STATISTICS WITH JMP: HYPOTHESIS TESTS, ANOVA AND REGRESSION

STATISTICS WITH JMP
HYPOTHESIS TESTS, ANOVA AND REGRESSION

Peter Goos

University of Leuven and University of Antwerp, Belgium

David Meintrup

University of Applied Sciences Ingolstadt, Germany

Library of Congress Cataloging-in-Publication Data

Names: Goos, Peter. | Meintrup, David.
Title: Statistics with JMP : hypothesis tests, ANOVA, and regression / Peter
    Goos, David Meintrup.
Description: Chichester, West Sussex : John Wiley & Sons, Inc., 2016. |
    Includes index.
Identifiers: LCCN 2015039990 (print) | LCCN 2015047679 (ebook) | ISBN 9781119097150 (cloth) |
    ISBN 9781119097044 (Adobe PDF) | ISBN 9781119097167 (ePub)
Subjects: LCSH: Probabilities–Data processing. | Mathematical statistics–Data processing. |
    Regression analysis. | JMP (Computer file)
Classification: LCC QA273.19.E4 G68 2016 (print) | LCC QA273.19.E4 (ebook) |
    DDC 519.50285/53–dc23
LC record available at http://lccn.loc.gov/2015039990

A catalogue record for this book is available from the British Library.

1   2016

To Marijke, Bas, Loes, and Mien
To Béatrice and Werner

Contents

Preface

This book is the result of a thorough revision of the lecture notes for the course "Statistics for Business and Economics 2" that were developed by Peter Goos at the Faculty of Applied Economics of the University of Antwerp in Belgium. Encouraged by the success of the Dutch version of this book (entitled *Verklarende Statistiek: Schatten and Toetsen*, published in 2014 by Acco Leuven/Den Haag), we joined forces to create an English version. The new book builds on our first joint work, *Statistics with JMP: Graphs, Descriptive Statistics and Probability* (Wiley, 2015), which adopts the same philosophy, and uses the same software package, JMP. Hence, it can be regarded as a sequel, but it can also be read as a stand-alone book.

In this book, we give a detailed introduction to point estimators, interval estimators, hypothesis tests, analysis of variance, and simple linear regression. Compared with other introductory textbooks on inferential statistics, we cover several additional topics. For example, considerable attention is paid to nonparametric tests, such as the sign test, the signed-rank test, and the Kruskal–Wallis test. In addition, we discuss tests and confidence intervals for the Pearson and Spearman correlation coefficients, introduce the concept of equivalence tests, and include a chapter on the principles of optimal design of experiments. For nonparametric tests, exact p-values are discussed in detail, alongside the better-known approximate p-values. Throughout the book, we discuss different versions of tests and confidence intervals, which includes the construction of confidence intervals for a proportion and approximate p-values for certain tests. The book also incorporates recent insights from the literature, such as a more detailed table with critical values for the Shapiro–Wilk test and a recent table with critical values for the Kruskal–Wallis test.

As in our first book, we pay equal attention to mathematical aspects, the interpretation of all the statistical concepts that are introduced, and their practical application. In order to facilitate the understanding of the methods and to appreciate their usefulness, the book contains many examples involving real-life data. To demonstrate the broad applicability of statistics and probability, these examples have been taken from various fields of application, including business, economics, sports, engineering, and the natural sciences.

Our motivation in writing this book was twofold. First, we wanted to provide students and teachers with a resource that goes beyond other textbooks of similar scope in its technical and mathematical content. It has become increasingly fashionable for authors and statistics teachers to sweep technicalities and mathematical derivations under the carpet. We decided against this, because we feel that students should be encouraged to apply their mathematical knowledge, and that doing so deepens their understanding of statistical methods. Reading this book obviously requires some knowledge of mathematics. In most countries, students are taught mathematics in secondary or high school and the required mathematical concepts are revisited in introductory mathematics courses at university. Therefore, we are convinced that many university students have a sufficiently strong mathematical background to appreciate and benefit from the more thorough nature of this book. In the various derivations, we have tried to include all the intermediate steps, in order to keep the book readable.

Our second motivation was to ensure that the concepts introduced in the book can be successfully put into practice. To this end, we show how to generate estimates, carry out hypothesis tests, and perform regression analyses using the statistical software package JMP (pronounced "jump"). We chose JMP as supporting software because it is powerful yet easy to use, and suitable for a wide range of statistically oriented courses (including descriptive statistics, hypothesis testing, regression, analysis of variance, design of experiments, reliability, multivariate methods, and statistical and predictive modeling). We believe that introductory courses in statistics should use such software wherever possible. Indeed, we find that, because of the way in which students can easily interact with JMP, it can actually spark enthusiasm for statistics in class. The probability that a student will use statistics in his or her future professional career is far greater if the statistics classes were more pleasurable than painful.

In summary, our approach to teaching statistics combines theoretical and mathematical depth, detailed and clear explanations, numerous practical examples, and the use of a user-friendly and yet very powerful statistical package.

Software

As mentioned, we use JMPas enabling software.With the purchase of a hard copy of this book, you receive a one-year license for JMP's Student Edition. The license period starts when you activate your copy of the software using the code included with this book (to receive an access code, please visit www.wiley.com/go/statsjmp). To download JMP's Student Edition, visit http://www.jmp.com/wiley. For students accessing a digital version of the book, your lecturer may contact Wiley in order to procure unique codes with which to download the free software. For more information about JMP, go to http://www.jmp.com. JMP is available for the Windows and Mac operating systems. This book is based on JMP version 12 for Windows.

In our examples, we do not assume any familiarity with JMP: the step-by-step instructions are detailed and accompanied by screenshots. For more explanations and descriptions, www.jmp.com offers a substantial amount of free material, including many video demonstrations. In addition, there is a JMP Academic User Community where you can access content, discuss questions, and collaborate with other JMP users worldwide: instructors can share teaching resources and best practices, students can ask questions, and everyone can access the latest resources provided by the JMP Academic Team. To join the community, go to http://community .jmp.com/academic.

Data Files

Throughout the book, various data sets are used. We strongly encourage everyone who wants to learn statistics to actively try things out using data. JMP files containing the data sets as well as JMP scripts to reproduce figures, tables, and analyses can be downloaded from the publisher's companion web site to this book:

www.wiley.com/go/goosandmeintrup/JMP

There, we also provide some additional supporting files.

Peter Goos
David Meintrup

Acknowledgments

We consulted plenty of sources during the preparation of this book, and we would like to acknowledge at least the most important ones. A source on nonparametric techniques that we found extremely valuable, given its depth and comprehensive explanations, is the book *Nonparametric Statistical Methods* by M. Hollander, D.A. Wolfe, and E. Chicken. A more general book that we found very helpful is the *Handbook of Parametric and Nonparametric Statistical Procedures* by David J. Sheskin. The same applies to *Biostatistical Analysis* by J.H. Zar.

We would like to thank numerous people who have made the publication of this book possible. The first author, Peter Goos, is very grateful to Professor Willy Gochet of the University of Leuven, who introduced him to the topics of statistics and probability. Professor Gochet allowed Peter to use his lecture notes as a backbone for his own course material, which later developed into this book.

The authors are very grateful for the support and advice offered by several people from the JMP Division of SAS: Brady Brady, Ian Cox, Bradley Jones, Volker Kraft, John Sall, and Mia Stephens. It is Volker who brought the two authors together and encouraged them to work on a series of English books on statistics with JMP (the first book is entitled *Statistics with JMP: Graphs, Descriptive Statistics and Probability*). A very special word of thanks goes to Ian, whose suggestions substantially improved this book, and to José Ramirez for generously sharing an example and a data set. The authors would also like to thank Eva Angels, Kris Annaert, Stefan Becuwe, Hilde Bemelmans, Marco Castro, Filip De Baerdemaeker, Hajar Hamidouche, Jérémie Haumont, Roselinde Kessels, Ida Ruts, Bagus Sartono, Daniel Palhazi Cuervo, Evelien Stoffels, Anja Struyf, Utami Syafitri, Yahri Tillmans, Ellen Vandervieren, Katrien Van Driessen, Kristel Van Rompay, Alan Vazquez Alcocer, Diane Verbiest, Tom Vermeire, Nha Vo-Thanh, Sara Weyns, Peter Willemé, and Simone Willis for their detailed comments and constructive suggestions, and for their technical assistance in creating figures and tables.

Finally, we thank Liz Wingett, Baljinder Kaur, Heather Kay, Audrey Koh, and Geoffrey D. Palmer at John Wiley & Sons.

Part One

Estimators and Tests

1

Estimating Population Parameters

I don't know how long I stand there. I don't believe I've ever stood there mourning faithfully in a downpour, but statistically speaking it must have been spitting now and then, there must have been a bit of a drizzle once or twice.

(from *The Misfortunates*, Dimitri Verhulst, pp. 125–126)

A major goal in statistics is to make statements about populations or processes. Often, the interest is in specific parameters of the distributions or densities of the populations or processes under study. For instance, researchers in political science want to make statements about the proportion of a population that votes for a certain political party. Industrial engineers want to make statements about the proportion of defective smartphones produced by a production process. Bioscience engineers are interested in comparing the mean amounts of growth resulting from applying two or more different fertilizers. Economists are interested in income inequality and may want to compare the variance in income across different groups.

To be able to make such statements, the proportions, means, and variances under study need to be quantified. In statistical jargon, we say that these parameters need to be estimated. It is also important to quantify how reliable each of the estimates is, in order to judge the confidence we can have in any statement we make. This chapter discusses the properties of the most important sample statistics that are used to make statements about population and process means, proportions, and variances.

1.1 Introduction: Estimators Versus Estimates

In practice, population parameters such as μ, σ^2, π, and λ (see our book *Statistics with JMP: Graphs, Descriptive Statistics and Probability*) are rarely known. For example, if we study the arrival times of the customers of a bank, we know that the number of arrivals per unit of time often follows a Poisson[1] distribution. However, we do not know the exact value of the

---

[1] The Poisson distribution is commonly used for random variables representing a certain number of events per unit of time, per unit of length, per unit of volume, and so on. The Poisson distribution has one parameter λ, which is the average number of events per unit of time, per unit of length, per unit of volume, and so on. For more details, see *Statistics with JMP: Graphs, Descriptive Statistics and Probability*.

Statistics with JMP: Hypothesis Tests, ANOVA and Regression, First Edition. Peter Goos and David Meintrup.
© 2016 John Wiley & Sons, Ltd. Published 2016 by John Wiley & Sons, Ltd.
Companion Website: http://www.wiley.com/go/goosandmeintrup/JMP

distribution's parameter λ. One way or another, we therefore need to estimate this parameter. This estimate will be based on a number of measurements or observations, $x_1, x_2, \ldots, x_n$, that we perform in the bank; in other words, on the sample data we collect.

The **estimate** for the unknown λ will be a function of the sample values $x_1, x_2, \ldots, x_n$; for example, the sample mean $\bar{x}$. Every researcher who faces the same problem, studying the arrival pattern of customers, will obtain different sample values, and thus a different sample mean and another estimate. The reason for this is that the number of arrivals in the bank in a given time interval is a random variable. We can express this explicitly by using uppercase letters X_1, $X_2, \ldots, X_n$ for the sample observations. The fact that each researcher obtains another estimate for λ can also be made more explicit by using a capital letter to denote the sample mean: $\bar{X}$. The sample mean is interpreted as a random variable, and then it is called an **estimator** instead of an estimate. In short, an estimate is always a real number, while an estimator is a random variable the value of which is not yet known.

The sample mean is, of course, only one of many possible functions of the sample observations $X_1, X_2, \ldots, X_n$, and thus only one of many possible estimators. Obviously, a researcher is not interested in an arbitrary function of the sample observations, but he wants to get a good idea of the unknown parameter. In other words, the researcher wishes to obtain an estimate that, on average, is equal to the unknown parameter, and that, ideally, is guaranteed to be close to the unknown parameter. Statisticians translate these requirements into "the estimator should be unbiased" and "the estimator should have a small variance". These requirements will be clarified in the next section.

1.2　Estimating a Mean Value

The requirements for a good estimator can best be illustrated by means of two simulation studies. The first study simulates data from a normally distributed population, while the second one simulates data from an exponentially distributed population.

1.2.1　The Mean of a Normally Distributed Population

We first assume that a normally distributed population with mean $\mu = 3000$ and standard deviation $\sigma = 100$ is studied by 1000 (fictitious) students. The students are unaware of the μ value and wish to estimate it. To this end, each of these students performs five measurements. A first option to estimate the unknown value μ is to calculate the sample mean. In this way, we obtain 1000 sample means, shown in the histogram in Figure 1.1, at the top left. The mean of these 1000 sample means is 2998.33, while the standard deviation is 43.38.

Another possibility to estimate the unknown μ is to calculate the median. For a normally distributed population, both the median and the expected value are equal to the parameter μ, so that this makes sense. Based on the samples that the students have gathered, the 1000 medians can also be calculated and displayed in a histogram. The resulting histogram is shown in Figure 1.1, at the top right[2]. The attentive reader will notice immediately that the second histogram is

---

[2] Outputs as in Figures 1.1 and 1.2 can be created in JMP with the "Distribution" option in the "Analyze" menu.

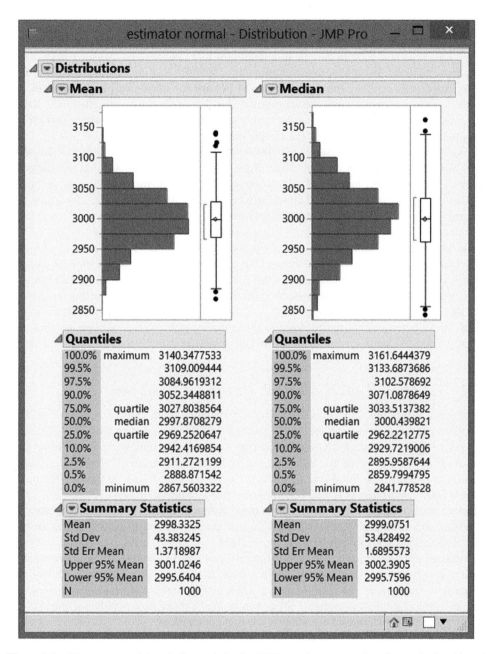

Figure 1.1 Histograms and descriptive statistics for 1000 sample means and medians calculated based on samples of five observations from a normally distributed population with mean 3000 and standard deviation 100.

just a bit wider than the first. Among other things, this is reflected by the fact that the standard deviation of the 1000 medians is 53.43. The mean of the 1000 medians is equal to 2999.08. In Figure 1.1, it can also be seen that the minimum (2841.78) and the first quartile (2962.22) of the sample medians are smaller than the minimum (2867.56) and the first quartile (2969.25) of the sample means. Also, the maximum (3161.64) and the third quartile (3033.51) of the sample medians are greater than the maximum (3140.35) and the third quartile (3027.80) of the sample means. This suggests that the sample medians are, in general, further away from the population mean $\mu = 3000$ than the sample means.

It is striking that both the mean of the 1000 sample means (2998.33) and that of the 1000 medians (2999.08) are very close to 3000. If the number of samples is raised significantly (theoretically, an infinite number of samples could be taken), the mean of the sample means and that of the sample medians will converge to the unknown $\mu = 3000$. Therefore, both the sample mean and the sample median are called **unbiased estimators** of the mean of a normally distributed population.

The fact that the range, the interquartile range, the standard deviation, and the variance of the 1000 sample means are smaller than those of the 1000 sample medians means that the sample mean is a more reliable estimator of the unknown population mean than the sample median. The larger variance of the medians indicates that the medians are generally further away from $\mu = 3000$ than the sample means. In short, a researcher should have more confidence in the sample mean because it is usually closer to the unknown μ. In such a case, we say that one estimator (here, the sample mean) is more **efficient** or **precise** than the other (here, the median).

1.2.2 The Mean of an Exponentially Distributed Population

We now investigate an exponentially distributed population with parameter $\lambda = 1/100$. The "unknown" population mean is therefore $\mu = 1/\lambda = 100$ (see *Statistics with JMP: Graphs, Descriptive Statistics and Probability*). Each of the 1000 fictitious students performs five measurements. A first option to estimate the unknown value μ is again to calculate the sample mean. A histogram of the 1000 sample means is shown in Figure 1.2, at the top left. The mean of these 1000 sample means is 99.2417, while the standard deviation is 44.10.

Based on the samples that the students have gathered, the 1000 medians can also be calculated and displayed in a histogram. This histogram is shown in Figure 1.2, at the top right. The mean of the 1000 medians is only 77.0114.

These calculations indicate that the population mean $\mu = 1/\lambda = 100$ can be approximated fairly well by using the sample means, with a mean of 99.2417. This is not the case for the medians, the mean value of which is far away from μ. This remains the case if the number of samples is increased. In this example, for an exponentially distributed population, the median is not an unbiased but a **biased** estimator of the population mean.

In addition, Figure 1.2 also shows that the standard deviation of the sample medians (46.13) is greater than that of the sample means (44.10).

1.3 Criteria for Estimators

Key properties of estimators are their expected values and their variances. These statistics are related to the concepts of bias and efficiency, respectively.

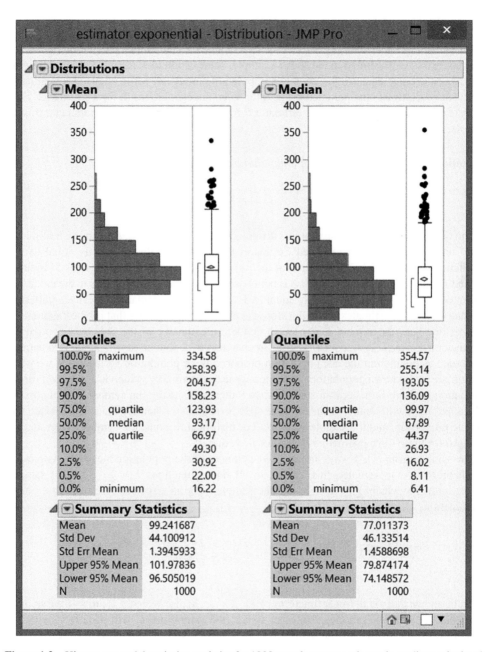

Figure 1.2 Histograms and descriptive statistics for 1000 sample means and sample medians calculated based on samples of five observations from an exponentially distributed population with parameter $\lambda = 1/100$.

1.3.1 Unbiased Estimators

An ideal estimator that always produces the exact value of an unknown population parameter does not exist. As illustrated in the above example, some estimators, namely unbiased estimators, are on average equal to the unknown population parameter, while others systematically under- or overestimate the parameter. The latter is an undesirable result for a researcher. Formally, the definition of an unbiased estimator $\hat{\theta}$ for an unknown population parameter θ is as follows:

Definition 1.3.1 *An estimator $\hat{\theta}$ of a population parameter θ is **unbiased** if*

$$E(\hat{\theta}) = \theta.$$

The **bias** of an estimator is the absolute difference $V(\hat{\theta}) = |E(\hat{\theta}) - \theta|$. An unbiased estimator has a bias of zero. For an unbiased estimator, the expected value is exactly equal to the population parameter. The histograms for the sample means on the left-hand sides of Figures 1.1 and 1.2 show that, once sample data is being used, the estimate will be close to the unknown population parameter, but not exactly equal to it. So, for any particular sample, even unbiased estimators result in estimates that differ from the population parameter that is being estimated.

Note that here the symbol $\hat{\theta}$ is used to denote an estimator of the unknown population parameter θ. As usual in statistics, we use Greek letters to denote unknown population parameters such as population means, population proportions, or population variances. If we want to estimate an unknown population parameter, we use an estimator, which is a synonym for an estimation method. In general in statistics, we indicate this using the symbol $\hat{\theta}$ (pronounced "theta hat"). We will mainly focus on three specific estimators, namely the sample mean, the sample proportion, and the sample variance. For historical reasons, the symbols $\overline{X}$, $\hat{P}$, and S^2 are used for these three estimators instead of $\hat{\mu}$, $\hat{\pi}$, and $\hat{\sigma}^2$.

The sample mean $\overline{X}$ is always an unbiased estimator of the population mean (this is proven in Theorem 1.5.1). Actually, this applies to all linear functions $Y = \sum_{i=1}^{n} \alpha_i X_i$ of sample observations for which $\sum_{i=1}^{n} \alpha_i = 1$, and the sample mean is a special case of such a linear combination, where each $\alpha_i = 1/n$:

$$\overline{X} = \frac{1}{n} \sum_{i=1}^{n} X_i = \frac{1}{n}(X_1 + X_2 + \cdots + X_n) = \frac{1}{n}X_1 + \frac{1}{n}X_2 + \cdots + \frac{1}{n}X_n.$$

It can be shown that, of all linear functions of $X_1, X_2, \ldots, X_n$, for which $\sum_{i=1}^{n} \alpha_i = 1$ the sample mean has the smallest variance[3]. In other words, the sample mean will usually provide an estimate that is closer to the population mean than any other linear function Y of $X_1, X_2, \ldots, X_n$.

In Theorem 1.7.1, we prove that the sample variance

$$S^2 = \frac{1}{n-1} \sum_{i=1}^{n} (X_i - \overline{X})^2$$

---

[3] Therefore, the sample mean is called the "best linear unbiased estimator", abbreviated as "BLUE".

is an unbiased estimator of a population variance σ^2. This theorem also explains why we divide by $n - 1$ when computing the sample variance, and not by n. It is important to note that the sample standard deviation S is a biased estimator of the population standard deviation σ.

Finally, in Section 1.6, we will see that a sample proportion $\hat{P}$ is a special case of a sample mean. Its expected value is equal to the population proportion π, so that $\hat{P}$ is an unbiased estimator of π.

1.3.2   The Efficiency of an Estimator

It is desirable that an estimator is as reliable as possible and yields estimates that are close to the unknown population parameter under investigation. In short, the estimator should have a small variance or standard deviation. An estimator with a small variance is called an efficient estimator.

If $\hat{\theta}_1$ and $\hat{\theta}_2$ are two unbiased estimators of the same unknown population parameter θ, the relative efficiency of $\hat{\theta}_2$ compared to $\hat{\theta}_1$ is computed as $\text{var}(\hat{\theta}_1)/\text{var}(\hat{\theta}_2)$.

Sometimes, we have the choice between an estimator that is unbiased but has a large variance, and an estimator that is biased but has a small variance. In this case, it is not immediately clear which estimator should be used. To make a decision in such situations, one can pick the estimator that has the smaller mean squared error, $\text{MSE}(\hat{\theta})$:

Definition 1.3.2   *The **mean squared error** of an estimator $\hat{\theta}$ is the sum of its variance and the square of its bias:*

$$MSE(\hat{\theta}) = \text{var}(\hat{\theta}) + [V(\hat{\theta})]^2.$$

Finally, it is also desirable that the precision of an estimator increases with the number of observations. More observations provide more information, so that better estimates can be expected. For example, Theorem 1.5.2 shows that the variance of the sample mean is equal to σ^2/n. The variance decreases as the sample size n increases. The precision of the sample mean is thus improved when more data is used.

1.4   Methods for the Calculation of Estimators

Finding estimators with good properties is not always easy. In the statistical literature[4], three methods are frequently used:

(1) the method of moments;
(2) the method of least squares; and
(3) the maximum likelihood method.

These general methods are beyond the scope of this book. This book primarily focuses on the following estimators: sample means, sample proportions, and sample variances. In the remainder of this chapter, each of these estimators is shown to be unbiased, and the probability density of each estimator is discussed.

---

[4] See *Statistics with JMP: Linear and Generalized Linear Models.*

1.5 The Sample Mean

1.5.1 The Expected Value and the Variance

If the sample mean is considered as an estimator and thus as a random variable, we can determine the expected value, the variance, and even the probability density. The sample mean is then written using a capital letter,

$$\overline{X} = \frac{1}{n} \sum_{i=1}^{n} X_i,$$

to indicate this explicitly. We consider the sample mean as an estimator or a random variable as long as we have no data; that is, the individual observations $X_1, X_2, \ldots, X_n$ are not known. Once the data has been collected, we use lowercase letters for the individual observations: $x_1, x_2, \ldots, x_n$. For the sample mean that we compute based on the observed values $x_1, x_2, \ldots, x_n$, we also use a lowercase letter:

$$\overline{x} = \frac{1}{n} \sum_{i=1}^{n} x_i.$$

Theorem 1.5.1   *For a random sample from a population with expected value μ, we have*

$$E(\overline{X}) = \mu.$$

Proof.

$$E(\overline{X}) = E\left(\frac{1}{n} \sum_{i=1}^{n} X_i\right),$$

$$= \frac{1}{n} \sum_{i=1}^{n} E(X_i),$$

$$= \frac{1}{n}(\mu + \mu + \cdots + \mu),$$

$$= \frac{n\mu}{n},$$

$$= \mu.$$ ∎

This theorem states that before sample data is obtained, the expected value of the sample mean is equal to the population mean. In other words, the theorem states that the sample mean is an unbiased estimator of the population mean.

Once we have sample observations $x_1, x_2, \ldots, x_n$, the sample mean is $\overline{x}$. Of course, this sample mean will not be exactly equal to μ. This was already illustrated in Section 1.2, where each student obtained a different sample mean. To get an idea of the size of the possible deviation between the sample mean $\overline{X}$ and the population mean μ, one should study the variance and standard deviation of $\overline{X}$. Figure 1.1 showed that the standard deviation of the sample means of

1000 (fictitious) students was equal to 43.38, while the original population of the individual values had a standard deviation of 100. In general, the standard deviation of a sample mean is lower than the standard deviation of the population studied. The same is true for the variance. The next theorem tells us how much smaller the variance of a sample mean is.

Theorem 1.5.2　*For a random sample of n observations from a population with variance σ^2, we have*

$$\sigma_{\overline{X}}^2 = \text{var}(\overline{X}) = \frac{\sigma^2}{n}$$

and

$$\sigma_{\overline{X}} = \frac{\sigma}{\sqrt{n}}.$$

Proof.

$$\sigma_{\overline{X}}^2 = \text{var}(\overline{X}) = \text{var}\left(\frac{1}{n}\sum_{i=1}^{n}X_i\right),$$

$$= \frac{1}{n^2}\sum_{i=1}^{n}\text{var}(X_i),$$

$$= \frac{1}{n^2}(\sigma^2 + \sigma^2 + \cdots + \sigma^2),$$

$$= \frac{n\sigma^2}{n^2},$$

$$= \frac{\sigma^2}{n}.$$

In the second step of this proof, it is assumed that the covariance between two different sample observations, X_i and X_j, is equal to zero. In that case, the variance of a linear combination of random variables is equal to a linear combination of the variances, with squared coefficients[5]. ■

---

[5] In *Statistics with JMP: Graphs, Descriptive Statistics and Probability*, we show that

$$\text{var}(aX + bY) = a^2\text{var}(X) + b^2\text{var}(Y) + 2ab\text{cov}(X, Y),$$

which can be simplified to

$$\text{var}(aX + bY) = a^2\text{var}(X) + b^2\text{var}(Y)$$

if the random variables X and Y are independent or uncorrelated (and thus $\text{cov}(X, Y) = 0$). This result can be generalized to scenarios involving more than two random variables.

The proof shows that the variance of the sample mean decreases linearly when the sample size n increases. This means that as the sample size increases, the probability that the sample mean $\bar{x}$ is close to (the unknown) μ increases as well.

The square root of the variance, namely $\sigma_{\bar{X}}$, is called the **standard error**. The estimated version of this statistic, namely $s/\sqrt{n}$, can be found in the reports of statistical packages. Figures 1.1 and 1.2 illustrate that the standard error[6] is also reported in JMP, namely as "Std Err Mean". It is not difficult to verify that the standard error in these figures is a factor of $\sqrt{n} = \sqrt{1000} = 31.62$ smaller than the corresponding standard deviation (abbreviated as "Std Dev").

1.5.2   The Probability Density of the Sample Mean for a Normally Distributed Population

If the sample is drawn from a normally distributed population, we can use the following theorem.

Theorem 1.5.3   *Let $X_1, X_2, \ldots, X_k$ be independent normally distributed random variables with expected values $E(X_1) = \mu_1, E(X_2) = \mu_2, \ldots, E(X_k) = \mu_k$ and variances $\mathrm{var}(X_1) = \sigma_1^2, \mathrm{var}(X_2) = \sigma_2^2, \ldots, \mathrm{var}(X_k) = \sigma_k^2$. Then, the linear function $Y = \alpha_0 + \sum_{i=1}^{k} \alpha_i X_i$ is also normally distributed, with expected value $E(Y) = \alpha_0 + \sum_{i=1}^{k} \alpha_i \mu_i$ and variance $\mathrm{var}(Y) = \sum_{i=1}^{k} \alpha_i^2 \sigma_i^2$.*

It follows from this theorem that the mean of a number of normally distributed random variables with the same mean μ and the same variance σ^2 is also normally distributed. Indeed, the mean of the variables $X_1, X_2, \ldots, X_n$ is a linear function with $\alpha_0 = 0$ and $\alpha_i = 1/n$ for $i \geq 1$. In this case, the sample mean $\bar{X}$ is normally distributed with mean μ and variance σ^2/n. We denote this by

$$\bar{X} \sim N\left(\mu, \frac{\sigma^2}{n}\right).$$

This result is valid for any sample size – also for a small one – and is illustrated in Section 1.5.4.

1.5.3   The Probability Density of the Sample Mean for a Nonnormally Distributed Population

If the population under investigation has an unknown probability density or probability distribution, the probability distribution of the sample mean typically cannot be determined exactly.

---

[6] The standard error can be calculated in JMP using the menu "Analyze" and the option "Distribution". An alternative way to obtain the standard error for a particular variable in a data table is to use the "Summary" option in the "Tables" menu, and to choose "Std Err" in the "Statistics" drop-down menu that then becomes available.

In this case, however, a large sample size may help, because the central limit theorem can be used for large samples. One version of this theorem, namely Theorem 1.5.6, indeed indicates that the sample mean is approximately normally distributed for large n, with mean μ and variance σ^2/n.

The **central limit theorem** is one of the main theorems of statistics. This theorem also explains to a large extent why the normal probability density is so crucial in statistics. There are different versions of this theorem.

Theorem 1.5.4 *Let $X_1, X_2, \ldots, X_n$ be independent random variables with expected values $E(X_i) = \mu_i$ and variances $\mathrm{var}(X_i) = \sigma_i^2$. Then, under very general conditions and for a sufficiently large value of n:*

(1) the random variable $Y = \sum_{i=1}^n X_i$ is approximately normally distributed with mean $\mu_Y = \sum_{i=1}^n \mu_i$ and variance $\sigma_Y^2 = \mathrm{var}(Y) = \sum_{i=1}^n \sigma_i^2$;
(2) and, consequently, the random variable

$$\frac{Y - \sum_{i=1}^n \mu_i}{\sqrt{\sum_{i=1}^n \sigma_i^2}}$$

approximately follows a standard normal distribution.

The general conditions mentioned in the theorem refer to the fact that none of the individual variances σ_i^2 makes a dominant contribution to the total variance of Y. In many practical applications of the central limit theorem, all random variables $X_1, X_2, \ldots, X_n$ have the same distribution or density, and therefore the same variance. In that case, this condition is automatically met. If all random variables $X_1, X_2, \ldots, X_n$ have the same distribution or density, the central limit theorem can be rewritten as follows:

Theorem 1.5.5 *Let $X_1, X_2, \ldots, X_n$ be independent random variables with expected value $E(X_i) = \mu$ and variance $\mathrm{var}(X_i) = \sigma^2$. Then, for a sufficiently large value of n:*

(1) the random variable $Y = \sum_{i=1}^n X_i$ is approximately normally distributed with mean $\mu_Y = n\mu$ and variance $\sigma_Y^2 = \mathrm{var}(Y) = n\sigma^2$;
(2) and, consequently, the random variable

$$\frac{Y - n\mu}{\sigma\sqrt{n}}$$

approximately follows a standard normal distribution.

The central limit theorem can also be formulated in terms of the sample mean $\overline{X} = Y/n$:

Theorem 1.5.6   *Let $X_1, X_2, \ldots, X_n$ be independent random variables with expected value $E(X_i) = \mu$ and variance $\text{var}(X_i) = \sigma^2$. Then, for a sufficiently large value of n:*

(1) the random variable $\overline{X} = \frac{Y}{n} = \frac{\sum_{i=1}^{n} X_i}{n}$ is approximately normally distributed with mean μ and variance $\frac{\sigma^2}{n}$;

(2) and, consequently, the random variable

$$\frac{\overline{X} - \mu}{\frac{\sigma}{\sqrt{n}}}$$

approximately follows a standard normal distribution.

An important practical question is how large the sample size n must be before one can apply the central limit theorem. There is no general answer to this question. The required size of n depends on the distribution or density of the individual random variables X_i:

- If the probability density of X_i is similar to the normal density, $n = 5$ is sufficient.
- If the probability density of X_i does not show any pronounced peaks – such as, for example, the uniform density – then $n = 12$ should be sufficient.
- If the probability distribution or density of X_i shows pronounced peaks, it is difficult to specify a value of n. A value of $n = 100$ will usually suffice. An example of a distribution with a peak is $P(X = 1) = 0.06$ and $P(X = 10) = 1 - P(X = 1) = 0.94$.
- For continuous variables that appear in practice, typically $n = 25$ or $n = 30$ is sufficient.

In the next section, the third version of the central limit theorem (Theorem 1.5.6) is illustrated in detail, using simulations.

1.5.4   An Illustration of the Central Limit Theorem

Suppose that some students are interested in the value of the Euro Stoxx 50 index, which summarizes the performance of the 50 most important stocks inside the eurozone. Student 1 will take a sample of n observations of the Euro Stoxx 50 index and calculate the mean, namely $\overline{X}_1$. Student 2 will also take a sample of n observations. Since the Euro Stoxx 50 index changes from minute to minute, Student 2 will obviously observe different values of the Euro Stoxx 50 index (unless, by coincidence, the two students make their observations at exactly the same times). Student 2 also calculates the mean of his sample: $\overline{X}_2$. In the same way, all students collect n observations and calculate their sample means. If there are 200 students, we finally obtain 200 sample means: $\overline{X}_1, \overline{X}_2, \ldots, \overline{X}_{200}$.

The third version of the central limit theorem now states that these 200 means have a distribution that is very similar to that of the normal density. With a histogram of these 200 means, this is easy to verify.

This is exactly what we will do in this section. We will not use real students, but we will simulate the scenario sketched above in JMP. To this end, we will create 200 samples of n observations (one for each hypothetical student), calculate the mean for each sample, and create a histogram of the 200 sample means. This simulation requires that we specify a probability distribution or probability density in JMP for the generation of the observations.

We start with a normal distribution. Hence, we first assume that the Euro Stoxx 50 index behaves like a normally distributed random variable. We use $\mu = 3000$ as the mean of this normal distribution (more or less the value of the index when work on this book was started in March 2014), and we choose $\sigma = 100$ as the standard deviation. We assume that all of the students' observations are independent of each other.

Normally Distributed X

First, suppose that each student collects a sample of five observations; in other words, that $n = 5$. In this scenario, we need to simulate 200 sets of five observations using JMP. To this end, we create a JMP data table with 200 rows and five columns, filled with pseudo-random numbers from a normal probability density with parameters $\mu = 3000$ and $\sigma = 100$. The formula we use for each of the five columns is "Random Normal(3000, 100)". We calculate the mean of the five observations in each row, and then display all means in a histogram. If we create a second data table in the same way, this corresponds to a second group of 200 hypothetical students who also collect samples of five observations. Two possible histograms obtained in this way are shown in Figure 1.3. The resulting histograms are quite bell-shaped, indicating that the sample means are normally distributed, as predicted by Theorem 1.5.6 (and also Theorem 1.5.3, because, here, we assume that the observations are normally distributed).

The fastest way to generate 200 new samples is to ask JMP to recalculate the formula "Random Normal(3000, 100)". This is done using the command "Rerun Formulas", which appears when you click on the hotspot (red triangle) menu next to the name of the data table. This is illustrated in Figure 1.4.

If each student takes samples of 20 instead of five observations, the histograms have a different shape: they are still bell-shaped but they are significantly narrower. Two histograms

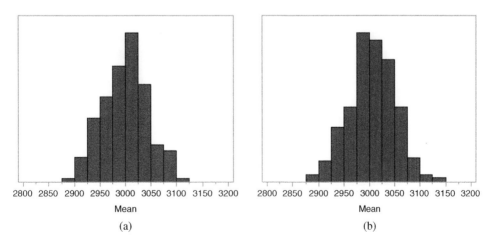

Figure 1.3 Two histograms of 200 sample means for normally distributed data and samples of five observations.

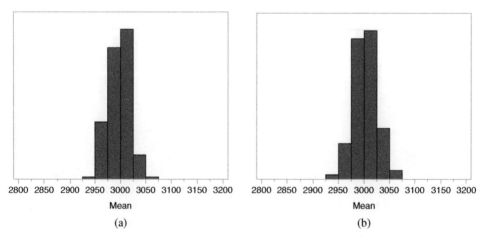

Student	X 1	X 2	X 3	X 4	X 5	Gemiddelde
1	3002.6685849	2754.8475711	2915.1166936	3113.2692434	3116.3843194	2980.4572825
2	3138.4333315	3040.6978943	2997.7014831	3044.5665303	3003.3665967	3044.9531672
3	3083.6358485	2803.3275928	2983.7722675	3023.7965761	3013.465223	2981.5995016
4	3012.652982	3182.8646996	3102.903273	2983.8800801	2866.6021016	3029.7806273
5	2992.1312803	3122.1013141	3119.9502417	2976.1198246	2979.1901827	3037.8985687
6	3016.475296	2874.642051	3045.5307397	2904.1367408	3101.4027804	2988.4375216
7	3093.5120276	2930.3191128	2802.936002	3002.8337279	2973.0109294	2960.5223599
8	2928.7451567	2733.6422314	2949.5822286	2985.5309564	2996.9011808	2918.8803508
9	2972.1380577	3007.9092767	2953.1749961	2839.8548	3134.6064487	2981.5367158
10	3047.1200541	3186.2084688	2861.5540236	2808.6086328	2987.1588664	2978.1300091
11	3056.0086781	2992.3071469	3042.7047188	3003.665274	3089.75286	3036.8877356
12	2804.5323296	3044.8696397	2956.7823256	3013.0488495	2915.1761182	2946.8818525
13	3027.4855952	3112.4078005	2874.6057682	2826.7155787	2960.9388176	2960.430712
14	3069.4612224	2876.08195	2926.7753377	3049.9048368	2842.5903386	2952.9627371
15	2965.7742713	2931.9980201	3155.3385954	2996.6162531	2938.869673	2997.7193626
16	3007.4087116	3141.857531	2977.4923338	2789.0545745	2945.3022324	2972.2230767
17	3005.689361	2931.5480681	2836.1739186	2899.6186354	3003.3939463	2935.2847859
18	2947.6201434	3005.7704412	2945.2295183	2880.3853891	3133.9229707	2982.5856926
19	2873.1393501	3014.3197318	3125.0422273	2915.9663006	2888.0201263	2963.2975472
20	2891.5391115	3228.6519175	2958.9416133	3065.6038004	3144.6497196	3057.8772325
21	3058.4785455	2843.3741273	2949.0198178	3000.4095136	3030.5744576	2976.3712924
22	2919.230771	3140.3024044	3100.414135	3001.5744445	2796.442942	2991.5929394
23	3083.1242778	2899.3117875	2861.4850361	2996.5462235	2953.9607325	2958.8856115
24	3006.8061398	3103.0460201	2931.4580164	2845.9086494	3105.3125958	2998.5062843

Figure 1.4   Generating new pseudo-random observations in JMP with the option "`Rerun Formulas`".

for 200 sample means of samples with 20 observations are shown in Figure 1.5. The bell shape tells us that the sample means are still normally distributed. The fact that the histograms are narrower should not come as a surprise, since the central limit theorem and Theorem 1.5.3 imply that the variance of the sample mean is equal to σ^2/n. As a consequence, sample means of 20 observations have a variance that is four times smaller than the variance of sample means of five observations.

Uniformly Distributed X

Suppose that the value of the Euro Stoxx 50 index is not normally distributed, but is uniformly distributed between 2800 and 3200. First, suppose again that each student takes a sample of

Mean

(a)

Mean

(b)

Figure 1.5   Two histograms of 200 sample means for normally distributed data and samples of 20 observations.

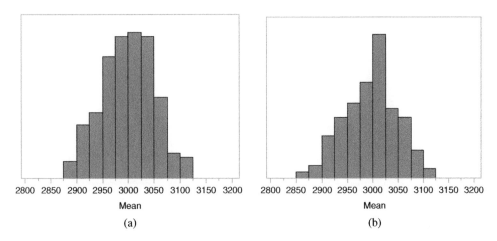

Figure 1.6 Two histograms of 200 sample means for uniformly distributed data and samples of five observations.

five observations. For this new scenario involving the uniform density, we again simulate 200 samples of five observations using JMP. To this end, we need to enter the formula "`Random Uniform(2800, 3200)`" in five columns of a data table with 200 rows. For each sample of five observations, we calculate the mean, and then we display all means in a histogram. Two possible histograms obtained in this way are shown in Figure 1.6. It is striking that, again, the histograms are quite bell-shaped, indicating that the sample means are still approximately normally distributed, even though the original data is uniformly distributed.

When the students take samples of 20 instead of five observations, the corresponding bell-shaped histograms are significantly narrower. Two histograms for 200 means of samples of 20 observations are shown in Figure 1.7.

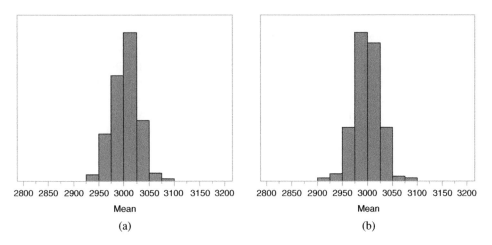

Figure 1.7 Two histograms of 200 sample means for uniformly distributed data and samples of 20 observations.

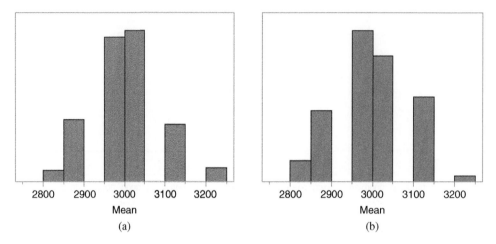

Figure 1.8 Two histograms of 200 sample means for Bernoulli distributed data and samples of five observations.

Bernoulli Distributed X

Now, suppose that the value of the Euro Stoxx 50 index is Bernoulli distributed, with a 50% chance that its value is 2800 and a 50% chance that its value is 3200. First, suppose again that each student takes a sample of five observations. We again need to simulate 200 samples of five observations using JMP. This time, we need to enter the formula "2800 + 400 * Random Binomial(1, 0.5)" in five columns of a data table with 200 rows. For each sample of five observations, we calculate the mean, and display the resulting 200 means in a histogram. Two possible histograms obtained in this way are shown in Figure 1.8. This time, the histograms are not bell-shaped. It is clearly visible that the original data comes from a discrete distribution, namely the Bernoulli distribution. The central limit theorem does not seem to work for the Bernoulli distribution and a sample of five observations.

When, however, the students take samples of 20 instead of five observations, the histograms look totally different. Although the histograms still do not exhibit a perfect bell shape, it is no longer obvious that the original data had a discrete probability distribution. Two possible histograms for 200 sample means of samples with 20 observations are shown in Figure 1.9. In order to obtain an even better bell shape, a slightly larger sample size is required.

This last example demonstrates that the central limit theorem is very powerful. Even probability distributions or probability densities that are quite different from the normal density still lead to distributions of sample means that are approximately normal, provided that the number of observations is sufficiently large.

1.6   The Sample Proportion

A sample proportion is a special case of the sample mean. Oftentimes, a variable under study can only take the values 0 or 1. Examples of such variables are gender (male/female) or

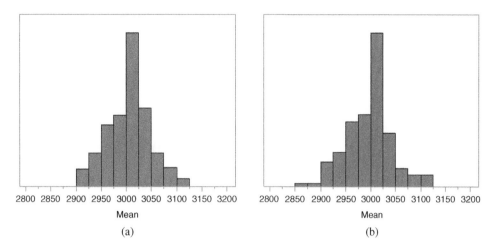

Figure 1.9 Two histograms of 200 sample means for Bernoulli distributed data and samples of 20 observations.

quality (defective/not defective). In general, the terms "success" and "failure" are used. The unknown population proportion is denoted by the letter π. This unknown population proportion indicates the proportion of successes in the entire population and is estimated using the sample proportion, which is simply the relative frequency of successes in a sample:

Definition 1.6.1   *The **sample proportion** is the number of successes in a sample divided by the number of observations.*

We can consider the sample proportion as a random variable or as a computed real number. We consider the sample proportion as a random variable as long as no sample data has been obtained. In that case, we use the symbol $\hat{P}$ to denote the sample proportion. If data is available and the sample proportion has been calculated, we use the symbol $\hat{p}$. For the sample proportion as random variable, we also use the symbol $\hat{P}$ to avoid confusion with a probability, which is typically denoted by the letter P.

If we adopt the convention of assigning the value "1" to the random variable X_i when the ith observation is a success, and the value "0" in the event of a failure, then we have

$$\hat{P} = \frac{1}{n} \sum_{i=1}^{n} X_i = \overline{X},$$

which shows that a sample proportion is actually a sample mean. Hence, the central limit theorem can be used: for large samples, the sample proportion is approximately normally distributed. The expected value and variance of the sample proportion are easily determined. Indeed, the sample proportion is a sum (and thus a linear combination) of n independent Bernoulli distributed random variables X_i with parameter π. As the expected value of a linear combination of random variables is equal to the linear combination of the expected values, we obtain

$$E(\hat{P}) = E\left(\frac{1}{n}\sum_{i=1}^{n}X_i\right),$$

$$= \frac{1}{n}E(X_1 + X_2 + \cdots + X_n),$$

$$= \frac{1}{n}\left(E(X_1) + E(X_2) + \cdots + E(X_n)\right),$$

$$= \frac{1}{n}(\pi + \pi + \cdots + \pi),$$

$$= \frac{1}{n}n\pi,$$

$$= \pi.$$

Since the variance of a linear combination of independent random variables is the linear combination of the variances with squared coefficients, we have

$$\text{var}(\hat{P}) = \text{var}\left(\frac{1}{n}\sum_{i=1}^{n}X_i\right),$$

$$= \frac{1}{n^2}\text{var}(X_1 + X_2 + \cdots + X_n),$$

$$= \frac{1}{n^2}\left(\text{var}(X_1) + \text{var}(X_2) + \cdots + \text{var}(X_n)\right),$$

$$= \frac{1}{n^2}(\pi(1-\pi) + \pi(1-\pi) + \cdots + \pi(1-\pi)),$$

$$= \frac{1}{n^2}n\pi(1-\pi),$$

$$= \frac{\pi(1-\pi)}{n}.$$

The expected value of the sample proportion thus proves to be equal to the population proportion π, so that the sample proportion is an unbiased estimator of the population proportion. The variance of the sample proportion is equal to $\pi(1-\pi)/n$. This variance decreases linearly with the number of observations, n. In other words, if you collect more data, the sample proportion will give you a more precise estimate of the population proportion. All this is summarized in the following theorem:

Theorem 1.6.2   *Let $X_1, X_2, \ldots, X_n$ be independent random variables with only two possible outcomes, 0 (failure) or 1 (success), and with a probability of success equal to π. Then, for a sufficiently large value of n:*

(1) the sample proportion $\hat{P} = \frac{\sum_{i=1}^{n}X_i}{n}$ is approximately normally distributed, with expected value π and variance $\pi(1-\pi)/n$;

(2) and, consequently, the random variable

$$\frac{\hat{P} - \pi}{\sqrt{\dfrac{\pi(1 - \pi)}{n}}}$$

approximately follows a standard normal distribution.

As a rule of thumb for using the normal density as an approximation, the conditions $n\pi > 5$ and $n(1 - \pi) > 5$ must be met. If one of these conditions is not valid, the normal distribution cannot be used for the sample proportion. In that case, the binomial distribution needs to be used instead of the normal probability density. Indeed, the number of successes in a sample can be described by a binomially distributed random variable with parameters n and π, if the probability of success for each individual observation is equal to π.

Figures 1.10 and 1.11 illustrate why the normal density can be used if $n\pi > 5$ and $n(1 - \pi) > 5$. Figure 1.10 shows three binomial distributions for $n = 25$. Figure 1.10a shows the binomial distribution for $\pi = 0.5$. This distribution is nicely symmetrical and almost perfectly bell-shaped. In this case, the binomial distribution seems similar to a normal density. Figure 1.10b shows the binomial distribution for $\pi = 0.25$. While this distribution is not perfectly symmetrical, it still looks bell-shaped and resembles a normal probability density. The same applies to the binomial distribution with $\pi = 0.75$ in Figure 1.10c. In all these cases, $n\pi > 5$ and $n(1 - \pi) > 5$.

Figure 1.11 shows two other binomial distributions, again with $n = 25$. Figure 1.11a shows the binomial distribution for $\pi = 0.05$. This distribution is far from symmetrical. In this case, the binomial probability distribution does not look like a normal probability density. Figure 1.11b shows the binomial distribution for $\pi = 0.95$. Again, the binomial distribution is neither symmetrical nor bell-shaped. In other words, the binomial probability distribution does not look like a normal probability density. For $\pi = 0.05$, the value of $n\pi$ is smaller than 5. For $\pi = 0.95$, the value of $n(1 - \pi)$ is smaller than 5.

1.7   The Sample Variance

The sample variance can be viewed as an estimator for the population variance. Thus, just like the sample mean and the sample proportion, it can also be treated as a random variable. In this case, we use capital letters to define the sample variance:

$$S^2 = \frac{1}{n - 1} \sum_{i=1}^{n} (X_i - \overline{X})^2. \tag{1.1}$$

If we consider the sample variance as an estimate and therefore as a real number, we denote it by

$$s^2 = \frac{1}{n - 1} \sum_{i=1}^{n} (x_i - \overline{x})^2.$$

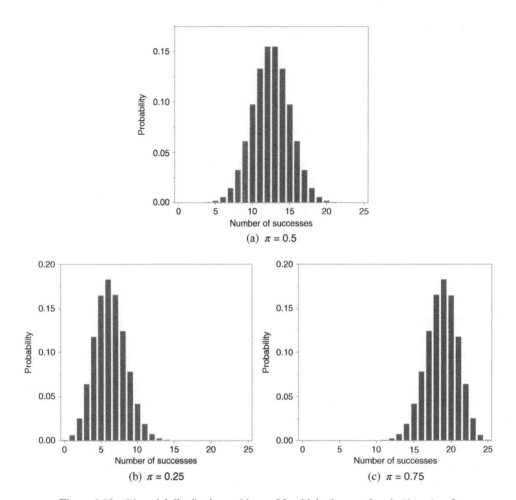

Figure 1.10   Binomial distributions with $n = 25$, with both $n\pi > 5$ and $n(1 - \pi) > 5$.

This second notation is used in descriptive statistics (see *Statistics with JMP: Graphs, Descriptive Statistics and Probability*). Once again, we use capital letters as long as there is no data, and lowercase letters when sample data has been used.

Like any other random variable, the random variable S^2 has an expected value, a variance, and a probability density.

1.7.1   The Expected Value

Theorem 1.7.1   *For a random sample from a population with variance σ^2, we have*

$$E(S^2) = \sigma^2.$$

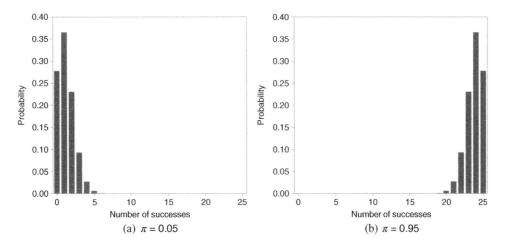

Figure 1.11   Binomial distributions with $n = 25$, with either $n\pi$ or $n(1 - \pi)$ less than 5.

Proof.

$$E(S^2) = E\left\{ \frac{1}{n-1} \sum_{i=1}^{n} (X_i - \overline{X})^2 \right\},$$

$$= \frac{1}{n-1} E\left\{ \sum_{i=1}^{n} (X_i - \overline{X})^2 \right\},$$

$$= \frac{1}{n-1} E\left\{ \sum_{i=1}^{n} (X_i - \mu + \mu - \overline{X})^2 \right\},$$

$$= \frac{1}{n-1} E\left\{ \sum_{i=1}^{n} (X_i - \mu)^2 + 2 \sum_{i=1}^{n} (X_i - \mu)(\mu - \overline{X}) + \sum_{i=1}^{n} (\mu - \overline{X})^2 \right\},$$

$$= \frac{1}{n-1} E\left\{ \sum_{i=1}^{n} (X_i - \mu)^2 + 2(\mu - \overline{X}) \sum_{i=1}^{n} (X_i - \mu) + n(\mu - \overline{X})^2 \right\},$$

$$= \frac{1}{n-1} E\left\{ \sum_{i=1}^{n} (X_i - \mu)^2 + 2(\mu - \overline{X})(n\overline{X} - n\mu) + n(\mu - \overline{X})^2 \right\},$$

$$= \frac{1}{n-1} E\left\{ \sum_{i=1}^{n} (X_i - \mu)^2 - 2n(\mu - \overline{X})^2 + n(\mu - \overline{X})^2 \right\},$$

$$= \frac{1}{n-1} E\left\{ \sum_{i=1}^{n} (X_i - \mu)^2 - n(\mu - \overline{X})^2 \right\},$$

$$= \frac{1}{n-1} \left[\sum_{i=1}^{n} E\{(X_i - \mu)^2\} - nE\{(\mu - \overline{X})^2\} \right],$$

$$= \frac{1}{n-1} \left[\sum_{i=1}^{n} E\{(X_i - \mu)^2\} - nE\{(\overline{X} - \mu)^2\} \right],$$

$$= \frac{1}{n-1} \left\{ \sum_{i=1}^{n} \text{var}(X_i) - n \, \text{var}(\overline{X}) \right\},$$

$$= \frac{1}{n-1} \left(\sum_{i=1}^{n} \sigma^2 - n\frac{\sigma^2}{n} \right),$$

$$= \frac{1}{n-1} (n\sigma^2 - \sigma^2),$$

$$= \sigma^2. \qquad\qquad\blacksquare$$

The theorem proves that the sample variance S^2 is an unbiased estimator of the population variance σ^2, regardless of the probability distribution or probability density of the population. In addition, the theorem explains why it is important that we divide by $n-1$ instead of n in the computation of the sample variance. It is important to note that the unbiasedness is a property of the sample variance, but not of the sample standard deviation S. In general,

$$E(S) \neq \sigma.$$

This implies that the sample standard deviation S is a biased estimator of the population standard deviation. We briefly discuss the sample standard deviation in Section 1.8. Before we discuss the distribution of the sample variance S^2, we introduce a new probability density, namely the χ^2-distribution (pronounced "chi-square").

1.7.2   The χ^2-Distribution

Apart from the normal distribution, the χ^2-distribution[7], which is derived from normal distributions, is a very important family of probability densities. This family has one parameter k, which is called the **degrees of freedom**. The probability density of this distribution is

$$f_X(x; k) = \frac{x^{\frac{k}{2}-1} e^{-x/2}}{\Gamma\left(\frac{k}{2}\right) 2^{\frac{k}{2}}}, \text{ for } x > 0.$$

---

[7] The name of the χ^2-distribution derives from the fact that the square of a standard normally distributed random variable has a χ^2-distribution: a random variable is typically denoted by an X and its square by X^2, and X is the capital form of the Greek letter χ.

In this expression, $\Gamma()$ is the so-called gamma function[8]. The χ^2-distribution is a special case of a gamma distribution (see *Statistics with JMP: Graphs, Descriptive Statistics and Probability*). The expected value and variance of a χ^2-distributed random variable X with k degrees of freedom are

$$\mu_X = E(X) = k$$

and

$$\sigma_X^2 = \text{var}(X) = 2k,$$

respectively. The median is approximately equal to

$$k\left(1 - \frac{2}{9k}\right)^3.$$

To indicate that a random variable has a χ^2-distribution with k degrees of freedom – that is, with parameter k – we use the notation

$$X \sim \chi_k^2.$$

χ^2-distributions with two, four, eight, and 12 degrees of freedom are shown in Figure 1.12.

Probabilities based on the χ^2-distribution can, of course, be computed using JMP. The probability

$$P(\chi_k^2 \leq x),$$

where χ_k^2 is a χ^2-distributed random variable with k degrees of freedom, can be calculated using the formula "ChiSquare Distribution(x, k)". If you want to find a quantile or percentile of a χ^2-distributed random variable with k degrees of freedom, the function "ChiSquare Quantile(p, k)" is available. Some specific quantiles can also be found in the table in Appendix C.

1.7.3   The Relation Between the Standard Normal and the χ^2-Distribution

A sum of k squared independent standard normally distributed random variables $X_1, X_2, \ldots, X_k$ is χ^2-distributed with k degrees of freedom. In other words, the random variable

$$Y = \sum_{i=1}^{k} X_i^2,$$

---

[8] The gamma function is an extension of the factorial function for integers n: $\Gamma(n+1) = n! = n \times (n-1) \times \cdots \times 2 \times 1$. If z is a positive real number, then $\Gamma(z) = (z-1)\Gamma(z-1) = \int_0^{+\infty} e^{-t} t^{z-1} dt$. A special case is $\Gamma(1/2) = \sqrt{\pi}$. In addition, $\Gamma(2) = \Gamma(1) = 1! = 0! = 1$, and $\Gamma(n) = (n-1)!$ for all integers n.

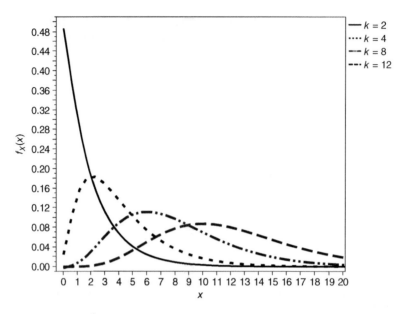

Figure 1.12 χ^2-distributions with two, four, eight, and 12 degrees of freedom.

where the X_i are standard normally distributed, is χ^2-distributed. The number of degrees of freedom of the resulting χ^2-distribution is equal to the number of independent variables involved in the sum of squares.

This assertion can easily be verified using JMP. To do so, generate 10 000 sets of eight pseudo-random numbers from the standard normal density. Next, square all numbers in each set of eight and sum the squares. Finally, create a histogram of the 10 000 sums of squares. You can then compare the shape of the histogram with that of the χ^2-distribution with eight degrees of freedom in Figure 1.12. Alternatively, you can compute the mean and the variance of the 10 000 sums of squares that you obtained. The mean should lie close to 8, while the variance should be close to 16. A histogram that was obtained in this way is shown in Figure 1.13. As the JMP output in Figure 1.14 shows, the mean of the 10 000 values is 7.9755, while the variance is 15.8786. Obviously, if you repeat this exercise in JMP by yourself, you will obtain other pseudo-random numbers and therefore a (slightly) different mean and a (slightly) different variance.

As the number of degrees of freedom of the χ^2-distribution increases, the distribution looks more and more like a normal probability density. This is a consequence of the central limit theorem: a χ^2-distributed random variable with a large number of degrees of freedom is a sum of a large number of independent random variables (the square of a standard normally distributed random variable is, of course, also a random variable), and the central limit theorem tells us that such a sum is approximately normally distributed. Figure 1.15 compares the χ^2-distributions with 25 and 30 degrees of freedom with the normal probability density with $\mu = 25$ and $\sigma^2 = 50$ and the normal probability density with $\mu = 30$ and $\sigma^2 = 60$, respectively. The figure clearly shows that, for a large number of degrees of freedom, there is a strong resemblance between the χ^2-distribution and the normal distribution. Figure 1.12 shows that this resemblance is absent if the number of degrees of freedom is small.

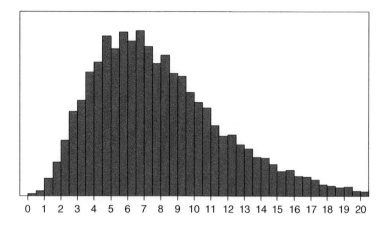

Figure 1.13 The histogram obtained by generating 10 000 sets of eight (pseudo-random) draws from a standard normal density, squaring them, and summing the squares.

Summary Statistics	
Mean	7.9754589
Std Dev	3.984791
Std Err Mean	0.0398479
Upper 95% Mean	8.0535688
Lower 95% Mean	7.897349
N	10000
Variance	15.878559

Figure 1.14 The descriptive statistics of the 10 000 (pseudo-random) numbers shown in Figure 1.13.

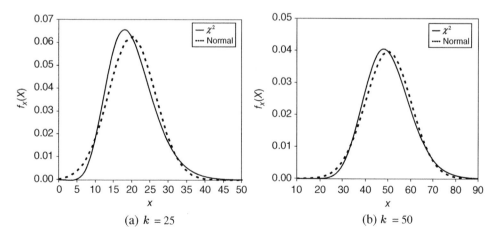

(a) $k = 25$ (b) $k = 50$

Figure 1.15 A comparison of a χ^2-distribution with k degrees of freedom and the corresponding normal probability density with $\mu = k$ and $\sigma^2 = 2k$.

1.7.4   The Probability Density of the Sample Variance

Starting with the definition of a sample variance in Equation (1.1), it is not difficult to see that there is a connection between the sample variance and the χ^2-distribution. Dividing both sides of Equation (1.1) by σ^2 and multiplying by $(n-1)$ yields

$$\frac{(n-1)S^2}{\sigma^2} = \frac{\sum_{i=1}^{n}(X_i - \overline{X})^2}{\sigma^2},$$

$$= \sum_{i=1}^{n}\left(\frac{X_i - \overline{X}}{\sigma}\right)^2. \tag{1.2}$$

Replacing the sample mean $\overline{X}$ in the right-hand side of this expression by the population mean μ, we obtain

$$\sum_{i=1}^{n}\left(\frac{X_i - \mu}{\sigma}\right)^2.$$

If $X_1, X_2, \ldots, X_n$ are independent normally distributed random variables with expected value μ and variance σ^2, then this expression is a sum of squared independent standard normally distributed random variables. Accordingly, this expression has a χ^2-distribution with n degrees of freedom.

In Equation (1.2), however, the population mean μ is estimated by the sample mean $\overline{X}$. As a consequence, the n terms in the sum

$$\sum_{i=1}^{n}\frac{(X_i - \overline{X})^2}{\sigma^2}$$

are not independent. This is due to the fact that

$$\overline{X} = \frac{1}{n}\sum_{i=1}^{n}X_i, \tag{1.3}$$

which implies that the X_i are subject to the following constraint:

$$\sum_{i=1}^{n}(X_i - \overline{X}) = 0.$$

As a result of this one constraint, the sum in Equation (1.3) contains only $n-1$ independent terms. Therefore, the χ^2-distribution of $(n-1)S^2/\sigma^2$ has only $n-1$ degrees of freedom.

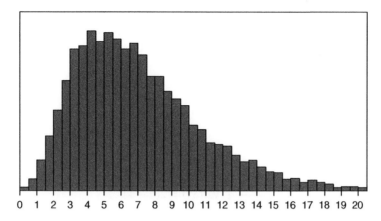

Figure 1.16   The histogram of 10 000 (pseudo-random) values of the random variable $(n-1)S^2/\sigma^2$.

Theorem 1.7.2   *Let $X_1, X_2, \ldots, X_n$ be independent normally distributed random variables with variance σ^2, and $n \geq 2$. Then, the random variable*

$$\frac{(n-1)S^2}{\sigma^2}$$

has a χ^2-distribution with $n-1$ degrees of freedom.

These theoretical results can also be verified using JMP. To do so, we can generate 10 000 sets of eight numbers from a normal density. We determine the sample variance for each of these 10 000 sets, multiply it by $n-1=7$, and divide the result by σ^2. A histogram of 10 000 values obtained in this way is shown in Figure 1.16. The mean of all these numbers is 7.0280 (see Figure 1.17), which indicates that the underlying χ^2-distribution has only seven degrees of freedom instead of eight. The histogram in Figure 1.16 looks a lot like the one in Figure 1.13 (generated from a χ^2-distribution with eight degrees of freedom). However, the difference between the two histograms is that the maximum is reached slightly earlier in Figure 1.16 than in the histogram in Figure 1.13, as a result of the smaller number of degrees of freedom of the distribution in Figure 1.16.

Summary Statistics	
Mean	7.0280156
Std Dev	3.7366969
Std Err Mean	0.037367
Upper 95% Mean	7.1012624
Lower 95% Mean	6.9547688
N	10000
Variance	13.962903

Figure 1.17   The descriptive statistics of the 10 000 (pseudo-random) values of the random variable $(n-1)S^2/\sigma^2$ shown in Figure 1.16.

Now that the probability density of the sample variance is known, we can determine the variance of the sample variance. We know that the variance of a χ^2-distributed random variable with k degrees of freedom is equal to $2k$. Thus a χ^2-distributed random variable with $n-1$ degrees of freedom has a variance of $2(n-1)$. Therefore,

$$\text{var}\left\{ \frac{(n-1)S^2}{\sigma^2} \right\} = 2(n-1).$$

Hence,

$$\left(\frac{n-1}{\sigma^2} \right)^2 \text{var}(S^2) = 2(n-1),$$

and

$$\text{var}(S^2) = 2(n-1) \left(\frac{\sigma^2}{n-1} \right)^2 = \frac{2\sigma^4}{n-1}.$$

The variance of the sample variance S^2 thus decreases with the number of observations in the sample, n. In other words, the sample variance is a more precise and efficient estimator when a larger amount of data is used. The same applies to the sample mean and the sample proportion.

Finally, it should be emphasized that the use of the χ^2-distribution in the context of sample variances is only valid for normally distributed variables $X_1, X_2, \ldots, X_k$. In other words, if you study a nonnormally distributed population, you should not assume that $(n-1)S^2/\sigma^2$ has a χ^2-distribution. Therefore, the above expression for the variance of the sample variance, $\text{var}(S^2)$, is also invalid for nonnormally distributed populations.

1.8   The Sample Standard Deviation

The sample standard deviation S is the square root of the sample variance S^2:

$$S = \sqrt{\frac{1}{n-1} \sum_{i=1}^{n} (X_i - \overline{X})^2}.$$

In Section 1.7.1, we showed that the sample variance is an unbiased estimator of the population variance. However, the sample standard deviation turns out to be a biased estimator of the population standard deviation. It can be shown that

$$E(S) < \sigma,$$

so that the sample standard deviation S generally underestimates the population standard deviation σ.

If the population under investigation is normally distributed, a bias correction can be applied to eliminate the systematic underestimation. This correction is based on the fact that

$$E(S) = \sigma \sqrt{\frac{2}{n-1}} \cdot \frac{\Gamma\left(\frac{n}{2}\right)}{\Gamma\left(\frac{n-1}{2}\right)}, \tag{1.4}$$

where $\Gamma()$ again represents the gamma function. The bias correction factor

$$\sqrt{\frac{2}{n-1}} \cdot \frac{\Gamma\left(\frac{n}{2}\right)}{\Gamma\left(\frac{n-1}{2}\right)}$$

is referred to as c_4 in the statistical literature.

It follows from the expression for the expected value of S in Equation (1.4) that the corrected sample standard deviation

$$S^* = \frac{S}{c_4} = S \sqrt{\frac{n-1}{2}} \cdot \frac{\Gamma\left(\frac{n-1}{2}\right)}{\Gamma\left(\frac{n}{2}\right)}$$

Table 1.1 Values of the factor c_4 and the extent to which the population standard deviation is underestimated by the sample standard deviation for a normally distributed population.

n	c_4 Exact	Approximation	Underestimation (%)
2	$\sqrt{\frac{2}{\pi}}$	0.7979	20.21
3	$\frac{\sqrt{\pi}}{2}$	0.8862	11.38
4	$2\sqrt{\frac{2}{3\pi}}$	0.9213	7.87
5	$\frac{3}{4}\sqrt{\frac{\pi}{2}}$	0.9400	6.00
6	$\frac{8}{3}\sqrt{\frac{2}{5\pi}}$	0.9515	4.85
7	$\frac{5\sqrt{3\pi}}{16}$	0.9594	4.06
8	$\frac{16}{5}\sqrt{\frac{2}{7\pi}}$	0.9650	3.50
9	$\frac{35\sqrt{\pi}}{64}$	0.9693	3.07
10	$\frac{128}{105}\sqrt{\frac{2}{\pi}}$	0.9727	2.73
100	$\frac{6511077190}{1570338389}\sqrt{\frac{2}{11\pi}}$	0.9975	0.25

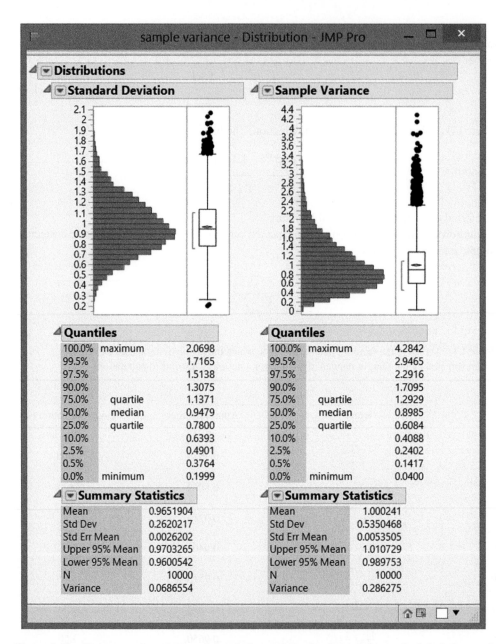

Figure 1.18 Histograms, box plots, and descriptive statistics for the sample standard deviations and variances of 10 000 samples of eight (pseudo-random) values drawn from a standard normally distributed population.

is an unbiased estimator of the population standard deviation σ. The extent to which σ is underestimated is larger for a small number of observations. The factor c_4 is approximately 0.80 or 80% if $n = 2$. This means that the sample standard deviation yields estimates that are generally 20% smaller than the population standard deviation. If n is equal to 10, the factor c_4 is approximately 0.97. In that case, the sample standard deviation provides estimates that are, on average, 3% too low. If the number of observations increases up to 100, then c_4 is nearly equal to 1, and the sample standard deviation is almost an unbiased estimator. Table 1.1 contains a list of values of the factor c_4 and the degree of underestimation of the population standard deviation by the sample standard deviation. Note that in this table, π is not a success probability, but the circle constant $3.1415\ldots$.

To illustrate all this, we have used JMP to generate 10 000 samples consisting of eight pseudo-random numbers drawn from the standard normal probability distribution. As a result, both the population variance and the population standard deviation are equal to 1 because the data is generated from the standard normal probability density. For each sample, the sample standard deviation and the sample variance are computed. Histograms, box plots, and descriptive statistics for the 10 000 sample standard deviations and sample variances are shown in Figure 1.18. The mean sample standard deviation is 0.9652, so that the population standard deviation is underestimated by about 3.5%. This corresponds to the value of c_4 and the underestimation displayed in Table 1.1. The mean sample variance is 1.0002, which is very close to the population variance of 1. This illustrates that the sample variance is an unbiased estimator of the population variance. The variance of the 10 000 sample variances is 0.2863, which is close to the theoretical value of $2\sigma^4/(n-1) = 2 \times 1^4/(8-1) = 2/7 = 0.2857$.

1.9 Applications

The derived probability densities for the sample mean, the sample proportion, and the sample variance are the cornerstones for the remaining chapters of the book. The normal (and therefore also the standard normal) distribution, and the χ^2-distribution will be used to establish so-called confidence intervals for the population mean, the population proportion, and the population variance. In addition, two new probability densities, namely the t-distribution and the F-distribution, will be derived. These new distributions will be needed for other confidence intervals and hypothesis tests.

2

Interval Estimators

If this fermata would have allowed for a breeze, I might have seen the stones swaying lazily on their strings, circumscribing the confidence interval of their trajectories.

(from *Omega Minor*, Paul Verhaeghen, p. 123)

2.1 Point and Interval Estimators

In the previous chapter, we explained how to use the sample mean $\overline{X}$ as an estimator for the unknown population mean μ, and how the sample variance S^2 can be used as an estimator of the population variance. Since these estimators result in exactly one value, they are also called **point estimators**.

One problem with these (point) estimators is that they do not contain any information on how reliable the estimation of the unknown population parameter is. For example, for a given estimate such as the sample mean, a researcher would like to get an idea of the extent to which it may differ from the unknown population mean. The calculations in Chapter 1 revealed that the variance of the sample mean depends on the variance σ^2 of the unknown population, and the number of observations in the sample. Indeed, the variance of the sample mean $\overline{X}$ is equal to σ^2/n (see the central limit theorem and Theorem 1.5.2). By using so-called **interval estimators**, we can assign a certain confidence to a point estimator. The computing of an interval estimate involves the calculation of a lower limit and an upper limit based on sample data. The intention is that this interval contains the unknown estimated parameter with a high probability. Such an interval is called a **confidence interval**. A confidence interval depends on the desired confidence level.

Definition 2.1.1 *The **confidence level** $1 - \alpha$ of a confidence interval $[L, U]$ for an unknown parameter θ of a population or a process is the probability that the interval contains the parameter θ:*

$$P(L \leq \theta \leq U) = 1 - \alpha.$$

Statistics with JMP: Hypothesis Tests, ANOVA and Regression, First Edition. Peter Goos and David Meintrup.
© 2016 John Wiley & Sons, Ltd. Published 2016 by John Wiley & Sons, Ltd.
Companion Website: http://www.wiley.com/go/goosandmeintrup/JMP

The notation L and U is derived from the terms used for the bounds of the interval, namely **lower bound** and **upper bound**. As the confidence level $1 - \alpha$ represents a probability, it has to lie between 0 and 1. Correspondingly, α needs to be between 0 and 1 as well.

Strictly speaking, the probability $1 - \alpha$ in the above definition should not be interpreted as the probability that the unknown parameter θ lies in the interval. In fact, the unknown population parameter does not vary. Instead, it is the limits of the confidence interval that are random variables and that vary from one data set to another. As a result, each sample leads to a different lower and upper bound, and hence to a different confidence interval. This is why the lower and upper bounds of the interval, L and U, are denoted by uppercase letters. Approximately, a fraction of $1 - \alpha$ of these intervals will contain the parameter θ. The interpretation of this is as follows: if 1000 data sets are collected independently by 1000 researchers to study a certain parameter θ and if each of these researchers constructs his own 95% confidence interval for θ, then about 950 of the researchers will obtain an interval that includes θ. The remaining researchers will obtain an interval that does not include θ. The latter researchers have bad luck, in the sense that their particular data set contained misleading information concerning θ, resulting in either an underestimation or an overestimation.

Ideally, confidence intervals have the following properties:

- The confidence level should be high. Typically, one opts for a confidence level $1 - \alpha$ of 90%, 95%, or even 99%.
- The interval should be as narrow as possible. Indeed, a wide interval contains information that is not very precise.

These two requirements are difficult to fulfill simultaneously unless the sample size n is increased.

2.2 Confidence Intervals for a Population Mean with Known Variance

For simplicity, we assume in this section that the population mean μ is unknown, but the variance σ^2 of the population is known. This assumption is not realistic, but we use it for educational purposes.

2.2.1 The Percentiles of the Standard Normal Density

The construction of confidence intervals requires some new notation. For any α value between 0 and 1, we denote by z_α the value that satisfies

$$P(Z \geq z_\alpha) = \alpha,$$

where Z represents a standard normally distributed random variable. Due to the symmetry of the standard normal probability density,

$$z_{1-\alpha} = -z_\alpha$$

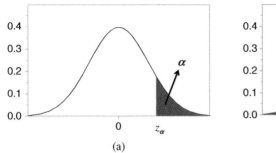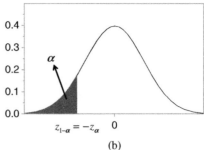

(a)   (b)

Figure 2.1   The graphical representation of z_α and $z_{1-\alpha} = -z_\alpha$.

and

$$P(Z \leq -z_\alpha) = \alpha.$$

The values $z_{1-\alpha} = -z_\alpha$ and z_α equal the $\alpha \times 100$th percentile (or quantile) and the $(1 - \alpha) \times$ 100th percentile (or quantile), respectively. The meaning of z_α and $z_{1-\alpha} = -z_\alpha$ is illustrated in Figure 2.1. In later chapters, the percentiles $-z_\alpha$, z_α, $-z_{\alpha/2}$, and $z_{\alpha/2}$ will be used as **critical values** in various hypothesis tests. The percentiles $z_{1-\alpha/2} = -z_{\alpha/2}$ and $z_{\alpha/2}$ are shown in Figure 2.2. For these percentiles, we have

$$P(z_{1-\alpha/2} \leq Z \leq z_{\alpha/2}) = P(-z_{\alpha/2} \leq Z \leq z_{\alpha/2}) = 1 - \alpha.$$

Note that, for any value of $1 - \alpha$ between 0 and 1, the interval $[-z_{\alpha/2}, z_{\alpha/2}] = [z_{1-\alpha/2}, z_{\alpha/2}]$ is the narrowest possible interval that contains a fraction $1 - \alpha$ of the values that a standard normally distributed random variable can take.

The percentiles of the standard normal probability density can be derived from the table in Appendix B. Alternatively, they can be calculated in JMP. For example, z_α can be computed using the command "Normal Quantile$(1 - \alpha)$". If one chooses α to be 10% or 0.10, then, according to the table in Appendix B, z_α is approximately 1.28. The command "Normal Quantile(0.9)" in JMP gives us a more precise value, namely 1.281552.

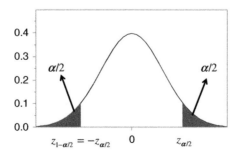

Figure 2.2   The graphical representation of $z_{1-\alpha/2} = -z_{\alpha/2}$ and $z_{\alpha/2}$.

2.2.2   Computing a Confidence Interval

A consequence of the central limit theorem (Theorem 1.5.6) is that a sample mean is approximately normally distributed regardless of the probability distribution or density of the population under investigation, provided that the sample is sufficiently large. For small samples, the sample mean can also be normally distributed, but this requires that the population under investigation is normally distributed. This follows from Theorem 1.5.3. In both cases, we have that

$$\overline{X} \sim N\left(\mu, \frac{\sigma^2}{n}\right)$$

whenever the sample data $X_1, X_2, \dots, X_n$ comes from a probability distribution or density with mean μ and variance σ^2. Consequently,

$$Z = \frac{\overline{X} - \mu}{\frac{\sigma}{\sqrt{n}}} \sim N(0, 1),$$

and hence

$$P\left(-z_{\alpha/2} \le \frac{\overline{X} - \mu}{\frac{\sigma}{\sqrt{n}}} \le z_{\alpha/2}\right) = 1 - \alpha.$$

This expression can then be turned into a confidence interval for μ. The required intermediate steps to do so are the following:

$$P\left(-z_{\alpha/2}\frac{\sigma}{\sqrt{n}} \le \overline{X} - \mu \le z_{\alpha/2}\frac{\sigma}{\sqrt{n}}\right) = 1 - \alpha,$$

$$P\left(-z_{\alpha/2}\frac{\sigma}{\sqrt{n}} \le \mu - \overline{X} \le z_{\alpha/2}\frac{\sigma}{\sqrt{n}}\right) = 1 - \alpha,$$

and

$$P\left(\overline{X} - z_{\alpha/2}\frac{\sigma}{\sqrt{n}} \le \mu \le \overline{X} + z_{\alpha/2}\frac{\sigma}{\sqrt{n}}\right) = 1 - \alpha.$$

Now, the confidence interval is

$$\left[\overline{X} - z_{\alpha/2}\frac{\sigma}{\sqrt{n}} \; , \; \overline{X} + z_{\alpha/2}\frac{\sigma}{\sqrt{n}}\right]. \tag{2.1}$$

Because the interval $[-z_{\alpha/2}, z_{\alpha/2}]$ is the narrowest possible interval that contains a fraction $1 - \alpha$ of the values that a standard normally distributed random variable can take, the confidence interval in Expression (2.1) is also the narrowest one possible. Therefore, it provides the best possible idea of the magnitude of the population mean μ.

2.2.3   The Width of a Confidence Interval

The bounds of the confidence interval in Expression (2.1) depend on the sample mean and therefore also on the sample data. In addition, the standard deviation σ, the sample size n, and the value chosen for α each play a role. The width of the confidence interval is

$$B = \overline{X} + z_{\alpha/2}\frac{\sigma}{\sqrt{n}} - \left(\overline{X} - z_{\alpha/2}\frac{\sigma}{\sqrt{n}}\right) = 2z_{\alpha/2}\frac{\sigma}{\sqrt{n}}. \tag{2.2}$$

This expression shows that the width of the interval:

- decreases with the sample size n;
- increases if a higher confidence level $1 - \alpha$ (and thus a smaller α value) is selected; and
- increases with the variance σ^2 of the population under investigation.

Generally, the value of the variance σ^2 is beyond the control of the researcher. Only the sample size n and the value of α are determined by the researcher.

Figure 2.3 shows the extent to which the sample size n, the confidence level $1 - \alpha$, and the standard deviation σ affect the width of a confidence interval. For a confidence level of

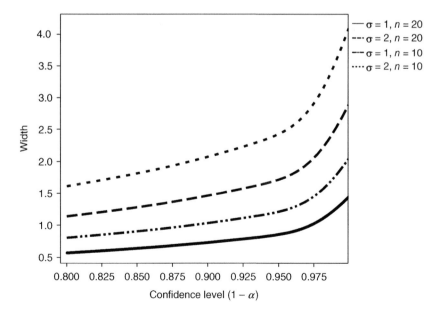

Figure 2.3   The influence of the sample size n, the confidence level $1 - \alpha$, and the standard deviation σ on the width of a confidence interval for a population with mean μ and known standard deviation σ.

100% or 1, the confidence interval is infinitely wide. Therefore, this value is not included in the figure.

In JMP, two scripts are available for determining a confidence interval or interval estimate, namely "CI for Mean from Data" and "CI for Mean from Summary Statistics". The former script is useful if you have a data table with all data points. The latter script is useful if you only have access to summary statistics, but not to the original data.

Example 2.2.1   *The temperature at which pure milk freezes is normally distributed with mean $\mu = -0.545\,°C$ and standard deviation $\sigma = 0.008\,°C$. In order to determine whether a batch of milk from a certain supplier was diluted with water, we investigate a sample of 5 bottles from the batch. The mean freezing point of the milk samples is $\bar{x} = -0.535\,°C$. A 99% confidence interval (correspondingly, α is $1 - 0.99 = 0.01$, so that $\alpha/2 = 0.005$) for the average freezing point of the milk batch is given by*

$$\left[-0.535 - z_{0.005}\frac{0.008}{\sqrt{5}} \ , \ -0.535 + z_{0.005}\frac{0.008}{\sqrt{5}} \right].$$

With the formula "`Normal Quantile(1 - 0.005)`*" or, alternatively, "*`Normal Quantile (0.995)`*", we find that $z_{0.005} = 2.5758$, which is slightly more precise than the value of 2.58 from the table in Appendix B. We therefore obtain the confidence interval*

$$\left[-0.535 - 2.5758\frac{0.008}{\sqrt{5}} \ , \ -0.535 + 2.5758\frac{0.008}{\sqrt{5}} \right],$$

or

$$[-0.5442 \ , \ -0.5258].$$

In a similar way, a 95% confidence interval ($\alpha = 1 - 0.95 = 0.05$, so that $\alpha/2 = 0.025$) for the freezing point of the batch of milk can be calculated:

$$[-0.5420 \ , \ -0.5280].$$

First, it is useful to note that the 95% confidence interval is narrower than the 99% confidence interval. In addition, it is striking that neither of the two intervals contains the population mean $\mu = -0.545$. This indicates that the milk supplier most likely (100% certainty is impossible) diluted the milk with water.

Figure 2.4 shows how the script "CI for Mean from Summary Statistics" can be used for the computation of the 99% confidence interval in this example.

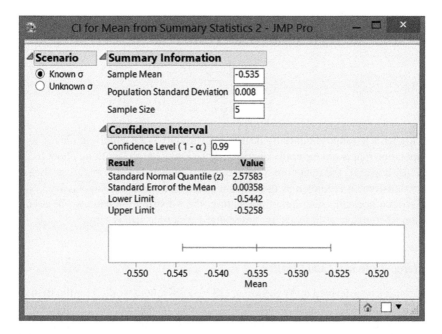

Figure 2.4 The use of the script "CI for Mean from Summary Statistics" in JMP for Example 2.2.1 involving a known σ.

2.2.4 The Margin of Error

Half of the width B of a confidence interval is called the **margin of error** of the point estimate of the unknown population mean. In other words, the margin of error is

$$b = \frac{B}{2} = z_{\alpha/2} \frac{\sigma}{\sqrt{n}}. \tag{2.3}$$

Just like the width of a confidence interval, the margin of error depends on the confidence level chosen, the number of observations, and the variance or standard deviation.

2.3 Confidence Intervals for a Population Mean with Unknown Variance

In practice, it is unlikely that a researcher will know the population variance σ^2. In general, the variance must be estimated, and, of course, this is done using the sample variance S^2. In that case, the confidence interval from the previous section can no longer be used and the symbol σ in the expression

$$Z = \frac{\overline{X} - \mu}{\frac{\sigma}{\sqrt{n}}},$$

which formed the starting point for the construction of the confidence interval, needs to be replaced by S. When doing so, we obtain

$$T = \frac{\overline{X} - \mu}{\frac{S}{\sqrt{n}}},$$

which no longer is a standard normally distributed variable. For a small sample $X_1, X_2, \ldots, X_n$ from a population that is not normally distributed, we know hardly anything about the random variable T. If, however, the data comes from a normally distributed population, then T has a probability density that is known as the Student t-distribution. If the data $X_1, X_2, \ldots, X_n$ does not come from a normally distributed population, the t-distribution can also be used for the computation of confidence intervals, provided that the sample size is large.

2.3.1   The Student t-Distribution

The t-distribution was derived by W.S. Gosset (1876–1937), who worked for the Irish brewery Guinness from 1899. To inspect the quality of the beer produced, he developed a number of statistical methods based on the t-distribution. Because Gosset's employer did not allow scientific publications, Gosset published his findings under the pseudonym of Student.

A t-distributed random variable T is defined as the quotient of a standard normally distributed random variable and the square root of a χ^2-distributed random variable, divided by its degrees of freedom. The standard normal and the χ^2-distribution have to be independent. Let X be a χ^2-distributed random variable with k degrees of freedom, and Z a standard normally distributed random variable. Then, a t-distributed random variable T is obtained as

$$T = \frac{Z}{\sqrt{\frac{X}{k}}}.$$

Just like a standard normally distributed random variable, a t-distributed random variable can take values between $-\infty$ and $+\infty$. The t-distribution has one parameter, namely the degrees of freedom k of the χ^2-distributed random variable, and is given by

$$f(t; k) = \frac{\Gamma\left(\frac{k+1}{2}\right)}{\Gamma\left(\frac{k}{2}\right)\sqrt{k\pi}} \left(1 + \frac{t^2}{k}\right)^{-\frac{k+1}{2}},$$

where $\Gamma()$ again represents the gamma function and π is the circle constant $3.1415 \ldots$. The expected value of a t-distributed random variable is zero, as the probability density is symmetric around zero and resembles the standard normal distribution. However, the t-distribution has a slightly lower peak than the standard normal probability density, and hence the probabilities for values further away from zero are slightly greater. However, as the number of degrees of

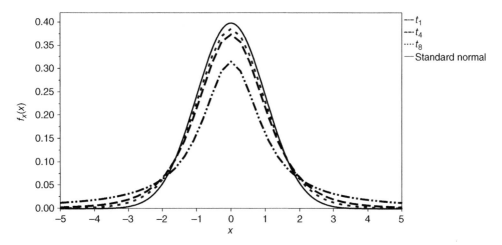

Figure 2.5 The graphical representation of the probability densities of t-distributed random variables with one, four, and eight degrees of freedom, and the standard normal density.

freedom k increases, the t-distribution increasingly resembles the standard normal distribution. All this is illustrated in Figure 2.5, where densities of t-distributed random variables with one, four, and eight degrees of freedom are compared to the standard normal density. We denote by t_k a t-distributed random variable with k degrees of freedom. Note that here, in order to be consistent with the traditional notation of a t-distributed random variable in the statistical literature, a small letter is exceptionally used to indicate a random variable. To indicate that a random variable X has a t-distribution, we use the notation

$$X \sim t_k,$$

where k represents the degrees of freedom.

Probabilities for the t-distribution can be calculated in JMP. Percentiles or quantiles of the t-distribution, which later will appear in confidence intervals and in hypothesis tests, can be determined by using JMP or by looking them up in the table in Appendix D. For example, the probability that a t-distributed random variable with five degrees of freedom is smaller than or equal to 2, $P(t_5 \leq 2)$, can be calculated in JMP using the command "t Distribution (2, 5)", and is equal to 0.9490. The first argument of the formula "t Distribution" is the value 2; the second one is the number of degrees of freedom of the t-distribution.

If we want to determine the values $t_{\alpha;k}$ for which

$$P(t_k \geq t_{\alpha;k}) = \alpha,$$

then, for α values of 0.4, 0.25, 0.15, 0.10, 0.05, 0.025, 0.01, 0.005, 0.001, and 0.0005, we can consult the table in Appendix D. For example, $t_{0.025;5} = 2.5706$, which means that

$$P(t_5 \geq 2.5706) = 0.025.$$

Table 2.1 The percentiles or quantiles $t_{\alpha;k}$ of four different t-distributions and z_α of the standard normal distribution.

	Degrees of freedom, k, of t-distribution				Standard normal
$1 - \alpha$	1	2	4	8	
0.005	−63.656741	−9.924843	−4.604095	−3.355387	−2.575829
0.01	−31.820516	−6.964557	−3.746947	−2.896459	−2.326348
0.025	−12.706205	−4.302653	−2.776445	−2.306004	−1.959964
0.05	−6.313752	−2.919986	−2.131847	−1.859548	−1.644854
0.1	−3.077684	−1.885618	−1.533206	−1.396815	−1.281552
0.9	3.077684	1.885618	1.533206	1.396815	1.281552
0.95	6.313752	2.919986	2.131847	1.859548	1.644854
0.975	12.706205	4.302653	2.776445	2.306004	1.959964
0.99	31.820516	6.964557	3.746947	2.896459	2.326348
0.995	63.656741	9.924843	4.604095	3.355387	2.575829

This value can also be calculated using JMP. The required formula is "t Quantile($1 - \alpha$, k)". For the calculation of $t_{0.025;5}$, this becomes "t Quantile(0.975, 5)". Note that $t_{0.025;5}$ is the 97.5th percentile of a t-distributed random variable with five degrees of freedom. In later chapters, we call such percentiles "critical values".

Table 2.1 lists the main percentiles or quantiles for four different t-distributions and the standard normal distribution. Note that the percentiles of the t-distributions are more extreme than those of the standard normal distribution. In other words, the percentiles of a t-distributed random variable are further away from zero than the percentiles of a standard normal random variable.

The t-distribution is symmetrical; hence $t_{1-\alpha;n-1} = -t_{\alpha;n-1}$. For example, the 10th percentile, $t_{0.90;8} = -1.3968$, is the negative of the 90th percentile, $t_{0.10;8} = 1.3968$. In other words, $t_{0.90;8} = -t_{0.10;8}$. A graphical representation of these statistics is given in Figure 2.6. The

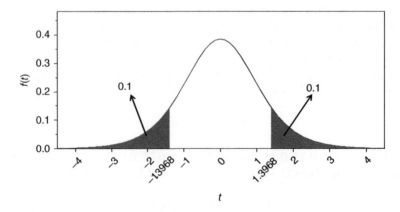

Figure 2.6 The graphical representation of $t_{0.90;8} = -t_{0.10;8} = -1.3968$ and $t_{0.10;8} = 1.3968$.

white-colored area under the curve in the figure is the probability that a t-distributed random variable with eight degrees of freedom takes values between $-t_{0.10;8}$ and $t_{0.10;8}$:

$$P\left(-t_{0.10;8} \leq t_8 \leq t_{0.10;8}\right) = P\left(-1.3968 \leq t_8 \leq 1.3968\right) = 0.80.$$

2.3.2 The Application of the t-Distribution to Construct Confidence Intervals

It is not obvious that

$$T = \frac{\overline{X} - \mu}{\frac{S}{\sqrt{n}}}$$

is a t-distributed random variable. However, this becomes clear if we rewrite the expression for T as

$$T = \frac{\frac{\overline{X} - \mu}{\sigma}}{\frac{S}{\sigma\sqrt{n}}} = \frac{\frac{\overline{X} - \mu}{\sigma/\sqrt{n}}}{\frac{S}{\sigma}} = \frac{\frac{\overline{X} - \mu}{\sigma/\sqrt{n}}}{\sqrt{\frac{S^2}{\sigma^2}}} = \frac{\frac{\overline{X} - \mu}{\sigma/\sqrt{n}}}{\sqrt{\frac{(n-1)S^2}{(n-1)\sigma^2}}}. \tag{2.4}$$

If we assume that the sample data, $X_1, X_2, \ldots, X_n$, is normally distributed with expected value μ and variance σ^2, then $\overline{X}$ is normally distributed with mean μ and variance σ^2/n, and $\frac{\overline{X}-\mu}{\sigma/\sqrt{n}}$ has a standard normal distribution. Moreover, we know from Section 1.7.4 that $(n-1)S^2/\sigma^2$ is χ^2-distributed with $n-1$ degrees of freedom if the sample data is normally distributed. In that case,

$$\sqrt{\frac{(n-1)S^2}{(n-1)\sigma^2}}$$

is the square root of a χ^2-distributed random variable divided by its degrees of freedom. In conclusion, we have shown that T is a standard normally distributed random variable divided by the square root of a χ^2-distributed random variable divided by its degrees of freedom. It can also be shown that these two probability densities are independent, so that we can conclude that T is a t-distributed random variable.

In short,

$$T = \frac{\overline{X} - \mu}{\frac{S}{\sqrt{n}}} \sim t_{n-1},$$

so that

$$P\left(-t_{\alpha/2;n-1} \leq \frac{\overline{X} - \mu}{\frac{S}{\sqrt{n}}} \leq t_{\alpha/2;n-1}\right) = 1 - \alpha.$$

In order to construct a $(1 - \alpha) \times 100\%$ confidence interval for μ starting from this expression, we need the following intermediate steps:

$$P\left(-t_{\alpha/2;n-1}\frac{S}{\sqrt{n}} \leq \overline{X} - \mu \leq t_{\alpha/2;n-1}\frac{S}{\sqrt{n}}\right) = 1 - \alpha,$$

$$P\left(-t_{\alpha/2;n-1}\frac{S}{\sqrt{n}} \leq \mu - \overline{X} \leq t_{\alpha/2;n-1}\frac{S}{\sqrt{n}}\right) = 1 - \alpha,$$

and

$$P\left(\overline{X} - t_{\alpha/2;n-1}\frac{S}{\sqrt{n}} \leq \mu \leq \overline{X} + t_{\alpha/2;n-1}\frac{S}{\sqrt{n}}\right) = 1 - \alpha.$$

Therefore, the confidence interval is

$$\left[\overline{X} - t_{\alpha/2;n-1}\frac{S}{\sqrt{n}} , \ \overline{X} + t_{\alpha/2;n-1}\frac{S}{\sqrt{n}}\right].$$

The JMP scripts "CI for Mean from Data" and "CI for Mean from Summary Statistics" can also be used to compute confidence intervals with unknown variance σ^2. You should always explicitly specify that you want to use the t-distribution instead of the standard normal distribution, by selecting the option "Unknown σ".

Example 2.3.1   *A sample of $n = 20$ observations shows that a filling machine on average puts $\overline{x} = 49.54$ d of a certain liquid in a bottle. The sample variance is $(0.4294\,\mathrm{d})^2$. A 95% confidence interval ($\alpha = 1 - 0.95 = 0.05$, so that $\alpha/2 = 0.025$) is given by*

$$\left[49.54 - t_{0.025;19}\frac{0.4294}{\sqrt{20}} , \ 49.54 + t_{0.025;19}\frac{0.4294}{\sqrt{20}}\right].$$

The required percentile $t_{0.025;19}$ is 2.0930, so that we obtain the interval

$$\left[49.54 - 2.0930\frac{0.4294}{\sqrt{20}} , \ 49.54 + 2.0930\frac{0.4294}{\sqrt{20}}\right] = [49.3390, 49.7410].$$

Figure 2.7 illustrates the use of the script "CI for Mean from Summary Statistics" for the computation of this interval.

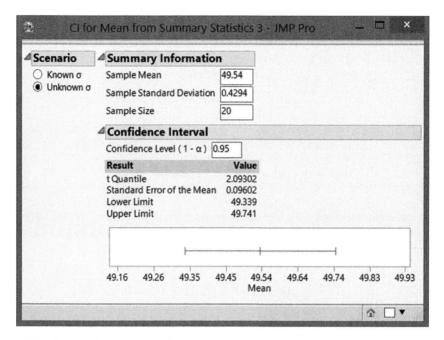

Figure 2.7   The use of the script "CI for Mean from Summary Statistics" in JMP for Example 2.3.1.

2.4   Confidence Intervals for a Population Proportion

We can compute interval estimates not only for population means μ, but also for population proportions π. There are several methods for computing confidence intervals for a population proportion. The most common methods start from the fact that the sample proportion $\hat{P}$ is approximately normally distributed with expected value π and variance $\pi(1 - \pi)/n$. Another approach is based on the binomial distribution.

2.4.1   A First Interval Estimator Based on the Normal Distribution

The sample proportion $\hat{P}$ is approximately normally distributed with expected value π and variance $\pi(1 - \pi)/n$ (see Theorem 1.6.2). As a result,

$$P\left(-z_{\alpha/2} \leq \frac{\hat{P} - \pi}{\sqrt{\dfrac{\pi(1 - \pi)}{n}}} \leq z_{\alpha/2}\right) = 1 - \alpha.$$

Now, the inequality

$$-z_{\alpha/2} \leq \frac{\hat{P} - \pi}{\sqrt{\dfrac{\pi(1 - \pi)}{n}}} \leq z_{\alpha/2}$$

can be rewritten as

$$\frac{(\hat{P} - \pi)^2}{\dfrac{\pi(1 - \pi)}{n}} \leq z_{\alpha/2}^2$$

and

$$\left(n + z_{\alpha/2}^2\right)\pi^2 - \left(2n\hat{P} + z_{\alpha/2}^2\right)\pi + n\hat{P}^2 \leq 0.$$

This is a quadratic inequality for the unknown population proportion π. All values of π satisfying this inequality form a confidence interval for π. First, we have to find the roots of the quadratic equation

$$\left(n + z_{\alpha/2}^2\right)\pi^2 - \left(2n\hat{P} + z_{\alpha/2}^2\right)\pi + n\hat{P}^2 = 0.$$

The discriminant of this equation is

$$D = \left(2n\hat{P} + z_{\alpha/2}^2\right)^2 - 4n\hat{P}^2\left(n + z_{\alpha/2}^2\right) = z_{\alpha/2}^4 + 4n\hat{P}z_{\alpha/2}^2 - 4n\hat{P}^2 z_{\alpha/2}^2$$

$$= z_{\alpha/2}^2\left(z_{\alpha/2}^2 + 4n\hat{P} - 4n\hat{P}^2\right),$$

which is always positive. Indeed, $\hat{P}$ lies between 0 and 1, so that $\hat{P} \geq \hat{P}^2$, $\hat{P} - \hat{P}^2 \geq 0$, and finally $4n\hat{P} - 4n\hat{P}^2 \geq 0$.

The roots of the quadratic equation are

$$\frac{\left(2n\hat{P} + z_{\alpha/2}^2\right) \pm \sqrt{z_{\alpha/2}^2\left(z_{\alpha/2}^2 + 4n\hat{P} - 4n\hat{P}^2\right)}}{2\left(n + z_{\alpha/2}^2\right)}$$

$$= \frac{\left(2n\hat{P} + z_{\alpha/2}^2\right) \pm z_{\alpha/2}\sqrt{z_{\alpha/2}^2 + 4n\hat{P} - 4n\hat{P}^2}}{2\left(n + z_{\alpha/2}^2\right)},$$

which can be rewritten as

$$\frac{\left(2n\hat{P} + z_{\alpha/2}^2\right) \pm z_{\alpha/2}\sqrt{z_{\alpha/2}^2 + 4n\hat{P}(1 - \hat{P})}}{2\left(n + z_{\alpha/2}^2\right)}.$$

These two roots are the lower and upper limits of the confidence interval for the population proportion π. Therefore, the confidence interval is

$$
\left[\frac{\left(2n\hat{P} + z_{\alpha/2}^2\right) - z_{\alpha/2}\sqrt{z_{\alpha/2}^2 + 4n\hat{P}(1 - \hat{P})}}{2\left(n + z_{\alpha/2}^2\right)}, \right.
$$

$$
\left. \frac{\left(2n\hat{P} + z_{\alpha/2}^2\right) + z_{\alpha/2}\sqrt{z_{\alpha/2}^2 + 4n\hat{P}(1 - \hat{P})}}{2\left(n + z_{\alpha/2}^2\right)} \right]. \tag{2.5}
$$

This confidence interval is called the **Wilson score interval**, after the American mathematician Edwin Bidwell Wilson.

Wilson's approach works well for small as well as large numbers of observations, and for extremely large or small probabilities. The Wilson confidence interval is symmetric around the value

$$
\frac{\left(2n\hat{P} + z_{\alpha/2}^2\right)}{2\left(n + z_{\alpha/2}^2\right)} = \frac{\left(\hat{P} + z_{\alpha/2}^2/2n\right)}{\left(1 + z_{\alpha/2}^2/n\right)}.
$$

In some cases, the lower limit of the confidence interval in Equation (2.5) is negative, or the upper limit is greater than 1. If the lower limit in Equation (2.5) is negative, it is replaced by 0. If the upper limit in Equation (2.5) is greater than 1, it is replaced by the value 1.

2.4.2   A Second Interval Estimator Based on the Normal Distribution

In some practical applications:

- n is much larger than $z_{\alpha/2}^2$, so that we can replace $n + z_{\alpha/2}^2$ by n;
- $2n\hat{P}$ is much larger than $z_{\alpha/2}^2$, so that we can replace $2n\hat{P} + z_{\alpha/2}^2$ by $2n\hat{P}$;
- $4n\hat{P}(1 - \hat{P})$ is much larger than $z_{\alpha/2}^2$, so that we can replace $z_{\alpha/2}^2 + 4n\hat{P}(1 - \hat{P})$ by $4n\hat{P}(1 - \hat{P})$.

The confidence interval in Expression (2.5) then simplifies to

$$
\left[\frac{2n\hat{P} - z_{\alpha/2}\sqrt{4n\hat{P}(1 - \hat{P})}}{2n}, \frac{2n\hat{P} + z_{\alpha/2}\sqrt{4n\hat{P}(1 - \hat{P})}}{2n} \right],
$$

which can be rewritten as

$$
\left[\hat{P} - z_{\alpha/2}\sqrt{\frac{\hat{P}(1 - \hat{P})}{n}}, \hat{P} + z_{\alpha/2}\sqrt{\frac{\hat{P}(1 - \hat{P})}{n}} \right]. \tag{2.6}
$$

This confidence interval, which was already derived in 1812 by Laplace in a different way, heavily relies on the assumption that the sample size n is large enough. The use of this interval is not recommended, but it does form the basis for determining the required sample size for the estimation of a population proportion. This topic is discussed in Section 2.7.2.

In some cases, the lower limit of the confidence interval in Expression (2.6) is negative, or the upper limit is greater than 1. If the lower limit is negative, it is replaced by 0. If the upper limit is greater than 1, it is replaced by the value 1.

The width of the confidence interval in Expression (2.6) for a population proportion is

$$B = 2\, z_{\alpha/2} \sqrt{\frac{\hat{P}(1 - \hat{P})}{n}},$$

while the margin of error is

$$b = \frac{B}{2} = z_{\alpha/2} \sqrt{\frac{\hat{P}(1 - \hat{P})}{n}}. \tag{2.7}$$

2.4.3 An Interval Estimator Based on the Binomial Distribution

If the sample size is small, the sample proportion is not approximately normally distributed. In that case, a confidence interval for a population proportion can be constructed based on the binomial distribution. This kind of construction is called **exact** because it does not make use of any approximation. The method dates back to 1934 and was named after the authors Clopper and Pearson. The exact derivation of the Clopper–Pearson confidence interval is complicated, but there is a relatively simple calculation method for both the lower limit and the upper limit of the confidence interval. Suppose that the total number of observations is equal to n, the number of "successes" in the sample is equal to s, and the number of "failures" in the sample is equal to $n - s$. Then, the lower and upper limits can be calculated in the following way:

- The lower limit of the interval is equal to the $(\alpha/2) \times 100\%$ percentile or quantile of a beta-distributed random variable with parameters s and $n - s + 1$.
- The upper limit of the interval is equal to the $(1 - \alpha/2) \times 100\%$ percentile or quantile of a beta-distributed random variable with parameters $s + 1$ and $n - s$.

These percentiles or quantiles can be calculated in JMP using the formulas "Beta Quantile $(\alpha/2,\ s,\ n - s + 1)$" and "Beta Quantile $(1 - \alpha/2,\ s + 1,\ n - s)$". As a beta-distributed random variable can only take values between 0 and 1, the beta distribution is suitable for the construction of a confidence interval for proportions, which, by definition, also lie between 0 and 1.

There is also an alternative method for calculating the limits of the Clopper–Pearson confidence interval, which uses the F-distribution[1] that we will discuss in detail in Section 8.3:

$$\text{lower limit} = \left(1 + \frac{n - s + 1}{s F_{1 - \frac{\alpha}{2} ; 2s ; 2(n - s + 1)}} \right)^{-1}$$

and

$$\text{upper limit} = \left(1 + \frac{n - s}{(s + 1) F_{\frac{\alpha}{2} ; 2(s + 1) ; 2(n - s)}} \right)^{-1},$$

where $F_{1 - \alpha/2 ; 2s ; 2(n-s+1)}$ and $F_{\alpha/2 ; 2(s+1) ; 2(n-s)}$ are percentiles or quantiles of an F-distributed random variable. These quantiles can be obtained in JMP using the formulas "F Quantile$(\alpha/2,\ 2*s,\ 2*(n - s + 1))$" and "F Quantile$(1 - \alpha/2,\ 2*(s + 1),\ 2*(n - s))$". The lower limit of the Clopper–Pearson interval is never negative, and the upper limit is never greater than 1.

The fact that the Clopper–Pearson confidence interval is based on the binomial distribution and not on the normal approximation of the binomial distribution is the reason why many researchers prefer this interval. Quite recently, however, the statisticians Agresti and Coull demonstrated that the Wilson score interval in Expression (2.5) is the better[2] one in almost all cases.

To construct confidence intervals with JMP, the scripts "CI for Proportion from Summary Statistics" and "CI for Proportion from Data" are available. When running these scripts, you can either opt for a confidence interval based on Expression (2.6) and the normal approximation or for a (Clopper–Pearson) confidence interval based on the binomial distribution. To calculate the Wilson score interval, you can use JMP directly. This is explained in Section 2.6.

Example 2.4.1 *A production line manufactures toys that are packaged in boxes of five. During a quality control exercise, 100 boxes were examined. There were 31 boxes without any defective piece, 42 boxes containing one defective piece, 18 boxes containing two defective pieces, six boxes containing three defective pieces, two boxes with four defective pieces, and one box with five defective pieces. We compute a 90% confidence interval for the proportion of defects for this production line.*

As a total of $100 \times 5 = 500$ toys were inspected and 109 defective products were found, the sample proportion of defective products is

$$\hat{p} = \frac{109}{500} = 0.218.$$

---

[1] The alternative calculation is due to the fact that a specific function of a beta-distributed random variable follows an F-distribution.

[2] A confidence interval is better than another confidence interval if the probability that the interval contains the unknown population parameter is closer to the confidence level $1 - \alpha$.

With $1 - \alpha = 0.90$ and, hence, $\alpha = 0.10$, we obtain the following confidence interval using the normal density and Equation (2.6):

$$\left[0.218 - 1.644854\sqrt{\frac{0.218(1 - 0.218)}{500}}, 0.218 + 1.644854\sqrt{\frac{0.218(1 - 0.218)}{500}}\right],$$

or

$$[0.18763, 0.24837].$$

If we use the binomial distribution, we obtain the Clopper–Pearson confidence interval

$$[0.18793, 0.25058].$$

Finally, Equation (2.5) yields the Wilson score interval

$$[0.18919, 0.24985].$$

In this example, there are no big differences between the three methods for determining a confidence interval for the population proportion under investigation. This is due to the very large number of observations.

The use of the script "CI for Proportion from Summary Statistics" in JMP is illustrated in Figure 2.8.

2.5   Confidence Intervals for a Population Variance

We explained in Section 1.7.4 that, for normally distributed data,

$$\frac{(n - 1)S^2}{\sigma^2} \sim \chi^2_{n-1}.$$

We can use this result to determine a confidence interval for the population variance σ^2. To this end, we have to find a lower bound l and an upper bound u so that

$$P\left(l \leq \frac{(n - 1)S^2}{\sigma^2} \leq u\right) = 1 - \alpha.$$

A logical choice is to select the values for l and u that satisfy

$$P\left(\frac{(n - 1)S^2}{\sigma^2} \leq l\right) = \frac{\alpha}{2}$$

and

$$P\left(\frac{(n - 1)S^2}{\sigma^2} \geq u\right) = \frac{\alpha}{2}.$$

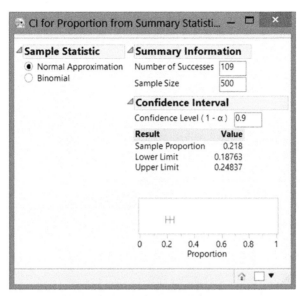

(a) Normal approximation

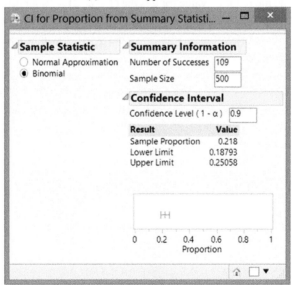

(b) Binomial distribution

Figure 2.8   The use of the script "CI for Proportion from Summary Statistics" in JMP for determining a confidence interval for a population proportion with an approximation based on the normal probability density, and the exact method based on the binomial distribution.

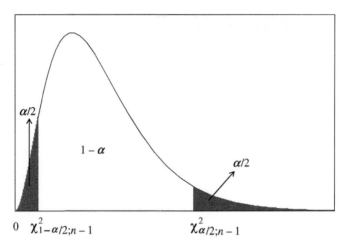

Figure 2.9 The graphical representation of $\chi^2_{1-\alpha/2;n-1}$ and $\chi^2_{\alpha/2;n-1}$.

The required value for l then, of course, is the $(\alpha/2) \times 100$th percentile of a χ^2-distributed random variable, while the required value for u is the $(1 - \alpha/2) \times 100$th percentile. We denote these percentiles by the symbols $\chi^2_{1-\alpha/2;n-1}$ and $\chi^2_{\alpha/2;n-1}$. A graphical representation of these values can be found in Figure 2.9. Later, $\chi^2_{1-\alpha/2;n-1}$ and $\chi^2_{\alpha/2;n-1}$ will be used as critical values in the chapters on hypothesis tests.

Note that in contrast to the standard normal distribution and the t-distribution, the χ^2-distribution is not symmetric around 0. As a result, $\chi^2_{1-\alpha/2;n-1}$ is not the negative of $\chi^2_{\alpha/2;n-1}$. Since a χ^2-distributed random variable can only take positive values, each of its percentiles or quantiles is positive. Therefore, both $\chi^2_{1-\alpha/2;n-1}$ and $\chi^2_{\alpha/2;n-1}$ are positive.

The confidence interval can be constructed in the following way:

$$P\left(\chi^2_{1-\alpha/2;n-1} \leq \frac{(n-1)S^2}{\sigma^2} \leq \chi^2_{\alpha/2;n-1}\right) = 1 - \alpha,$$

$$P\left(\frac{1}{\chi^2_{\alpha/2;n-1}} \leq \frac{\sigma^2}{(n-1)S^2} \leq \frac{1}{\chi^2_{1-\alpha/2;n-1}}\right) = 1 - \alpha,$$

$$P\left(\frac{(n-1)S^2}{\chi^2_{\alpha/2;n-1}} \leq \sigma^2 \leq \frac{(n-1)S^2}{\chi^2_{1-\alpha/2;n-1}}\right) = 1 - \alpha,$$

so that an interval with confidence level $1 - \alpha$ for σ^2 is given by

$$\left[\frac{(n-1)S^2}{\chi^2_{\alpha/2;n-1}}, \frac{(n-1)S^2}{\chi^2_{1-\alpha/2;n-1}}\right]. \tag{2.8}$$

The JMP scripts "CI for Variance from Data" and "CI for Variance from Summary Statistics" are available for the computation of confidence intervals for population variances.

Example 2.5.1   *In order to evaluate the reliability of a manufacturing process for plastic film, one option is to measure the thickness of the film at 51 places. The mean thickness of all these measurements is $\bar{x} = 23.46$ μm, while the sample variance is $s^2 = 1.24$ μm^2. A 90% confidence interval for the population variance is given by*

$$\left[\frac{(51-1)(1.24)}{\chi^2_{0.05;50}} \, , \, \frac{(51-1)(1.24)}{\chi^2_{0.95;50}} \right].$$

The required percentiles $\chi^2_{0.05;50} = 67.505$ and $\chi^2_{0.95;50} = 34.764$ can be looked up in Appendix C. Alternatively, the JMP formulas "ChiSquare Quantile(0.95, 50)" and "ChiSquare Quantile(0.05, 50)" can be used. With these values, the interval is

$$\left[\frac{(51-1)(1.24)}{67.505} \, , \, \frac{(51-1)(1.24)}{34.764} \right] = [0.9185, 1.7834].$$

Figure 2.10 shows how the JMP script "CI for Variance from Summary Statistics" can be used for the determination of this interval.

An important difference between confidence intervals for variances and confidence intervals for population means is that the former are not symmetrical around s^2, while the latter are

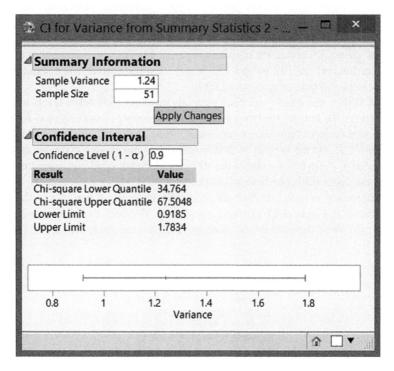

Figure 2.10   The use of the JMP script "CI for Variance from Summary Statistics" for Example 2.5.1.

symmetrical around $\bar{x}$. This is due to the fact that the χ^2-distribution is not symmetrical, whereas both the normal distribution and the t-distribution are.

Finally, it is useful to note that the confidence interval for a population variance σ^2 can easily be converted to a confidence interval for a population standard deviation σ, simply by taking the square root of the lower and upper limits of the interval for σ^2 in Expression (2.8). In other words,

$$\left[\sqrt{\frac{(n-1)S^2}{\chi^2_{\alpha/2;n-1}}} \, , \, \sqrt{\frac{(n-1)S^2}{\chi^2_{1-\alpha/2;n-1}}} \, \right]$$

is a confidence interval with confidence level $1 - \alpha$ for the population standard deviation σ.

2.6 More Confidence Intervals in JMP

So far, we have used a few JMP scripts to illustrate how to determine interval estimates or confidence intervals in JMP. These scripts have been programmed for students, so that they can easily compute different types of confidence intervals, but they are not contained in a standard version of JMP. In this section, we examine the options that JMP offers by default for interval estimators and confidence intervals.

First, each univariate analysis of a quantitative variable via the "Distribution" platform in the "Analyze" menu automatically provides a 95% confidence interval for the population mean μ of the variable under study. We illustrate this by means of a data table on cereals. If, for example, we study the variable "Calories" and we choose the "Distribution" option in the "Analyze" menu, we obtain the output in Figure 2.11. The lower and upper limits of the 95% confidence interval are 129.19 and 151.86 calories, respectively. The standard error of the sample mean is also listed and equals 5.69.

You can ask JMP to compute confidence intervals for other confidence levels too. This can be done by clicking the hotspot (red triangle) next to "Summary Statistics" in the output. If you choose the option "Customize Summary Statistics", you will see the window shown in Figure 2.12. At the bottom of that window, you can change the value of $1 - \alpha$. For example, if we set $1 - \alpha$ to 0.9, we obtain the 90% confidence interval, the lower limit of which is 131.05 and the upper limit of which is 150.00.

It is also interesting to note that JMP automatically generates a box plot of the data. The box plot, as shown in Figure 2.13, contains a diamond. The center of the diamond indicates the sample mean, while the ends of the diamond indicate the lower and upper limits of the

Summary Statistics	
Mean	140.52632
Std Dev	49.608997
Std Err Mean	5.6905423
Upper 95% Mean	151.86246
Lower 95% Mean	129.19017
N	76

Figure 2.11 The standard output of the "Distribution" platform in the "Analyze" menu in JMP.

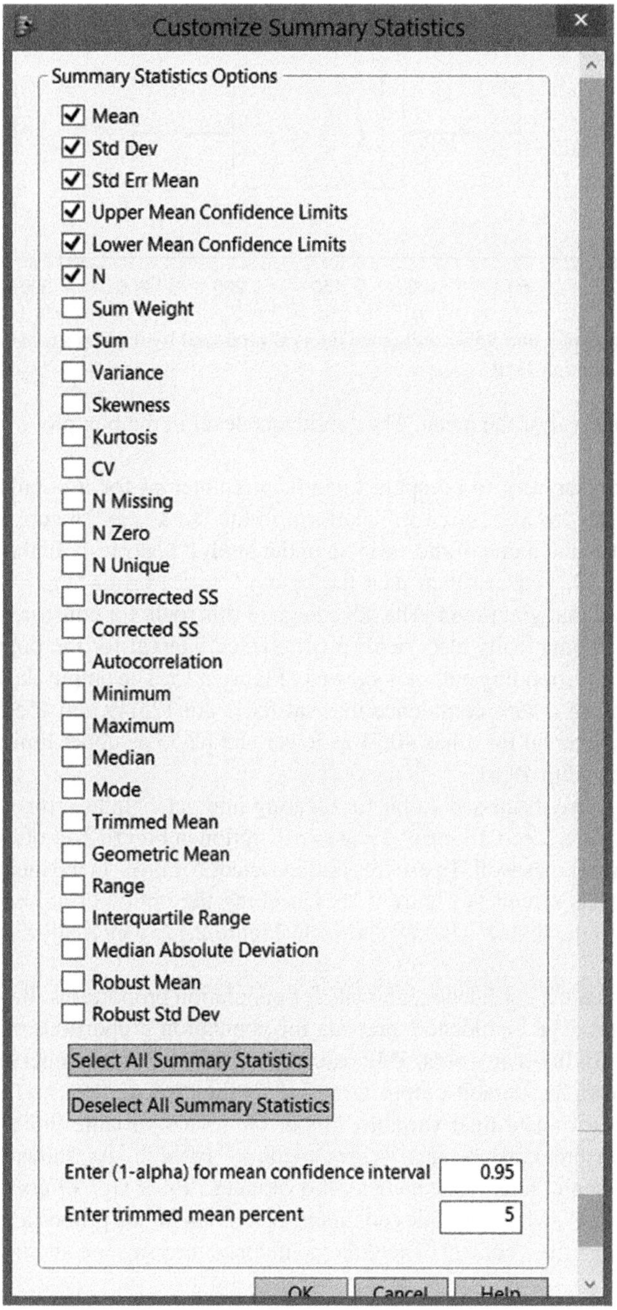

Figure 2.12   The JMP screen that allows you to set $1 - \alpha$.

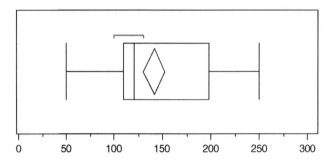

Figure 2.13   The box plot and 95% confidence interval produced by the "`Distribution`" platform in the "`Analyze`" menu in JMP.

95% confidence interval of the mean. The confidence level in the box plot cannot be changed and is always 95%.

There is yet another way to compute a confidence interval for a population mean μ in JMP. After using the "`Distribution`" platform in the "`Analyze`" menu, click the hotspot (red triangle) next to the name of the variable under study, "Calories". In the resulting menu, shown in Figure 2.14, you can then pick the option "`Confidence Interval`" as well as the confidence level that you prefer. The advantage of this route for constructing a confidence interval is that it automatically also yields a confidence interval for the population standard deviation σ. The corresponding output is shown in Figure 2.15. The output shows that the lower and the upper limit of a 99% confidence interval for μ are 125.49 and 155.57, respectively. A 99% confidence interval for σ has 40.91 as lower and 62.53 as upper limit, while the point estimate for σ is equal to 49.61.

The options we have discussed so far for creating interval estimates for μ in JMP assume that σ is unknown. The "`Confidence Interval`" option in Figure 2.14 offers the possibility of dealing with known σ as well. To use this option, select "`Other`" in the menu in Figure 2.14. This will produce the screen in Figure 2.16. Checking the option "`Use known Sigma`" in that screen leads to the dialog window shown in Figure 2.17, which allows you to enter the value of σ.

JMP can also compute confidence intervals for population proportions. It does this only for qualitative variables. The confidence intervals for population proportions calculated in JMP use Expression (2.5). In other words, JMP calculates the Wilson score interval.

Figure 2.18 shows the default output provided by the "`Distribution`" platform in the "`Analyze`" menu for an ordinal variable, "Fiber Gr". This variable indicates whether the amount of fiber in breakfast cereals is low, medium, or high. As shown in Figure 2.19, the hotspot (red triangle) next to the name of the variable, "Fiber Gr", offers the option "`Confidence Interval`", which provides confidence intervals for the proportions of cereals with low, medium, or high fiber content. The three confidence intervals are shown in Figure 2.20.

2.7   Determining the Sample Size

The collection of sample data is usually expensive and time-consuming. Therefore, researchers try to use statistics to determine the minimum number of data points they need. This requires choosing a confidence level $1 - \alpha$, and then fixing either the maximum width $B_{\max}$ of a

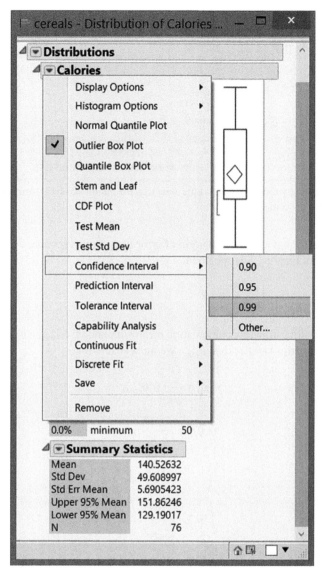

Figure 2.14 The option for creating confidence intervals for the population mean μ and the population standard deviation σ of quantitative variables.

Confidence Intervals			
Parameter	Estimate	Lower CI	Upper CI	1-Alpha
Mean	140.5263	125.4863	155.5663	0.990
Std Dev	49.609	40.91021	62.53055	0.990

Figure 2.15 The output of the option "Confidence Interval" in Figure 2.14.

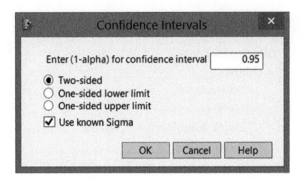

Figure 2.16   The dialog window obtained after selecting the option "`Confidence Interval`" in Figure 2.14, and then choosing "`Other`".

confidence interval, or the maximum margin of error b_{max}. In this section, we assume that the maximum margin of error b_{max} is specified.

2.7.1   The Population Mean

The margin of error for estimating a population mean with known σ is given in Equation (2.3). In order to keep this error smaller than b_{max}, we must ensure that

$$z_{\alpha/2}\frac{\sigma}{\sqrt{n}} \le b_{max},$$

$$\sqrt{n} \ge \frac{z_{\alpha/2}\sigma}{b_{max}},$$

and, hence,

$$n \ge \frac{z_{\alpha/2}^2\sigma^2}{b_{max}^2}.$$

Thus, the determination of the sample size requires not only choosing a maximum margin of error and a confidence level, but also a value for the population standard deviation σ.

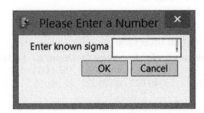

Figure 2.17   The dialog window for entering the value of σ.

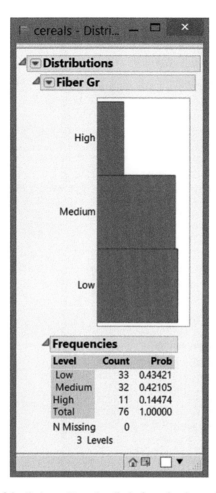

Figure 2.18 The output of the "Distribution" platform for the qualitative variable "Fiber Gr".

The JMP script "Sample Size For Mean" can be used to determine the sample size for estimating a population mean, so that the confidence interval does not exceed the desired maximum width of $2b_{\max}$. The use of this script is illustrated in the following example.

Example 2.7.1 *Suppose that you wish to set up a study for estimating a population mean* μ. *The population under investigation has a standard deviation of* $\sigma = 5$. *You would like to compute an interval estimate for* μ *with a confidence level of 90% and a margin of error of at most 2, and you wonder how many observations this requires.*

Since the margin of error is at most 2, we have $b_{max} = 2$. *The confidence level of 90% means that* $\alpha = 0.10$. *As a result, the required sample size has to satisfy the inequality*

$$n \geq \frac{z_{0.10/2}^2 5^2}{2^2} = \frac{z_{0.05}^2 5^2}{2^2} = \frac{(1.644854)^2 5^2}{2^2} = 16.9097.$$

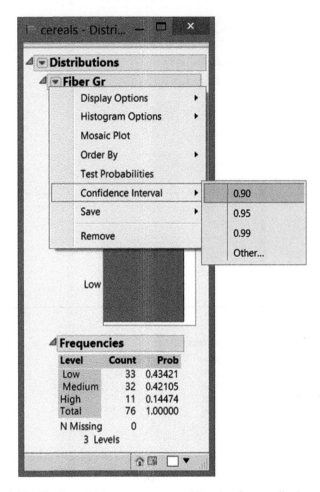

Figure 2.19   The "Confidence Interval" option for a qualitative variable.

Confidence Intervals					
Level	Count	Prob	Lower CI	Upper CI	1-Alpha
Low	33	0.43421	0.344547	0.528397	0.900
Medium	32	0.42105	0.332186	0.515347	0.900
High	11	0.14474	0.090583	0.223315	0.900
Total	76				

Note: Computed using score confidence intervals.

Figure 2.20   The confidence intervals produced by the option "Confidence Interval" for the qualitative variable "Fiber Gr".

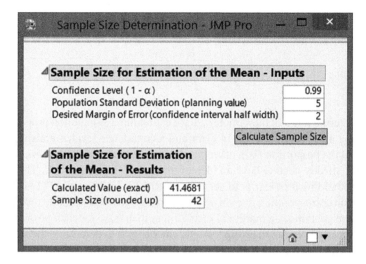

Figure 2.21 The use of the JMP script "Sample Size For Mean" for Example 2.7.1.

Of course, the number of observations n needs to be an integer. The smallest integer that satisfies the inequality is 17, *so that* 17 *is the minimum number of observations required to keep the margin of error below* 2.

If you want a higher confidence level for your interval estimate – for example, 99% – *then, of course,* $1 - \alpha = 0.99$ *and* $\alpha = 0.01$. *The sample size then has to satisfy the inequality*

$$n \geq \frac{z_{0.01/2}^2 5^2}{2^2} = \frac{z_{0.005}^2 5^2}{2^2} = \frac{(2.575829)^2 5^2}{2^2} = 41.4681.$$

The smallest integer that satisfies this inequality is 42, *so that* 42 *is the required sample size to keep the margin of error below* 2 *for a confidence level of* 99%.

Figure 2.21 shows how the script "Sample Size For Mean" can be used for the determination of a sample size for a confidence interval with a certain maximum margin of error.

2.7.2 The Population Proportion

The calculation of the required sample size based on the Wilson score interval in Expression (2.5) or the Clopper–Pearson interval is quite complex. For this reason, we usually start from the simplest confidence interval for a population proportion, which is the one in Expression (2.6). The margin of error when estimating a population proportion using that interval is given in Equation (2.7). In order for this margin of error to be less than b_{max}, we must ensure that

$$z_{\alpha/2}\sqrt{\frac{\hat{P}(1 - \hat{P})}{n}} \leq b_{max},$$

$$\sqrt{n} \geq \frac{z_{\alpha/2}\sqrt{\hat{P}(1 - \hat{P})}}{b_{max}},$$

and hence

$$n \geq \frac{z_{\alpha/2}^2 \hat{P}(1 - \hat{P})}{b_{max}^2}.$$

Therefore, the determination of the sample size for estimating a population proportion not only requires choosing a maximum margin of error and a confidence level, but also indicating the expected value of the proportion $\hat{P}$. In other words, to determine the sample size for estimating a proportion, we already need to have an idea of the size of this proportion. The most careful way to work around this problem is to set $\hat{P}$ equal to $1/2$. In this case, $\hat{P}(1 - \hat{P})$ reaches its maximum, and the corresponding margin of error is the largest one possible too. The resulting sample size n then guarantees a margin of error of less than b_{max} for each possible value of $\hat{P}$.

The JMP script "Sample Size For Proportion" can be used to determine the sample size for estimating a population proportion so that the maximum margin of error is not exceeded.

Example 2.7.2 *Suppose that you want to set up a study to estimate the proportion of left-handers in your country. You choose the confidence level to be 99.9%, with a maximum margin of error of 1%. In other words, you set $1 - \alpha = 0.999$, $\alpha = 0.001$, and $b_{max} = 0.01$. You wonder how many observations this requires. If you assume that half of the population is left-handed (i.e., if, beforehand, you believe that $\hat{P}$ will be 0.5), then the required sample size is*

$$n \geq \frac{z_{\alpha/2}^2 \hat{P}(1 - \hat{P})}{b_{max}^2} = \frac{z_{0.0005}^2 0.5(1 - 0.5)}{(0.01)^2}$$

$$= \frac{(3.29053)^2 \times 0.5(1 - 0.5)}{(0.01)^2} = 27\,068.9.$$

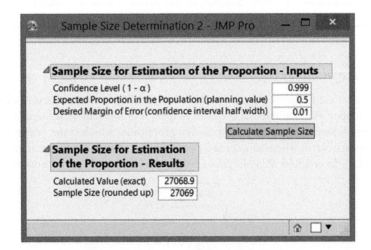

Figure 2.22   The use of the JMP script "Sample Size For Proportion" for Example 2.7.2.

The smallest integer that satisfies this inequality is 27 069, so that 27 069 is the minimum number of observations needed to keep the margin of error below 0.01. This is quite a large number of observations.

Figure 2.22 shows how the JMP script "Sample Size For Proportion" can be used to determine the sample size in this example.

One way to lower the required number of observations is to better prepare your study. For example, you can refer to the Internet to inquire about the number of left-handers. Then, you would see that the proportion of left-handers varies between 10% and 20%. If, for example, you decide that a good a priori estimate of the proportion is 15%, then the required sample size is

$$n \geq \frac{z_{\alpha/2}^2 \hat{P}(1-\hat{P})}{b_{max}^2} = \frac{z_{0.0005}^2 0.15(1-0.15)}{(0.01)^2} = 13\,805.1.$$

In other words, in this case only 13 806 observations are necessary.

3

Hypothesis Tests

It was an understandable response. But my strategy was to minimise the chance of making a type-one error – wasting time on an unsuitable choice. Inevitably, that increased the risk of a type-two error – rejecting a suitable person. But this was an acceptable risk as I was dealing with a very large population.

(from *The Rosie Project*, Graeme Simsion, p. 32)

In this chapter, the principles of statistical **hypothesis testing** are explained in detail. The purpose of a hypothesis test is to determine whether a statement or claim about a population or a process is right or wrong. This statement can concern one or more population parameters, or the probability distribution or density of the population or the process. The judgment of the statement is based on sample data, and provides an indication of how likely or unlikely it is that the statement is right or wrong.

3.1 Key Concepts

To introduce the main concepts of hypothesis testing in this chapter, we restrict our attention to a hypothesis test for a population or process mean μ. For simplicity, we assume in this introductory chapter that the population variance σ^2 is known. The following examples illustrate that hypotheses can also relate to population or process parameters other than a mean μ, and to more than one parameter.

Example 3.1.1 *In Example 2.2.1, milk samples were tested for their freezing point. Using a sample, we can test whether or not a supplier's claim that a batch of milk has an average freezing temperature of $\mu = -0.545\,°C$ (and, hence, was not diluted) is correct.*

Example 3.1.2 *Filling machines are widely used in various industrial applications. Typically, these machines have a (large) number of filling heads, to fill several bottles at the same time. The calibration of each of the filling heads is often a difficult task. Ideally, the calibration of all filling heads would be exactly the same, so that they deposit the same average amount in the bottles, and so that all of them are equally reliable. A possible statement or hypothesis*

Statistics with JMP: Hypothesis Tests, ANOVA and Regression, First Edition. Peter Goos and David Meintrup.
© 2016 John Wiley & Sons, Ltd. Published 2016 by John Wiley & Sons, Ltd.
Companion Website: http://www.wiley.com/go/goosandmeintrup/JMP

that could be tested is: "The average filling weight of the k filling heads of the machine is the same". Another possible hypothesis is: "The variances of the k filling heads are all equal".

Example 3.1.3   *The pattern in the number of goals scored by the teams participating in the football World Cup in 1998 is very similar to a Poisson distribution. A possible statement that could be tested is: "The number of goals is Poisson distributed with parameter $\lambda = 1.25$".*

The approach for testing hypotheses that we follow here was developed by Neyman and Pearson. For this approach, we need two clear hypotheses about the studied population or the process under consideration. A first hypothesis is called the **null hypothesis**. This hypothesis will be indicated by H_0. The second hypothesis is called the **alternative hypothesis** and is denoted by H_a. This hypothesis will be accepted if the null hypothesis is rejected.

Example 3.1.4   *In Example 3.1.1, one possible hypothesis is that the supplier did not add water to the milk. This hypothesis is the null hypothesis:*

$$H_0 : \mu = -0.545\,^\circ C.$$

The alternative hypothesis states that the supplier added water to the milk, so that the freezing temperature increased:

$$H_a : \mu > -0.545\,^\circ C.$$

Example 3.1.5   *In Example 3.1.2, a possible null hypothesis is*

$$H_0 : \mu_1 = \mu_2 = \cdots = \mu_k,$$

where $\mu_1, \mu_2, \ldots, \mu_k$ represent the mean filling weights of the k filling heads. The alternative hypothesis states that at least one filling head has a different mean:

$$H_a : \text{not } H_0.$$

If one is more interested in the variances of the filling heads than in their means, then the following null hypothesis is suitable:

$$H_0 : \sigma_1^2 = \sigma_2^2 = \cdots = \sigma_k^2.$$

A possible alternative hypothesis is

$$H_a : \text{not } H_0.$$

Example 3.1.6   *An obvious null hypothesis for Example 3.1.3 is*

$$H_0 : \text{The number of goals is Poisson distributed with parameter } \lambda = 1.25.$$

The alternative hypothesis is

H_a :   *The number of goals is not Poisson distributed with parameter* $\lambda = 1.25$.

In the formulation of the null hypothesis, one usually starts from a situation in which there is no effect, no difference, or no relation. This is illustrated in Examples 3.1.4 and 3.1.5. If, for example, the goal of a study is to find out whether a certain treatment reduces the number of bacteria in a sample, the null hypothesis will state that there is no decrease due to the treatment, while the alternative hypothesis will state that there is an effect of the treatment. The null hypothesis does not necessarily reflect the situation that is desirable. More often, the desired situation is the content of the alternative hypothesis.

The alternative hypothesis is sometimes called the **research hypothesis**, because it often contains a claim or statement that a researcher seeks to prove. Typically, a researcher wants to demonstrate that "something new" is better than "something existing", that a particular product does not comply with a legal requirement, and so on.

Based on sample data, the researcher attempts to take a decision about the null and the alternative hypotheses. In order to use the sample data for deciding between the two proposed hypotheses, the researcher uses a so-called **decision rule**. Making a decision means that the null hypothesis is either **accepted** or **rejected**.

Of course, when making a decision, it is possible to make a mistake. One possible error is to falsely reject the null hypothesis. This error is called a **type I error** or α-error. A second erroneous type of decision is to accept the null hypothesis when in fact the alternative hypothesis reflects the truth. This error is called a **type II error** or β-error. These type I and type II errors are summarized in Table 3.1.

Type I and type II errors are due to a combination of two factors: the nature of the data and the decision rule used. The data is sample data, which, by definition, offers an incomplete picture of a population or a process. Drawing a conclusion based on a subset of all elements of a population or a subset of all elements produced by a process inevitably involves a risk. It is possible to have bad luck when collecting data, so that the data sample leads you to an incorrect decision. The decision rule used to choose between the null hypothesis and the alternative hypothesis also has an influence. Some decision rules favor the null hypothesis, while others favor the alternative hypothesis. Some decision rules are liberal in the sense that they lead to a rejection of the null hypothesis too often, while other decision rules are conservative and lead to an acceptance of the null hypothesis too often.

Table 3.1   Possible decisions and error types when testing hypotheses.

	True situation	
Decision based on sample	H_0 is true	H_a is true
Accept H_0	Correct decision, probability $1 - \alpha$	Type II error, probability β
Accept H_a	Type I error, probability α	Correct decision, probability $1 - \beta$

The probability of a type I error is typically represented using the symbol α. This probability is also known as the **significance level of the test**[1]. The probability of a type II error is represented by β. The complement of this probability, $1 - \beta$, is called the **power** of the test. The power is the probability that a decision rule correctly rejects the null hypothesis H_0, which means that the null hypothesis is rejected if it is indeed false. The α- and β-probabilities are also shown in Table 3.1. Clearly, both α and β are conditional probabilities:

$$\alpha = P(H_0 \text{ is rejected} \mid H_0 \text{ is true})$$

and

$$\beta = P(H_0 \text{ is accepted} \mid H_a \text{ is true}).$$

This implies that

$$1 - \alpha = P(H_0 \text{ is accepted} \mid H_0 \text{ is true})$$

and

$$1 - \beta = P(H_a \text{ is accepted} \mid H_a \text{ is true}).$$

There is an interesting analogy between hypothesis tests and jurisdiction. In court, the applied null hypothesis is that a defendant is innocent. The alternative hypothesis states that the defendant is guilty. If the court decides that a defendant is guilty while this is not the case, then the court is making a type I error. If the defendant is actually guilty but is acquitted, the court is making a type II error. The information on which the court relies to make a decision with regard to guilt or innocence is similar to the information contained within a data sample in a statistical study. The information available in court may also be incomplete and may point in the wrong direction.

If a tested hypothesis contains only one value of a parameter of a population, then it is called a **simple** hypothesis. Otherwise, it is called a **composite** hypothesis. Often, a simple null hypothesis is tested against a composite alternative hypothesis. For example, this is the case in Example 3.1.4, where the unknown parameter has exactly one value under the null hypothesis, and many different values under the alternative hypothesis.

Finally, we distinguish between **left-tailed**, **right-tailed**, and **two-tailed** hypotheses. The alternative hypothesis H_a in Example 3.1.4 is a right-tailed hypothesis. In general, a right-tailed hypothesis is of the form $H_a : \mu > \mu_0$. Here, μ_0 represents a predetermined real number. In Example 3.1.4, $\mu_0 = -0.545\,°C$. A left-tailed alternative hypothesis is of the form $H_a : \mu < \mu_0$, while a two-tailed hypothesis is of the type $H_a : \mu \neq \mu_0$.

---

[1] The probability $1 - \alpha$ is called the confidence level of the test.

3.2 Testing Hypotheses About a Population Mean

In this section, we introduce tests with right-tailed, left-tailed, and two-tailed alternative hypotheses. We discuss three approaches for testing each of these hypotheses.

3.2.1 The Right-Tailed Test

For the data from Examples 2.2.1, 3.1.1, and 3.1.4, the hypotheses tested are

$$H_0 : \mu = -0.545\,°\text{C}$$

and

$$H_a : \mu > -0.545\,°\text{C}.$$

More generally, we can write these hypotheses as

$$H_0 : \mu = \mu_0$$

and

$$H_a : \mu > \mu_0.$$

A First Approach: A Critical Value for the Sample Mean

An intuitive approach for choosing between the two hypotheses is a decision rule of the following form:

- Reject the null hypothesis H_0 (and accept the alternative hypothesis H_a) if the sample mean $\bar{x}$ is large compared to $\mu_0 = -0.545\,°\text{C}$; in particular, if $\bar{x}$ exceeds a certain **critical value** c.
- Accept the null hypothesis if the sample mean $\bar{x}$ is smaller than or equal to the critical value c.

The purpose of introducing critical values is to make sure that the null hypothesis is not rejected if the sample mean $\bar{x}$ is only "a little" larger than $\mu_0 = -0.545\,°\text{C}$. The reason for this is that a small difference between $\bar{x}$ and $\mu_0 = -0.545\,°\text{C}$ can be due to pure chance. As explained earlier, the sample mean is a random variable, so that the exact value that a researcher obtains for the mean is subject to randomness.

The obvious question now is how to choose the critical value c. The choice will influence both the probability α of a type I error and the probability β of a type II error.

In practice, one typically fixes a value for the significance level α. Usually, the value chosen for α is 0.10, 0.05, or 0.01. In any case, by fixing the significance level α, one chooses the critical value c so that

$$\alpha = P(H_0 \text{ is rejected} \mid H_0 \text{ is true}),$$
$$= P(\overline{X} > c \mid H_0 \text{ is true}),$$
$$= P(\overline{X} > c \mid \mu = \mu_0).$$

We know that the sample mean $\overline{X}$ is normally distributed if the individual sample observations are normally distributed, and that it is approximately normally distributed if the data is not normally distributed, but the sample size is sufficiently large. In both cases, we can rewrite the probability of a type I error as

$$\alpha = P\left(\frac{\overline{X}-\mu}{\sigma/\sqrt{n}} > \frac{c-\mu}{\sigma/\sqrt{n}} \;\middle|\; \mu = \mu_0\right),$$

$$= P\left(\frac{\overline{X}-\mu_0}{\sigma/\sqrt{n}} > \frac{c-\mu_0}{\sigma/\sqrt{n}} \;\middle|\; \mu = \mu_0\right),$$

$$= P\left(Z > \frac{c-\mu_0}{\sigma/\sqrt{n}}\right),$$

where Z is standard normally distributed. We know from Section 2.2.1 that

$$\alpha = P(Z \geq z_\alpha) = P(Z > z_\alpha),$$

so that we can find the desired critical value c by solving the equation

$$\frac{c-\mu_0}{\sigma/\sqrt{n}} = z_\alpha.$$

Consequently, the critical value is

$$c = \mu_0 + z_\alpha \frac{\sigma}{\sqrt{n}}. \tag{3.1}$$

This leads to the following decision rule:

- Reject the null hypothesis H_0 and accept the alternative hypothesis H_a if

$$\overline{x} > \mu_0 + z_\alpha \frac{\sigma}{\sqrt{n}}.$$

- Accept the null hypothesis if

$$\overline{x} \leq \mu_0 + z_\alpha \frac{\sigma}{\sqrt{n}}.$$

The number $\mu_0 + z_\alpha \sigma/\sqrt{n}$ is the critical value of the test. The interval $]c, +\infty[$ is called the **critical region** or **rejection region**. This interval includes all values of the sample mean $\overline{x}$ for which the null hypothesis is rejected. The interval $]-\infty, c]$ contains the values of $\overline{x}$ for which the null hypothesis is accepted. This interval is called the **acceptance region**.

Example 3.2.1 *For Example 2.2.1 (see also Examples 3.1.1 and 3.1.4) and $\alpha = 0.05$, we obtain the following decision rule:*

- *Reject the null hypothesis H_0 and accept the alternative hypothesis H_a if*

$$\bar{x} > -0.545 + 1.645\frac{0.008}{\sqrt{5}} = -0.5391.$$

- *Accept the null hypothesis if*

$$\bar{x} \leq -0.545 + 1.645\frac{0.008}{\sqrt{5}} = -0.5391.$$

A graphical representation of the normal density of the sample mean, the critical value, the acceptance region, and the critical or rejection region is shown in Figure 3.1.

The observed sample mean is $\bar{x} = -0.535$, which is larger than the critical value -0.5391. Consequently, the null hypothesis H_0 can be rejected and the alternative hypothesis can be accepted. This means that we have a strong indication that the milk supplier did manipulate the milk.

In this example, one says that the population mean μ (the freezing point of the batch of milk) is **significantly** larger than $\mu_0 = -0.545\,°\mathrm{C}$ at a **significance level** of $\alpha = 0.05 = 5\%$. The use of a significance level of 5% implies that, in this example, there is a probability of 5% that the milk supplier will be accused even if he does not manipulate the milk.

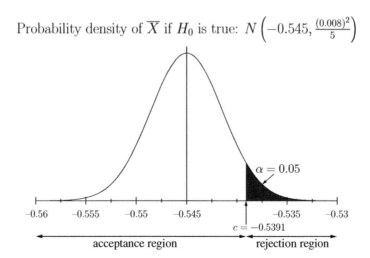

Figure 3.1 An illustration of the first approach for testing a hypothesis on the population mean μ with known σ (Example 3.2.1).

A Second Approach: A Critical Value for the Test Statistic

From the above reasoning, it is clear that the decision rule for accepting or rejecting the null hypothesis can be rewritten as follows:

- Reject the null hypothesis H_0 and accept the alternative hypothesis H_a if

$$z = \frac{\bar{x} - \mu_0}{\sigma/\sqrt{n}} > z_\alpha.$$

- Accept the null hypothesis if

$$z = \frac{\bar{x} - \mu_0}{\sigma/\sqrt{n}} \leq z_\alpha.$$

In this approach, the sample mean $\bar{x}$ is standardized to a **test statistic** z, which is compared to the critical value z_α.

Example 3.2.2   *For Example 2.2.1 (see also Examples 3.1.1, 3.1.4, and 3.2.1) and $\alpha = 0.05$, the test statistic is*

$$z = \frac{-0.535 - (-0.545)}{0.008/\sqrt{5}} = 2.795.$$

This value is considerably larger than the critical value $z_\alpha = z_{0.05} = 1.645$, so that the null hypothesis H_0 is rejected at a significance level $\alpha = 0.05$.

This second approach for right-tailed hypothesis tests is illustrated graphically in Figure 3.2. The figure shows the standard normal density of the test statistic, the critical value $z_{0.05}$, and the acceptance and rejection regions.

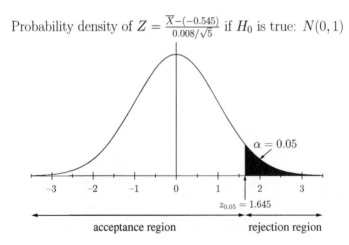

Probability density of $Z = \frac{\bar{X} - (-0.545)}{0.008/\sqrt{5}}$ if H_0 is true: $N(0, 1)$

Figure 3.2   An illustration of the second approach for testing a hypothesis on the population mean μ with known σ (Example 3.2.2).

A Third Approach: The p-Value

There is a third approach for performing a hypothesis test, namely the approach that makes use of the so-called p-value. This approach is undoubtedly the most informative one. The p-value of a right-tailed hypothesis test about a population mean μ is the probability that $\overline{X}$ takes a value that exceeds the actually observed sample mean $\overline{x}$ if the null hypothesis H_0 is true. Therefore, the p-value can be interpreted as an indication of how likely or unlikely it is that the null hypothesis is correct.

For a right-tailed test, the p-value can be calculated in the following way:

$$p = P(\overline{X} > \overline{x} \mid H_0 \text{ is true}),$$

$$= P(\overline{X} > \overline{x} \mid \mu = \mu_0),$$

$$= P\left(\frac{\overline{X} - \mu}{\sigma/\sqrt{n}} > \frac{\overline{x} - \mu}{\sigma/\sqrt{n}} \,\Bigg|\, \mu = \mu_0\right),$$

$$= P\left(\frac{\overline{X} - \mu_0}{\sigma/\sqrt{n}} > \frac{\overline{x} - \mu_0}{\sigma/\sqrt{n}} \,\Bigg|\, \mu = \mu_0\right),$$

$$= P(Z > z),$$

where

$$z = \frac{\overline{x} - \mu_0}{\sigma/\sqrt{n}}$$

again represents the test statistic from the previous approach. The decision rule for accepting or rejecting the null hypothesis is as follows:

- Reject the null hypothesis H_0 and accept the alternative hypothesis H_a if $p = P(Z > z) < \alpha$.
- Accept the null hypothesis if $p = P(Z > z) \geq \alpha$.

This decision rule leads to exactly the same conclusion as the decision rule based on the critical value c for the sample mean $\overline{x}$ and the decision rule based on the critical value z_α for the test statistic z.

The idea behind the use of the p-value is that we reject the null hypothesis in only $\alpha \times 100\%$ of the cases if the null hypothesis is true. In other words, we reject the null hypothesis when the observed sample mean $\overline{x}$ belongs to the $\alpha \times 100\%$ largest possible sample means, assuming that the null hypothesis is true. The p-value indicates what fraction of the possible sample means is more extreme than the observed one, under the assumption that the null hypothesis is true. If this fraction is smaller than α, it means that the observed sample mean belongs to the $\alpha \times 100\%$ largest possible sample means.

Example 3.2.3 *For Example 2.2.1 (see also Examples 3.1.1, 3.1.4, 3.2.1, and 3.2.2), the p-value is as follows:*

$$p = P(\overline{X} > -0.535 \mid H_0 \text{ is true}),$$

$$= P(\overline{X} > -0.535 \mid \mu = -0.545),$$

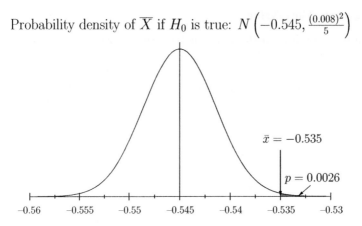

Probability density of $\overline{X}$ if H_0 is true: $N\left(-0.545, \frac{(0.008)^2}{5}\right)$

Figure 3.3 An illustration of the third approach for testing a hypothesis on the population mean μ with known σ (Example 3.2.3).

$$= P\left(\frac{\overline{X} - \mu}{\sigma/\sqrt{n}} > \frac{-0.535 - \mu}{\sigma/\sqrt{n}} \,\middle|\, \mu = -0.545\right),$$

$$= P\left(\frac{\overline{X} - (-0.545)}{0.008/\sqrt{5}} > \frac{-0.535 - (-0.545)}{0.008/\sqrt{5}} \,\middle|\, \mu = -0.545\right),$$

$$= P(Z > 2.795),$$

$$= 0.0026.$$

This value is considerably smaller than $\alpha = 0.05$, so that, at a significance level α of 5%, the null hypothesis H_0 is rejected. It is clear that the null hypothesis would also be rejected for smaller values of α; for example, for $\alpha = 0.01$ or even for $\alpha = 0.0027$.

This third approach for right-tailed hypothesis tests is illustrated graphically in Figures 3.3 and 3.4. Figure 3.3 shows how the p-value can be calculated starting from the probability density of $\overline{X}$ when the null hypothesis is true. Figure 3.4 shows how the same p-value can be obtained starting from the probability density of the test statistic Z when the null hypothesis is true.

The example shows that the p-value corresponds to the smallest value of the significance level α, for which the sample data does not allow the null hypothesis to be rejected. This third approach to hypothesis testing provides information on the weight of the evidence against the null hypothesis: p-values of 0.049 and 0.0026 both indicate that the null hypothesis can be rejected at a significance level of $\alpha = 0.05$, but a p-value of 0.0026 is a much stronger indication that the null hypothesis might be wrong than a p-value of 0.049.

The p-value thus provides us with more detailed information than that needed for merely rejecting or accepting the null hypothesis. The only disadvantage of the approach is that an exact calculation of p-values generally requires the use of a statistical software package. Statistical software usually reports the p-value of a test by default, and therefore adopts the

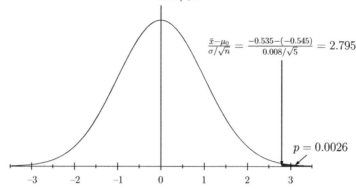

Figure 3.4 An alternative illustration of the third approach for testing a hypothesis on the population mean μ with known σ (Example 3.2.3).

third approach for testing hypotheses. The interpretation of the reported p-value is usually left to the researcher.

Researchers who only have access to tables with critical values need to resort to the second, less informative approach, involving critical values for test statistics. However, it is important to note that the three approaches to testing hypotheses invariably lead to the same decisions concerning the acceptance or rejection of the null hypothesis.

3.2.2 The Left-Tailed Test

The process for a left-tailed hypothesis test is entirely analogous to that for a right-tailed test. In a left-tailed test, the tested hypotheses can generally be written as

$$H_0 : \mu = \mu_0$$

and

$$H_a : \mu < \mu_0.$$

The First Approach

- Reject the null hypothesis H_0 and accept the alternative hypothesis H_a if

$$\bar{x} < \mu_0 - z_\alpha \frac{\sigma}{\sqrt{n}}.$$

- Accept the null hypothesis if

$$\bar{x} \geq \mu_0 - z_\alpha \frac{\sigma}{\sqrt{n}}.$$

The Second Approach

- Reject the null hypothesis H_0 and accept the alternative hypothesis H_a if

$$z = \frac{\bar{x} - \mu_0}{\sigma/\sqrt{n}} < -z_\alpha.$$

- Accept the null hypothesis if

$$z = \frac{\bar{x} - \mu_0}{\sigma/\sqrt{n}} \geq -z_\alpha.$$

The Third Approach

- Reject the null hypothesis H_0 and accept the alternative hypothesis H_a if

$$p = P(Z < z) = P\left(\frac{\bar{X} - \mu}{\sigma/\sqrt{n}} < \frac{\bar{x} - \mu}{\sigma/\sqrt{n}} \,\middle|\, \mu = \mu_0\right) < \alpha.$$

- Accept the null hypothesis if

$$p = P(Z < z) = P\left(\frac{\bar{X} - \mu}{\sigma/\sqrt{n}} < \frac{\bar{x} - \mu}{\sigma/\sqrt{n}} \,\middle|\, \mu = \mu_0\right) \geq \alpha.$$

3.2.3 The Two-Tailed Test

In a two-tailed test, the null hypothesis and the alternative hypothesis have the following forms:

$$H_0 : \mu = \mu_0$$

and

$$H_a : \mu \neq \mu_0.$$

Example 3.2.4 *The purpose of a filling process is to inject* 33 cl *of lemonade into a can. The mean filling content of the filling machine can be set by the operator on the factory floor, but the standard deviation of* 0.5 cl *cannot be manipulated. The machine is set to an average fill volume of* 34 cl *to avoid a large number of cans being underfilled. Despite this conservative machine setting, the quality assurance department wants to ensure that the cans neither contain too little lemonade nor too much. Therefore, a sample of* 64 cans *is taken on a regular basis, and the content of the cans in the sample is carefully measured.*

Suppose that the mean content of the 64 cans *in a given sample is equal to* 33.89 cl *and that the fill content of the cans is normally distributed. The management wants to determine whether the mean fill content is indeed equal to the set* 34 *cl, and so is interested in testing the hypotheses*

$$H_0 : \mu = 34 \ \text{cl}$$

and

$$H_a : \mu \neq 34 \ \text{cl}.$$

The First Approach

Again, it is logical to reject the null hypothesis when the observed sample mean $\bar{x}$ deviates too much from the hypothesized value μ_0 (34 cl in the example). This time, the null hypothesis needs to be rejected in two cases. It needs to be rejected when the sample mean is substantially smaller than μ_0, but also when the sample mean is substantially larger than μ_0. In other words, the null hypothesis will be rejected when $\bar{x}$ is smaller than a certain lower critical value c_L or larger than another, upper, critical value c_U. Obviously, the lower critical value has to be smaller than μ_0, while the upper critical value has to be larger than μ_0.

Typically, the critical values c_L and c_U are chosen to be symmetrical around μ_0. In that case, we can write that $c_L = \mu_0 - \Delta$ and $c_U = \mu_0 + \Delta$, with $\Delta > 0$. The decision rule can then be written as follows:

- Reject the null hypothesis H_0 and accept the alternative hypothesis H_a if

$$\bar{x} < c_L = \mu_0 - \Delta$$

or if

$$\bar{x} > c_U = \mu_0 + \Delta.$$

- Accept the null hypothesis if

$$c_L = \mu_0 - \Delta \leq \bar{x} \leq c_U = \mu_0 + \Delta.$$

To determine the value of Δ, and hence c_L and c_U, for a two-tailed test, the significance level α is again fixed at a small value. The α value again represents the probability that the null hypothesis will be rejected when it is actually true:

$$\alpha = P[(\overline{X} < c_L) \text{ or } (\overline{X} > c_U) \mid H_0 \text{ is true}],$$

$$= P[(\overline{X} < \mu_0 - \Delta) \text{ or } (\overline{X} > \mu_0 + \Delta) \mid H_0 \text{ is true}],$$

$$= P[(\overline{X} < \mu_0 - \Delta) \text{ or } (\overline{X} > \mu_0 + \Delta) \mid \mu = \mu_0],$$

$$= 1 - P(\mu_0 - \Delta \leq \overline{X} \leq \mu_0 + \Delta \mid \mu = \mu_0),$$

$$= 1 - P\left(\frac{\mu_0 - \Delta - \mu_0}{\sigma/\sqrt{n}} \leq \frac{\overline{X} - \mu_0}{\sigma/\sqrt{n}} \leq \frac{\mu_0 + \Delta - \mu_0}{\sigma/\sqrt{n}} \,\middle|\, \mu = \mu_0\right),$$

$$= 1 - P\left(\frac{-\Delta}{\sigma/\sqrt{n}} \leq Z \leq \frac{\Delta}{\sigma/\sqrt{n}}\right).$$

Therefore,

$$1 - \alpha = P\left(\frac{-\Delta}{\sigma/\sqrt{n}} \leq Z \leq \frac{\Delta}{\sigma/\sqrt{n}}\right),$$

$$= 1 - P\left(Z \leq \frac{-\Delta}{\sigma/\sqrt{n}}\right) - P\left(Z \geq \frac{\Delta}{\sigma/\sqrt{n}}\right),$$

$$= 1 - 2P\left(Z \geq \frac{\Delta}{\sigma/\sqrt{n}}\right),$$

and hence

$$P\left(Z \geq \frac{\Delta}{\sigma/\sqrt{n}}\right) = \frac{\alpha}{2}.$$

Therefore, the required value for Δ can be obtained by solving the equation

$$\frac{\Delta}{\sigma/\sqrt{n}} = z_{\frac{\alpha}{2}},$$

which leads to

$$\Delta = z_{\frac{\alpha}{2}} \frac{\sigma}{\sqrt{n}}.$$

As a consequence, the acceptance region of the two-tailed hypothesis test is

$$[c_L, c_U] = [\mu_0 - \Delta, \mu_0 + \Delta] = \left[\mu_0 - z_{\frac{\alpha}{2}} \frac{\sigma}{\sqrt{n}}, \mu_0 + z_{\frac{\alpha}{2}} \frac{\sigma}{\sqrt{n}}\right],$$

while the critical region (or rejection region) consists of the intervals

$$]-\infty, c_L[\quad \text{and} \quad]c_U, +\infty[.$$

The corresponding decision rule is as follows:

- Reject the null hypothesis H_0 and accept the alternative hypothesis H_a if

$$\bar{x} < \mu_0 - z_{\frac{\alpha}{2}} \frac{\sigma}{\sqrt{n}}$$

or if

$$\bar{x} > \mu_0 + z_{\frac{\alpha}{2}} \frac{\sigma}{\sqrt{n}}.$$

- Accept the null hypothesis if

$$\mu_0 - z_{\frac{\alpha}{2}} \frac{\sigma}{\sqrt{n}} \leq \bar{x} \leq \mu_0 + z_{\frac{\alpha}{2}} \frac{\sigma}{\sqrt{n}}.$$

Example 3.2.5 *In Example 3.2.4, the sample mean was $\bar{x} = 33.89$ cl for a sample of size $n = 64$. The standard deviation σ is 0.5 cl and the population mean μ_0 given in the null hypothesis is 34 cl. If we pick an α value of 0.05, then $\alpha/2 = 0.025$ and*

$$\Delta = z_{0.025} \frac{0.5}{\sqrt{64}} = 1.96 \frac{0.5}{\sqrt{64}} = 0.1225,$$

so that the acceptance region is

$$[c_L, c_U] = [\mu_0 - \Delta, \mu_0 + \Delta] = [34 - 0.1225, 34 + 0.1225] = [33.8775, 34.1225].$$

The acceptance region and the corresponding critical region for the sample mean $\bar{x}$ are illustrated graphically in Figure 3.5.

The sample mean $\bar{x} = 33.89$ cl falls within the acceptance region, so that the null hypothesis cannot be rejected at a significance level of $\alpha = 0.05$. The conclusion is that the population mean μ is not significantly different from 34 cl at a significance level of 5%.

The Second Approach

The above decision rule for accepting or rejecting the null hypothesis in a two-tailed test can be rewritten in terms of the test statistic z:

- Reject the null hypothesis H_0 and accept the alternative hypothesis H_a if

$$z = \frac{\bar{x} - \mu_0}{\sigma/\sqrt{n}} < -z_{\frac{\alpha}{2}}$$

or if

$$z = \frac{\bar{x} - \mu_0}{\sigma/\sqrt{n}} > z_{\frac{\alpha}{2}}.$$

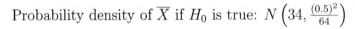

Probability density of $\overline{X}$ if H_0 is true: $N\left(34, \frac{(0.5)^2}{64}\right)$

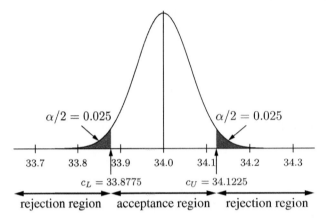

Figure 3.5 An illustration of the first approach for testing the two-tailed alternative hypothesis of Example 3.2.5.

- Accept the null hypothesis if

$$-z_{\frac{\alpha}{2}} \leq z = \frac{\overline{x} - \mu_0}{\sigma/\sqrt{n}} \leq z_{\frac{\alpha}{2}}.$$

In this approach, the sample mean is again standardized to a test statistic z, which is compared to the critical values $-z_{\frac{\alpha}{2}}$ and $z_{\frac{\alpha}{2}}$.

Example 3.2.6 *In Example 3.2.4, the sample mean was $\overline{x} = 33.89$ cl for a sample of size $n = 64$. The standard deviation σ is 0.5 cl, and the population mean μ_0 given in the null hypothesis is 34 cl. Therefore, the test statistic is*

$$z = \frac{33.89 - 34}{0.5/\sqrt{64}} = -1.76.$$

When choosing a significance level of $\alpha = 0.05$, the test statistic value has to be compared to $-z_{0.025} = -1.96$ and $z_{0.025} = 1.96$. The test statistic z lies between these two values, and hence within the acceptance region. Therefore, we accept the null hypothesis at a significance level of 5%.

The acceptance region and the critical region for a two-tailed hypothesis test, expressed in terms of the test statistic, are graphically shown in Figure 3.6.

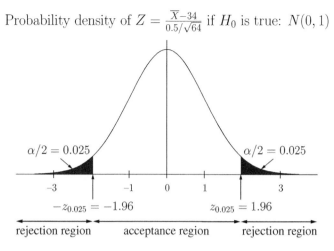

Probability density of $Z = \frac{\overline{X}-34}{0.5/\sqrt{64}}$ if H_0 is true: $N(0,1)$

$\alpha/2 = 0.025$ $\alpha/2 = 0.025$

-3 -1 0 1 3

$-z_{0.025} = -1.96$ $z_{0.025} = 1.96$

rejection region   acceptance region   rejection region

Figure 3.6 An illustration of the second approach for testing the two-tailed alternative hypothesis of Example 3.2.6.

The Third Approach

The computation of the p-value for a two-tailed hypothesis test is a bit more complicated than for a one-tailed test. The reason for this is that, in a two-tailed test, we have to consider deviations in both directions. If the observed sample mean $\overline{x}$ is smaller than μ_0, then finding the p-value requires calculating the probability $P(\overline{X} < \overline{x} \mid \mu = \mu_0)$ that the sample mean is smaller than the observed value $\overline{x}$, assuming that the null hypothesis is correct. This probability then has to be compared to $\alpha/2$. This yields the same result as comparing $2P(\overline{X} < \overline{x} \mid \mu = \mu_0)$ to α. If, however, $\overline{x}$ is larger than μ_0, then finding the p-value requires calculating the probability $P(\overline{X} > \overline{x} \mid \mu = \mu_0)$ that the sample mean exceeds the observed sample mean $\overline{x}$, assuming that the null hypothesis is correct. This probability also has to be compared to $\alpha/2$. This gives the same result as comparing $2P(\overline{X} > \overline{x} \mid \mu = \mu_0)$ to α. Therefore, for a two-tailed test, the p-value is calculated in the following way:

- If $\overline{x} \leq \mu_0$, then

$$p = 2P(\overline{X} < \overline{x} \mid \mu = \mu_0) = 2P(Z < z).$$

- If $\overline{x} > \mu_0$, then

$$p = 2P(\overline{X} > \overline{x} \mid \mu = \mu_0) = 2P(Z > z).$$

The interpretation of the p-value does not change: it still provides an indication of how likely or unlikely it is that the null hypothesis is correct. A small p-value (less than α) means that the null hypothesis is probably incorrect, while a large p-value (greater than α) indicates that the sample data contains no convincing indication that the null hypothesis is wrong.

Example 3.2.7   *In Example 3.2.4, we had $\bar{x} = 33.89$ cl, $n = 64$, $\sigma = 0.5$ cl, and $\mu_0 = 34$ cl. As $\bar{x} < \mu_0$, the p-value is calculated as*

$$p = 2P(\overline{X} < 33.89 \mid \mu = 34) = 2P(Z < -1.76) = 0.0784.$$

This probability is greater than the chosen significance level of 5%, so that the null hypothesis should be accepted. The p-value is quite close to 0.05. Despite the fact that the null hypothesis is accepted here, the p-value suggests that the average filling content may differ from 34 cl.

A slightly shorter calculation method for the p-value uses the fact that $\overline{X} - \mu_0$ is normally distributed around zero when the null hypothesis holds. For a sample mean with $\bar{x} > \mu_0$, $\bar{x} - \mu_0$ is positive and the p-value is

$$2P(\overline{X} > \bar{x} \mid \mu = \mu_0) = 2P(\overline{X} - \mu_0 > \bar{x} - \mu_0 \mid \mu = \mu_0),$$
$$= 2P\{(\overline{X} - \mu_0 > |\bar{x} - \mu_0|) \mid \mu = \mu_0\},$$

while, for a sample mean with $\bar{x} < \mu_0$, $\bar{x} - \mu_0$ is negative and the p-value can be rewritten as

$$2P(\overline{X} < \bar{x} \mid \mu = \mu_0) = 2P(\overline{X} - \mu_0 < \bar{x} - \mu_0 \mid \mu = \mu_0),$$
$$= 2P\{(\overline{X} - \mu_0 > |\bar{x} - \mu_0|) \mid \mu = \mu_0\}.$$

As a conclusion, the p-value in a two-tailed test can be calculated as

$$p = 2P\{(\overline{X} - \mu_0 > |\bar{x} - \mu_0|) \mid \mu = \mu_0\},$$

regardless of the size of $\bar{x}$.

Making use of the test statistic z, the p-value can be obtained as

$$2P(Z > z) = 2P(Z > |z|)$$

when z is positive, and as

$$2P(Z < z) = 2P(Z > |z|)$$

when z is negative. As a result, the p-value in a two-tailed test can be calculated as

$$p = 2P(Z > |z|),$$

regardless of the sign of the test statistic z.

The Relation Between a Two-Tailed Hypothesis Test and a Confidence Interval

There is a close relation between the $(1 - \alpha) \times 100\%$ confidence interval from Section 2.2.2 and the two-tailed hypothesis test with significance level α described above. In fact, the $(1 - \alpha) \times 100\%$ confidence interval for μ will not contain the hypothesized value μ_0 if the null hypothesis $H_0 : \mu = \mu_0$ is rejected in favor of the two-tailed alternative hypothesis $H_a : \mu \neq \mu_0$. If the confidence interval does contain the value μ_0, then the null hypothesis is accepted. This shows that the terms "confidence level" and "significance level" are closely related, and it explains why we have used the Greek letter α in both the previous and the current chapter: when α represents a hypothesis test's significance level, $1 - \alpha$ is the confidence level of the test and the corresponding confidence interval.

Example 3.2.8 *Consider again the data from Example 3.2.4, where $\bar{x} = 33.89$ cl, $n = 64$, $\sigma = 0.5$ cl, and $\mu_0 = 34$ cl. A 95% confidence interval (thus, $1 - \alpha = 0.95$ and $\alpha = 0.05$) for μ is*

$$\left[33.89 - z_{0.025} \frac{0.5}{\sqrt{64}} \ , \ 33.89 + z_{0.025} \frac{0.5}{\sqrt{64}} \right]$$

$$= \left[33.89 - 1.96 \frac{0.5}{8} \ , \ 33.89 + 1.96 \frac{0.5}{8} \right] = [33.7675, 34.0125].$$

This interval contains the hypothesized value $\mu_0 = 34$ cl, which means that the null hypothesis should be accepted at a significance level of $\alpha = 0.05$.

We recommend that, whenever possible, the report of a statistical analysis should not only include the tested hypotheses and associated p-values, but also the confidence intervals for the population parameters under investigation. After all, confidence intervals provide a clear picture of the uncertainty concerning the values of the population parameters under study.

3.3 The Probability of a Type II Error and the Power

In Section 3.1, a type II error was defined as wrongly accepting the null hypothesis when in fact the alternative hypothesis is correct. We illustrate how the probability β of a type II error can be calculated for a right-tailed hypothesis test. Suppose that the null hypothesis

$$H_0 : \mu = \mu_0$$

is wrong, and that the alternative hypothesis

$$H_a : \mu > \mu_0$$

is correct. More specifically, suppose that the actual population mean μ is equal to μ_1, with $\mu = \mu_1 > \mu_0$. This situation is depicted in Figure 3.7.

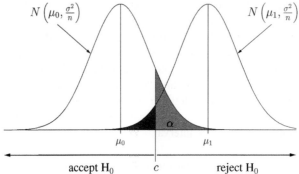

Probability density of $\overline{X}$ if H_0 is true: Probability density of $\overline{X}$ if H_0 is not true:

$N\left(\mu_0, \frac{\sigma^2}{n}\right)$ $N\left(\mu_1, \frac{\sigma^2}{n}\right)$

μ_0 α μ_1

accept H_0 c reject H_0

Figure 3.7 A graphical representation of the probability of a type I error and the probability of a type II error.

For a right-tailed test, the null hypothesis is accepted if $\overline{x}$ is smaller than the critical value $c = \mu_0 + z_\alpha \frac{\sigma}{\sqrt{n}}$. If the null hypothesis is incorrect and, instead, we have that $\mu = \mu_1$, then the probability of this happening is

$$\beta = P(\overline{X} < c \mid \mu = \mu_1),$$

$$= P\left(\frac{\overline{X} - \mu_1}{\sigma/\sqrt{n}} < \frac{c - \mu_1}{\sigma/\sqrt{n}} \ \middle| \ \mu = \mu_1\right),$$

$$= P\left(Z < \frac{c - \mu_1}{\sigma/\sqrt{n}}\right).$$

Indeed, under the assumption that $\mu = \mu_1$, the random variable $\overline{X}$ is normally distributed with expected value μ_1 and variance σ^2/n, and

$$\frac{\overline{X} - \mu_1}{\sigma/\sqrt{n}}$$

is standard normally distributed.

The above derivation shows that the probability β of a type II error depends on several factors. It is not difficult to see that β increases:

- if the significance level α is lowered (indeed, this leads to an increase of z_α and thus also of the critical value c);
- if μ_1 gets closer μ_0;
- if the sample size n is lowered; and
- if the variance σ^2 of the population or process under study increases.

Each of these results has a logical explanation. If the significance level α is reduced, this means that the researcher wants to make fewer type I errors; in other words, that the researcher wants to falsely reject the null hypothesis less often. As a result, the null hypothesis will be accepted more often, even when it is wrong; that is, there will be more type II errors.

If μ_1 and μ_0 get closer, the true population mean gets closer to μ_0 and the normal densities of Figure 3.7 move towards each other. It then becomes more difficult to distinguish μ_1 from μ_0. Hence, it becomes harder to distinguish whether the null hypothesis holds or whether the alternative hypothesis is true. The null hypothesis will therefore be accepted more often, even if it is not correct.

Finally, the probability of a type II error will increase if there is less information in the sample. Less information means more uncertainty, which will be reflected in wider densities in Figure 3.7 and a larger critical value c. One cause for a smaller amount of information is that the sample involves fewer observations and therefore a smaller value of n. A second reason is that the data exhibits more variability, meaning that σ^2 is larger. This results in less precise statements concerning the sample mean, which has a variance of σ^2/n, and so in more erroneous conclusions.

Example 3.3.1   *Suppose that the milk supplier from Example 2.2.1 (see also Examples 3.1.1, 3.1.4, 3.2.1, 3.2.2, and 3.2.3) has manipulated the milk so that its mean freezing temperature in reality is $\mu_1 = -0.535$. The probability of a type II error is the probability that this manipulation goes undetected. In other words, it is the probability that the sample mean $\overline{X}$ takes a value smaller than the critical value $c = -0.545 + 1.645\frac{0.008}{\sqrt{5}} = -0.5391$ (see Example 3.2.1), despite the manipulation. This probability can be calculated in the following way:*

$$\beta = P(\overline{X} < -0.5391 \mid \mu = -0.535),$$

$$= P\left(\frac{\overline{X} - (-0.535)}{0.008/\sqrt{5}} < \frac{-0.5391 - (-0.535)}{0.008/\sqrt{5}} \,\middle|\, \mu = -0.535\right),$$

$$= P(Z < -1.146),$$

$$= 0.1259.$$

So, there is a pretty good chance that the milk supplier's manipulation will remain undetected if a sample with only five observations is used. It is a useful exercise to demonstrate that a sample of size 10 decreases the probability of a type II error from 12.59% to 1.05%. A sample of 11 units is sufficient to reduce the probability of a type II error to less than 1%.

The "power" of a significance test is the complement of the probability of a type II error. For a right-tailed test, it is calculated in the following way:

$$1 - \beta = P(\overline{X} \geq c \mid \mu = \mu_1),$$

$$= P\left(\frac{\overline{X} - \mu_1}{\sigma/\sqrt{n}} \geq \frac{c - \mu_1}{\sigma/\sqrt{n}} \,\middle|\, \mu = \mu_1\right),$$

$$= P\left(Z \geq \frac{c - \mu_1}{\sigma/\sqrt{n}}\right).$$

Example 3.3.2 *Suppose that the milk supplier from Example 2.2.1 (see also Examples 3.1.1, 3.1.4, 3.2.1, 3.2.2, 3.2.3, and 3.3.1) has manipulated the milk so that the mean freezing temperature in reality is* $\mu_1 = -0.535$. *In that case, the power of the right-tailed hypothesis test, with a sample of five observations, is*

$$1 - \beta = P(\overline{X} \geq -0.5391 \mid \mu = -0.535),$$

$$= P\left(\frac{\overline{X} - (-0.535)}{0.008/\sqrt{5}} \geq \frac{-0.5391 - (-0.535)}{0.008/\sqrt{5}} \,\middle|\, \mu = -0.535\right),$$

$$= P(Z \geq -1.146),$$

$$= 0.8741.$$

As a result, the probability that the milk supplier will be accused is 0.8741 *if a sample of five observations is used. A sample size of* 10 *would increase the power to* 98.95%.

3.4 Determination of the Sample Size

When preparing a study on a population mean μ, a researcher must decide which sample size n he will use. In Section 2.7, we studied how to determine the required sample size if the intent of the study is to compute a confidence interval. Here, we determine the sample size for a hypothesis test. We start with a right-tailed hypothesis test for the mean μ of a normally distributed population with known σ. For a right-tailed test, the hypotheses tested are

$$H_0 : \mu = \mu_0$$

and

$$H_a : \mu > \mu_0.$$

Suppose a researcher believes that the true population mean is equal to μ_1 and $\mu_1 > \mu_0$; that is, the researcher believes that the alternative hypothesis is true. The researcher now wants to determine the sample size so that there is a large probability that the null hypothesis will be be rejected. In other words, the researcher wants the hypothesis test to have a large power. Suppose that the desired power is equal to $1 - \beta^*$. We therefore seek to determine the sample size n required to obtain a power of $1 - \beta^*$.

The power is the probability that the null hypothesis is rejected, given that the alternative hypothesis is true; that is,

$$P(\overline{X} > c \mid \mu = \mu_1).$$

For a right-tailed test, this probability is equal to $1 - \beta^*$ if

$$P\left(\overline{X} > \mu_0 + z_\alpha \frac{\sigma}{\sqrt{n}} \,\middle|\, \mu = \mu_1\right) = 1 - \beta^*,$$

which is equivalent to

$$P\left(\frac{\overline{X} - \mu_1}{\sigma/\sqrt{n}} \geq \frac{\mu_0 + z_\alpha \frac{\sigma}{\sqrt{n}} - \mu_1}{\frac{\sigma}{\sqrt{n}}} \,\middle|\, \mu = \mu_1\right) = 1 - \beta^*$$

and

$$P\left(Z \geq \frac{\mu_0 + z_\alpha \frac{\sigma}{\sqrt{n}} - \mu_1}{\frac{\sigma}{\sqrt{n}}}\right) = 1 - \beta^*.$$

This equality implies that

$$\frac{\mu_0 + z_\alpha \frac{\sigma}{\sqrt{n}} - \mu_1}{\frac{\sigma}{\sqrt{n}}} = z_{1-\beta^*},$$

and that

$$\mu_0 + z_\alpha \frac{\sigma}{\sqrt{n}} - \mu_1 = z_{1-\beta^*} \frac{\sigma}{\sqrt{n}},$$

$$z_\alpha \frac{\sigma}{\sqrt{n}} - z_{1-\beta^*} \frac{\sigma}{\sqrt{n}} = \mu_1 - \mu_0,$$

$$(z_\alpha - z_{1-\beta^*}) \frac{\sigma}{\sqrt{n}} = \mu_1 - \mu_0,$$

$$\frac{(z_\alpha - z_{1-\beta^*})\sigma}{\mu_1 - \mu_0} = \sqrt{n},$$

so that, finally,

$$n = \frac{(z_\alpha - z_{1-\beta^*})^2 \sigma^2}{(\mu_1 - \mu_0)^2}.$$

This formula gives the minimum number of observations required to obtain the desired power. The required number of observations increases with the desired power and with the variance σ^2, and when μ_1 gets closer to μ_0.

The difference between μ_1 and μ_0 is usually called δ, so that the required sample size is also written as

$$n = \frac{(z_\alpha - z_{1-\beta^*})^2 \sigma^2}{\delta^2}.$$

The symbol δ can be interpreted as the smallest difference that you would like to detect as a researcher.

Example 3.4.1   *A representative of the food control agency inspects the quality of milk and visits some milk suppliers each day. The agent likes to be strict and wants to increase the probability of catching unfair suppliers. He uses a significance level of 5% and wants a probability of at least 99.9% to catch suppliers who manipulate their milk so that its freezing temperature is in reality $\mu_1 = -0.535\,°C$. The agent assumes that the freezing temperature of milk is a normally distributed random variable, and that $\sigma = 0.008$.*

Statistically speaking, the food control agent wants a power of 99.9%. In other words, the agent wants the probability that the null hypothesis

$$H_0 : \mu = \mu_0 = -0.545\,°C$$

is rejected to be equal to 99.9%, if the true population mean is equal to $-0.535\,°C$. As $\alpha = 0.05$, $1 - \beta^* = 0.999$ and thus $\beta^* = 0.001$, the required sample size is

$$n = \frac{(z_\alpha - z_{1-\beta^*})^2 \sigma^2}{(\mu_1 - \mu_0)^2},$$

$$= \frac{(z_{0.05} - z_{0.999})^2 (0.008)^2}{(-0.535 - (-0.545))^2},$$

$$= \frac{(1.6449 - (-3.0902))^2 (0.008)^2}{(0.01)^2},$$

$$= 14.3495.$$

Therefore, we need 15 observations for the probability of rejecting the null hypothesis to be at least 0.999. The exact power for $n = 15$ is 0.9993. With $n = 14$, the power would only be 0.9988.

These probabilities can be computed in JMP using the formulas

"`1-Normal Distribution`(−0.545
                    + `Normal Quantile`(0.95)*0.008/`Sqrt`(15),
                                    −0.535, 0.008/ `Sqrt`(15))"

and

"`1-Normal Distribution`(−0.545
                    + `Normal Quantile`(0.95)*0.008/`Sqrt`(14),
                                    −0.535, 0.008/`Sqrt`(14))".

The first argument of the function "`Normal Distribution`" *is the JMP formula for calculating the critical value c. The second argument is the population mean assuming that the alternative hypothesis is true, namely $-0.535\,°C$. The last argument is the standard deviation of $\overline{X}$.*

3.5   JMP

The hypothesis test we have discussed in this chapter, namely the hypothesis test for a population mean μ with known σ, can also be performed in JMP. One option to do so is to use

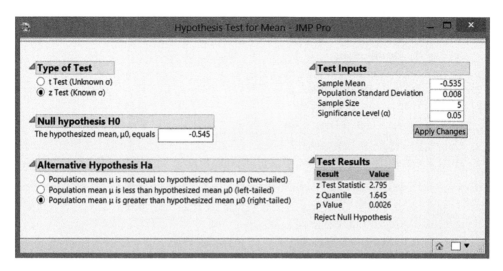

Figure 3.8 The JMP script "Hypothesis Test for Mean Using Summary Statistics" applied to the example of the milk samples.

the script "Hypothesis Test for Mean Using Summary Statistics". This script allows you to indicate that you know the population standard deviation σ or the population variance σ^2. You can do this by choosing the option "z Test (Known σ)", as shown in Figure 3.8.

The script "Hypothesis Test for Mean Using Summary Statistics" allows you to enter the sample mean, the population standard deviation, the sample size, and the significance level α. Also, you can enter the value for the hypothesized mean μ_0 in the field "Hypothesized Mean". Finally, you can choose a two-tailed test ("Population mean μ is not equal to hypothesized mean μ_0 (two-tailed)"), a left-tailed test ("Population mean μ is less than hypothesized mean μ_0 (left-tailed)"), or a right-tailed test ("Population mean μ is greater than hypothesized mean μ_0 (right-tailed)").

The output consists of the computed value for the test statistic ("z Test Statistic"), the required percentile(s) of the standard normal distribution (the critical value(s), "z Quantile(s)"), the p-value ("p Value"), and the decision. The decision is either "Fail to Reject Null Hypothesis" or "Reject Null Hypothesis".

Figure 3.8 shows the input and the output of the script "Hypothesis Test for Mean Using Summary Statistics" for Example 3.2.3. The output shows one more time the p-value of 0.0026, and the decision to reject the null hypothesis.

If you have the original data available, you can also use JMP directly. To do so, you need to first select the "Distribution" option in the "Analyze" menu, and indicate the variable(s) you want to study. Next, you can perform a hypothesis test for the mean using the option "Test Mean" in the hotspot (red triangle) menu next to the variable name. This option is shown in Figure 3.9 for the data on the amount of calories in breakfast cereals (see Section 2.6, where we have already discussed this example). Suppose that we want to test the following hypotheses:

$$H_0 : \mu = 120$$

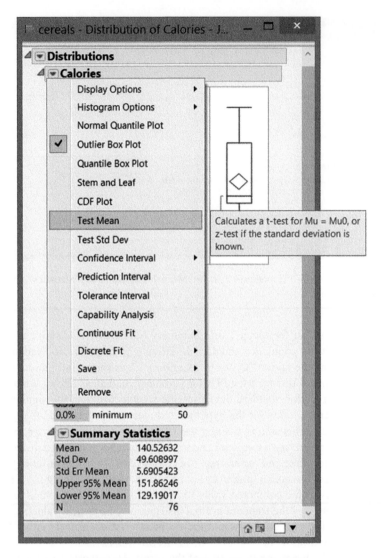

Figure 3.9   The option "Test Mean" in JMP.

and

$$H_a : \mu > 120.$$

In other words, we want to test whether the number of calories in breakfast cereals is greater than 120.

After selecting the option "Test Mean", the dialog window in Figure 3.10 shows up. In that window, you can enter the value for μ_0 (the "Hypothesized Mean" in the null hypothesis), as well as the value of σ. In the figure, the value 120 was specified for μ_0, and the value 50

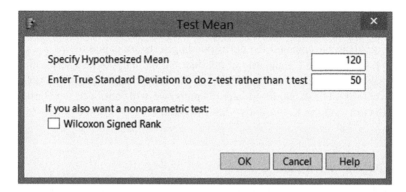

Figure 3.10 The dialog window obtained after selecting the option "Test Mean" in Figure 3.9.

was entered for σ. After clicking on "OK", you obtain the additional output for the hypothesis test shown in Figure 3.11.

The output contains three different p-values: a p-value for the two-tailed test ("Prob > |z|"), a p-value for the right-tailed test ("Prob > z"), and a p-value for the left-tailed test ("Prob < z"). The p-value corresponding to the right-tailed test is 0.0002, which is considerably smaller than the usual significance level of 5%. Therefore, we conclude that the number of calories is significantly greater than 120. The test statistic z is 3.5789, which is the result of the formula

$$z = \frac{\bar{x} - \mu_0}{\frac{\sigma}{\sqrt{n}}} = \frac{140.52632 - 120}{\frac{50}{\sqrt{76}}}.$$

3.6 Some Important Notes Concerning Hypothesis Testing

In this section, we will make some remarks on several issues that arise in the context of hypothesis tests. We start with a comment on the names of hypothesis tests. Every hypothesis test comes with a test statistic. The test statistic is a random variable with a distribution that

Test Mean

Hypothesized Value	120
Actual Estimate	140.526
DF	75
Std Dev	49.609
Sigma given	50

	z Test
Test Statistic	3.5789
Prob > |z|	0.0003*
Prob > z	0.0002*
Prob < z	0.9998

Figure 3.11 The output of the option "Test Mean" in JMP.

is known when the null hypothesis is true. Very often, the name of the distribution of the test statistic determines the name of the hypothesis test. Typical distributions of test statistics are the t-, F-, or χ^2-distributions, and the corresponding tests are called the t-, F-, or χ^2-tests. As Z is the typical letter for a standard normal distribution, a hypothesis test with a standard normally distributed test statistic is called a z-test. Some hypothesis tests, however, are named after their inventor(s). For example, in later chapters, we will introduce the Bartlett test, the Kruskal–Wallis test, and the Kolmogorov–Smirnov test.

3.6.1   Fixing the Significance Level

The critical value in the above hypothesis test corresponds to a choice of significance level α. The significance level is always chosen to be small. Since α indicates the probability of a type I error, choosing a small value for it means that we prefer to have few type I errors when we carry out hypothesis tests.

It is, of course, logical that we prefer to have few type I errors. However, in Section 3.3, we learnt that lowering the significance level α results in a larger probability β for the type II error, all other things being equal. As a result, there is a downside to choosing a very low significance level: it entails a large risk that we will also accept the null hypothesis when it is false.

A wise choice for the significance level strikes a balance between the probability of a type I error and the probability of a type II error. On the one hand, we do not want to reject the null hypothesis too quickly (i.e., when it is correct), but, on the other hand, we also do not want to accept it too often (i.e., when it is false).

It is sometimes recommended to consider the cost of making a wrong decision concerning the null hypothesis and the alternative hypothesis when choosing the significance level. If erroneously rejecting the null hypothesis is likely to result in a large cost, you may want to avoid it and pick a small α value. If erroneously accepting the null hypothesis and rejecting the alternative hypothesis is likely to result in a large cost, it is advisable to pick a larger α value. In many courses, textbooks, and software packages, the default significance level α is 0.05 or 5%. This does certainly not mean that this particular significance level is appropriate in every possible practical application.

It is, for example, common to use an α value of 10% in the context of very small, expensive industrial experiments, where the researchers do not want to miss any potential effect. In that context, the researchers do not mind rejecting the null hypothesis too often, since they only incur a limited cost. In fact, follow-up experimentation, which yields additional data, will allow them to correct their error. However, accepting the null hypothesis in a case where the alternative hypothesis is true may cost them hundreds of thousands of euros. Accepting a null hypothesis in this context usually means stopping a study, in which case the mistake can never be rectified.

When a lot of data is available, almost every difference or every effect, however tiny, is significant when $\alpha = 0.05$ or $\alpha = 0.10$. Since small differences or effects are uninteresting, there is no point in performing a statistical analysis that identifies these small differences. Therefore, in the case of a large data set, the significance level may be lowered to 0.01 or less, to ensure that only practically important effects are signaled as statistically significant. For example, in the analysis of gene expressions, tens of thousands of genes are tested simultaneously. In order to control the false discovery rate, significance levels of 0.001 or smaller are common. This point is related to the discussion in Section 3.6.3.

Regarding the fact that α is fixed in advance when conducting a hypothesis test, two things need to be mentioned. First, fixing α implies that you start from a scenario in which the null hypothesis is true when conducting a hypothesis test. In other words, you start from the assumption that the null hypothesis is true. The critical value and the p-value are computed based on a distribution that was derived under the assumption that the null hypothesis is true. Second, fixing α at a small value and assuming that the null hypothesis is true is similar to what happens in court. There, the defendant is considered innocent (null hypothesis), unless he or she is proven to be guilty (alternative hypothesis). In court, the small α value means that it is considered unacceptable that an innocent defendant is considered guilty and hence convicted.

3.6.2   A Note on the "Acceptance" of the Null Hypothesis

As mentioned above, when performing a right-tailed test, the interval $]-\infty, c]$ is called the "acceptance region". It is therefore common to use the phrase "accepting the null hypothesis", if the sample mean $\bar{x}$ lies within the acceptance region.

It is important to stress that accepting the null hypothesis is something completely different than proving the null hypothesis to be correct. No hypothesis test can ever prove a null hypothesis to be correct, because all hypothesis tests are built on the assumption that the null hypothesis is true. The precise meaning of "accepting the null hypothesis" is that there is no evidence to suggest that the null hypothesis should be rejected. Not being able to reject a certain statement is obviously a much weaker conclusion than proving it to be correct.

This can be clarified using the analogy with jurisdiction. If a court fails to demonstrate the guilt (alternative hypothesis) of a defendant, this by no means implies that the defendant actually is innocent (null hypothesis). All it means is that there was not enough evidence to reject the assumption of innocence.

This being said, we will occasionally use the phrase "accepting the null hypothesis", because it is convenient and because it is in line with the terms "acceptance region" and "rejection region". However, always bear in mind that this does not mean we have shown the null hypothesis to be true. We will revisit this topic in more detail at the beginning of Chapter 16, which deals with equivalence tests.

3.6.3   Statistical and Practical Significance

When testing hypotheses, the alternative hypothesis is the research hypothesis: this is typically the hypothesis that the researcher wants to "prove". Therefore, a researcher will usually hope that he can reject the null hypothesis and accept the alternative hypothesis. If this is the case, the result is called "statistically significant". If, for example, the alternative hypothesis in Example 3.2.4 could be accepted, then the researcher would say that the fill content of the cans of lemonade differs significantly from 34 cl. However, Example 3.2.7 shows that the null hypothesis cannot be rejected. As a result, the fill content is not significantly different from 34 cl. Similarly, in Example 3.2.1, one can say that the freezing temperature of the milk is significantly higher than $-0.545\,°C$, because the right-tailed alternative hypothesis of Example 3.1.4 was accepted.

For researchers, a statistically significant result is very important, because without statistical evidence for a hypothesis, their work will generally not be published in scientific journals and will not be picked up by the press. Similarly, without statistical evidence and hence without statistically significant results, a manufacturer of dietary supplements cannot claim that the dietary supplements have a beneficial health impact. Therefore, researchers often make major efforts to try to get statistically significant results; that is, to reject the null hypothesis and accept the alternative hypothesis (which is their research hypothesis). This can be done in different ways, both ethical and nonethical. It is therefore important that the research results of one researcher or one laboratory can be confirmed by other independent laboratories.

One way to obtain statistically significant results is to increase the number of observations n considerably. If a researcher has a sufficiently large number of observations, then the null hypothesis is rejected every single time. Indeed, if the number of observations is large enough, then the power $1 - \beta$ of the hypothesis test is so big that the null hypothesis will never be accepted.

In doing so, it is certainly possible that researchers may obtain statistically significant results that at first glance look spectacular, but actually have very little practical importance. In an article in a Belgian magazine, a study was mentioned where 50 000 people had to be monitored for eight years to show that a certain yogurt is good for your health. The only reason to carry out such a large study is that the positive health effect is extremely small, so small that it can only be declared significant if the sample size is huge.

In terms of Figure 3.7, this means that μ_1 (the health of a person eating the yogurt) is just a bit greater than μ_0 (the health of someone who does not use the yogurt). In that case, since the two normal probability densities overlap almost completely, the probability of a type II error, β, is large, and the power, $1 - \beta$, is small. This means that there is a small chance of rejecting the null hypothesis, even if the alternative hypothesis is true. However, if the number of observations n is increased, the normal probability densities become narrower and the critical value c shifts to the left. This lowers the probability of a type II error, β, and leads to an increase of the power, $1 - \beta$. Therefore, if the difference between μ_1 and μ_0 is small, then a large number of observations is required to have a substantial probability of rejecting the null hypothesis.

In summary, one can say that a statistically significant result should always be looked at with some skepticism. Often, statistically significant differences are merely the result of a large number of observations, and do not show a difference that is of any practical importance.

Part Two

One Population

4

Hypothesis Tests for a Population Mean, Proportion, or Variance

As Sherlock Holmes once said, "It is a capital mistake to theorize before one has data, … "
(from *The Eight*, Katherine Neville, p. 116)

Just as some statistics are meaningful only for certain types of data, not all hypothesis tests can be used for each type of variable. However, the basic principle of each of the hypothesis tests for choosing between a null hypothesis and an alternative hypothesis is the same in each case. First, a probability density is constructed for a certain test statistic, assuming that the null hypothesis is true. Based on this probability density, a p-value is calculated that provides an indication of how likely or unlikely it is that the null hypothesis is correct. An alternative approach makes use of critical values, which are often derived from a binomial, normal, t-, χ^2-, or F-distribution.

4.1 Hypothesis Tests for One Population Mean

In the previous chapter, all hypothesis tests were constructed for a population mean μ. This was done under the assumption that the population variance σ^2 was known. This assumption is obviously not very realistic, because there is no reason to suppose that the population variance would be known if the population mean were not. In this section, we construct a hypothesis test for the more practical situation in which the population variance σ^2 is estimated by means of the sample variance s^2. This hypothesis test does not use the normal or the standard normal distribution, but the t-distribution. After the introduction of the t-distribution for the construction of a confidence interval for a population mean, in Section 2.3, this should not come as a surprise. A prerequisite for the use of the t-distribution is that a normally distributed population is studied, or a process that generates normally distributed data. In

Statistics with JMP: Hypothesis Tests, ANOVA and Regression, First Edition. Peter Goos and David Meintrup.
© 2016 John Wiley & Sons, Ltd. Published 2016 by John Wiley & Sons, Ltd.
Companion Website: http://www.wiley.com/go/goosandmeintrup/JMP

Sections 4.1.1–4.1.3, we assume that this is indeed the case. The hypothesis tests for a population mean with unknown σ^2 are then based on the fact that the test statistic

$$T = \frac{\overline{X} - \mu_0}{S/\sqrt{n}}$$

is t-distributed with $n-1$ degrees of freedom if the null hypothesis $\mu = \mu_0$ is true. Two approaches can be distinguished: the first approach is based on critical values (either for the sample mean or for the associated test statistic), while the second method involves a p-value. Again, the details of these approaches differ for right-tailed, left-tailed, and two-tailed tests.

4.1.1   The Right-Tailed Test

For a right-tailed hypothesis test for a population mean, the null hypothesis and the alternative hypothesis are

$$H_0 : \mu = \mu_0$$

and

$$H_a : \mu > \mu_0,$$

respectively. The critical or rejection region can be expressed in terms of the sample mean $\bar{x}$, on the one hand, and in terms of the test statistic t, on the other. In the first case, the critical region is the interval

$$]\mu_0 + t_{\alpha;n-1}\frac{s}{\sqrt{n}}, +\infty[.$$

Determining the critical region for $\bar{x}$ is done in the same fashion as in Section 3.2.1. For a right-tailed test, we reject the null hypothesis if the sample mean $\bar{x}$ exceeds a critical value c. If the null hypothesis is true, we do not want this to happen too often. We ensure this by choosing a small value for the significance level α. The significance level is the probability that the null hypothesis is rejected when it is actually correct. In mathematical notation, this conditional probability, which we also call the probability of a type I error, is

$$\alpha = P(\overline{X} > c \mid \mu = \mu_0),$$

$$= P\left(\frac{\overline{X} - \mu}{S/\sqrt{n}} > \frac{c - \mu}{S/\sqrt{n}} \,\middle|\, \mu = \mu_0\right),$$

$$= P\left(\frac{\overline{X} - \mu_0}{S/\sqrt{n}} > \frac{c - \mu_0}{S/\sqrt{n}} \,\middle|\, \mu = \mu_0\right),$$

$$= P\left(t_{n-1} > \frac{c - \mu_0}{S/\sqrt{n}}\right),$$

where t_{n-1} represents a t-distributed random variable with $n-1$ degrees of freedom. In Section 2.3.1, we learnt that

$$\alpha = P(t_{n-1} \geq t_{\alpha;n-1}) = P(t_{n-1} > t_{\alpha;n-1}),$$

so that we can determine the requested critical value c by solving the equation

$$\frac{c - \mu_0}{S/\sqrt{n}} = t_{\alpha;n-1}.$$

Hence, the critical value is

$$c = \mu_0 + t_{\alpha;n-1}\frac{S}{\sqrt{n}}.$$

The critical value is a function of the sample standard deviation, which is unknown until the data has been collected. The critical value is thus itself a random variable. Once the data is collected, we can replace S in the above formula by the computed sample standard deviation s, and we obtain the computed critical value

$$c = \mu_0 + t_{\alpha;n-1}\frac{s}{\sqrt{n}}.$$

In terms of the test statistic

$$t = \frac{\bar{x} - \mu_0}{s/\sqrt{n}},$$

the critical region is $]t_{\alpha;n-1}, +\infty[$. In other words, we reject the null hypothesis if the computed test statistic t is greater than the critical value $t_{\alpha;n-1}$. If the test statistic value does not exceed $t_{\alpha;n-1}$, then we cannot reject the null hypothesis.

The p-value of the right-tailed test is

$$p = P\left(t_{n-1} > \frac{\bar{x} - \mu_0}{s/\sqrt{n}}\right) = P(t_{n-1} > t).$$

The p-value is the probability that a t-distributed random variable with $n-1$ degrees of freedom exceeds the test statistic. The null hypothesis is rejected if the p-value is smaller than the chosen significance level α.

4.1.2   The Left-Tailed Test

For a left-tailed hypothesis test for a population mean, the null hypothesis and the alternative hypothesis are

$$H_0 : \mu = \mu_0$$

and

$$H_a : \mu < \mu_0,$$

respectively. The critical region for the sample mean $\bar{x}$ is the interval

$$]-\infty, \mu_0 - t_{\alpha;n-1} \frac{s}{\sqrt{n}}[.$$

The null hypothesis $H_0 : \mu = \mu_0$ is thus rejected (and the alternative hypothesis $H_a : \mu < \mu_0$ accepted) if the computed sample mean $\bar{x}$ is smaller than the computed critical value

$$c = \mu_0 - t_{\alpha;n-1} \frac{s}{\sqrt{n}}.$$

In terms of the test statistic

$$t = \frac{\bar{x} - \mu_0}{s/\sqrt{n}},$$

the critical region is $]-\infty, -t_{\alpha;n-1}[$. In other words, we reject the null hypothesis if the computed test statistic t is smaller than the critical value $-t_{\alpha;n-1}$.

The p-value of the test is

$$p = P\left(t_{n-1} < \frac{\bar{x} - \mu_0}{s/\sqrt{n}} \right) = P(t_{n-1} < t).$$

The null hypothesis is rejected if the p-value is smaller than α.

4.1.3   The Two-Tailed Test

For a two-tailed hypothesis test for a population mean, the null hypothesis and the alternative hypothesis are

$$H_0 : \mu = \mu_0$$

and

$$H_a : \mu \neq \mu_0,$$

respectively. The critical region for the two-tailed test, expressed in terms of the sample mean $\bar{x}$, consists of the intervals

$$]-\infty, \mu_0 - t_{\alpha/2;n-1}\frac{s}{\sqrt{n}}[\quad \text{and} \quad]\mu_0 + t_{\alpha/2;n-1}\frac{s}{\sqrt{n}}, +\infty[.$$

The null hypothesis $H_0 : \mu = \mu_0$ is rejected (and the alternative hypothesis $H_a : \mu \neq \mu_0$ accepted) if the computed sample mean $\bar{x}$ lies in one of these intervals. In other words, the null hypothesis is rejected if $\bar{x}$ is smaller than the computed left critical value

$$c_L = \mu_0 - t_{\alpha/2;n-1}\frac{s}{\sqrt{n}}$$

or larger than the computed right critical value

$$c_R = \mu_0 + t_{\alpha/2;n-1}\frac{s}{\sqrt{n}}.$$

In terms of the test statistic

$$t = \frac{\bar{x} - \mu_0}{s/\sqrt{n}},$$

the critical region is given by the two intervals $]-\infty, -t_{\alpha/2;n-1}[$ and $]t_{\alpha/2;n-1}, +\infty[$. The null hypothesis is thus rejected if the computed test statistic is less than $-t_{\alpha/2;n-1}$ or greater than $t_{\alpha/2;n-1}$. A shorter way to put this is that the null hypothesis is rejected if the absolute value of the computed test statistic, $|t|$, is larger than $t_{\alpha/2;n-1}$.

The p-value of the test is

$$p = 2P\left(t_{n-1} > \left|\frac{\bar{x} - \mu_0}{s/\sqrt{n}}\right|\right) = 2P(t_{n-1} > |t|).$$

The null hypothesis is rejected if the p-value is smaller than α.

Example 4.1.1 *On average, a filling machine deposits 50.5 g (grams) of rice into a plastic bag and generates normally distributed weights. Both too much and too little weight is perceived as a problem, so that, on a regular basis, data is collected to test the hypotheses*

$$H_0 : \mu = 50.5 \text{ g}$$

and

$$H_a : \mu \neq 50.5 \text{ g},$$

using a significance level of $\alpha = 0.05$.

Assume that a sample of 20 measurements provides a sample mean of $\bar{x} = 50.2525$ g and a sample standard deviation of $s = 0.5326$ g. For this two-tailed test, the test statistic

$$t = \frac{\bar{x} - \mu_0}{s/\sqrt{n}} = \frac{50.2525 - 50.5}{0.5326/\sqrt{20}} = -2.078$$

needs to be compared with the critical values $-t_{\alpha/2;n-1} = -t_{0.025;19} = -2.093$ and $t_{\alpha/2;n-1} = t_{0.025;19} = 2.093$. The test statistic lies between these critical values (and thus outside the critical region), so that the null hypothesis cannot be rejected at a significance level of 5%. The p-value of the hypothesis test is

$$p = 2P\left(t_{n-1} > \left|\frac{50.2525 - 50.5}{0.5326/\sqrt{20}}\right|\right) = 2P(t_{19} > 2.078) = 0.0515,$$

which is slightly larger than $\alpha = 0.05$, so that the p-value leads to acceptance of the null hypothesis.

*The p-value in this example can be calculated in JMP as $2(1 - P(t_{19} \leq 2.078))$ by means of the formula "2 * (1 − t Distribution(2.078, 19))". The first argument in the function "t Distribution (2.078, 19)" is the absolute value of the test statistic, $|t|$. The second argument is the number of degrees of freedom, $n - 1 = 19$. To determine the critical value $t_{0.025;19}$, we can rely on Appendix D or use JMP. The required formula in JMP is "t Quantile (0.975, 19)". In general, the required JMP formula for a quantile $t_{\alpha;n-1}$ is equal to "t Quantile (1 − α, n − 1)".*

For the hypothesis test in this example, you can also use the script "Hypothesis Test for Mean Using Summary Statistics". This script allows you to specify whether or not you know the population variance σ^2 or the population standard deviation σ. In this example, σ^2 is estimated by means of the sample variance, so that a test based on the t-distribution is required. In the dialog window, we can specify this together with the summary statistics, the significance level, and the value of μ_0 ("Hypothesized Mean"). For this example, where we want to perform a two-tailed test, we need to select the option "Population mean μ is not equal to hypothesized mean μ_0 (two-tailed)". The output of the script consists of the computed value for the test statistic ("t Test Statistic"), the percentiles or quantiles from the t-distribution ($-t_{0.025;19}$ and $t_{0.975;19}$), the p-value, and the decision. The decision in this example is "Fail to Reject Null Hypothesis". Figure 4.1 shows the input and the output of the script "Hypothesis Test for Mean Using Summary Statistics" for Example 4.1.1. The output shows the p-value of 0.0515 and the decision not to reject the null hypothesis.

It is recommended to also compute the confidence interval for μ, and not merely report the result of the hypothesis test. In this example, the 95% confidence interval is

[50.0032, 50.5018].

This interval just includes the hypothesized value $\mu_0 = 50.5$ g of the null hypothesis. The interval also shows that we cannot reject the null hypothesis that $\mu = 50.5$ g, but that it is

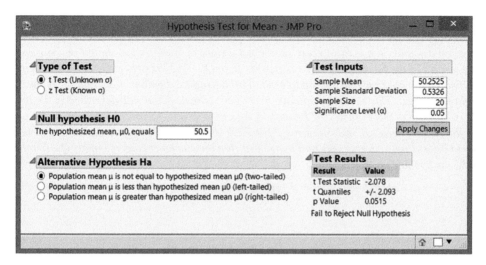

Figure 4.1 The JMP script "Hypothesis Test for Mean Using Summary Statistics" applied to Example 4.1.1.

also quite possible that μ is close to 50 *g. Given this result, it would be useful to monitor the filling machine closely, even though the hypothesis test did not lead to a rejection of the null hypothesis. Figure 4.2 shows how the confidence interval was determined using the JMP script "CI for Mean from Summary Statistics".*

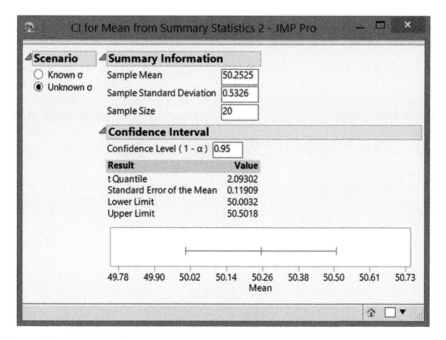

Figure 4.2 The JMP script "CI for Mean from Summary Statistics" applied to Example 4.1.1.

Example 4.1.2   *A Belgian law of January 1, 1980 describes how control of the content of packaged goods should be undertaken by employees of the Ministry of Economic Affairs. The law stipulates that, in a control exercise, the net weight of 50 packages should be measured. If the mean $\bar{x}$ of the 50 measurements is smaller than $Q_n - 0.379s$, where Q_n is the net weight specified on the package and s is sample standard deviation, then the claim that Q_n is the net weight is rejected. This example explains the origin of the decision rule described in the law.*

It is clear that the Ministry of Economic Affairs is trying to protect the consumer, and therefore it is interested in testing the left-tailed alternative hypothesis that the net mean content is less than Q_n versus the null hypothesis that the mean net content equals Q_n. In other words, the hypotheses tested are

$$H_0 : \mu = Q_n$$

and

$$H_a : \mu < Q_n.$$

The null hypothesis is rejected in favor of the left-tailed alternative hypothesis if

$$\bar{x} < Q_n - 0.379s.$$

A decision rule of the same type can be constructed based on Section 4.1.2, where the null hypothesis was rejected if

$$t = \frac{\bar{x} - \mu_0}{s/\sqrt{n}} < -t_{\alpha;n-1}.$$

If μ_0 and n are replaced by Q_n and 50, respectively, we obtain

$$t = \frac{\bar{x} - Q_n}{s/\sqrt{50}} < -t_{\alpha;49},$$

which can be rewritten as

$$\bar{x} < Q_n - t_{\alpha;49}\frac{s}{\sqrt{50}}.$$

This decision rule is identical to that in the 1980 law if

$$\frac{t_{\alpha;49}}{\sqrt{50}} = 0.379,$$

or

$$t_{\alpha;49} = 0.379\sqrt{50} = 2.6799.$$

By means of the JMP formula "1 − t Distribution(2.6799, 49)", which computes the probability $1 - P(t_{49} \leq 2.6799) = P(t_{49} > 2.6799)$, it is not difficult to verify that this *corresponds to an α value of 0.005. Thus, the significance level used by the Belgian authorities is α = 0.005, so that there is only a very small probability that the manufacturer will be falsely accused of fabricating packages that are too light. The small value of α leaves substantial room for manipulating the net weight. A small α implies that the probability of a type II error is quite large. In other words, it is quite possible that a producer is systematically cheating on consumers, but that this will not be detected.*

4.1.4 Nonnormal Data

An important assumption when applying the t-distribution for testing hypotheses regarding a population mean with unknown variance σ^2 in Sections 4.1.1–4.1.3 is that the population under study is normally distributed or that the examined process generates normally distributed data. In practice, this assumption is rarely met exactly. In many cases, however, the data will be more or less normally distributed. As we will see in Chapter 6, there exist hypothesis tests that allow us to determine whether or not data is normally distributed (and, thus, whether or not a hypothesis test based on the t-distribution may be used).

An important question in this context is how much the distribution or density of the data may differ from the normal distribution before the use of the t-distribution leads to wrong decisions about the population mean. Detailed simulation studies have shown that:

- for samples with less than 15 observations, hypothesis tests based on the t-distribution should preferably not be used if the distribution of the population is significantly different from the normal distribution;
- for samples of at least 15 observations, hypothesis tests based on the t-distribution may be used, unless extreme values occur in the data set, or the distribution of the population or process is heavily skewed (see the book *Statistics with JMP: Graphs, Descriptive Statistics and Probability*);
- for samples of more than 30 observations, t-tests can almost always be used.

In practice, the following rule of thumb for hypothesis tests on a population mean with unknown σ^2 is recommended:

- If the sample contains at least 30 observations, then the t-distribution can be used. The same tests as in Sections 4.1.1–4.1.3 can thus be used for nonnormally distributed data, if the sample is sufficiently large. As for such "large" data sets, the number of degrees of freedom of the t-distribution is high, this distribution hardly differs from the standard normal probability density. The reason for this is that, for large samples, the sample standard deviation S is a nearly unbiased estimator of the population standard deviation of σ.
- If the sample contains between 15 and 30 observations, check whether the data is normally distributed (see Chapter 6). If so, then confidence intervals can be constructed and a hypothesis test can be performed based on the t-distribution. If not, then the best option is to use so-called nonparametric hypothesis tests (see Chapter 5).
- If the number of observations is less than 15, then it is safer to use nonparametric hypothesis tests instead of the t-test (see Chapter 5).

In Section 6.3, we explain the reasons for this practical recommendation in more detail.

4.1.5   The Use of JMP

You do not need the JMP script "Hypothesis Test for Mean Using Summary Statistics" to perform a *t*-test for a population mean in JMP. The following example illustrates this.

Example 4.1.3   *In the coffee room for the professors of the Faculty of Applied Economics at a German university, the results of the previous examination period are being discussed. The professors debate whether the group of students that they had in the past academic year was a strong group. The statistics professor says: "If I want to know whether a group of students is strong or not, I look at the group average and test whether it is better or worse than my students in the past. On average, students in the past scored 10 out of 20 in my exam. If a group of students scores significantly better than a 10, then I speak of a strong group. If a group scores significantly less than a 10, then I speak of a weak group. If the score is not significantly different from 10, I speak of an average group". As his colleagues are nodding, the statistics professor opens up his laptop with a JMP data table containing the scores of the students for his subject. He informs his colleagues that he uses a significance level of 5% and that the hypotheses tested are as follows:*

$$H_0 : \mu = 10 \text{ (The group of students is an average group)}$$

and

$$H_a : \mu \neq 10 \text{ (The group of students is either weak or strong)}.$$

Here, the hypothesized mean $\mu_0 = 10$ is the students' score in the previous examination periods. In the "Analyze" menu, the professor selects the option "Distribution", and indicates in the resulting dialog window that he wants to study the variable "Final mark". The output is shown in Figure 4.3. As the histogram is not perfectly symmetrical, a few fellow professors suggest that the students' score is not normally distributed. The statistics professor explains that even if the variable "Final mark" is indeed not normally distributed, there still is absolutely no problem because there are 132 observations in total. As a result, it is unproblematic to assume that the sample mean is normally distributed, and perform a hypothesis test for the mean score using the t-distribution with 131 degrees of freedom.

To perform this hypothesis test, the professor subsequently clicks the hotspot (red triangle) next to the word "Final mark" at the top of the output, and then selects the option "Test Mean" (see Figure 4.4). This results in the dialog window shown in Figure 4.5. In this window, he indicates in the input field "Specify Hypothesized Mean" that he wants to test whether the population mean μ deviates significantly from 10, and he clicks "OK". This produces the additional piece of output in Figure 4.6.

The p-value for the two-tailed hypothesis test is denoted by "Prob > |t|" in JMP. The p-value equals 0.5867, which is considerably higher than the significance level of 5% that the statistics professor uses. Therefore, the null hypothesis cannot be rejected. Hence the group of students is neither strong nor weak.

Just before he closes his laptop, the statistics professor points at the 95% confidence interval that JMP reports. The lower limit of the interval is 9.58, while the upper limit is equal to 10.74 (see Figure 4.3). The professor explains that it is always a good idea to also report the

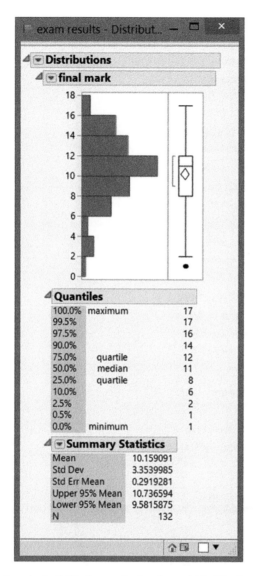

Figure 4.3 The initial output in JMP when analyzing the variable "Final mark" in Example 4.1.3.

confidence interval when performing a two-tailed hypothesis test: "The confidence interval expresses the fact that we have observed a mean of 10.16, but that it is equally possible that the unknown population mean is 9.58, or 10.74. Note that the hypothesized value of $\mu = 10$ in the null hypothesis is contained within the confidence interval. This is always the case when the null hypothesis of a two-tailed test cannot be rejected".

An alternative way to conduct the data analysis in Example 4.1.3 is to make use of the script "Hypothesis Test for Mean Using Data". Here, you must first indicate which variable you

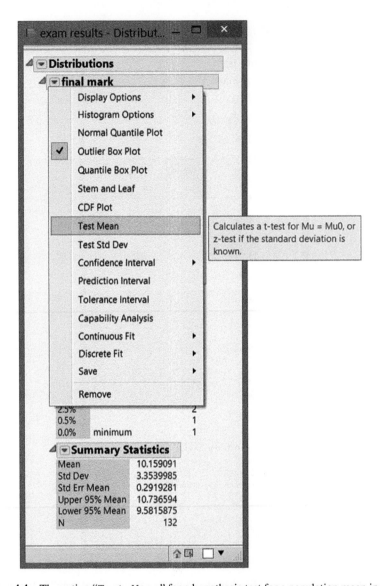

Figure 4.4   The option "Test Mean" for a hypothesis test for a population mean in JMP.

want to study, using the button "Pick A Column" (see Figure 4.7). Next, the value μ_0 and the significance level α must be specified. Using the button "Apply Changes", you obtain the result displayed in Figure 4.8.

4.2   Hypothesis Tests for a Population Proportion

In general, there are two ways to test a population proportion. The first method relies on the central limit theorem and the approximate normal distribution of the sample proportion. The

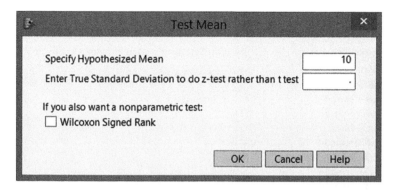

Figure 4.5   The dialog window for the option "Test Mean" for a hypothesis test for a population mean in JMP.

second method makes use of the binomial distribution, because the number of successes in a series of independent observations with success probability π is binomially distributed.

4.2.1   Tests Based on the Normal Distribution

In Section 1.6, we explained that the sample proportion $\hat{P}$ is a special case of a sample mean. Therefore, due to the central limit theorem (Theorem 1.5.6), $\hat{P}$ is approximately normally distributed with an expected value equal to the population proportion π and variance $\pi(1 - \pi)/n$:

$$\hat{P} \sim N\left(\pi, \frac{\pi(1 - \pi)}{n} \right).$$

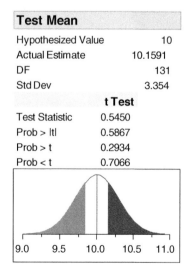

Figure 4.6   Additional JMP output containing the outcome of the hypothesis test for the mean of the variable "Final mark" in Example 4.1.3.

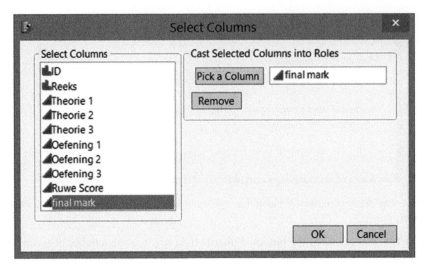

Figure 4.7 The dialog window of the script "Hypothesis Test for Mean Using Data" for selecting a variable.

This approximate probability density is, of course, only valid if the sample is sufficiently large. Most textbooks recommend that the conditions $n\hat{p} > 5$ and $n(1 - \hat{p}) > 5$ should be met to justify the use of the normal distribution.

Once again, we distinguish between hypothesis tests with right-tailed, left-tailed, or two-tailed alternative hypotheses. There are three different approaches: two methods involve critical values (either for the sample proportion or for a test statistic), while the third method makes use of a p-value.

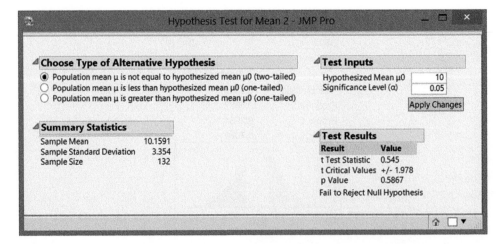

Figure 4.8 The output generated by the script "Hypothesis Test for Mean Using Data" for Example 4.1.3.

The Right-Tailed Test

In a right-tailed test for a population proportion π, the hypotheses tested are

$$H_0 : \pi = \pi_0$$

and

$$H_a : \pi > \pi_0.$$

The null hypothesis is rejected if the sample proportion $\hat{p}$ is larger than a critical value c. This critical value can be computed as soon as a significance level α has been chosen. The computation starts in the following way:

$$\alpha = P(\hat{P} > c \mid \pi = \pi_0),$$

$$= P\left(\frac{\hat{P} - \pi_0}{\sqrt{\dfrac{\pi_0(1 - \pi_0)}{n}}} > \frac{c - \pi_0}{\sqrt{\dfrac{\pi_0(1 - \pi_0)}{n}}} \;\middle|\; \pi = \pi_0 \right),$$

$$= P\left(Z > \frac{c - \pi_0}{\sqrt{\dfrac{\pi_0(1 - \pi_0)}{n}}} \right).$$

This implies that

$$z_\alpha = \frac{c - \pi_0}{\sqrt{\dfrac{\pi_0(1 - \pi_0)}{n}}},$$

which leads to the following critical value:

$$c = \pi_0 + z_\alpha \sqrt{\frac{\pi_0(1 - \pi_0)}{n}}.$$

The null hypothesis is rejected if the sample proportion $\hat{p}$ is greater than the critical value c. This is equivalent to rejecting the null hypothesis if

$$z = \frac{\hat{p} - \pi_0}{\sqrt{\dfrac{\pi_0(1 - \pi_0)}{n}}} > z_\alpha.$$

The p-value of the right-tailed test is calculated as

$$p = P\left(Z > \frac{\hat{p} - \pi_0}{\sqrt{\frac{\pi_0(1 - \pi_0)}{n}}}\right).$$

If the p-value is smaller than the significance level α, then the null hypothesis is rejected in favor of the alternative hypothesis.

The Left-Tailed Test

In a left-tailed test for a population proportion π, the hypotheses tested are

$$H_0 : \pi = \pi_0$$

and

$$H_a : \pi < \pi_0.$$

In that case, the null hypothesis is rejected if the sample proportion $\hat{p}$ is smaller than the critical value

$$c = \pi_0 - z_\alpha \sqrt{\frac{\pi_0(1 - \pi_0)}{n}},$$

or if

$$z = \frac{\hat{p} - \pi_0}{\sqrt{\frac{\pi_0(1 - \pi_0)}{n}}} < -z_\alpha.$$

For a left-tailed test, the p-value is calculated as

$$p = P\left(Z < \frac{\hat{p} - \pi_0}{\sqrt{\frac{\pi_0(1 - \pi_0)}{n}}}\right).$$

If the p-value is smaller than the significance level α, then the null hypothesis is rejected and the alternative hypothesis is accepted.

The Two-Tailed Test

In a two-tailed test for a population proportion π, the tested hypotheses are

$$H_0 : \pi = \pi_0$$

and

$$H_a : \pi \neq \pi_0.$$

The null hypothesis is rejected if the sample proportion $\hat{p}$ is smaller than the lower critical value

$$c_L = \pi_0 - z_{\alpha/2} \sqrt{\frac{\pi_0(1 - \pi_0)}{n}}$$

or larger than the upper critical value

$$c_U = \pi_0 + z_{\alpha/2} \sqrt{\frac{\pi_0(1 - \pi_0)}{n}}.$$

This decision rule is equivalent to rejecting the null hypothesis if

$$|z| = \left| \frac{\hat{p} - \pi_0}{\sqrt{\frac{\pi_0(1 - \pi_0)}{n}}} \right| > z_{\alpha/2}.$$

The p-value for a two-tailed test can be calculated as

$$p = 2 P\left(Z > \left| \frac{\hat{p} - \pi_0}{\sqrt{\frac{\pi_0(1 - \pi_0)}{n}}} \right| \right).$$

If the p-value is smaller than the significance level α, then the null hypothesis is rejected and the alternative hypothesis is accepted.

Example 4.2.1   *A detective who specializes in white-collar crime investigates so-called insider trading in the stock market. Insider trading means that a person who has internal information about a company is buying or selling stocks before this information is made public. Because all investors should have equal access to information that might influence stock prices, insider trading is prohibited.*

The detective analyzes 576 purchases of stocks by a number of senior managers. He notes that the stock price increased on the day after purchase in 327 cases. An obvious question for the detective is whether this data proves that the buyers did exploit insider information.

In order to answer this question, the null hypothesis that the senior managers do not use insider information must be tested against the alternative hypothesis that they do. If the null hypothesis is true, then it would be natural that, in about half of the cases, the stock price would increase after purchase, and, in the other half of the cases, the price would decrease. If the alternative hypothesis is true, then the price of more than half of the purchased stocks should rise. If π represents the proportion of all purchases of stocks the price of which rose the next day, then the null hypothesis can be written as

$$H_0 : \pi = 0.5,$$

while the alternative hypothesis is

$$H_a : \pi > 0.5.$$

Suppose that the detective is willing to take a risk of 5% that he is going to falsely accuse the senior managers. In such a case, he is going to choose a significance level of $\alpha = 0.05$.

The sample proportion in this example is $\hat{p} = 327/576 = 0.5677$. This sample proportion must be compared with the critical value

$$c = 0.5 + z_\alpha \sqrt{\frac{0.5(1-0.5)}{576}} = 0.5 + (1.6449)(0.0208) = 0.5343.$$

Clearly, $\hat{p} > c$, so that the null hypothesis has to be rejected. The test statistic is

$$z = \frac{\hat{p} - \pi_0}{\sqrt{\dfrac{\pi_0(1-\pi_0)}{n}}} = \frac{0.5677 - 0.5}{\sqrt{\dfrac{0.5(1-0.5)}{576}}} = 3.25,$$

which is larger than $z_\alpha = 1.6449$. The p-value is

$$p = P(Z > 3.25) = 0.000577,$$

which is smaller than the significance level. The data thus leads to a rejection of the null hypothesis and thereby suggests that insider trading does take place.

The hypothesis test in this example can be performed in JMP using the scripts "Hypothesis Test for Proportion Using Data" and "Hypothesis Test for Proportion Using Summary Statistics". When using the former script, you must indicate that you are interested in rising stock prices. This can be done by indicating in the dialog window shown in Figure 4.9 that a "success" (this is the event that interests us) corresponds to the value "Up" in the data set. Next, as shown in Figure 4.10, we need to indicate that we want to carry out a right-tailed hypothesis test for $\pi_0 = 0.5$ at a significance level of 0.05. Finally, clicking the button "Apply Changes" tells us that the p-value is equal to approximately 0.0006, and that the null hypothesis should therefore be rejected.

The 95% confidence interval for the population proportion π, computed using Equation (2.5), is [0.5269, 0.6076]. This interval confirms our suspicion that insider trading does take

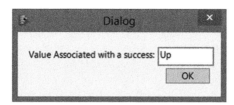

Figure 4.9 The dialog window of the script "Hypothesis Test for Proportion Using Data" for a hypothesis test for a population proportion in JMP.

place: our statistical analysis suggests that between 52.69% and 60.76% of all purchases of stocks by senior managers are followed by an immediate price increase, while this would be the case in only 50% of the cases if there were no insider trading.

4.2.2   Tests Based on the Binomial Distribution

When we collect sample data on "successes" and "failures", each individual observation can be seen as a Bernoulli experiment with a certain success probability, namely the population proportion π. The number of successes in a sample of n observations is binomially distributed with parameters n and π (see *Statistics with JMP: Graphs, Descriptive Statistics and Probability*). The binomial distribution can therefore be used to calculate a p-value for a test on a population proportion. We denote the number of successes in a sample by the random variable S.

The test based on the binomial distribution is especially important in cases where the sample is small and we cannot rely on the central limit theorem. The binomial test is said to be **exact**, because the binomial distribution is not approximated by the normal distribution in the calculation of the p-value. If a null hypothesis is rejected using the binomial distribution, this is stronger evidence against the null hypothesis than if it is done using the normal density approach from Section 4.2.1. Therefore, many researchers prefer the test based on the binomial distribution.

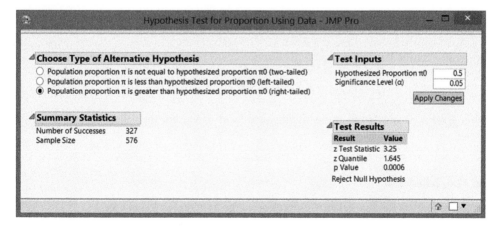

Figure 4.10 The JMP script "Hypothesis Test for Proportion Using Data" applied to Example 4.2.1.

In contrast to the tests based on the normal distribution in Section 4.2.1, for the binomial test, we do not use critical values, but only p-values. As usual, we assume that the null hypothesis is true when calculating the p-values.

The Right-Tailed Test

In a right-tailed test for a population proportion π, the hypotheses tested are

$$H_0 : \pi = \pi_0$$

and

$$H_a : \pi > \pi_0.$$

If the null hypothesis is true, the number of successes in the sample, S, is binomially distributed with parameters n and π_0. Now, suppose that the observed number of successes is equal to s. Then, the p-value for the right-tailed hypothesis test is

$$p = P(S \geq s).$$

If this p-value is smaller than the significance level α chosen, then the null hypothesis is rejected.

The Left-Tailed Test

In a left-tailed test for a population proportion π, the hypotheses tested are

$$H_0 : \pi = \pi_0$$

and

$$H_a : \pi < \pi_0.$$

The p-value for the left-tailed hypothesis test is

$$p = P(S \leq s).$$

If this p-value is smaller than the significance level α chosen, then the null hypothesis is rejected.

The Two-Tailed Test

In a two-tailed test for a population proportion π, the hypotheses tested are

$$H_0 : \pi = \pi_0$$

and

$$H_a : \pi \neq \pi_0.$$

The p-value for this two-tailed test is equal to the minimum of

$$2 \, P(S \geq s)$$

and

$$2 \, P(S \leq s).$$

4.2.3   Testing Proportions in JMP

You do not need the JMP scripts "Hypothesis Test for Proportion Using Data" and "Hypothesis Test for Proportion Using Summary Statistics" to conduct a test for a population proportion in JMP. You can also carry out the test after selecting the "Distribution" platform in the "Analyze" menu. You can choose a left- or right-tailed test, or a two-tailed test. If you opt for a one-tailed test, JMP calculates the p-value based on the binomial distribution (as in Section 4.2.2). JMP then uses the exact test. However, if you opt for a two-tailed test, JMP calculates two p-values based on a χ^2-distribution with one degree of freedom. The p-value labeled "Pearson" is the p-value we would obtain if we used the two-tailed test based on the normal distribution[1] (as in Section 4.2.1). The p-value labeled "Likelihood Ratio" is based on a completely different method[2], which is not discussed in this book. The interpretation of both p-values is identical: p-values smaller than the significance level α lead to a rejection of the null hypothesis.

JMP uses different methods for one-tailed and two-tailed tests for a population proportion. This seems to be inconsistent. However, the reason for this is that the power of the two-tailed test based on the binomial distribution is considerably smaller than the power of the two-tailed test based on the χ^2-distribution or on the normal distribution.

We illustrate the use of JMP by revisiting the hypothesis test on insider trading from Example 4.2.1.

Example 4.2.2   *The null hypothesis in Example 4.2.1 was*

$$H_0 : \pi = 0.5,$$

while the alternative hypothesis was

$$H_a : \pi > 0.5.$$

---

[1] A χ^2-distributed random variable with one degree of freedom is the square of a standard normally distributed random variable. Therefore, the two distributions contain the same kind of information and result in the same p-value. See also Section 1.7.3.

[2] Likelihood ratio tests are used very often in statistics. They start from a specific type of estimators, namely maximum likelihood estimators, which are not studied here.

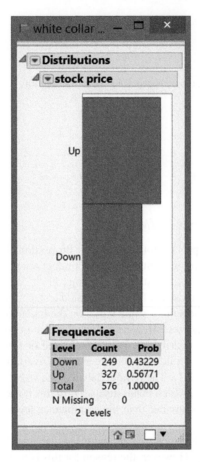

Figure 4.11 The initial JMP output for the analysis of the data on insider trading in Examples 4.2.1 and 4.2.2.

To perform the test in JMP, you have to use the "`Distribution`*" platform in the "*`Analyze`*" menu after opening the data table. This produces the output shown in Figure 4.11. The output indicates that there are 327 successes (rising stock prices, labeled "Up") in the sample of 576 observations.*

Next, you need to click on the hotspot (red triangle) next to the name of the variable under study, "Stock price". In the resulting menu, shown in Figure 4.12, you have to select the option "`Test Probabilities`*", and, if you also desire a confidence interval, the option "*`Confidence Interval`*" as well. The option "*`Test Probabilities`*" leads to the dialog window shown in Figure 4.13. In this window, next to the label "Up", we can see that the sample proportion of rising stock prices is equal to 0.56771.*

In this example, we want to perform a test for the proportion of rising stock prices, and the hypothesized value, π_0, is 0.5. Therefore, we need to enter the value 0.5 next to the label "Up". We must also indicate that we want to perform a right-tailed test by selecting the option "`Probability greater than hypothesized value (one-sided binomial test)`*" (see Figure 4.14). Finally, clicking "Done" produces the result shown in Figure 4.15.*

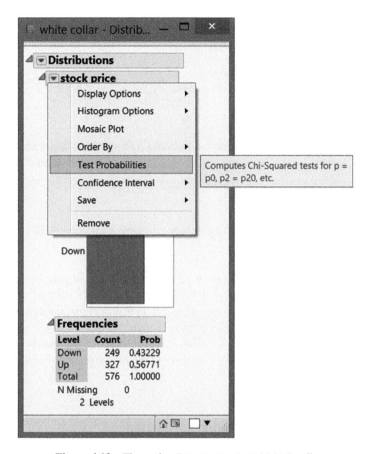

Figure 4.12 The option "Test Probabilities".

Test Probabilities

Level	Estim Prob	Hypoth Prob
Down	0.43229	
Up	0.56771	

Click then Enter Hypothesized Probabilities.

Select an alternative hypothesis for testing probabilities.

- ⦿ probabilities not equal to hypothesized value (two-sided chi-square test)
- ◯ probability greater than hypothesized value (exact one-sided binomial test)
- ◯ probability less than hypothesized value (exact one-sided binomial test)

[Done] [Help]

Figure 4.13 The dialog window for a hypothesis test for a population proportion in JMP.

⊿ Test Probabilities

Level	Estim Prob	Hypoth Prob
Down	0.43229	.
Up	0.56771	0.50000

Click then Enter Hypothesized Probabilities.

Select an alternative hypothesis for testing
probabilities.

⊙ probabilities not equal to hypothesized value (two-sided chi-square test)

◉ probability greater than hypothesized value (exact one-sided binomial test)

⊙ probability less than hypothesized value (exact one-sided binomial test)

| Done | Help |

Figure 4.14   The input required for conducting the right-tailed hypothesis test in JMP for the population proportion in Examples 4.2.1 and 4.2.2.

The p-value that JMP reports in Figure 4.15 equals 0.0007. The p-value differs slightly from the p-value of Example 4.2.1 because, for a one-tailed test, JMP uses the exact approach based on the binomial distribution for a one-tailed test. In this example, the difference between the p-values is minimal, because we have a large number of observations. Therefore, both $n\hat{p}$ and $n(1 - \hat{p})$ are considerably greater than 5, so that the normal approximation of the binomial distribution is excellent. The JMP formula to obtain the p-value is jmp 1 − Binomial Distribution (0.5, 576, 326).

4.3   Hypothesis Tests for a Population Variance

As a result of the relation between the sample variance S^2 and the χ^2-distribution (see Sections 1.7.5 and 2.5), the hypothesis test for a population variance σ^2 will be based on the χ^2-distribution. Of course, for the test to be valid, the population studied has to be normally distributed.

Once again, we distinguish between hypothesis tests involving right-tailed, left-tailed, or two-tailed alternative hypotheses. Again, three different approaches exist for each of these tests. Two of the approaches are based on critical values (either for the sample variance or for the test statistic), while the third one uses a *p*-value.

⊿ Test Probabilities

Level	Estim Prob	Hypoth Prob
Down	0.43229	0.50000
Up	0.56771	0.50000

Binomial Test	Level Tested	Hypoth Prob (p1)	p-Value
Ha: Prob(p > p1)	Up	0.50000	0.0007*

Figure 4.15   The JMP output for the right-tailed hypothesis test for the population proportion in Examples 4.2.1 and 4.2.2.

4.3.1   The Right-Tailed Test

The hypotheses tested are

$$H_0 : \sigma^2 = \sigma_0^2$$

and

$$H_a : \sigma^2 > \sigma_0^2.$$

A logical decision is to reject the null hypothesis H_0 in favor of the alternative hypothesis H_a as soon as the calculated sample variance s^2 is larger than a certain critical value c. This critical value depends on the significance level α chosen, and can be determined using the following result from Section 1.7.4:

$$\frac{(n-1)S^2}{\sigma^2} \sim \chi_{n-1}^2.$$

The critical value is then calculated starting from

$$\alpha = P(S^2 > c \mid \sigma^2 = \sigma_0^2),$$

$$= P\left(\frac{(n-1)S^2}{\sigma_0^2} > \frac{(n-1)c}{\sigma_0^2} \,\middle|\, \sigma^2 = \sigma_0^2 \right),$$

$$= P\left(\chi_{n-1}^2 > \frac{(n-1)c}{\sigma_0^2} \right),$$

where χ_{n-1}^2 represents a χ^2-distributed random variable with $n-1$ degrees of freedom. This implies that

$$\frac{(n-1)c}{\sigma_0^2} = \chi_{\alpha;n-1}^2,$$

so that

$$c = \frac{\sigma_0^2}{n-1} \chi_{\alpha;n-1}^2.$$

Consequently, we reject the null hypothesis if

$$s^2 > \frac{\sigma_0^2}{n-1} \chi_{\alpha;n-1}^2,$$

or if

$$\frac{(n-1)s^2}{\sigma_0^2} > \chi^2_{\alpha;n-1}.$$

The value

$$\frac{(n-1)s^2}{\sigma_0^2}$$

is called the test statistic of the hypothesis test for a population variance.

The third approach uses the p-value and leads to a rejection of the null hypothesis whenever

$$p = P\left(\chi^2_{n-1} > \frac{(n-1)s^2}{\sigma_0^2} \right) < \alpha.$$

These three conditions are equivalent: they all lead to same decision.

4.3.2   The Left-Tailed Test

The hypotheses tested are

$$H_0 : \sigma^2 = \sigma_0^2$$

and

$$H_a : \sigma^2 < \sigma_0^2.$$

The null hypothesis is rejected if

$$s^2 < \frac{\sigma_0^2}{n-1}\chi^2_{1-\alpha;n-1}$$

or, in terms of the test statistic, if

$$\frac{(n-1)s^2}{\sigma_0^2} < \chi^2_{1-\alpha;n-1}.$$

The third approach uses the p-value and leads to a rejection of the null hypothesis whenever

$$p = P\left(\chi^2_{n-1} < \frac{(n-1)s^2}{\sigma_0^2} \right) < \alpha.$$

4.3.3   The Two-Tailed Test

The hypotheses tested are

$$H_0 : \sigma^2 = \sigma_0^2$$

and

$$H_a : \sigma^2 \neq \sigma_0^2.$$

The null hypothesis is rejected if

$$s^2 < \frac{\sigma_0^2}{n-1}\chi_{1-\alpha/2;n-1}^2 \quad \text{or} \quad s^2 > \frac{\sigma_0^2}{n-1}\chi_{\alpha/2;n-1}^2,$$

or, in terms of the test statistic, if

$$\frac{(n-1)s^2}{\sigma_0^2} < \chi_{1-\alpha/2;n-1}^2 \quad \text{or} \quad \frac{(n-1)s^2}{\sigma_0^2} > \chi_{\alpha/2;n-1}^2.$$

To calculate the p-value of the two-tailed hypothesis test, we need to distinguish between the cases in which $s^2 > \sigma_0^2$ and $s^2 < \sigma_0^2$. The reason for this is that the χ^2-distribution is not symmetrical. When $s^2 > \sigma_0^2$, the p-value is calculated as

$$p = 2\,P\left(\chi_{n-1}^2 > \frac{(n-1)s^2}{\sigma_0^2}\right).$$

In the opposite case – that is, when $s^2 < \sigma_0^2$ – the p-value is

$$p = 2\,P\left(\chi_{n-1}^2 < \frac{(n-1)s^2}{\sigma_0^2}\right).$$

Alternatively, to calculate the p-value for a two-tailed hypothesis test, one can determine the minimum of the p-values of the left-tailed and right-tailed hypothesis tests, and then multiply this minimum by 2.

Two JMP scripts are available for performing hypothesis tests for a population variance, namely the scripts "Hypothesis Test for Variance Using Data" and "Hypothesis Test for Variance Using Summary Statistics".

Example 4.3.1   *A sample of 20 observations from a filling machine has a standard deviation s of 0.4294 cl, and thus a variance s^2 of $(0.4294)^2$ cl^2. We test the hypotheses*

$$H_0 : \sigma^2 = 0.1 \text{ cl}^2$$

and

$$H_a : \sigma^2 \neq 0.1 \text{ cl}^2.$$

The computed test statistic is

$$\frac{(n-1)s^2}{\sigma_0^2} = \frac{(20-1)(0.4294)^2}{0.1} = 35.0330.$$

For the two-sided hypothesis test, this quantity should be compared with the critical values $\chi^2_{1-\alpha/2;n-1}$ *and* $\chi^2_{\alpha/2;n-1}$. *For the usual significance level* $\alpha = 0.05$, *and given that* $n = 20$, *these critical values are* $\chi^2_{0.975;19} = 8.9065$ *and* $\chi^2_{0.025;19} = 32.8523$. *The test statistic does not lie between these two critical values, so that the null hypothesis can be rejected. As* $s^2 = (0.4294)^2 = 0.1844$ *is larger than* $\sigma_0^2 = 0.1$, *the p-value is*

$$p = 2\, P(\chi^2_{n-1} > 35.0330) = 2(0.0138) = 0.0277.$$

The fact that the p-value is smaller than $\alpha = 0.05$ *confirms the decision to reject the null hypothesis. The p-value can be calculated in JMP using the formula "2 * (1 − ChiSquare Distribution (35.0330, 19))". The critical values* $\chi^2_{0.975;19}$ *and* $\chi^2_{0.025;19}$ *can be obtained in JMP using the formulas "ChiSquare Quantile(0.025, 19)" and "ChiSquare Quantile(0.975, 19)". These two values can also be found in the table in Appendix C, but only up to three digits after the decimal point.*

The JMP script "Hypothesis Test for Variance Using Summary Statistics" also allows the hypothesis test in this example to be performed. Figure 4.16 shows the result produced by this script.

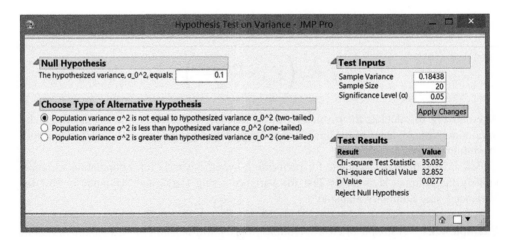

Figure 4.16 The JMP script "Hypothesis Test for Variance Using Summary Statistics" applied to Example 4.3.1.

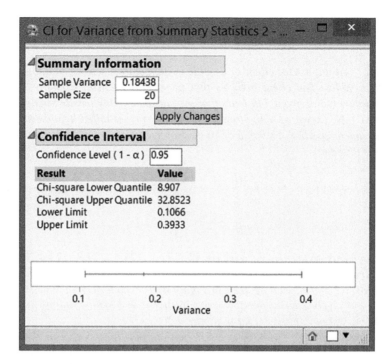

Figure 4.17   The JMP script "CI for Variance from Summary Statistics" applied to Example 4.3.1.

It is always recommended that the confidence interval for σ^2 is also reported, and not merely the result of the hypothesis test. In this example, the 95% confidence interval for σ^2 is

$$[0.1066, 0.3933].$$

This interval does not include the hypothesized value $\sigma_0^2 = 0.1$. Therefore, the interval shows that we can reject the null hypothesis $\sigma^2 = 0.1$ if we use a significance level of 5%. Figure 4.17 shows how the confidence interval was determined using the JMP script "CI for Variance from Summary Statistics".

4.3.4   The Use of JMP

In JMP, it is not possible to perform a hypothesis test for a population variance σ^2 as described above. However, JMP provides an alternative, namely a hypothesis test for a population standard deviation σ. The calculations in JMP are identical to those that were described above. The only difference is that you need to specify the value of σ_0 instead of the value of σ_0^2. If you use JMP for this hypothesis test, you obtain three *p*-values. The *p*-value for the two-tailed test given in JMP is labeled "Min PValue". The *p*-value for the right-tailed hypothesis test is labeled "`Prob > ChiSq`", while the p-value for the left-tailed hypothesis test is labeled "`Prob < ChiSq`".

Example 4.3.2   *An industrial client of a cable manufacturer is worried about the quality of the supplied cables. The client believes that the average tensile strength of the cables fulfills the minimum requirement, but fears that the variance of the tensile strength is greater than the agreed* 1 N^2 *(newton). Therefore, he performs a right-tailed hypothesis test for the population variance based on a sample of* 10 *measurements, with a significance level of* 5%. *The hypotheses tested are*

$$H_0 : \sigma^2 = 1 \ N^2$$

and

$$H_a : \sigma^2 > 1 \ N^2.$$

The 10 *measurements of the tensile strength are* 172.1, 174.5, 172.8, 173.4, 175.4, 176.1, 173.5, 174.1, *and* 172 *N, so that the sample standard deviation s is* 1.3812 *N and the sample variance is* 1.9077 N^2. *This can be seen in Figure 4.18, which also shows how the option* "`Test Std Dev`" *can be carried out. If you specify that* $\sigma_0 = 1$ *in the resulting dialog window (see Figure 4.19), you obtain the output in Figure 4.20.*

   From the output labeled "`Prob > ChiSq`", *we see that the p-value of the right-tailed hypothesis test is* 0.0461. *The p-value is smaller than the significance level used. The client therefore concludes that the variance is significantly larger than* 1 N^2. *To further support his conclusion, he also calculates the* 95% *confidence interval for the population standard deviation and the population variance. He obtains the interval* [0.9500, 2.5215] *for the standard deviation, and* [0.9026, 6.3580] *for the variance.*

   Note that the confidence intervals for the standard deviation and the variance contain the hypothesized values $\sigma_0 = \sigma_0^2 = 1$, *while the null hypothesis that* $\sigma_0 = \sigma_0^2 = 1$ *is rejected. As a result, there is seemingly a contradiction between the hypothesis test and the confidence intervals. However, as was explained in Section 3.2.3, the correspondence between a confidence interval and a hypothesis test is valid for a two-tailed test, but not for a one-tailed test such as the one carried out in this example.*

4.4   The Probability of a Type II Error and the Power

To determine the probability of a type II error and the power of a hypothesis test, we consider the situation where the alternative hypothesis H_a is correct. This requires us to determine the probability density of the test statistic when the alternative hypothesis is true. In Section 3.3, we studied the probability of a type II error and the power of a test for a population mean with known σ^2, and we assumed that, under the alternative hypothesis, $\overline{X}$ was normally distributed with expected value μ_1 instead of μ_0. The use of the normal distribution in that section (both when the null hypothesis was true and when the alternative hypothesis was true) was only possible because σ^2 was known. The normal distribution is no longer appropriate when σ is unknown.

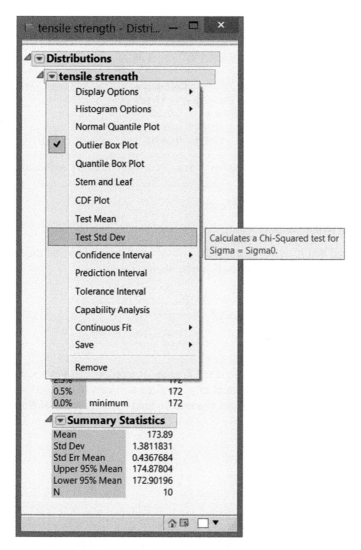

Figure 4.18 The option "`Test Std Dev`" in JMP for a hypothesis test for a population standard deviation (see Example 4.3.2).

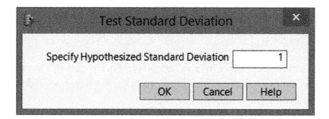

Figure 4.19 Specifying the value of σ_0 when using the option "`Test Std Dev`" in JMP (see Example 4.3.2).

Test Standard Deviation

Hypothesized Value	1
Actual Estimate	1.38118
DF	9

Test	**ChiSquare**
Test Statistic	17.169
Min PValue	0.0923
Prob < ChiSq	0.9539
Prob > ChiSq	0.0461*

Figure 4.20 The output of the option "`Test Std Dev`" in JMP (see Example 4.3.2).

4.4.1 Tests for a Population Mean

The Noncentral t-Distribution

Earlier, we argued that if we study a normally distributed population and the null hypothesis that the population mean μ is equal to μ_0 is true, then

$$T = \frac{\overline{X} - \mu_0}{\frac{S}{\sqrt{n}}}$$

is a t-distributed random variable with $n-1$ degrees of freedom. This is because the random variable T can be written as the quotient of a standard normally distributed variable and the square root of a χ^2-distributed random variable divided by its degrees of freedom. In other words,

$$T = \frac{Z}{\sqrt{\frac{X}{v}}}, \tag{4.1}$$

where Z is a standard normally distributed random variable, and X is a χ^2-distributed random variable with v degrees of freedom. The t-distribution, as defined in Equation (4.1), is symmetrical around zero. Therefore, it is called the central t-distribution.

However, there is also a noncentral t-distribution, defined as

$$T = \frac{Z + \delta}{\sqrt{\frac{X}{v}}}, \tag{4.2}$$

with δ a real number that can be positive or negative. This parameter δ is the noncentrality parameter of the noncentral t-distribution. Thus, besides the number of degrees of freedom, the noncentral t-distribution has a second parameter.

Unlike the central t-distribution, the noncentral t-distribution is skewed. It is skewed to the right if δ is positive, and skewed to the left if δ is negative. Figure 4.21 shows the central t-distribution with three degrees of freedom, and three noncentral t-distributions, also with

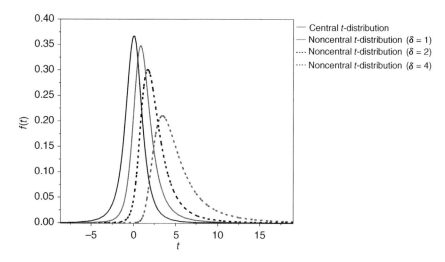

Figure 4.21   The central and three noncentral t-distributions with three degrees of freedom.

three degrees of freedom. The three noncentral t- distributions have noncentrality parameters 1, 2, and 4. The figure shows that the distribution shifts to the right when the noncentrality parameter goes up, and that the skewness of the distributions increases with the noncentrality parameter.

Note that the central t-distribution is a special case of the noncentral t-distribution. Indeed, if the noncentrality parameter δ is equal to zero, then the noncentral t-distribution reduces to the central t-distribution.

In JMP, we can use the functions "t Distribution", "t Quantile", and "t Density" to calculate probabilities and quantiles of t-distributed random variables, and for creating graphs of their probability densities. When we use these functions, we can either enter one or two parameters for the t- distribution. The parameter that JMP always requires is the number of degrees of freedom. The entering of a second parameter, the noncentrality parameter, is optional. If we do not enter a noncentrality parameter, JMP automatically uses a central t-distribution (and thereby sets the noncentrality parameter to zero). For example, to calculate the probability that a central t-distributed random variable with three degrees of freedom takes values smaller than 2, we can use the formula "t Distribution(2, 3)". On the other hand, to calculate the probability that a noncentral t-distributed random variable with three degrees of freedom and a noncentrality parameter of 4 takes values smaller than 2, we have to use the formula "t Distribution(2, 3, 4)". These two JMP formulas return the probabilities 0.9303 and 0.0476, respectively.

The Use of the Noncentral t-Distribution

Suppose that the population under study is normally distributed and that we want to test the following hypotheses:

$$H_0 : \mu = \mu_0$$

and

$$H_a : \mu > \mu_0.$$

The decision rule for this hypothesis test states that we reject the null hypothesis if the computed test statistic

$$t = \frac{\bar{x} - \mu_0}{\frac{s}{\sqrt{n}}}$$

is larger than t_α.

If the null hypothesis is false and the population mean is $\mu_1 > \mu_0$, then the test statistic

$$T = \frac{\bar{X} - \mu_0}{\frac{S}{\sqrt{n}}}$$

follows a noncentral t-distribution with $n - 1$ degrees of freedom and noncentrality parameter

$$\frac{\mu_1 - \mu_0}{\frac{\sigma}{\sqrt{n}}}.$$

To see this, it is useful to rewrite the test statistic as

$$T = \frac{\bar{X} - \mu_0}{\frac{S}{\sqrt{n}}} = \frac{\bar{X} - \mu_1 + \mu_1 - \mu_0}{\frac{S}{\sqrt{n}}} = \frac{\frac{\bar{X} - \mu_1 + \mu_1 - \mu_0}{\sigma}}{\frac{S}{\sigma\sqrt{n}}}$$

$$= \frac{\frac{\bar{X} - \mu_1 + \mu_1 - \mu_0}{\frac{\sigma}{\sqrt{n}}}}{\frac{S}{\sigma}} = \frac{\frac{\bar{X} - \mu_1}{\frac{\sigma}{\sqrt{n}}} + \frac{\mu_1 - \mu_0}{\frac{\sigma}{\sqrt{n}}}}{\sqrt{\frac{(n-1)S^2}{\sigma^2}}{n-1}}.$$

The numerator in this equation is the sum of a standard normally distributed random variable

$$\frac{\bar{X} - \mu_1}{\frac{\sigma}{\sqrt{n}}}$$

and a constant

$$\frac{\mu_1 - \mu_0}{\frac{\sigma}{\sqrt{n}}},$$

while the denominator is the square root of a χ^2-distributed random variable, namely

$$\frac{(n-1)S^2}{\sigma^2},$$

divided by its number of degrees of freedom, $n-1$. As a result, if $\mu = \mu_1$, then the test statistic follows a noncentral t-distribution with $n-1$ degrees of freedom and noncentrality parameter

$$\delta = \frac{\mu_1 - \mu_0}{\frac{\sigma}{\sqrt{n}}}.$$

The probability of a type II error is the probability that the test statistic does not exceed the critical value $t_{\alpha;n-1}$, given that $\mu = \mu_1$. This is the probability that a noncentral t-distributed random variable with $n-1$ degrees of freedom and noncentrality parameter δ takes values smaller than $t_{\alpha;n-1}$.

In order to determine the probability of a type II error and its complement, the power, we have to select a value for μ_1 and have an idea of the magnitude of σ and of the number of observations n. The values of μ_1, n, and σ are needed to calculate the noncentrality parameter δ of the required noncentral t-distribution. It follows that if we want to estimate the probability of a type II error or the power of a test for a population mean μ with unknown σ, we have to pretend that we do know σ.

Example 4.4.1   *Suppose that you want to perform a right-tailed hypothesis test for a population mean at a significance level of 5%, and that you believe the true population mean to be $\mu_1 = \mu_0 + 2$. Your guess is that the population standard deviation σ is equal to 4. The budget for your research only allows a sample size n of five observations.*

In order to determine the probability of a type II error and the power, you first have to calculate the noncentrality parameter δ of the required noncentral t-distribution. This parameter is

$$\delta = \frac{\mu_1 - \mu_0}{\frac{\sigma}{\sqrt{n}}} = \frac{2}{\frac{4}{\sqrt{5}}} = 1.1180.$$

The number of degrees of freedom is $n - 1 = 4$, while the critical value for the test statistic, $t_{\alpha;n-1} = t_{0.05;4}$, is equal to 2.1318. The probability of a type II error is equal to the probability that a noncentral t-distributed random variable with four degrees of freedom and noncentrality parameter 1.1180 is less than 2.1318. This probability can be calculated in JMP using the formula "t Distribution(2.1318, 4, 1.1180)", or "t Distribution(t

`Quantile(0.95, 4)`, `4`, `2/4 * Sqrt(5))`", *if the critical value and the noncentrality parameter are not calculated in advance. The result is equal to* 0.7610. *Its complement,* 0.2390, *is the power.*

A power of 0.2390 *is on the low side. A researcher who seeks to reject the null hypothesis when the true population mean is* $\mu_1 = \mu_0 + 2$ *and* $\sigma = 4$ *has no choice other than to increase the number of observations n. In order to illustrate this, suppose that the budget allows for* 20 *observations instead of five. In that case, the noncentrality parameter is*

$$\delta = \frac{\mu_1 - \mu_0}{\frac{\sigma}{\sqrt{n}}} = \frac{2}{\frac{4}{\sqrt{20}}} = \frac{2}{\frac{4}{2\sqrt{5}}} = \sqrt{5} = 2.2361.$$

The number of degrees of freedom is then $n - 1 = 19$, *and the probability of a type II error is equal to* 0.3049. *As a result, the power is* 0.6951. *This is substantially larger than the power for* $n = 5$.

Suppose that you do not perform a right-tailed hypothesis test, but a two-tailed test. The hypotheses tested are then

$$H_0 : \mu = \mu_0$$

and

$$H_a : \mu \neq \mu_0.$$

The decision rule for this hypothesis test states that we reject the null hypothesis if the calculated test statistic is smaller than $t_{1-\alpha/2;n-1} = -t_{\alpha/2;n-1}$, or larger than $t_{\alpha/2;n-1}$.

Therefore, the probability of a type II error is the probability that a noncentral t-distributed random variable with $n - 1$ degrees of freedom and noncentrality parameter δ takes values between $t_{1-\alpha/2;n-1} = -t_{\alpha/2;n-1}$ and $t_{\alpha/2;n-1}$.

Example 4.4.2   *Suppose that you want to perform a two-tailed hypothesis test for a population mean at a significance level of* 5% , *and that you believe the true population mean to be* $\mu_1 = \mu_0 + 2$. *Your guess for the population standard deviation* σ *is* 4. *The sample size n is* 5. *As shown in Example 4.4.1, the noncentrality parameter* δ *is* 1.1180 *in this case, and there are four degrees of freedom.*

The two critical values for the test statistic, $t_{1-\alpha/2;n-1} = t_{0.975;4} = -t_{\alpha/2;n-1} = -t_{0.025;4}$ *and* $t_{\alpha/2;n-1} = t_{0.025;4}$, *are* -2.7764 *and* 2.7764. *The probability of a type II error – that is, the probability of erroneously accepting the null hypothesis – is the probability that a noncentral t-distributed random variable takes values between* -2.7764 *and* 2.7764. *With the formulas*

"`t Distribution(2.7764, 4, 1.1180)`
              `-t Distribution(-2.7764, 4, 1.1180)`"

and

"t Distribution(t Quantile(0.975, 4), 4, 2/4*Sqrt(5))
              −t Distribution(t Quantile(0.025, 4), 4, 2/4*Sqrt(5))",

we can calculate in JMP that this probability is equal to 0.8595. The complement, 0.1405, is the power. This low power can be increased by using a larger number of observations n.

In this example, we calculated the power for $\mu_1 = \mu_0 + 2$. This means that we were interested in detecting a difference of $\mu_1 - \mu_0 = 2$. The low power indicates that it is difficult to detect such a small difference when the standard deviation σ is as large as 4. If we are interested in a difference of $\mu_1 - \mu_0 = 10$ (which is substantially larger than σ), then the noncentrality parameter is

$$\delta = \frac{\mu_1 - \mu_0}{\frac{\sigma}{\sqrt{n}}} = \frac{10}{\frac{4}{\sqrt{5}}} = 5.5902.$$

With the same number of observations ($n = 5$), the probability of a type II error drops to 0.0025 and the power increases to 0.9975.

This calculation shows that it is easier to detect a difference that is large relative to σ than a difference that is small relative to σ. In other words, the more the reality (μ_1) deviates from the null hypothesis (μ_0), the larger is the probability that the null hypothesis is rejected.

JMP

JMP allows an easy calculation of the probability of a type II error and the power of a *t*-test for a population mean via the menu "DOE" and the option "Sample Size and Power". In the menu that appears (see Figure 4.22), you need to select the option "One Sample Mean" for a *t*-test for one population mean. This will open a dialog window in which you can enter all the relevant information about your problem. To calculate the power in Example 4.4.2, we first need to fill the fields "Alpha", "Std Dev", "Difference to detect", and "Sample Size". In these fields, we enter the significance level α, the (expected) population standard deviation σ, the difference between μ_1 and μ_0 that we wish to detect, and the sample size n, respectively. For the kind of analysis we perform here, the value for "Extra Parameters" should remain zero. If you leave the field called "Power" empty, and click on "Continue", then JMP calculates the power. This is illustrated in Figure 4.23. Note that JMP assumes that you are interested in performing a two-tailed test.

Instead of the field "Power", you can also leave the field "Sample Size" blank. If you then click on "Continue", JMP calculates the required sample size to obtain the desired power. Figure 4.24 shows how we can determine the required number of observations using JMP if we aim for a power of 90%. It turns out that 44 observations are required.

In Figures 4.23 and 4.24, you can see a button called "Animation Script". Clicking this generates a graphical representation of the probability of a type I error and that of a type II error. The graph is shown in Figure 4.25. In this animation, you can read off both the

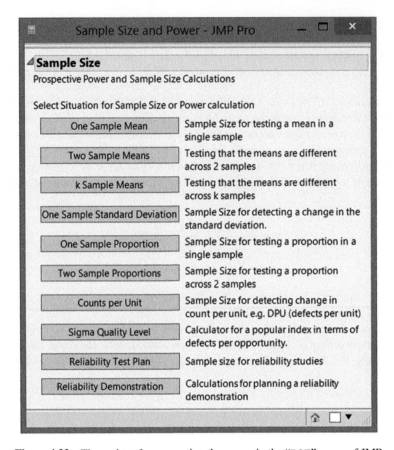

Figure 4.22   The options for computing the power in the "DOE" menu of JMP.

probability of a type II error and the power next to the words "Beta" and "Power". The dark gray areas together form the probability of a type I error – or, in other words, the significance level α. The light gray area is the probability of a type II error.

On the left, the figure shows the probability density of $\overline{X}$ when the null hypothesis is correct. This probability density is nicely symmetrical. On the right, the figure shows the probability density of $\overline{X}$ if the alternative hypothesis is correct. This probability density is skewed to the right and has a lower peak. The figure does not show the central and noncentral t-distributions, but a transformation of those distributions.

In JMP, the figure is interactive: the little squares in the figure can be dragged, so that the values of μ_0 ("Hypothesized Mean") and μ_1 ("True Mean") can be changed. When doing so, the corresponding probabilities of a type II error and the power are recalculated immediately. If you click on the button "High Side" (this button is not visible in Figure 4.25), you will see a similar graph, but now for the right-tailed test. This graph is shown in Figure 4.26. The probabilities shown in this figure were calculated in Example 4.4.1.

Finally, if you leave two fields empty in the dialog window shown in Figure 4.23, including the field "Power", and then click "Continue", you will obtain a graph showing the relationship

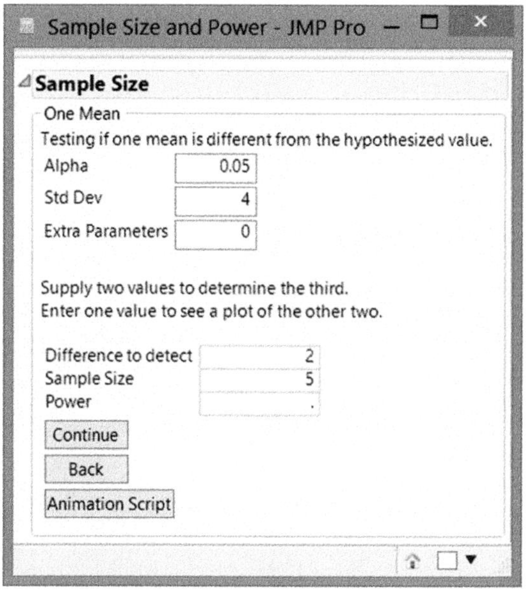

(a)

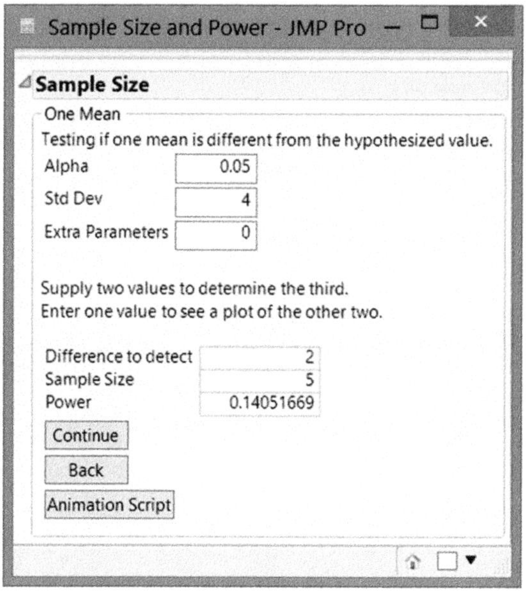

(b)

Figure 4.23 Determining the power of a hypothesis test for a population mean in JMP when σ is unknown.

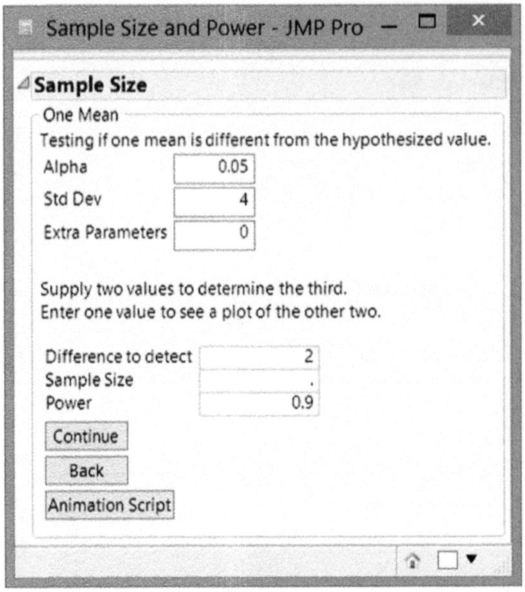

(a)

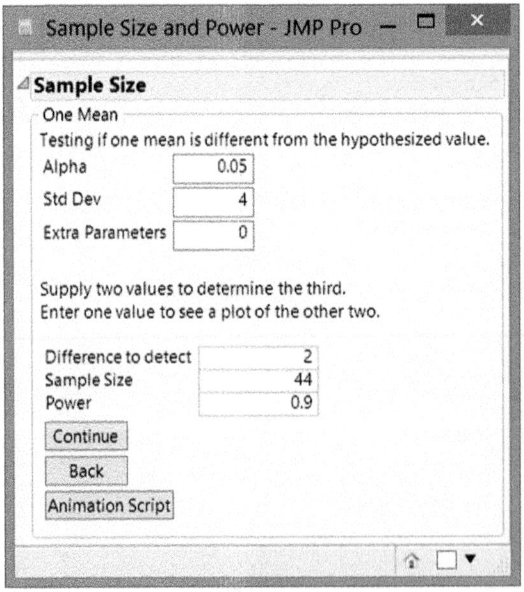

(b)

Figure 4.24 Determining the required sample size n for a hypothesis test for a population mean in JMP when σ is unknown.

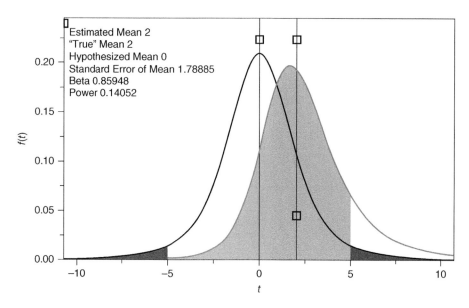

Figure 4.25 A graphical representation of the probabilities of type I and type II errors in the two-tailed test in Example 4.4.2. The probability of a type I error is shown by the dark gray areas. The probability of a type II error is represented by the light gray area.

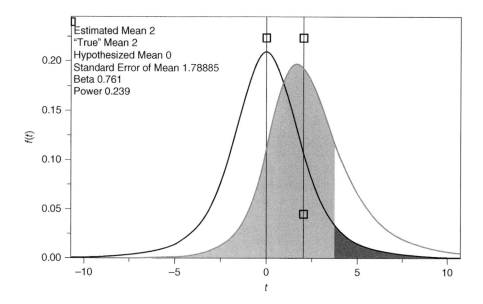

Figure 4.26 A graphical representation of the probabilities of type I and type II errors in the right-tailed test for the population mean in Example 4.4.1. The probability of a type I error is shown by the dark gray area. The probability of a type II error is represented by the light gray area.

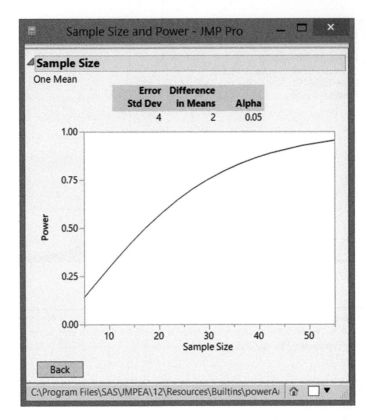

Figure 4.27 The relationship between the power and the sample size in the two-tailed test for the population mean in Example 4.4.2.

between either the power and the sample size, or between the power and the difference to be detected between μ_1 and μ_0. When creating the graph, JMP again assumes that you are interested in performing a two-tailed test. Figure 4.27 contains a graph that shows the relationship between the power and the sample size when the difference to be detected is 2 and the population standard deviation is equal to 4. For example, we can see in this graph that 44 observations are required for a power of 90%, which matches the result of our calculation in Figure 4.24.

4.4.2 Tests for a Population Proportion

The option "Sample Size and Power" in the "DOE" menu also allows the power of one-tailed or two-tailed hypothesis tests for a population proportion to be determined. To do so, select the option "One Sample Proportion" (see Figure 4.22). Using that option, one may also compute the sample size that is needed to achieve a certain power. Finally, the option also makes it possible to display the relationship between the power and the sample size graphically.

4.4.3   Tests for a Population Variance and Standard Deviation

The option "Sample Size and Power" in the "DOE" menu also allows the power of one-tailed or two-tailed hypothesis tests for a population standard deviation to be determined. To do so, select the option "One Sample Standard Deviation" (see Figure 4.22). This option also allows the determination of the required sample sizes and the graphical display of the relationship between the power and the sample size. JMP does not work with variances directly, but it is not difficult to convert variances into standard deviations.

5

Two Hypothesis Tests for the Median of a Population

He was deeply annoyed when people asked for him to provide a "gut response" before the data had provided any real direction: as far as Caston was concerned, to go on hunches was to go off half-cocked. It prevented one from analyzing things logically; it impeded the workings of reason and the rigorous techniques of probabilistic analysis.

(from *The Ambler Warning*, Robert Ludlum, p. 80)

We have already mentioned in Section 4.1.4 that the *t*-test for a population mean μ cannot be used for small samples of nonnormal data. Using the *t*-distribution in such cases may lead to a high probability of a type I error (larger than the significance level α), and reduced power. In short, for small samples and nonnormal data, the test for the population mean as described in the previous chapter cannot be used.

However, for the analysis of nonnormal data, there are many **nonparametric** or **distribution-free** techniques[1]. Traditional (parametric) tests have a larger power than nonparametric ones when the assumptions regarding the probability distribution or the probability density of the population are met. Nonparametric tests are therefore often only used as an emergency solution. However, in their book *Nonparametric Statistical Methods*, M. Hollander, D.A. Wolfe, and E. Chicken emphasize that the loss in power of nonparametric hypothesis tests compared to classical parametric hypothesis tests is quite small even when the assumptions with respect to the probability distribution or density of the population are met. If these assumptions are not met, then, in many cases, nonparametric hypothesis tests are much better than classic tests such as the *t*-test discussed in Section 4.1.

The sign test for a population median belongs to the category of nonparametric hypothesis tests. This test is used not only for small data sets with quantitative, nonnormally distributed

---

[1] Although, strictly speaking, there is a difference between nonparametric and distribution-free methods, these terms are typically used interchangeably. A possible definition of distribution-free or nonparametric tests is the following: statistical tests that do not depend on assumptions about the probability distribution or probability density of the population studied.

Statistics with JMP: Hypothesis Tests, ANOVA and Regression, First Edition. Peter Goos and David Meintrup.
© 2016 John Wiley & Sons, Ltd. Published 2016 by John Wiley & Sons, Ltd.
Companion Website: http://www.wiley.com/go/goosandmeintrup/JMP

data, but even for data sets with ordinal data. For quantitative data, the Wilcoxon signed-rank test is an alternative to the sign test. It is also a nonparametric hypothesis test. Unlike the sign test, the Wilcoxon signed-rank test cannot be used for ordinal data.

An important difference between the sign test, on the one hand, and the t-test from the previous chapter, on the other, is that the former test does not deal with the population mean, but with the population median. The Wilcoxon signed-rank test, as we will use it in this chapter, is designed for the median of populations with a symmetrical distribution. For symmetrical probability distributions and probability densities, however, the population median is equal to the population mean. Therefore, the signed-rank test for a single (symmetrically distributed) population can be seen as a test for the population mean and the population median simultaneously.

An important advantage of both the sign test and the signed-rank test is that they are insensitive to extreme values or outliers in the data. An advantage of the sign test over the signed-rank test is that it can always be used, regardless of whether or not the population studied is symmetrically distributed.

5.1 The Sign Test

The sign test is a hypothesis test for the median of a population. In contrast to the mean, the median is less sensitive to extreme values, to asymmetry, or to skewness in the data. Therefore, the median is suitable for a hypothesis test about the central location if the population under study is not normally distributed and the sample size is too small to rely on the central limit theorem. The sign test tests the null hypothesis

$$H_0 : \mathrm{Me} = \mathrm{Me}_0,$$

where Me represents the population median and Me_0 is a hypothesized value for it, against an alternative hypothesis. The alternative hypothesis is right- or left-tailed, or two-tailed. For a right-tailed hypothesis test, we have

$$H_a : \mathrm{Me} > \mathrm{Me}_0.$$

For a left-tailed test, the alternative hypothesis is

$$H_a : \mathrm{Me} < \mathrm{Me}_0,$$

and, for a two-tailed test, it is

$$H_a : \mathrm{Me} \neq \mathrm{Me}_0.$$

A characteristic of the sign test is that it only uses a decision rule based on the p-value. A correct, exact p-value can be calculated using the binomial distribution. Therefore, the sign test is very similar to the binomial test that we encountered in Section 4.2.2. When calculating the probabilities based on the binomial distribution is problematic, then the p-value can be approximated using the normal probability density. For this approach, however, the sample size should be sufficiently large.

5.1.1 The Starting Point of the Sign Test

Regardless of how the p-value is calculated, the sign test relies on the definition of the median, which states that 50% of the values in a population are smaller than the median, and the remaining 50% are larger than the median. Therefore, one can expect that a sample from any population will contain approximately as many observations that are smaller than the median as observations that are larger than the median.

If the null hypothesis of the sign test is correct, then one can expect that the sample will contain as many observations that are smaller than the hypothesized value Me_0 as observations that are larger than Me_0. If the null hypothesis is true, then, in a sample of size n, we expect about $n/2$ observations to be smaller than Me_0, and $n/2$ observations to be larger than Me_0[2].

One version of the sign test, discussed in Section 5.1.3, compares $n/2$, which we expect if the null hypothesis is true, to a test statistic s. For a right-tailed test, the test statistic s is the number of observations in the sample that are larger than Me_0. For a left-tailed test, s is the number of observations that are smaller than Me_0. For a two-tailed test, s is the maximum of these two numbers.

Figure 5.1 shows, for a sample of $n = 10$ observations, how the number of observations larger than Me_0 and the number of observations smaller than Me_0 vary as a function of the population median Me. If the null hypothesis is true, then $n/2 = 5$ observations (exactly half of all observations) are smaller than Me_0 and $n/2 = 5$ observations are greater than Me_0. If the null hypothesis is false and $Me > Me_0$, then more than $n/2 = 5$ observations (to be precise, seven observations) are larger than Me_0 and less than $n/2 = 5$ observations (to be precise, three observations) are smaller than Me_0. If, however, $Me < Me_0$, then less than $n/2 = 5$ observations (two observations, to be precise) are larger than Me_0 and more than $n/2 = 5$ observations (eight observations, to be precise) are smaller than Me_0.

It is useful to find the value of the test statistic s for the three versions of the sign test and the three possible scenarios in Figure 5.1:

- If we want to perform a right-tailed sign test and the scenario in Figure 5.1a occurs, then the test statistic s is equal to 5. If, however, we are in the scenario shown in Figure 5.1b, then $s = 7$. If the scenario in Figure 5.1c occurs in a right-tailed test, then $s = 2$.
- If we want to perform a left-tailed sign test and the scenario in Figure 5.1a occurs, then the test statistic s also equals 5. If, however, we are in the scenario shown in Figure 5.1b, then $s = 3$. For the scenario in Figure 5.1c, we have $s = 8$.
- If, finally, we want to perform a two-tailed sign test and the scenario in Figure 5.1a occurs, then the test statistic s is again equal to 5. For the scenario in Figure 5.1b, we have $s = 7$. For the scenario in Figure 5.1c, we have $s = 8$.

Just like the sample mean, the sample proportion, and the sample variance, the number s can be interpreted as a random variable, as long as no sample data has been collected. In that event, we use an uppercase letter for the test statistic: S.

---

[2] Some samples contain observations that are equal to Me_0. These observations should be omitted from the data set for the implementation of the sign test. The sample size n then needs to be reduced by the number of observations equal to Me_0

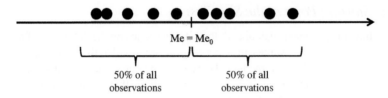

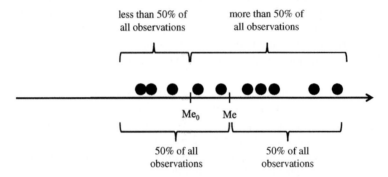

(a) Null hypothesis is true: Me = Me_0. In this case, 50% of all observations are smaller than Me_0, and the remaining 50% of all observations are larger than Me_0.

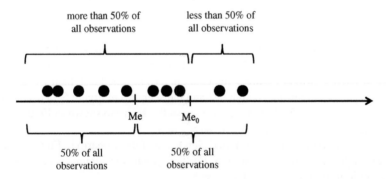

(b) Null hypothesis is false: Me > Me_0. In this case, more than 50% of all observations are larger than Me_0.

(c) Null hypothesis is false: Me < Me_0. In this case, more than 50% of all observations are smaller than Me_0.

Figure 5.1   A schematic representation of the procedure of the sign test, where Me represents the unknown population median, and Me_0 is the hypothesized value of the population median in the null hypothesis. In each situation, (approximately) 50% of the observations are below the population median Me and (approximately) 50% are above the population median Me.

Since the number of observations larger (or smaller) than Me_0 can be interpreted as a number of "successes"[3], it is not hard to see that S is binomially distributed.

If the null hypothesis is true, the parameter π of this binomial distribution is 0.5, because the probability of a success (an observation larger or smaller than Me_0) is exactly 50%. The binomial distribution with parameters n and $\pi = 0.5$ can then be used to compute a p-value for the hypothesis test.

5.1.2 Exact p-Values

The p-value of a hypothesis test provides an indication of how likely the test statistic's value is if the null hypothesis is true. If the p-value is smaller than the significance level chosen, we reject the null hypothesis and accept the alternative hypothesis.

For a one-tailed test, the p-value is

$$p = P(S \geq s),$$

where S represents a binomially distributed random variable with success probability $\pi = 0.5$. Here, the random variable S is the test statistic, while s represents the computed test statistic. It is not difficult to see that the p-value $P(S \geq s)$ makes sense.

Suppose, first, that we test a right-tailed alternative hypothesis, $Me > Me_0$. If this alternative hypothesis is true, then many observations will be larger than Me_0. In other words, the computed test statistic s (which for a right-tailed alternative hypothesis is equal to the number of observations greater than Me_0) will be large, and the probability $P(S \geq s)$ will be small. In short, the p-value will be small, which indicates that the null hypothesis is most likely incorrect.

For a left-tailed alternative hypothesis, $Me < Me_0$, we can set up a similar argument. If the left-tailed alternative hypothesis is true, then many observations will be smaller than Me_0. In other words, the computed test statistic s (which for a left-tailed alternative hypothesis is equal to the number of observations smaller than Me_0) will be large, and the probability $P(S \geq s)$ will be small. Therefore, the p-value will be small, which again indicates that the null hypothesis is most likely incorrect.

For a two-tailed test, the p-value is

$$p = 2\,P(S \geq s).$$

In a two-tailed test, we accept the alternative hypothesis if $Me > Me_0$, but also if $Me < Me_0$. In both cases, s will be large and $P(S \geq s)$ will be small. This results in a small p-value, indicating that the null hypothesis is probably incorrect and should be rejected.

The calculation of the p-values here differs from the approach used for the hypothesis tests for a population mean, variance, or proportion, because no distinction is made between

---

[3] Each individual observation is considered as a Bernoulli experiment, with the possible results "larger than Me_0" and "smaller than Me_0". A sample of n independent observations constitutes a set of n independent Bernoulli experiments. In the sign test, we are interested in the number of observations that exceed Me_0, or in the number of observations that are smaller than Me_0. These numbers correspond to the number of successes or failures in n Bernoulli experiments. The number of observations larger or smaller than Me_0 is therefore binomially distributed. More details about the Bernoulli and binomial distributions can be found in the book *Statistics with JMP: Graphs, Descriptive Statistics and Probability*.

a left-tailed and a right-tailed test. This is due to the fact that the test statistic S is defined differently for the two types of tests, so that, for both the right-tailed test and the left-tailed test, an exceedance probability has to be calculated. The interpretation of the p-value remains unchanged. The procedure of the sign test can best be illustrated by an example.

Example 5.1.1 *A manufacturer of DVD players knows that the median life span of his devices is 5250 hours. In order to evaluate this performance, he has tested 20 DVD players from a competitor. The shortest life span of the tested devices was 2523 hours, while the best device worked for 8120 hours. In total, 14 out of the 20 DVD players had a life span that was longer than 5250 hours, while the other six had a shorter life span. The manufacturer wonders whether this data supports the conclusion that the life span of the competitor's products exceeds that of his own devices.*
 It is clear that the hypotheses tested here are

$$H_0 : Me = 5250 \ hours$$

and

$$H_a : Me > 5250 \ hours.$$

The computed test statistic, the number of sample observations larger than 5250, is $s = 14$. The p-value of the right-tailed hypothesis test is

$$p = P(S \geq s) = P(S \geq 14) = 0.0577,$$

where S is binomially distributed with parameters $n = 20$ and $\pi = 0.5$. This p-value can be computed by means of the JMP formula "$1-\texttt{Binomial Distribution}(0.5, 20, 13)$", or found in Appendix A. Although the p-value is larger than the usual significance level of 5% and therefore the null hypothesis cannot be rejected, it points in the direction of the alternative hypothesis.

5.1.3 Approximate p-Values Based on the Normal Distribution

The normal probability density function is a good approximation for the binomial distribution for sufficiently large values n. This follows from the central limit theorem and requires that $n\pi > 5$ and $n(1 - \pi) > 5$ (where π represents the probability of success). For the sign test, $n > 10$ is sufficient, as π is equal to 0.5 when the null hypothesis is true (which is the starting point for computing the p-value). Due to the fact that $\pi = 0.5$, the random variable S in the sign test is approximately normally distributed, with expected value $n\pi = n/2$ and variance $n\pi(1 - \pi) = n\frac{1}{2}(1 - \frac{1}{2}) = \frac{n}{4}$. Consequently,

$$\frac{S - \frac{n}{2}}{\sqrt{n\frac{1}{2}\left(1 - \frac{1}{2}\right)}} = \frac{S - \frac{n}{2}}{\sqrt{\frac{n}{4}}} = \frac{S - \frac{n}{2}}{\frac{\sqrt{n}}{2}}.$$

has a standard normal distribution. This leads to the computed test statistic

$$z = \frac{s - \frac{n}{2}}{\sqrt{n \frac{1}{2}\left(1 - \frac{1}{2}\right)}} = \frac{s - \frac{n}{2}}{\frac{\sqrt{n}}{2}},$$

which can be compared with the critical value z_α for a one-tailed test, and with $z_{\alpha/2}$ for a two-tailed test.

Slightly more accurate results are obtained, however, if this test statistic is corrected by subtracting $1/2$ in the numerator:

$$z = \frac{s - \frac{1}{2} - \frac{n}{2}}{\sqrt{n \frac{1}{2}\left(1 - \frac{1}{2}\right)}} = \frac{s - \frac{1}{2} - \frac{n}{2}}{\frac{\sqrt{n}}{2}}.$$

This correction, which is called a continuity correction[4], is carried out because, in contrast to the normal distribution, the binomial distribution is a discrete probability distribution.

Taking the continuity correction into account, the approximate p-value for a one-tailed sign test is

$$p = P\left(Z > \frac{s - \frac{1}{2} - \frac{n}{2}}{\frac{\sqrt{n}}{2}}\right) = P(Z > z),$$

while the approximate p-value for a two-tailed test is

$$p = 2 P\left(Z > \frac{s - \frac{1}{2} - \frac{n}{2}}{\frac{\sqrt{n}}{2}}\right) = 2 P(Z > z).$$

Because of the different definitions of the test statistic s in the left-tailed, right-tailed, and two-tailed hypothesis tests, we always have to calculate an exceedance probability to determine the p-value. It is not difficult to see that a large value of s, which provides strong evidence against the null hypothesis, implies a large value for the test statistic z and a small p-value.

Example 5.1.2   *In Example 5.1.1, the manufacturer analyses $n = 20$ devices. If he wants to use the normal approximation when carrying out the sign test, the test statistic is*

$$z = \frac{s - \frac{1}{2} - \frac{n}{2}}{\frac{\sqrt{n}}{2}} = \frac{14 - \frac{1}{2} - \frac{20}{2}}{\frac{\sqrt{20}}{2}} = 1.5652.$$

---

[4] For a probability of the type $P(X \geq x)$, where X is binomially distributed, the approximation using the normal distribution is done by calculating the probability $P(X^* \geq x - \frac{1}{2})$, where X^* is normally distributed with expected value $n\pi$ and variance $n\pi(1 - \pi)$. A probability of the form $P(X \leq x)$ is approximated by $P(X^* \leq x + \frac{1}{2})$.

This value is smaller than $z_\alpha = z_{0.05} = 1.645$, so that the null hypothesis should be accepted. The corresponding p-value, $P(Z \geq 1.5652)$, is 0.0588. This p-value is larger than α and thus also leads to an acceptance of the null hypothesis.

The name of the sign test is inspired by the fact that one can indicate by means of a plus or a minus sign whether a sample observation is above or below the hypothesized median Me_0. Four JMP scripts are available for the sign test. The scripts named "Sign Test for Median Using Summary Statistics (Exact Test)" and "Sign Test for Median Using Data (Exact Test)" calculate p-values based on the binomial distribution. The scripts named "Sign Test for Median Using Summary Statistics (Normal Approximation)" and "Sign Test for Median Using Data (Normal Approximation)" calculate approximate p-values based on the normal distribution. Figure 5.2 shows the functioning of the scripts "Sign Test for Median Using Summary Statistics (Exact Test)" and "Sign Test for Median Using Summary Statistics (Normal Approximation)" for Examples 5.1.1 and 5.1.2.

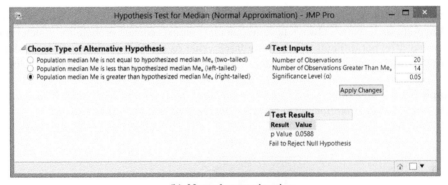

(a) Exact test

(b) Normal approximation

Figure 5.2 The JMP scripts "Sign Test for Median Using Summary Statistics (Exact Test)" and "Sign Test for Median Using Summary Statistics (Normal Approximation)" applied to Examples 5.1.1 and 5.1.2.

5.2   The Wilcoxon Signed-Rank Test

Although the Wilcoxon signed-rank test was originally introduced for the analysis of data from two (dependent) samples, the test can also be used as an alternative to the sign test for one sample. The signed-rank test requires a symmetrical data distribution. Like the sign test, the signed-rank test tests the null hypothesis

$$H_0 : \text{Me} = \text{Me}_0,$$

where Me represents the population median and Me_0 is the hypothesized median, against either a one-tailed alternative hypothesis or a two-tailed alternative hypothesis. As the name suggests, the signed-rank test uses ranks. Each rank is assigned a plus or a minus sign.

For symmetrically distributed populations, the median is obviously equal to the mean, so that we can also say that the signed-rank test tests the null hypothesis

$$H_0 : \mu = \mu_0$$

against a one- or two-sided alternative hypothesis. In summary, we can say that the signed-rank test tests the following null hypothesis:

$$H_0 : \text{Me} = \mu = \text{Me}_0 = \mu_0.$$

The corresponding alternative hypotheses are

$$H_a : \text{Me} = \mu > \text{Me}_0 = \mu_0,$$
$$H_a : \text{Me} = \mu < \text{Me}_0 = \mu_0$$

and

$$H_a : \text{Me} = \mu \neq \text{Me}_0 = \mu_0.$$

5.2.1   The Use of Ranks

To perform a signed-rank test, we proceed in the following way:

(1) Subtract the value Me_0 from all observations $x_1, x_2, \ldots, x_n$. This results in n differences $d_1 = x_1 - \text{Me}_0, d_2 = x_2 - \text{Me}_0, \ldots, d_n = x_n - \text{Me}_0$.
(2) Typically, some of the differences $d_1, d_2, \ldots, d_n$ are positive, while other differences are negative. If some differences are zero, then remove these differences and the corresponding observations from the data set[5]. You then have to carry out the hypothesis test with a smaller number of observations.
(3) Compute the absolute values of all differences, $|d_1|, |d_2|, \ldots, |d_n|$.

---

[5] Alternatively, one could rank the zeros along with the other differences, and then drop the ranks of the zeros. This so-called Pratt method is implemented in JMP 12.

(4) Order the absolute values $|d_1|, |d_2|, \ldots, |d_n|$ from small to large and, based on this ranking, assign rank numbers to all observations ranging from 1 to n. Ensure that the observation with the smallest $|d_i|$ value is given the smallest rank number and the observation with the largest $|d_i|$ value is given the largest rank number.

(5) If there are ties in the ranking, assign an average rank number to the corresponding observations. Denote by r_1 the final rank number of the first observation x_1, by r_2 the final rank number of the second observation $x_2, \ldots$, and by r_n the final rank number of the last observation x_n.

(6) Attach a plus sign to all rank numbers r_i that correspond to a positive difference $d_i = x_i - Me_0$, and attach a minus sign to all rank numbers r_i that correspond to a negative difference $d_i = x_i - Me_0$. Denote the resulting signed ranks as s_i.

(7) Compute the sum of all positive ranks and name this sum t_+.

(8) Compute the sum of the absolute values of all negative ranks and name this sum t_-.

(9) Check that the sum of t_+ and t_- is equal to $n(n+1)/2$. Indeed, the sum of the natural numbers from 1 to n is equal to $n(n+1)/2$. Hence, the sum of all ranks $r_1, r_2, \ldots, r_n$, and thus also the sum of t_+ and t_-, should be equal to $n(n+1)/2$.

Because the signed-rank test uses differences, this test is not suitable for ordinal data. This is due to the fact that there is no meaningful way in which to calculate differences for ordinal variables.

Example 5.2.1   *Table 5.1 shows the life span of the 20 DVD players in Examples 5.1.1 and 5.1.2. We denote these observations by $x_1, x_2, \ldots, x_{20}$. The table also shows how the differences d_i are calculated and how, subsequently, rank numbers are assigned to the individual observations. For the calculation of the differences d_i, the value $Me_0 = 5250$ is subtracted from all observations x_i, because the null hypothesis is*

$$H_0 : Me = 5250 \ hours.$$

In Table 5.1, neither ties nor zero differences occur. Dealing with ties is demonstrated in detail in later chapters. The sum of all positive ranks in the last column of Table 5.1, t_+, is 156, while the sum of the absolute values of all negative ranks, t_-, is 54. The sum of t_+ and t_- is 210, which is $n(n+1)/2 = 20(20+1)/2$.

5.2.2   The Starting Point of the Signed-Rank Test

If Me_0 is indeed the median of the population studied (i.e., if the null hypothesis is true), half of the sample observations will have a value that is smaller than Me_0 and the other half of the observations will be larger than Me_0. Accordingly, about half of the differences d_i will be positive, while the other half will be negative. If the probability distribution or probability density of the population studied is symmetrical, the magnitude of the positive ranks will be similar to the magnitude of the negative ranks[6]. Accordingly, the sum of all positive ranks, t_+,

---

[6] If the probability distribution or probability density of the studied variable is skewed to the right and the null hypothesis is true, then the largest ranks will almost all have a positive sign. For a left-skewed probability distribution

Table 5.1 The calculations needed for the determination of the ranks and their sign for the 20 observations in Example 5.2.1.

i	x_i	$d_i = x_i - Me_0$	$|d_i| = |x_i - Me_0|$	r_i	s_i
1	5689	439	439	10	10
2	5423	173	173	5	5
3	6243	993	993	15	15
4	5205	−45	45	2	−2
5	5173	−77	77	3	−3
6	5744	494	494	11	11
7	7087	1837	1837	18	18
8	5260	10	10	1	1
9	6020	770	770	14	14
10	5919	669	669	12	12
11	2523	−2727	2727	19	−19
12	5425	175	175	6	6
13	4915	−335	335	9	−9
14	4521	−729	729	13	−13
15	5374	124	124	4	4
16	5531	281	281	7	7
17	6325	1075	1075	16	16
18	4956	−294	294	8	−8
19	6627	1377	1377	17	17
20	8120	2870	2870	20	20

will be approximately equal to the sum of the negative ranks, t_-, if the null hypothesis that $Me = Me_0$ is true. Since the sum of t_+ and t_- is always equal to $n(n+1)/2$, both t_+ and t_- will then be close to $n(n+1)/4$.

If the null hypothesis is wrong and, for example, the population median Me is smaller than Me_0, then most of the sample observations will be smaller than Me_0. In that case, most of the differences d_i will be negative, and some of these differences will be large in absolute value. Consequently, most of the rank numbers, and also the largest rank numbers, will be negative. The sum of the absolute value of the negative ranks, t_-, will then be considerably larger than the sum of the positive ranks, t_+.

If, on the other hand, the population median Me is greater than Me_0, then most of the sample observations will be greater than Me_0. In that case, most of the differences d_i will be positive, and some of those differences will be large. As a result, most of the rank numbers, and also the largest rank numbers, will be positive, so that the sum of the positive ranks, t_+, will be considerably larger than the sum of the negative ranks, t_-.

Figure 5.3 graphically illustrates how the ranks, their signs, and the sums t_+ and t_- are influenced by the location of the population median Me relative to the hypothesized median Me_0. Figure 5.3a shows that, when the null hypothesis that $Me = Me_0$ is true, the number of

---

or probability density, almost all large ranks will have a negative sign. Thus, for a skewed distribution or density, the sum of all positive ranks will not be equal to the sum of the negative ranks, even if the null hypothesis is true. Therefore, the Wilcoxon signed-rank test cannot be used for skewed distributions or densities.

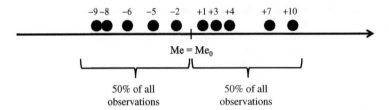

(a) Null hypothesis is true: Me = Me_0. In this case, half of the ranks are negative and the other half of the ranks is positive.

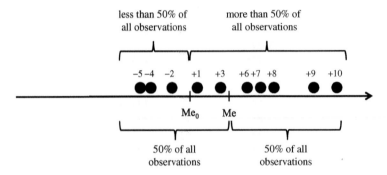

(b) Null hypothesis is false: Me > Me_0. In this case, more than 50% of all observations have a positive rank and less than 50% of all observations have a negative rank.

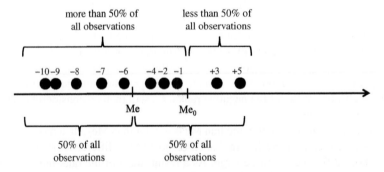

(c) Null hypothesis is false: Me < Me_0. In this case, more than 50% of all observations have a negative rank and less than 50% of all observations have a positive rank.

Figure 5.3   A schematic representation of the functioning of the Wilcoxon signed-rank test, where Me represents the unknown population median and Me_0 the hypothesized value for the population median in the null hypothesis. In all cases, (approximately) 50% of the observations are below the population median Me and (approximately) 50% are above the population median Me.

positive and negative ranks is the same. The sums t_+ and t_- are about the same size. More specifically, t_+ and t_- are equal to 25 and 30, respectively.

Figure 5.3b shows that, if $Me > Me_0$ so that the null hypothesis is false, the number of positive ranks is larger than the number of negative ranks. In addition, the largest ranks are positive. This results in a large value for t_+ and a small value for t_-. For the example in Figure 5.3b, the sums t_+ and t_- are equal to 44 and 11, respectively.

Finally, Figure 5.3c shows that if $Me < Me_0$, the number of negative ranks is larger than the number of positive ranks. In addition, the largest ranks (in absolute value) are negative. This leads to a large value for t_- and a small value for t_+. For the example in Figure 5.3c, the sums t_+ are t_- are equal to 8 and 47, respectively.

Figure 5.3 also illustrates how the ranks and their signs are determined. The point that is closest to Me_0 is given the rank number 1. If this point is to the right of Me_0, it is given a plus sign; otherwise, it is given a minus sign. The point that is the second closest to Me_0 is given the rank number 2. If this point is to the right of Me_0, it is given a plus sign; otherwise, it is given a minus sign. This procedure is repeated up to the point that is furthest from Me_0. That point is given the rank number 10, with a plus sign if the point is to the right of Me_0 and a minus sign if it is to the left of Me_0.

There are several ways to calculate p-values for the computed sums t_+ and t_-. The first method provides exact p-values. Therefore, it is the most correct one. However, it is also the most difficult method, and it can only be used for small samples. For large samples, there are alternative ways to calculate good p-values, based on the normal distribution and the t-distribution. In the following sections, we study all three methods.

For the determination of the p-values, we have to consider the ranks and the sums of positive and negative ranks as random variables, just as we do for individual observations. We emphasize this by using the notation T_+ and T_- instead of t_+ and t_-, whenever we consider these sums as random variables.

5.2.3   Exact p-Values

If there are no ties and if the number of observations is limited, exact p-values and critical values can be determined for the Wilcoxon signed-rank test. In this section, we study how to compute these values and we demonstrate their use. We prefer using exact p-values and the corresponding critical values whenever the number of observations n is smaller than or equal to 20.

Since, in the absence of ties, the random variables T_+ and T_- are sums of ranks and rank numbers are integers, T_+ and T_- can only take integer values. The random variables T_+ and T_- therefore have a discrete probability distribution. This discrete probability distribution is not a known probability distribution such as the binomial or Poisson distribution, but it can be calculated. We illustrate the calculation of the probability distribution of T_+ and T_- using a small example, with $n = 3$ observations.

If there are three observations, then there are only three possible ranks, namely 1, 2, and 3. Each of these ranks can be positive or negative, which leads to a total of $2^3 = 8$ possibilities. These possibilities are shown in Table 5.2. The table clearly shows that there are only seven possible values for the random variables T_+ and T_-. The values 0, 1, 2, 4, 5, and 6 all have a probability of 0.125, while the value 3 has a probability of 0.250. In fact, the value 3 is the

Table 5.2   Eight possible scenarios for the signs of the ranks in a sample with three observations without ties.

Number of minus signs	Signed ranks			Probability	t_+	t_-
0	1	2	3	0.125	6	0
1	−1	2	3	0.125	5	1
1	1	−2	3	0.125	4	2
1	1	2	−3	0.125	3	3
2	−1	−2	3	0.125	3	3
2	−1	2	−3	0.125	2	4
2	1	−2	−3	0.125	1	5
3	−1	−2	−3	0.125	0	6

only one that appears twice in the column for t_+ and in the column for t_- in Table 5.2. The probability distribution of T_+ is displayed in Table 5.3, along with the cumulative distribution function, $P(T_+ \leq t_+)$, and the probabilities of the type $P(T_+ \geq t_+)$. It is an easy exercise to establish the probability distribution of T_-.

Table 5.3 can be used to determine p-values for the signed-rank test where the number of observations is as low as three. Since it is time-consuming to determine the probability distribution of T_+, several researchers have created tables with probabilities of the type $P(T_+ \leq t_+)$ and of the type $P(T_+ \geq t_+)$. Appendix E contains such a table for $n \leq 20$. For the case where the number of sample observations n is equal to three, Appendix E contains the values that are printed in bold in Table 5.3.

JMP is able to compute exact p-values when $n \leq 20$.

The Right-Tailed Test

If the alternative hypothesis Me > Me_0 is correct, then t_+ will be large. To determine whether t_+ is large enough for the null hypothesis to be rejected in favor of the alternative hypothesis, we should consider how likely the observed value of t_+ is under the assumption that the null hypothesis is true. For this purpose, we can use Table 5.3. More specifically, we have to find the

Table 5.3   The probability distribution of T_+ for a sample with three observations without ties.

t_+	$P(T_+ = t_+)$	$P(T_+ \leq t_+)$	$P(T_+ \geq t_+)$
0	0.125	**0.125**	1.000
1	0.125	**0.250**	0.875
2	0.125	**0.375**	0.750
3	0.250	**0.625**	**0.625**
4	0.125	0.750	**0.375**
5	0.125	0.875	**0.250**
6	0.125	1.000	**0.125**

probability $P(T_+ \geq t_+)$. This probability is our p-value for the right-tailed test. If this p-value is smaller than the chosen significance level, we reject the null hypothesis.

If, for example, in a sample with three observations without ties we obtain a value of 5 for t_+, then the probability $P(T_+ \geq t_+) = P(T_+ \geq 5)$ equals 0.250. This p-value is larger than the significance levels of 10%, 5%, or 1% that we typically use.

The Left-Tailed Test

If the alternative hypothesis $Me < Me_0$ is correct, then t_+ will be small. To determine whether t_+ is small enough for the null hypothesis to be rejected in favor of the alternative hypothesis, we should consider how likely the observed value of t_+ is under the assumption that the null hypothesis is true. For this purpose, we can again use Table 5.3. More specifically, we have to find the probability $P(T_+ \leq t_+)$. This probability is our p-value for the left-tailed test. If this p-value is smaller than the chosen significance level, we reject the null hypothesis.

The Two-Tailed Test

If the alternative hypothesis that $Me \neq Me_0$ is correct, then there are two possibilities. One possibility is that $Me < Me_0$. In that case, t_+ will be small. The second possibility is that $Me > Me_0$. In that case, t_+ will be large. To decide whether the value of t_+ is sufficiently large or small for the null hypothesis to be rejected, we need to determine the corresponding p-value. How to determine the p-value depends on the exact value of t_+:

$$
p = \begin{cases} 2\,P(T_+ \leq t_+), & \text{if } t_+ \leq \dfrac{n(n+1)}{4}, \\ 2\,P(T_+ \geq t_+), & \text{if } t_+ \geq \dfrac{n(n+1)}{4}. \end{cases}
$$

Again, we reject the null hypothesis if the p-value is smaller than the chosen significance level.

Example 5.2.2   *Examples 5.1.1 and 5.1.2 dealt with a manufacturer of DVD players, who tested the following hypotheses using the sign test:*

$$H_0 : Me = 5250 \text{ hours}$$

and

$$H_a : Me > 5250 \text{ hours}.$$

The original observations are shown in Table 5.1, along with their ranks. In Example 5.2.1, we calculated that $t_+ = 156$ and $t_- = 54$.

To use the Wilcoxon signed-rank test instead of the sign test, the manufacturer must first verify that the observations come from a population with a symmetric probability density. For this purpose, he can construct a histogram of the 20 life spans in the data set. Figure 5.4 shows a histogram generated using JMP. The histogram is quite symmetrical, which suggests

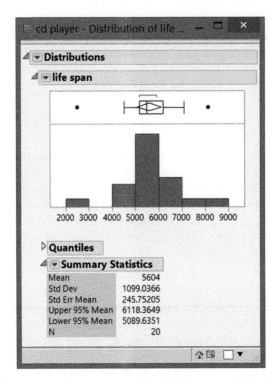

Figure 5.4   The histogram of the 20 life spans in Table 5.1.

that the life span of the DVD players has a symmetrical probability density. Consequently, the Wilcoxon signed-rank test can be used for testing the median.

As the manufacturer has used a right-tailed alternative hypothesis, the p-value of the signed-rank test is equal to $P(T_+ \geq t_+)$ for $n = 20$. Since $t_+ = 156$, we need to find the probability that T_+ takes values larger than or equal to 156. From Appendix E, we can see that this probability is equal to 0.0291. In this case, we can consult the table in Appendix E, because there are no ties in the data.

*The signed-rank test can also be conducted in JMP. To this end, start from the screen shown in Figure 5.4, and click on the hotspot (red triangle) next to the name of the variable "Life span". Next, you have to choose the option "*Test Mean*", as shown in Figure 5.5. This produces a dialog window where you can enter the value of Me_0, 5250. Note that the JMP input field is labeled as "*Specify Hypothesized Mean*". This is justified because the mean and the median are identical for any population with a symmetrical distribution. The final things you need to do are to check the box "*Wilcoxon Signed Rank*" and to click "*OK*" (see Figure 5.6). This results in the output shown in Figure 5.7, containing the p-value of 0.0291.*

Figure 5.7 reports the value 51 as a computed test statistic for the signed-rank test. This value is equal to neither t_+ nor t_-. The value that JMP uses as test statistic is $(t_+ - t_-)/2$. Therefore, JMP uses a different internal logic, but it arrives at the same p-value, 0.0291.

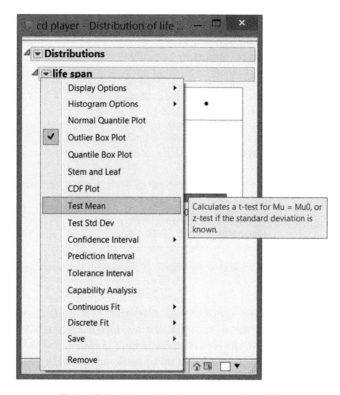

Figure 5.5   The option "Test Mean" in JMP.

Note that Figure 5.7 contrasts the t-test and the signed-rank test. The t-test yields a p-value of 0.0830 for the right-tailed hypothesis test, while the Wilcoxon signed-ranked test yields a p-value of 0.0291. According to the t-test, we cannot reject the null hypothesis, while we can reject it according to the signed-rank test. The reason for this is that there is one very small observation in the data, namely the life span of 2523. The sample mean and sample standard deviation are very sensitive to such an extreme value, which strongly influences the t-test. In contrast, the signed-rank test is insensitive to the precise value of the smallest observation.

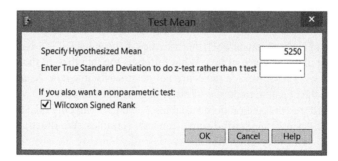

Figure 5.6   The dialog window for selecting the Wilcoxon signed-rank test for testing a population median or mean of a symmetrically distributed population.

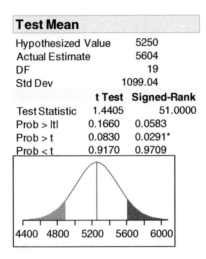

Figure 5.7   The output with p-values of the t-test and the signed-rank test for Example 5.2.2.

If the smallest life span had not been 2523 but 3300, the t-test would have led to a p-value of 0.0465, while the p-value of the signed-rank test would remain at 0.0291. This is due to the fact that the rank numbers of the observations remain unchanged if the smallest lifetime is 3300 instead of 2523, but the sample mean and sample standard deviation are influenced quite strongly.

The hypotheses in the DVD player example were first tested using the sign test (the p-values were 0.0577 and 0.0588) and then with the signed-rank test (the p-value was 0.0291) and the t-test (the p-value was 0.0830). Thus, if we use a significance level of 5%, the tests lead to different conclusions. This is an unfortunate situation that, as researchers, we prefer to avoid. The best that the manufacturer can do here is to conduct additional research, collect additional data, and perform new hypothesis tests. Due to the larger number of observations, the new hypothesis tests will be more powerful, so that the ambiguity regarding the null hypothesis and the alternative hypothesis can probably be eliminated.

5.2.4   Exact p-Values for Ties

The determination of exact p-values when there are ties in the values of $|d_i|$ is not very different from the procedure in the absence of ties. What differs is the fact that, in the presence of ties, we use average ranks. These average ranks are not necessarily integers. As a result, the sums of the positive and the negative ranks are also not necessarily integers.

However, it is still true that the random variables T_+ and T_- can only take a limited number of values, so that T_+ and T_- are still discrete random variables. We illustrate this by means of a sample with $n = 4$ observations, with two ties after ranking the absolute values of the differences d_i. Suppose that the ranks of the four observations are equal to 1.5, 1.5, 3, and 4. This means that the two smallest $|d_i|$ values are equal. These values would normally be assigned the rank numbers 1 and 2. Due to the tie, the average rank number 1.5 is used instead.

Table 5.4 The 16 possible scenarios for the signs of the ranks in a sample of four observations with a tie for the two smallest values of $|d_i|$, resulting in two average rank numbers of 1.5.

Number of minus signs	Signed ranks				Probability	t_+	t_-
0	1.5	1.5	3	4	0.0625	10	0
1	−1.5	1.5	3	4	0.0625	8.5	1.5
1	1.5	−1.5	3	4	0.0625	8.5	1.5
1	1.5	1.5	−3	4	0.0625	7	3
1	1.5	1.5	3	−4	0.0625	6	4
2	−1.5	−1.5	3	4	0.0625	7	3
2	−1.5	1.5	−3	4	0.0625	5.5	4.5
2	−1.5	1.5	3	−4	0.0625	4.5	5.5
2	1.5	−1.5	−3	4	0.0625	5.5	4.5
2	1.5	−1.5	3	−4	0.0625	4.5	5.5
2	1.5	1.5	−3	−4	0.0625	3	7
3	−1.5	−1.5	−3	4	0.0625	4	6
3	−1.5	−1.5	3	−4	0.0625	3	7
3	−1.5	1.5	−3	−4	0.0625	1.5	8.5
3	1.5	−1.5	−3	−4	0.0625	1.5	8.5
4	−1.5	−1.5	−3	−4	0.0625	0	10

Each of these ranks can be positive or negative. This leads to a total of $2^4 = 16$ options, all of which are listed in Table 5.4. Each of these 16 options is equally likely. Therefore, they all have probability $1/16 = 0.0625$.

The 16 rows of the table clearly show that there are only 10 possible values for the sums T_+ and T_-, namely 0, 1.5, 3, 4, 4.5, 5.5, 6, 7, 8.5, and 10. Some of these 10 values are more common than others. The probability distribution of T_+ is shown in Table 5.5.

Table 5.5 can now be used to calculate p-values. In the literature, no tables exist with p-values for data with ties in the ranking of the absolute d_i values. Therefore, in general, you

Table 5.5 The probability distribution of T_+ for a sample of four observations with a tie.

t_+	$P(T_+ = t_+)$	$P(T_+ \leq t_+)$	$P(T_+ \geq t_+)$
0	0.0625	**0.0625**	1.0000
1.5	0.1250	**0.1875**	0.9375
3	0.1250	**0.3125**	0.8125
4	0.0625	**0.3750**	0.6875
4.5	0.1250	**0.5000**	0.6250
5.5	0.1250	0.6250	**0.5000**
6	0.0625	0.6875	**0.3750**
7	0.1250	0.8125	**0.3125**
8.5	0.1250	0.9375	**0.1875**
10	0.0625	1.0000	**0.0625**

must either calculate the probability distribution of T_+ yourself, rely on statistical software to perform the computations required, or use an approximate p-value if you come across ties. The method for determining and interpreting the p-value is the same as in Section 5.2.3. For the example involving four observations and one tie, the possible p-values are printed in bold in Table 5.5.

5.2.5 Approximate p-Values Based on the Normal Distribution

For large samples, one can use the fact that T_+, the sum of the positive ranks, and T_-, the sum of the negative ranks, are approximately normally distributed with expected value

$$E(T_+) = E(T_-) = n(n+1)/4$$

and variance

$$\mathrm{var}(T_+) = \mathrm{var}(T_-) = n(n+1)(2n+1)/24.$$

This implies that the random variables

$$Z = \frac{T_+ - n(n+1)/4}{\sqrt{n(n+1)(2n+1)/24}} \tag{5.1}$$

and

$$Z = \frac{T_- - n(n+1)/4}{\sqrt{n(n+1)(2n+1)/24}}, \tag{5.2}$$

which act as test statistics, are approximately standard normally distributed. Based on this result, we can formulate decision rules for the signed-rank test.

Since $T_+ - n(n+1)/4$ and $T_- - n(n+1)/4$ are always complementary, the information contained in T_+ is identical to the information contained in T_-. Therefore, we can choose whether we work with T_+ or with T_-. As for the exact test, we work with T_+, but it is a good exercise to think about the consequences of working with T_- instead.

Many scientific papers suggest correcting the test statistics in Equations (5.1) and (5.2) in two ways. One correction is made in the numerator and is the so-called continuity correction, which accounts for the fact that T_+ and T_- are discrete random variables, in contrast to a (standard) normally distributed random variable. The second correction is in the denominator and is only needed when there are ties in the data. The corrected test statistic is

$$Z = \frac{T_+ - n(n+1)/4 - 1/2}{\sqrt{n(n+1)(2n+1)/24 - \sum_{i=1}^{g}\left(t_i^3 - t_i\right)/48}} \tag{5.3}$$

for a right-tailed test,

$$Z = \frac{T_+ - n(n+1)/4 + 1/2}{\sqrt{n(n+1)(2n+1)/24 - \sum_{i=1}^{g} \left(t_i^3 - t_i\right)/48}} \tag{5.4}$$

for a left-tailed test, and

$$Z = \frac{|T_+ - n(n+1)/4| - 1/2}{\sqrt{n(n+1)(2n+1)/24 - \sum_{i=1}^{g} \left(t_i^3 - t_i\right)/48}} \tag{5.5}$$

for a two-tailed test.

In all these expressions, g represents the number of groups of ties, and t_i is the number of tied observations in the ith group. In general, the correction for ties has little impact unless there are a lot of ties.

The Right-Tailed Test

If the alternative hypothesis that $Me > Me_0$ is correct, then t_+ will be considerably larger than $n(n+1)/4$. The computed test statistic in Equation (5.3) will then be strongly positive. In order to determine whether the computed test statistic z is sufficiently positive for the null hypothesis to be rejected in favor of the alternative hypothesis, we have to compare it with the critical value z_α. If $z > z_\alpha$, then we reject the null hypothesis in favor of the alternative hypothesis. If $z \leq z_\alpha$, then we cannot reject the null hypothesis.

The p-value for a right-tailed test is the probability that a standard normally distributed random variable Z takes values larger than the computed test statistic z. In other words, the p-value for a right-tailed test is $P(Z > z)$. If this p-value is smaller than the significance level chosen, we reject the null hypothesis. In mathematical terms, we reject the null hypothesis if

$$p = P\left(Z > \frac{t_+ - n(n+1)/4 - 1/2}{\sqrt{n(n+1)(2n+1)/24 - \sum_{i=1}^{g} \left(t_i^3 - t_i\right)/48}}\right) = P(Z > z) < \alpha.$$

The Left-Tailed Test

If the alternative hypothesis that $Me < Me_0$ is correct, then t_+ will be considerably smaller than $n(n+1)/4$. The test statistic in Equation (5.4) will then be strongly negative. In order to determine whether the computed test statistic z is sufficiently negative for the null hypothesis to be rejected in favor of the alternative hypothesis, we have to compare it with the critical value $-z_\alpha$. If $z < -z_\alpha$, then we reject the null hypothesis in favor of the alternative hypothesis. If $z \geq -z_\alpha$, then we cannot reject the null hypothesis.

The p-value for a left-tailed test is the probability that a standard normally distributed random variable takes values smaller than the computed test statistic z. In other words, the

p-value for a left-tailed test is $P(Z < z)$. We reject the null hypothesis if the p-value is smaller than the significance level chosen; that is, if

$$p = P\left(Z < \frac{t_+ - n(n+1)/4 + 1/2}{\sqrt{n(n+1)(2n+1)/24 - \sum_{i=1}^{g}\left(t_i^3 - t_i\right)/48}}\right) = P(Z < z) < \alpha.$$

The Two-Tailed Test

If the alternative hypothesis that $Me \neq Me_0$ is correct, then there are two possibilities. One possibility is that $Me < Me_0$. In that case, t_+ will be considerably smaller than $n(n+1)/4$; hence $t_+ - n(n+1)/4$ will be strongly negative and $|t_+ - n(n+1)/4|$ strongly positive. The whole test statistic in Equation (5.5) then takes a large positive value. The second possibility is that $Me > Me_0$. In that case, t_+ will be considerably larger than $n(n+1)/4$, and, hence, both $t_+ - n(n+1)/4$ and $|t_+ - n(n+1)/4|$ will be strongly positive. Clearly, the test statistic in Equation (5.5) will then also be strongly positive.

To determine whether the test statistic is sufficiently positive for the null hypothesis to be rejected, we need to compare it with $z_{\alpha/2}$. If the test statistic is larger than $z_{\alpha/2}$, then we reject the null hypothesis. Otherwise, we accept the null hypothesis. The p-value for a two-tailed test is equal to $2\,P(Z > z)$. If this p-value is smaller than the significance level α, then we reject the null hypothesis. In mathematical terms, we reject the null hypothesis if

$$p = 2\,P\left(Z > \frac{|t_+ - n(n+1)/4| - 1/2}{\sqrt{n(n+1)(2n+1)/24 - \sum_{i=1}^{g}\left(t_i^3 - t_i\right)/48}}\right) = 2\,P(Z > z) < \alpha.$$

Example 5.2.3   *If we use the approximation based on the normal distribution for determining the p-value of the signed-rank test in Example 5.2.2, then we should generally use Equation (5.3). However, as there are no ties, we have that $g = 0$, so that the denominator of Equation (5.3) is simplified. The computed value of the test statistic is then*

$$z = \frac{t_+ - n(n+1)/4 - 1/2}{\sqrt{n(n+1)(2n+1)/24}} = \frac{156 - 20(20+1)/4 - 1/2}{\sqrt{20(20+1)(2 \times 20+1)/24}} = 1.8853.$$

The p-value is $P(Z > 1.8853) = 0.0297$, which is a very good approximation of the exact p-value 0.0291, which we derived in Example 5.2.2.

5.2.6   *Approximate p-Values Based on the t-Distribution*

In 1974, Iman described an alternative method for performing a signed-rank test and for the calculation of an approximate p-value. He suggested testing whether the mean of all ranks is significantly different from zero, using a t-test. It can be shown that the required test statistic can be written as

$$t = \frac{t_+ - \frac{n(n+1)}{4}}{\sqrt{\frac{n^2(n+1)(2n+1)}{24(n-1)} - \frac{t_+ - \frac{n(n+1)}{4}}{n-1}}}. \tag{5.6}$$

This test statistic is approximately t-distributed with $n-1$ degrees of freedom.

This expression assumes that there are no ties, and that no observations have to be dropped because they are equal to Me_0. If some observations need to be dropped, then n should be lowered in the above expression. If there are ties, then the test statistic should be corrected as follows:

$$t = \frac{t_+ - \frac{n(n+1)}{4}}{\sqrt{\frac{n^2(n+1)(2n+1)}{24(n-1)} - \frac{t_+ - \frac{n(n+1)}{4}}{n-1} - c}}, \tag{5.7}$$

where

$$c = \frac{n}{48}\sum_{i=1}^{g} t_i(t_i+1)(t_i-1) = \frac{n}{48}\sum_{i=1}^{g}\left(t_i^3 - t_i\right),$$

g represents the number of groups of ties, and t_i is the number of tied observations in the ith group.

If the null hypothesis is true, then the sum of the positive ranks, t_+, will be close to $n(n+1)/4$. The numerator in Equations (5.6) and (5.7) will then be close to zero. This will result in a test statistic that is close to zero as well. When the null hypothesis is false, and t_+ will either be much larger or much smaller than $n(n+1)/4$, this will either result in a very positive or a very negative value for the test statistic.

JMP uses the approximate p-values based on the t-distribution when the number of observations n is larger than 20. The test statistic reported in JMP is $(t_+ - t_-)/2$.

The Right-Tailed Test

If the alternative hypothesis that $Me > Me_0$ is correct, then t_+ will be considerably larger than $n(n+1)/4$. The computed test statistic in Equations (5.6) and (5.7) will then be strongly positive. To determine whether the test statistic t is sufficiently positive for the null hypothesis to be rejected in favor of the alternative hypothesis, we have to compare it with the critical value $t_{\alpha;n-1}$. If $t > t_{\alpha;n-1}$, then we reject the null hypothesis in favor of the alternative hypothesis. If $t \le t_{\alpha;n-1}$, then we cannot reject the null hypothesis.

The p-value for a right-tailed test is the probability that a t-distributed random variable with $n-1$ degrees of freedom takes values larger than the computed test statistic t. In other words, the p-value for a right-tailed test is $P(t_{n-1} > t)$. If the p-value is smaller than the significance level chosen, we reject the null hypothesis.

The Left-Tailed Test

If the alternative hypothesis that $\text{Me} < \text{Me}_0$ is correct, then t_+ will be considerably smaller than $n(n+1)/4$. The computed test statistic in Equations (5.6) and (5.7) will then be strongly negative. To determine whether the test statistic t is sufficiently negative for the null hypothesis to be rejected, we have to compare it with the critical value $-t_{\alpha;n-1}$. If $t < -t_{\alpha;n-1}$, then we reject the null hypothesis in favor of the alternative hypothesis. If $t \geq -t_{\alpha;n-1}$, then we cannot reject the null hypothesis.

The p-value for a left-tailed test is the probability that a t-distributed random variable with $n-1$ degrees of freedom takes values smaller than the computed test statistic t. In other words, the p-value for a left-tailed test is $P(t_{n-1} < t)$. If the p-value is smaller than the significance level chosen, we reject the null hypothesis.

The Two-Tailed Test

If the alternative hypothesis that $\text{Me} \neq \text{Me}_0$ is correct, then there are two possibilities. One possibility is that $\text{Me} < \text{Me}_0$. In that case, t_+ will be considerably smaller than $n(n+1)/4$, and the test statistic t will be negative. The second possibility is that $\text{Me} > \text{Me}_0$. In that case, t_+ will be considerably larger than $n(n+1)/4$, and the test statistic will be positive. To determine whether the test statistic is sufficiently positive or sufficiently negative for the null hypothesis to be rejected, we need to compare it with $-t_{\alpha/2;n-1}$ and with $t_{\alpha/2;n-1}$. If the test statistic is smaller than $-t_{\alpha/2;n-1}$ or if it is larger than $t_{\alpha/2;n-1}$, then we reject the null hypothesis. Otherwise, we accept the null hypothesis. The p-value for a two-tailed test is equal to $2\,P(t_{n-1} > |t|)$. If this p-value is smaller than the significance level α, then we reject the null hypothesis.

Example 5.2.4   *If we use the approach based on the t-distribution for determining the p-value of the signed-rank test in Example 5.2.2, we have to use Expression (5.7). As there are no ties, we have that $g = 0$ and $c = 0$, so that the denominator of Equation (5.7) is simplified. The computed test statistic is then*

$$t = \frac{t_+ - \dfrac{n(n+1)}{4}}{\sqrt{\dfrac{n^2(n+1)(2n+1)}{24(n-1)} - \dfrac{t_+ - \frac{n(n+1)}{4}}{n-1}}}$$

$$= \frac{156 - \dfrac{20(20+1)}{4}}{\sqrt{\dfrac{20^2(20+1)(2 \times 20 + 1)}{24(20-1)} - \dfrac{156 - \frac{20(20+1)}{4}}{20-1}}} = 1.8591.$$

The p-value is $P(t_{n-1} > 1.8591) = 0.0393$. This p-value is a poorer approximation of the exact p-value, 0.0291, than the approximation based on the normal distribution in Example 5.2.3. Nevertheless, it leads to the same conclusion in this example.

6

Hypothesis Tests for the Distribution of a Population

Blomkvist worried about the coincidence but gradually ran out of questions. Bjurman must have known hundreds of people in his professional and social life. The fact that he happened to know someone who turned up in Svensson's material was neither improbable nor statistically unusual. Blomkvist was himself casually acquainted with a journalist who also appeared in the book.
(from *The Girl Who Played with Fire*, Stieg Larsson, p. 318)

When we obtain a set of observations from a continuous variable, it is usually not immediately clear whether these data points come from, for example, a normally distributed population or an exponentially distributed population. Similarly, if we obtain a set of observations from a discrete random variable, we do not know immediately whether these observations come from, for example, a Poisson process or a binomial process.

In this chapter, we explore how we can test the probability density or the probability distribution of a population or process under study.

6.1 Testing Probability Distributions

In order to verify whether qualitative or discrete quantitative data come from a population or process with a certain probability distribution, a χ^2-test can be performed. Several χ^2-tests exist. The possibility of conducting a χ^2-test based on the maximum likelihood approach has already been briefly mentioned in Chapter 4. In this section, we restrict ourselves to Pearson's simpler χ^2-test. However, both χ^2-tests are available in JMP.

The exact details of the Pearson χ^2-test depend on whether the parameters of the tested probability distribution are set in advance (i.e., before the data is collected) or estimated based on the data. We first consider the case where the parameters are fixed in advance. In this case, the parameters are treated as if they are "known".

Statistics with JMP: Hypothesis Tests, ANOVA and Regression, First Edition. Peter Goos and David Meintrup.
© 2016 John Wiley & Sons, Ltd. Published 2016 by John Wiley & Sons, Ltd.
Companion Website: http://www.wiley.com/go/goosandmeintrup/JMP

6.1.1 Known Parameters

When testing a probability distribution, a researcher sometimes has a good idea of the parameters of the underlying probability distribution. One possible source of inspiration for the parameter values is the scientific literature: by reading specialized academic papers, the researcher may obtain detailed information about the population or process that he is studying. Another possible source of inspiration is available historical data. This provides information about the size of the parameters in the past. By collecting new data, the researcher can then check whether the probability distribution and parameters of the past are still valid.

Example 6.1.1 *The pattern in the number of goals scored by the teams participating in the football World Cup in 1998 is very similar to that of a Poisson distribution. Suppose that an analysis of all previous World Cup soccer teams has taught us that, on average, 1.25 goals have been scored per team per game. Then, an obvious null hypothesis is*

H_0 : *The number of goals is Poisson distributed with parameter $\lambda = 1.25$.*

The alternative hypothesis is

H_a : *The number of goals is not Poisson distributed with parameter $\lambda = 1.25$.*

The 96 observations[1] for this example are listed in Table 6.1. The table shows that seven different outcomes were observed. In general, we denote the number of different outcomes by the letter k. When performing a χ^2-test, the frequencies of each of these outcomes are usually referred to as o_i (if the data has already been collected) or O_i (if the data has not yet been collected). This notation is derived from the term "observed frequency". For each outcome, the observed frequency is compared with the theoretical frequency expected if the data is indeed Poisson distributed with parameter $\lambda = 1.25$. These theoretical frequencies are referred to as e_i, which is short for "expected frequency". The theoretical or expected frequencies e_i can be calculated based on the Poisson distribution. To this end, one first has to calculate the probability for each outcome of the variable under study, using the Poisson distribution with parameter $\lambda = 1.25$. This provides the theoretical frequencies for a scenario with only one observation. Next, all probabilities must be multiplied by n = 96 to obtain the theoretical frequencies for 96 observations. The last two rows of Table 6.1 show how this is done.

Table 6.1 Observed frequencies o_i and theoretical frequencies e_i under the null hypothesis for the number of goals scored in the football World Cup in 1998.

x	0	1	2	3	4	5	≥ 6
o_i	26	34	24	8	1	2	1
$p_X(x; 1.25)$	0.2865	0.3581	0.2238	0.0933	0.0291	0.0073	0.0018[a]
$e_i = 96 \times p_X(x; 1.25)$	27.50	34.38	21.49	8.95	2.80	0.70	0.18

[a] The probability 0.0018 is the probability $P(X \geq 6)$ and not $P(X = 6)$.

---

[1] At the 1998 World Cup, 48 games were played by two teams. This results in $48 \times 2 = 96$ observations for the number of goals per team.

The required test statistic χ is

$$\chi = \sum_{i=1}^{k} \frac{(o_i - e_i)^2}{e_i},$$

$$= \frac{(26 - 27.50)^2}{27.50} + \frac{(34 - 34.38)^2}{34.38} + \frac{(24 - 21.49)^2}{21.49} + \frac{(8 - 8.95)^2}{8.95}$$

$$+ \frac{(1 - 2.80)^2}{2.80} + \frac{(2 - 0.70)^2}{0.70} + \frac{(1 - 0.18)^2}{0.18},$$

$$= 7.8986.$$

The formula for the test statistic shows that it compares the observed and expected frequencies by computing their differences. In this comparison, the (squared) differences between the two types of frequencies are weighted to indicate that a difference of one unit is more severe for a small expected frequency e_i than for a large one.

The test statistic χ is approximately χ^2-distributed when the null hypothesis is true. The number of degrees of freedom of the χ^2-distribution is $k - 1$, where k represents the number of different outcomes or classes. If the null hypothesis is true, then all differences $o_i - e_i$ are close to zero. Then, the weighted squares $(o_i - e_i)^2/e_i$ are all small positive numbers, so that the test statistic χ is not very large. However, when the null hypothesis is false, at least some of the differences $o_i - e_i$ will be strongly negative or strongly positive, so that some weighted squares $(o_i - e_i)^2/e_i$ will be large. Therefore, the test statistic χ will also take a large positive value when the null hypothesis is incorrect.

To decide whether the test statistic is sufficiently large for the null hypothesis to be rejected, we need a critical value. We obtain this critical value from the χ^2-distribution with $k - 1$ degrees of freedom, because this is the probability density of the test statistic under the null hypothesis. As a critical value, we use $\chi^2_{\alpha;k-1}$, where α represents the significance level. In the example, with $k - 1 = 6$, the critical value is $\chi^2_{\alpha;6} = \chi^2_{0.05;6} = 12.5916$, if we use a significance level of 5%.

We now have to compare the computed test statistic, 7.8986, with the critical value, 12.5916. As $7.8986 < 12.5916$, we accept the null hypothesis. If the computed test statistic had been larger than the critical value, we would have rejected the null hypothesis. The fact that the test statistic is smaller than the critical value indicates that the (weighted squared) deviations between the observed and theoretical frequencies are not large enough for the assumption that the number of goals is Poisson distribution with $\lambda = 1.25$ to be rejected.

In general, the p-value for the Pearson χ^2-test is calculated as

$$p = P\left(\chi^2_{k-1} > \chi\right).$$

For the example, we obtain

$$p = P\left(\chi^2_6 > 7.8986\right) = 0.2456.$$

The p-value is greater than the significance level of 5%, so that we accept the null hypothesis. The p-value can be calculated in JMP using the command "$1 -$ ChiSquare Distribution (7.8986, 6)", while the critical value can be determined using "ChiSquare Quantile (0.95, 6)".

The result of the hypothesis test is that the data does not allow us to conclude that the data does not come from a Poisson distribution with parameter $\lambda = 1.25$.

6.1.2   Unknown Parameters

In some situations, no literature can be consulted, and no historical data concerning the population or process under study is available. In that case, the researcher cannot specify parameter values in advance. The parameter(s) of the hypothesized probability distribution then cannot be specified in the null hypothesis.

The only alternative for the researcher is to estimate the unknown parameters based on the collected data. Afterwards, the researcher can use the same method as in Table 6.1 to compare the observed and theoretical frequencies. The only difference is that the theoretical frequencies must be calculated using the estimated parameter values.

The final test statistic is still approximately χ^2-distributed. However, the number of degrees of freedom is no longer $k - 1$, but $k - 1$ minus the number of estimated parameters. This is illustrated in the following example.

Example 6.1.2   *If an estimate for the parameter λ is calculated from the data in Example 6.1.1, then the null hypothesis and the alternative hypothesis are*

$$H_0 : \text{The number of goals is Poisson distributed}$$

and

$$H_a : \text{The number of goals is not Poisson distributed.}$$

The parameter λ of a Poisson distribution is the expected value or mean. Applied to our example, λ is the expected value or population mean of the number of goals scored by one team in one match. Therefore, we can calculate an unbiased estimate for the parameter λ by using the sample mean[2], which turns out to be $\overline{x} = 1.3125$. The observed frequencies o_i in this example and the theoretical or expected frequencies e_i for $\lambda = 1.3125$ are given in Table 6.2.

Table 6.2   Observed frequencies o_i and theoretical frequencies e_i under the null hypothesis for the number of goals scored in the football World Cup in 1998. For the computation of the theoretical frequencies, the estimated parameter value $\lambda = 1.3125$ was used.

x	0	1	2	3	4	5	≥ 6
o_i	26	34	24	8	1	2	1
$p_X(x; 1.3125)$	0.2691	0.3533	0.2318	0.1014	0.0333	0.0087	0.0023[a]
$e_i = 96 \times p_X(x; 1.3125)$	25.84	33.91	22.26	9.74	3.19	0.84	0.22

[a]The probability 0.0023 is the probability $P(X \geq 6)$ and not $P(X = 6)$.

---

[2] The sample mean can be calculated as $(0 \times 26 + 1 \times 34 + \cdots + 5 \times 2 + 6 \times 1)/96 = 1.3125$. See Example 6.1.1 for the original data.

The computed test statistic χ is

$$\chi = \sum_{i=1}^{k} \frac{(o_i - e_i)^2}{e_i},$$

$$= \frac{(26 - 25.84)^2}{25.84} + \frac{(34 - 33.91)^2}{33.91} + \frac{(24 - 22.26)^2}{22.26} + \frac{(8 - 9.74)^2}{9.74}$$

$$+ \frac{(1 - 3.19)^2}{3.19} + \frac{(2 - 0.84)^2}{0.84} + \frac{(1 - 0.22)^2}{0.22},$$

$$= 6.2439.$$

This statistic is approximately χ^2-distributed if the null hypothesis is true. The number of degrees of freedom of the χ^2-distribution is $k - 1 - 1$, where k represents the number of outcomes. The number of degrees of freedom is one unit smaller than in Example 6.1.1 because one parameter, namely λ, has been estimated from the data. The test statistic χ has to be compared with a critical value obtained from the χ^2-distribution with five degrees of freedom. For a significance level of 5%, the critical value is $\chi^2_{\alpha;5} = \chi^2_{0.05;5} = 11.0705$. As $6.2439 < 11.0705$, we accept the null hypothesis. If the computed test statistic had been larger than the critical value, we would have accepted the alternative hypothesis. The p-value for the χ^2-test can be calculated as

$$p = P\left(\chi^2_5 > 6.2439\right) = 0.2832.$$

It can be obtained by means of the JMP command "$1 - \texttt{ChiSquare Distribution}$ $(6.2439, 5)$", while the critical value can be found using the formula "$\texttt{ChiSquare}$ $\texttt{Quantile}(0.95, 5)$". The result of the hypothesis test is that the data does not allow us to conclude that the data does not come from a Poisson distribution.

A technical condition[3] for the χ^2-tests described above to work properly is that each outcome or class has an expected frequency of at least one, and that at most 20% of all outcomes or classes have an expected value that is less than 5. If this is not the case, then, ideally, some of the classes are merged. The calculation of the theoretical frequencies should then also take into account the merging of classes. In Table 6.2, two classes have an expected frequency of less than 1, and three of the seven classes have an expected frequency of less than 5. Ideally, these three classes (four goals, five goals, and at least six goals) are merged to meet the technical requirements. For the calculation of the theoretical frequency of the newly created class, one would use the formula $96 \times \{P(X = 4) + P(X = 5) + P(X \geq 6)\}$.

In JMP, the scripts "Chi Square Test for Poisson Distribution" and "Chi Square Test for Binomial Distribution" are available for the Pearson χ^2-test. The use of the first script for Example 6.1.2 is illustrated in Figures 6.1, 6.2, and 6.3. Figure 6.1 shows what the data table should look like. Figure 6.2 shows the required entries in the dialog window. Finally, Figure 6.3 contains the final result.

---

[3] Some textbooks state that this technical condition is perhaps too strict, and that the χ^2-test still works properly if there are some more small expected frequencies.

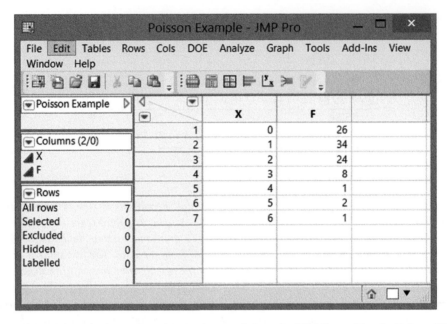

Figure 6.1 The required JMP data table for the use of the script "Chi Square Test for Poisson Distribution" in Example 6.1.2.

6.1.3 χ^2-Tests for Qualitative Variables

Example 6.1.3 *One of the University of Antwerp's professors gets a part-time appointment at the Erasmus University of Rotterdam and is supposed to work in Rotterdam one day a week. The professor wants to make the journey by train. However, the reliability of the international train service between Belgium and the Netherlands leaves a lot to be desired. The professor wants to minimize his loss of time and choose the weekday with the smallest frequency of delays on the Antwerp–Rotterdam–Antwerp round trip. For two months, the professor carefully records the weekdays on which the 8 o'clock Antwerp–Rotterdam train is delayed. The professor observes 23 delays. The distribution of the delays over the five weekdays is shown in Table 6.3.*

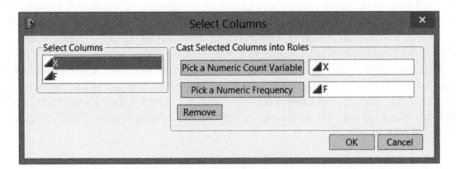

Figure 6.2 The dialog window of the script "Chi Square Test for Poisson Distribution" for Example 6.1.2.

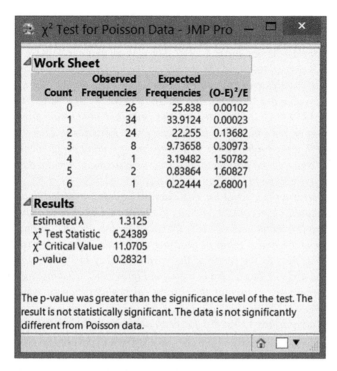

Figure 6.3   The final result of the script "Chi Square Test for Poisson Distribution" for Example 6.1.2.

The professor, however, is concerned that the delays are simply uniformly distributed over the days of the week. To verify whether this is indeed the case, he tests the following hypotheses:

H_0 : *The delays are uniformly distributed over the five weekdays.*

and

H_a : *The delays are not uniformly distributed over the five weekdays.*

If the null hypothesis is correct, then one fifth of all delays should happen on a Monday, one fifth on a Tuesday, and so on. The fraction 1/5, or 0.2, is the theoretical probability for each of the days, if the null hypothesis is correct. If we multiply the theoretical probability for each day by the number of observations, 23, we obtain the theoretical frequencies, e_i. The 5

Table 6.3   The data from Example 6.1.3 along with the calculations required for the Pearson χ^2-test.

Weekday	o_i	Theoretical probability	e_i	$(o_i - e_i)^2/e_i$
Monday	4	0.2	4.6	0.0783
Tuesday	4	0.2	4.6	0.0783
Wednesday	8	0.2	4.6	2.5130
Thursday	2	0.2	4.6	1.4696
Friday	5	0.2	4.6	0.0348
Sum	23	1.0	23	4.1739

theoretical frequencies in this example are all equal to $1/5 \times 23 = 4.6$. *The calculation of the test statistic* χ *is shown schematically in the right-hand columns of Table 6.3.*

The final value for the test statistic is 4.1739. The critical value is 9.4877, while the p-value is equal to 0.3830. Based on these values, we accept the null hypothesis and conclude that the delays are spread uniformly across the different weekdays. The required formulas for the critical value and the p-value are "ChiSquare Quantile(0.95, 4)" *and* "1 − ChiSquare Distribution(4.1739, 4)", *respectively. The practical conclusion for the professor is that it does not matter on which day he commutes to Rotterdam. There is no significant statistical evidence that the probabilities of a delay are spread unevenly over the 5 weekdays.*

Note that, in this example, no parameters have to be estimated from the data. Therefore, the number of degrees of freedom for the hypothesis test is $k - 1 = 5 - 1 = 4$.

You can easily perform the χ^2-*test for this example in JMP, provided that you have a data table with the individual observations, such as the table in Figure 6.4a. You can analyze the data in the table via the menu* "Analyze", *by choosing the option* "Distribution". *You will then see a bar chart and a table of absolute and relative frequencies. If you click the hotspot (red triangle) next to the name of the variable "Day of Week", you will see a new option:* "Test Probabilities". *This is shown in Figure 6.4b. Clicking this option will open a dialog window, where you can enter the theoretical probabilities for each outcome of the variable "Day of Week". This is shown in Figure 6.5.*

If we enter the probabilities $1/5 = 0.2$ *for each outcome and click the* "Done" *button, we obtain the output in Figure 6.6. The results for the Pearson* χ^2-*test in this output are perfectly consistent with the results we obtained above.*

Note that when you enter the days of the week in your data table in English, JMP orders them logically; that is, beginning with Monday and ending with Friday. If you enter the days in another language, then JMP will order them alphabetically. In that case, you can ensure that JMP orders the days logically by right-clicking on the header of the column "Day of Week", and then choosing the option "Column Info...". *Next, under* "Column Properties", *the option* "Value Ordering" *has to be selected. As shown in Figure 6.7, you can then indicate the desired order by using the buttons* "Move Up" *and* "Move Down".

Example 6.1.4   *The first lottery draw with 42 numbers in Belgium took place on April 30, 1984. Since then, some numbers have been drawn more often than others. Table 6.4 summarizes the results of 2422 draws (a total of* $2422 \times 7 = 16954$ *numbers have been drawn). For each number from 1 to 42, the table contains the (observed) frequency* o_i, *the relative frequency, and the date on which it was last drawn. A bar chart of the relative frequencies is shown in Figure 6.8.*

The National Lottery of Belgium claims that the lottery draws are completely random; that is, that each of the 42 numbers has the same probability of being drawn. In total, seven numbers are randomly selected in each draw. The numbers drawn are not put back into the drum, so the probability that a number is drawn in a lottery draw is $7/42 = 1/6$. *In total, there have been 2422 draws. The theoretical frequency for each of the numbers in 2422 draws is thus* $e_i = 2422 \times 1/6 = 403.6667$. *We can now test the following hypotheses:*

H_0 : *In the Belgian lottery,*
      the probability of drawing each of the 42 numbers is the same

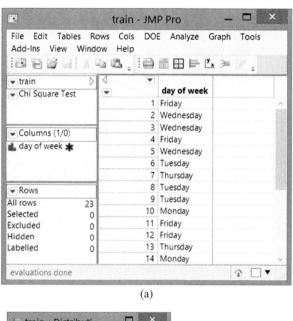

(a)

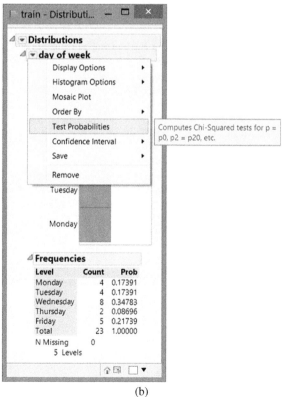

(b)

Figure 6.4 The data table and an intermediate step for the Pearson χ^2-test in JMP for Example 6.1.3.

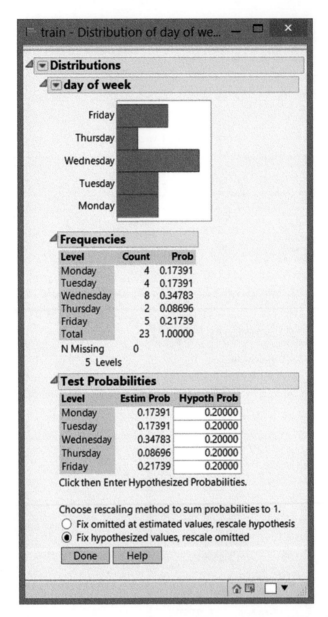

Figure 6.5 The screen for entering the theoretical probabilities for the Pearson χ^2-test in JMP for Example 6.1.3.

and

H_a : *In the Belgian lottery,*

the probability of drawing each of the 42 numbers is not the same.

Test Probabilities

Level	Estim Prob	Hypoth Prob
Monday	0.17391	0.20000
Tuesday	0.17391	0.20000
Wednesday	0.34783	0.20000
Thursday	0.08696	0.20000
Friday	0.21739	0.20000

Test	ChiSquare	DF	Prob> Chisq
Likelihood Ratio	4.1202	4	0.3900
Pearson	4.1739	4	0.3830

Method: Fix hypothesized values, rescale omitted

Figure 6.6   The JMP output for Example 6.1.3.

Using the Pearson χ^2-test, we can test whether the distribution for the 42 numbers is indeed uniform. The computed test statistic is

$$\chi = \frac{(406 - 403.6667)^2}{403.6667} + \frac{(416 - 403.6667)^2}{403.6667} + \cdots + \frac{(404 - 403.6667)^2}{403.6667},$$
$$= 46.8538.$$

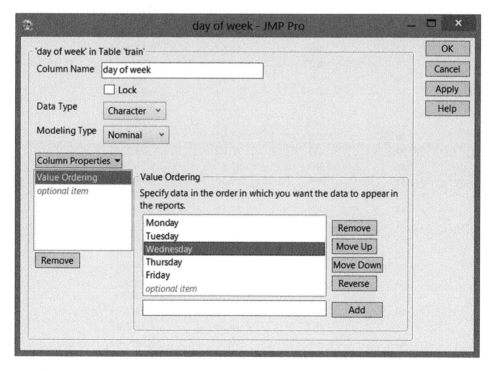

Figure 6.7   Ordering the values of the nominal variable "Day of Week" in Example 6.1.3.

Table 6.4 The data for the lottery draws in Belgium. Source: http://www.nationale-loterij.be/ (accessed January 4, 2012).

Number	Number of drawing	Relative frequency (%)	Date of most recent drawing
1	406	16.76	9/28/2011
2	416	17.18	8/27/2011
3	407	16.80	9/24/2011
4	416	17.18	9/21/2011
5	430	17.75	9/28/2011
6	396	16.35	9/24/2011
7	442	18.25	9/21/2011
8	363	14.99	9/17/2011
9	417	17.22	9/14/2011
10	405	16.72	9/3/2011
11	391	16.14	8/20/2011
12	438	18.08	8/20/2011
13	417	17.22	9/10/2011
14	418	17.26	9/24/2011
15	356	14.70	7/16/2011
16	433	17.88	8/24/2011
17	405	16.72	9/28/2011
18	379	15.65	9/28/2011
19	403	16.64	9/17/2011
20	376	15.52	9/10/2011
21	397	16.39	8/31/2011
22	449	18.54	9/17/2011
23	405	16.72	9/28/2011
24	439	18.13	9/14/2011
25	419	17.30	9/14/2011
26	385	15.90	9/28/2011
27	395	16.31	9/24/2011
28	411	16.97	9/24/2011
29	411	16.97	8/6/2011
30	383	15.81	9/21/2011
31	390	16.10	9/14/2011
32	385	15.90	9/7/2011
33	415	17.13	9/24/2011
34	401	16.56	9/17/2011
35	381	15.73	8/24/2011
36	395	16.31	8/17/2011
37	382	15.77	9/7/2011
38	430	17.75	9/3/2011
39	369	15.24	9/24/2011
40	392	16.18	8/24/2011
41	402	16.60	9/28/2011
42	404	16.68	8/24/2011

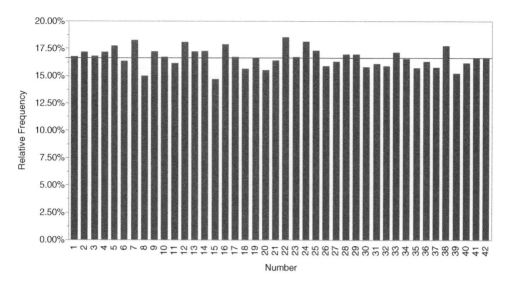

Figure 6.8 A bar chart of the relative frequencies of the 42 lottery numbers. The reference line represents the theoretical probability of $7/42 = 1/6$ that a specific number is drawn at any lottery draw.

If the null hypothesis is correct, the statistic is approximately χ^2-distributed with $k - 1 = 42 - 1 = 41$ degrees of freedom. The critical value, which is determined based on this distribution, is 56.9424. As a result, the value of the computed test statistic does not exceed the critical value, so that we accept the null hypothesis. The p-value is 0.2448, which, of course, also leads to acceptance of the null hypothesis. The required JMP formulas for the critical value and the p-value are, respectively, "`ChiSquare Quantile`*(0.95, 41)" and "*`1 - ChiSquare Distribution`*(46.8538, 41)".*

Starting from the data table in Figure 6.9, you can also perform Pearson's χ^2-test in JMP. To this end, it is important that you indicate in JMP that the column with all possible outcomes is a nominal variable. This can be done by right-clicking the header of the column labeled "Number", selecting "`Column Info...`*", and then choosing "Nominal" as "*`Modeling Type`*". If you then use the "*`Distribution`*" option in the "*`Analyze`*" menu, you can indicate in the resulting dialog window that you want to analyze the variable "Number", and that the observed frequencies of the outcomes appear in the column "Number of drawings". The required input is shown in Figure 6.10.*

If you then click "`OK`*", you obtain a chart with 42 bars and a table with 42 relative frequencies. To perform the χ^2-test, you have to click the hotspot (red triangle) next to the name of the variable, "Number". In the resulting menu, select the option "*`Test Probabilities`*" and enter the theoretical probability for each outcome $1/42 = 0.02381$. Alternatively, you can also enter another common value for each outcome, for example 1. JMP will then transform this common value into probabilities that are all of the same size (see Figure 6.11). If, finally, you click "*`Done`*", you obtain the result of the Pearson χ^2-test, together with the result of the alternative χ^2-test based on the maximum likelihood method.*

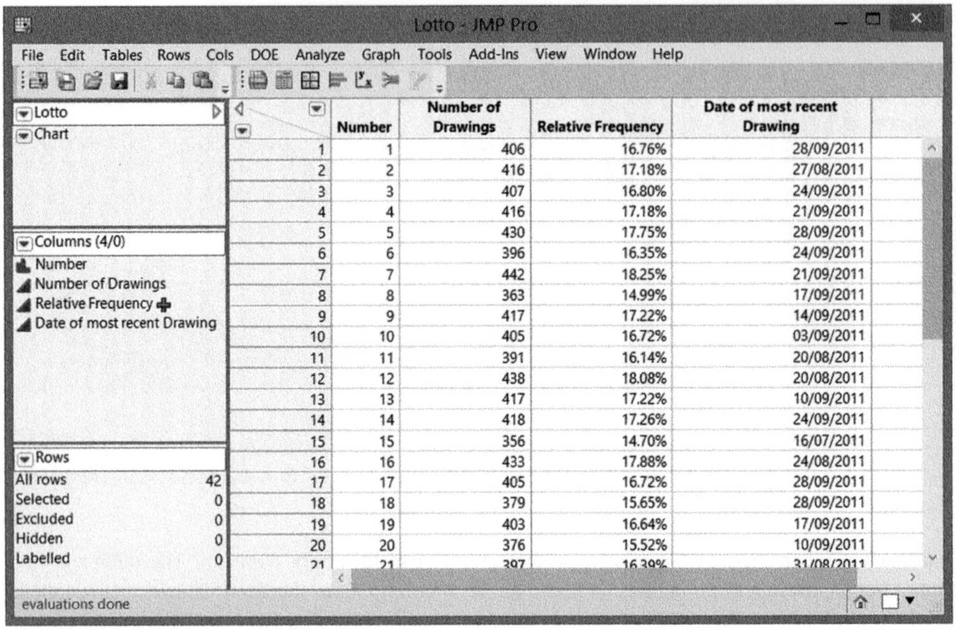

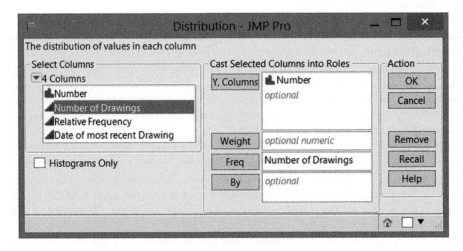

Figure 6.9   The JMP table with the data described in Example 6.1.4.

Figure 6.12 shows the final result of all these operations in JMP. The values entered for each outcome have been rescaled, as indicated by the message "Note: Hypothesized probabilities did not sum to 1. Probabilities have been rescaled". *in   the output.*

Figure 6.10   The dialog window for the hypothesis test in Example 6.1.4.

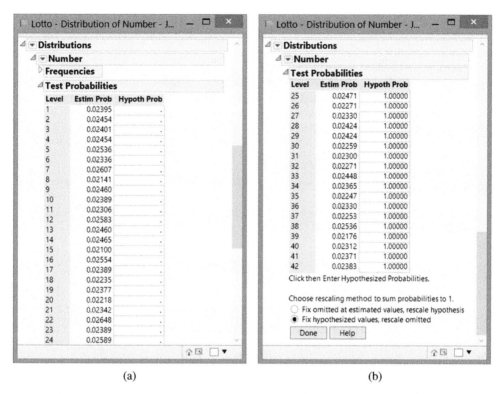

(a)   (b)

Figure 6.11   The intermediate steps for the Pearson χ^2-test in JMP for Example 6.1.4.

6.2   Testing Probability Densities

Testing whether a set of quantitative data points come from a population or a process with a certain probability density can also be done using the χ^2-test described in the previous section. For this purpose, however, the data first has to be divided into classes. Another approach involves generating a stem-and-leaf diagram, a histogram, or a frequency polygon (see the book *Statistics with JMP: Graphs, Descriptive Statistics and Probability*), and then comparing the shape of these graphical representations with the probability density under consideration. However, none of these approaches is ideal. The major reason for this is that

Test	ChiSquare	DF	Prob>Chisq
Likelihood Ratio	46.8952	41	0.2435
Pearson	46.8538	41	0.2448

Method: Fix hypothesized values, rescale omitted
Note: Hypothesized probabilities did not sum to 1. Probabilities have been rescaled.

Figure 6.12   The JMP output for Example 6.1.4.

quantitative data can be arranged in classes in many different ways, and each leads to a different test statistic value for the χ^2-test, and to a different look for the corresponding stem-and-leaf diagram, histogram, and frequency polygon.

Fortunately, in addition to these rudimentary graphical approaches, there exists a more objective visual method, namely the quantile diagram or Q–Q plot (quantile–quantile plot). Also, a number of more sophisticated methods exist, such as the Kolmogorov–Smirnov and Lilliefors tests and the Shapiro–Wilk test. These tests are available in JMP.

6.2.1   The Normal Probability Density

The Hypotheses Tested

In this section, the focus is on testing whether a data set contains normally distributed data. The hypotheses tested are therefore

$$H_0 : \text{The data is normally distributed}$$

and

$$H_a : \text{The data is not normally distributed.}$$

The test for normality of data is considered important for small data sets and quantitative variables, because the t-tests for one population mean (see Chapter 4) and for two population means (see Chapters 8 and 10) can only be used if the data is normally distributed. Also, for the use of the χ^2-test for a population variance, it should be verified that the data is normally distributed. We start by discussing a graphical approach to verifying the normality of a data set.

An Elementary Quantile Diagram

The purpose of a quantile diagram, also called a Q–Q plot or a quantile–quantile plot, is to graphically compare the actual observations with the ones we would expect based on the null hypothesis. A quantile diagram actually compares observed or empirical quantiles with theoretical quantiles. To gain a better understanding, we will use a small example.

Example 6.2.1   *Suppose that a sample of n = 10 observations is given: 61, 50, 26, 47, 38, 46, 60, 65, 62, and 43. We examine whether this data comes from a normally distributed population. To generate a quantile diagram of the data, we first have to arrange the data in ascending order. This was done in Table 6.5. The values in the table are called **order statistics**, and they are denoted by $x_{(1)}, x_{(2)}, \ldots, x_{(10)}$. Next, the cumulative relative frequencies have to be determined for the 10 order statistics. The cumulative relative frequency for the ith order statistic can be calculated as*

$$cf_i = \frac{i - 0.5}{n}, \tag{6.1}$$

Table 6.5 Calculations for the construction of a quantile diagram.

$x_{(i)}$	26	38	43	46	47	50	60	61	62	65
cf_i	0.05	0.15	0.25	0.35	0.45	0.55	0.65	0.75	0.85	0.95
z_{1-cf_i}	−1.64	−1.04	−0.67	−0.39	−0.13	0.13	0.39	0.67	1.04	1.64
$49.8 + 12.4z_{1-cf_i}$	29.40	36.95	41.44	45.02	48.24	51.36	54.58	58.16	62.65	70.20

where $n = 10$ in this example. The resulting cumulative relative frequencies $0.05, 0.15, \ldots, 0.95$ are shown in the second row of Table 6.5.

To understand why this specific formula is required, we first have to associate the quantiles or percentiles with the various observations in the sample. The definition of a quantile or percentile tells us that the 10th percentile or quantile of the data set with 10 observations lies between 26 and 38 (i.e., between the first order statistic $x_{(1)}$ and the second order statistic $x_{(1)}$), the 20th quantile between 38 and 43 (i.e., between the second order statistic $x_{(2)}$ and the third order statistic $x_{(3)}$), and so on. As a consequence, the 5th quantile will be near the value 26, the 15th quantile near the value 38, and so on. The numbers $5, 15, \ldots$ for the quantiles are those produced by Equation (6.1) for cf_i, up to a factor of 100. This explains the need to compute the cumulative relative frequencies cf_i: the cf_i values indicate with which quantile the ith order statistic (i.e., the ith observation after ranking the observations from small to large) needs to be compared.

If the null hypothesis that the data is normally distributed is correct and the expected value and variance are equal to μ and σ^2, then the 5th quantile[4] $\mu + z_{1-cf_i}\sigma = \mu + z_{0.95}\sigma$ of a normally distributed random variable will be close to the value 26, the 15th quantile $\mu + z_{0.85}\sigma$ close to the value 38, and so on. Generally, no values are hypothesized for the parameters μ and σ of the normal distribution. Therefore, these parameters are estimated from the data using the sample mean $\bar{x} = 49.8$ and the sample standard deviation $s = 12.4$. The theoretical quantiles based on these estimates are calculated in the third and fourth rows of Table 6.5.

The quantile diagram is simply a scatter plot, with the theoretical quantiles displayed on the horizontal axis and the sample observations (the observed or empirical quantiles) displayed on the vertical axis. If the null hypothesis is true, then the theoretical quantiles and the sample observations are very similar, and all points in the scatter plots virtually lie on a straight line, namely the bisector. The null hypothesis can therefore be accepted if the points are all close to this straight line. If a large number of points deviate from this line, the null hypothesis that the data is normally distributed should be rejected. The quantile diagram based on the results in Table 6.5 is shown in Figure 6.13. The two dashed lines indicate that the ninth point in the graph has the coordinates 62.65 and 62. The first coordinate is the theoretical quantile associated with the ninth order statistic, while the second coordinate is the ninth order statistic $x_{(9)}$ itself (i.e., the observed quantile). The points in the graph do not perfectly lie on the bisector. However, it is premature to conclude from this that the data is not normally

---

[4] In *Statistics with JMP: Graphs, Descriptive Statistics and Probability*, we explain in detail how to calculate quantiles or percentiles of normally distributed random variables with mean μ and standard deviation σ.

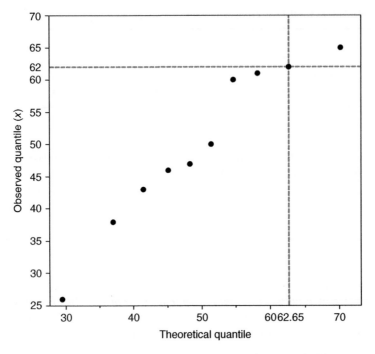

Figure 6.13   An elementary quantile diagram for Example 6.2.1.

distributed, because quantile diagrams only work well for data sets with larger numbers of observations.

An Improved Quantile Diagram

Although the construction of a quantile diagram as described above is perfectly logical, research has shown that it is better to calculate the theoretical quantiles using the following cumulative relative frequencies:

$$cf_i^* = \frac{i}{n+1}, \quad for\ i = 1, 2, \dots, n.$$

This provides the theoretical quantiles in Table 6.6. These theoretical quantiles are called the **van der Waerden scores**. *The corresponding quantile diagram is shown in Figure 6.14.*

Table 6.6   An improved construction of a quantile diagram using the van der Waerden scores.

$x_{(i)}$	26	38	43	46	47	50	60	61	62	65
cf_i^*	0.091	0.182	0.273	0.364	0.455	0.545	0.636	0.727	0.818	0.909
$z_{1-cf_i^*}$	−1.34	−0.91	−0.60	−0.35	−0.11	0.11	0.35	0.60	0.91	1.34
$49.8 + 12.4z_{1-cf_i^*}$	33.24	38.54	42.3	45.48	48.38	51.22	54.12	57.3	61.06	66.36

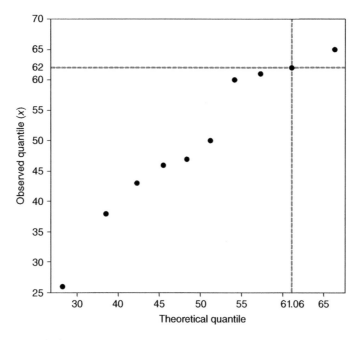

Figure 6.14   An improved quantile diagram with van der Waerden scores.

Quantile Plots in JMP

In the previous sections, we have explained how quantile diagrams can be constructed so that the observed and theoretical quantiles can be compared directly. Every software package constructs its own version of a quantile diagram. In this section, we study the quantile diagram that JMP constructs. Figure 6.15 shows JMP's quantile diagram for the data in Tables 6.5 and 6.6.

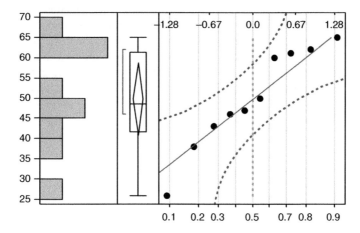

Figure 6.15   The quantile diagram created in JMP for the data in Tables 6.5 and 6.6.

One difference from the quantile diagrams in Figures 6.13 and 6.14 is that JMP shows the theoretical quantiles of the standard normal distribution on the horizontal axis at the top. This is not the case in Figures 6.13 and 6.14, where the theoretical quantiles on the horizontal axis correspond to a normal distribution with the same mean and the same variance as the data. This does not change the interpretation of the figure: if the data originates from a normally distributed population, all points more or less lie on a straight line. The numbers on the horizontal axis at the bottom of JMP's diagram in Figure 6.15 correspond to the cf_i^* values.

Another difference is that, in JMP, dotted lines are used to indicate the 95% Lilliefors confidence limits. If all the points in the chart lie within these limits, then there is no reason to reject the null hypothesis that the data is normally distributed. In Figure 6.15, all points fall within the confidence limits, so that there is no evidence in favor of the alternative hypothesis. If there are only a few observations, as is the case in Figure 6.15, the confidence limits are so wide that it virtually never happens that a point falls outside the confidence limits.

The Lilliefors confidence limits actually form a graphical hypothesis test. If all points lie within the limits, then we accept the null hypothesis that the studied population is normally distributed. As soon as there is a point outside the limits, we reject the null hypothesis.

Example 6.2.2   *In a study on the strength of fibers, the breaking strength of 50 samples was recorded. Figures 6.16a and 6.16b show the quantile plots of the variables "breaking strength" and "ln(breaking strength)", where "ln( )" denotes the natural logarithm. The graphs clearly show that the probability density of the variable "breaking strength" strongly deviates from the normal distribution, while the variable "ln(breaking strength)" corresponds pretty well to a normal distribution. This can also be seen from the histograms in Figures 6.16a and 6.16b.*

To generate a quantile diagram in JMP, you can use the "Distribution" option in the "Analyze" menu. If, in the resulting output, you click on the hotspot (red triangle) next to the name of the variable under study, one of the options that appear will be "Normal quantile plot". This is illustrated in Figure 6.17 for the data in Example 6.2.1.

Another interesting option in the same menu is "Continuous Fit" (see Figure 6.17 as well). With this option, you can, for example, ask JMP to identify the normal probability density that best fits your data. To this end, choose "Normal" after selecting the "Continuous Fit" option. This results in the output in Figure 6.18. In the upper part of the output, you see the estimated normal density together with the histogram. In the lower part of the output, the estimates of the parameters of the fitted density are shown, along with a 95% confidence interval. In the case of the normal probability density, two estimates are given, one for each parameter of the normal distribution (the mean μ and the standard deviation σ).

In the menu in Figure 6.17, there is also an option called "Discrete Fit". This option can be used to search for a discrete probability distribution that fits your data.

The Interpretation of Quantile Diagrams

If the quantile diagram contains a large number of observations, a quick look at it is generally sufficient to judge whether or not the data is normally distributed. However, if a variable is not normally distributed, the quantile diagram can also be used to say something about the nature of the distribution of the variable.

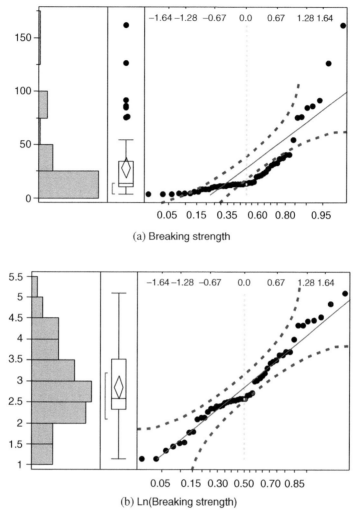

Figure 6.16 Quantile diagrams created in JMP for the variables "Breaking strength" and "ln(Breaking strength)" in Example 6.2.2.

Figure 6.19 contains two quantile diagrams for 1000 observations. The diagram in Figure 6.19a is a quantile plot for 1000 (simulated) observations from a standard normally distributed population, while the diagram in Figure 6.19b shows 1000 (simulated) observations from a population that is t-distributed with five degrees of freedom. Despite the fact that the t-distribution, just like the standard normal probability density, is bell-shaped and symmetrical around zero, the two quantile diagrams are very different. Figure 6.19b shows that the observed or empirical quantiles in the left-hand part of the diagram are smaller than the theoretical quantiles from the standard normal distribution. For example, the smallest observation (the smallest observed quantile) is near -8, while the smallest theoretical quantile is near -3. This suggests that the left tail of the distribution of the data is heavier or thicker

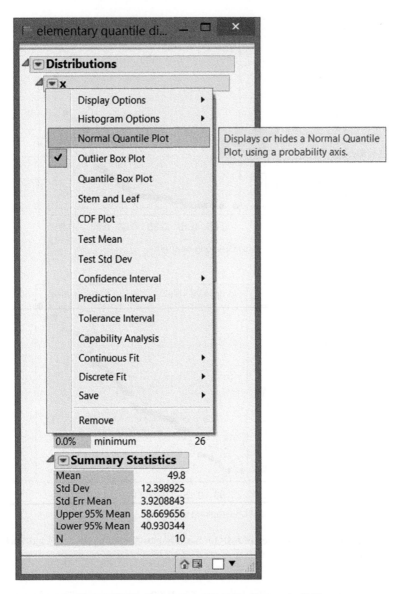

Figure 6.17   Generating a quantile diagram in JMP.

than the left tail of the normal density. In other words, the data set contains more very small observations than can be expected from a normal distribution. In the right-hand part of the diagram in Figure 6.19b, the observed or empirical quantiles are larger than the theoretical ones. For example, the largest observation (the largest observed quantile) is located in the neighborhood of +8, while the largest theoretical quantile is near +3. This indicates that the right tail of the distribution of the data is heavier or thicker than the right tail of the normal density. In other words, the data set contains more very large observations than can be expected from a normal distribution.

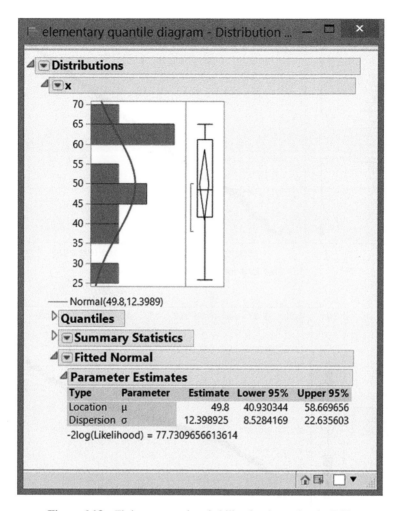

Figure 6.18 Fitting a normal probability density to data in JMP.

This should not come as a surprise: Figure 2.5 clearly shows that the tails of a t-distribution are thicker than those of the standard normal density. An immediate consequence of this is that $t_{\alpha;v} < z_\alpha$ when $\alpha > 0.5$ and $t_{\alpha;v} > z_\alpha$ when $\alpha < 0.5$, for any number of degrees of freedom v of the t-distribution. For example, the 5th quantile of a t-distributed random variable with five degrees of freedom is $t_{0.95;5} = -t_{0.05;5} = -2.015$, while that of a standard normally distributed random variable is $z_{0.95} = -z_{0.05} = -1.645$. The 95th quantile for the same t-distributed random variable is $t_{0.05;5} = 2.015$, while it is $z_{0.05} = 1.645$ for the standard normally distributed random variable. The fact that the quantiles of the t-distribution are more extreme than those of the standard normal probability density is thus reflected in the quantile diagram in Figure 6.19b.

Quantile diagrams also allow us to assess the skewness of a distribution. Figure 6.20 illustrates what a quantile diagram looks like when the data comes from a distribution that is

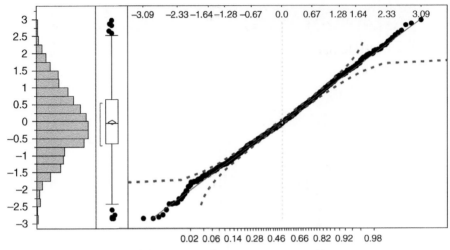

(a) Standard normally distributed population

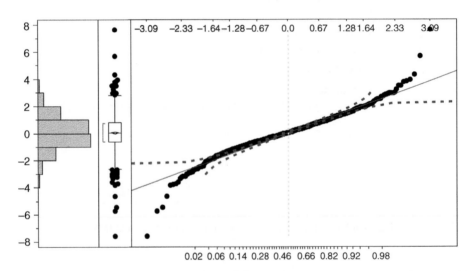

(b) t-distributed population with five degrees of freedom

Figure 6.19 The histograms, box plots, and corresponding quantile diagrams for 1000 observations from a standard normally distributed population and a t-distributed population.

skewed to the right or skewed to the left. The shape of the quantile diagram for the left-skewed data is clearly concave, while the shape of the quantile diagram for the right-skewed data is convex. For right-skewed distributions, the extreme observed quantiles are large compared to the theoretical quantiles derived from the normal distribution. For left-skewed distributions, the extreme observed quantiles are small compared to the theoretical quantiles from the normal distribution.

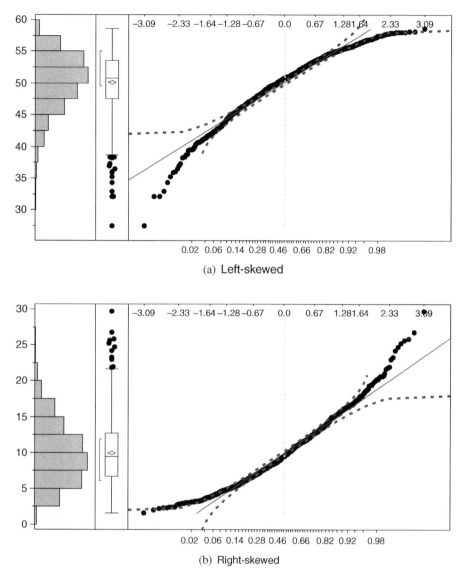

Figure 6.20 The histograms, box plots, and corresponding quantile diagrams of 1000 observations for populations with left-skewed and right-skewed distributions.

The Shapiro–Wilk Test

The Lilliefors confidence limits provide a graphical test to determine whether the population under study is normally distributed. However, several formal hypothesis tests also exist. Many of them do not work well. One example is the χ^2-test, which we have already discussed in Section 6.1 and which we used to test discrete probability distributions. As explained above, this test can also be used for continuous probability densities, provided that the observations are subdivided into classes. Another option is the so-called Kolmogorov–Smirnov test, which compares the observed or empirical cumulative distribution function with the theoretical cumulative distribution function. In a series of scientific papers, d'Agostino demonstrated that these techniques are fairly useless, because they have a low power when testing normality. In 1965, the researchers Shapiro and Wilk proposed a new test.

The Shapiro–Wilk test uses a regression technique to summarize the information in a quantile diagram. Just like the quantile diagram itself, it is based on a comparison of the observed and theoretical quantiles. With the regression technique, theoretical quantiles can be determined. These quantiles are called $a_1, a_2, \ldots, a_n$, with $a_1 < a_2 < \cdots < a_n$. They are compared with the observed quantiles; that is, the observations ranked from small to large. When we rank observations from small to large, we again obtain the so-called order statistics $x_{(1)}, x_{(2)}, \ldots, x_{(n)}$. Here, $x_{(1)}$ is the smallest number in the sample, $x_{(2)}$ the second smallest, and so on. However, as long as we have not yet collected the data, we use uppercase letters to denote the order statistics: $X_{(1)}, X_{(2)}, \ldots, X_{(n)}$.

The test statistic W in the Shapiro–Wilk test is the square of the Pearson correlation coefficient between the theoretical quantiles $a_1, a_2, \ldots, a_n$ and the observed quantiles or order statistics $X_{(1)}, X_{(2)}, \ldots, X_{(n)}$:

$$W = \frac{\left(\sum_{i=1}^{n} \left(X_{(i)} - \overline{X} \right) (a_i - \overline{a}) \right)^2}{\sum_{i=1}^{n} \left(X_{(i)} - \overline{X} \right)^2 \sum_{i=1}^{n} (a_i - \overline{a})^2} = \frac{\left(\sum_{i=1}^{n} X_{(i)} a_i \right)^2}{\sum_{i=1}^{n} \left(X_{(i)} - \overline{X} \right)^2}.$$

To prove this equality, it is important to note that the mean of all theoretical quantiles $a_1, a_2, \ldots, a_n$, namely $\overline{a}$, is equal to zero, that their sum is also zero, and that

$$\sum_{i=1}^{n} a_i^2 = \sum_{i=1}^{n} (a_i - \overline{a})^2 = 1.$$

Since the test statistic W is the square of a correlation coefficient, W always takes values between 0 and 1. The closer the value of the test statistic is to 1, the more the observed quantiles are in agreement with the theoretical quantiles. In other words, the closer the value of W is to 1, the more the data tends towards normality. The larger the difference between the value of W and the value 1, the more the data supports the alternative hypothesis that the population studied is not normally distributed.

To decide whether the null hypothesis can be rejected, we can compare the computed test statistic w with critical values $c_{\alpha,n}$. There are tables of critical values for different values of n and α. A detailed table of critical values for the Shapiro–Wilk test is given in Appendix F. If $w < c_{\alpha,n}$, then we reject the null hypothesis of normality at a significance level α. Otherwise, we cannot reject it.

One can also calculate approximate p-values for the Shapiro–Wilk test. The calculation method is based on a transformation of the original test statistic W. The transformation, proposed by Royston, is extremely complex and depends on the number of observations in the sample. Therefore, we do not go deeper into the calculation of the p-value. As always, small p-values point in the direction of the alternative hypothesis, while large p-values lead to an acceptance of the null hypothesis, which states that the population under study is normally distributed.

If you want to formally test a data set for normality, then JMP automatically performs the Shapiro–Wilk test when the number of observations is smaller than or equal to 2000. To perform the test, you first have to select "Continuous Fit", followed by "Normal", as explained earlier (see also Figure 6.17). You will then obtain an output like the one shown in Figure 6.18. If, in that output, you click on the hotspot (red triangle) next to the term "Fitted Normal", the option "Goodness of Fit" appears. This is illustrated in Figure 6.21.

Note that, in the menu in Figure 6.21, there is another option, "Diagnostic Plot". This option generates an alternative quantile diagram. In addition, the option "Save Fitted

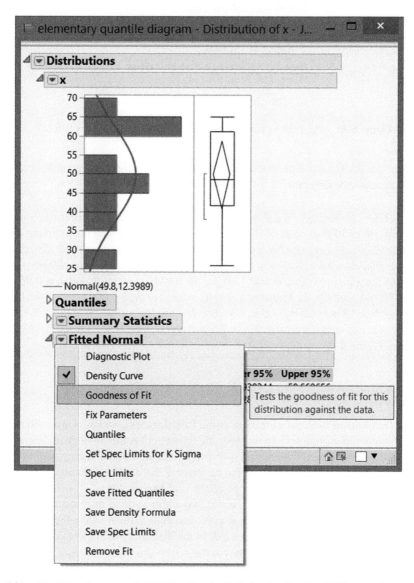

Figure 6.21 The "Goodness of Fit" option in JMP for the Shapiro–Wilk test when $n \leq 2000$, and for the Lilliefors test when $n > 2000$.

Goodness-of-Fit Test

Shapiro-Wilk W Test

W	Prob<W
0.700085	<.0001*

Note: Ho = The data is from the Normal distribution. Small p-values reject Ho.

(a) Breaking strength

Goodness-of-Fit Test

Shapiro-Wilk W Test

W	Prob<W
0.965386	0.1494

Note: Ho = The data is from the Normal distribution. Small p-values reject Ho.

(b) Ln(Breaking strength)

Figure 6.22   The JMP outputs for the Shapiro–Wilk tests in Example 6.2.3.

Quantiles" adds a new column to the data table, with the theoretical quantiles used for the creation of the quantile diagram.

Example 6.2.3   *In Example 6.2.2, two quantile diagrams were constructed, one for the breaking strength of 50 fibers and another for the natural logarithm of the breaking strengths. The quantile diagrams suggested that the breaking strength is not normally distributed, but that its logarithm does seem to be normally distributed. When we use the Shapiro–Wilk test, we obtain the test statistic values w = 0.7001 for the breaking strength and w = 0.9654 for its logarithm. The p-value for the breaking strength is smaller than 0.0001, so we have to reject the null hypothesis of normality for that variable. For the logarithm, the p-value is 0.1494, so that we can accept the null hypothesis of normality for it. The JMP outputs for the two tests are shown in Figure 6.22.*

The Lilliefors Test

One of the best-known tests for normality is the Lilliefors test, which is a modified version of the Kolmogorov–Smirnov test. In this test, the theoretical cumulative distribution function (which is assumed in the null hypothesis) is directly compared with the empirical or observed cumulative distribution function; that is, the cumulative distribution function of the data (see *Statistics with JMP: Graphs, Descriptive Statistics and Probability*).

The test statistic for the Lilliefors test is the maximum vertical distance between the two curves. If this maximum vertical distance exceeds a certain critical value, the null hypothesis that the data is normally distributed is rejected. In mathematical terms, the test statistic of the Lilliefors test and the Kolmogorov–Smirnov test is

$$D = \max |F^*(x) - F_X(x)|,$$

where $F^*(x)$ represents the empirical cumulative distribution function, and $F_X(x)$ is the theoretical cumulative distribution function. If we want to test for normality, the theoretical cumulative distribution function is, of course, the one corresponding to the normal density.

If the test statistic D takes a small value, then this suggests that the data could come from a normally distributed population. This makes sense: if the maximum vertical distance between the two functions is small, it means that, over the entire domain of the functions, the vertical distance is small. In that case, the two functions are very close to each other over their entire domain. The distribution of the data then looks very much like the distribution of normal data.

If, however, the test statistic D takes a large value, then the data probably does not come from a normally distributed population. The critical values to determine whether the test statistic is sufficiently large for the null hypothesis to be rejected are available in tables in the literature.

The difference between the Lilliefors test and the Kolmogorov–Smirnov test is that, for testing the normal distribution, Lilliefors has reduced the critical values of Kolmogorov and Smirnov, so that the null hypothesis of normality is rejected more easily. In fact, the Lilliefors test corrects the Kolmogorov–Smirnov test by taking into account the fact that the parameters of the theoretical normal distribution must be estimated from the data.

If you want to formally test the normality of a data set with more than 2000 observations for normality, then JMP automatically switches from the Shapiro–Wilk test to the Lilliefors test if you select the "Goodness of Fit" option in the hotspot next to "Fitted Normal" (see Figure 6.21). Figure 6.23 shows the result of the Lilliefors test in JMP, abbreviated to "KSL" (Kolmogorov–Smirnov–Lilliefors), for two artificial data sets with 3000 pseudo-random numbers from, on the one hand, a t-distribution with five degrees of freedom and, on the other, a standard normal distribution.

The calculations needed for performing the Kolmogorov–Smirnov test and the Lilliefors test are nicely summarized in the *Handbook of Parametric and Nonparametric Statistical Procedures* by David J. Sheskin. The calculation of p-values for the Lilliefors test is complex, and we do not pursue this further.

The Jarque–Bera Test

The last hypothesis test that we will mention here for testing normality is the Jarque–Bera test. This test is based on the third and fourth sample moments of the variable under study. The third and fourth moments contain information about the skewness and the kurtosis of the variable. The Jarque–Bera test compares the skewness and kurtosis of the sample data with those of the normal distribution that is assumed in the null hypothesis. The Jarque–Bera test is not available in JMP, because the test has a very low power in some cases.

6.2.2 Other Continuous Densities

It is not just the normal density for which quantile diagrams can be constructed. Via the option "Continuous Fit", JMP allows you to find the lognormal, gamma, beta, Weibull, or exponential distribution that best fits your data sample. JMP is able to estimate the parameters of these distributions, add the corresponding probability density to the histogram, and perform formal hypothesis tests for all these distributions.

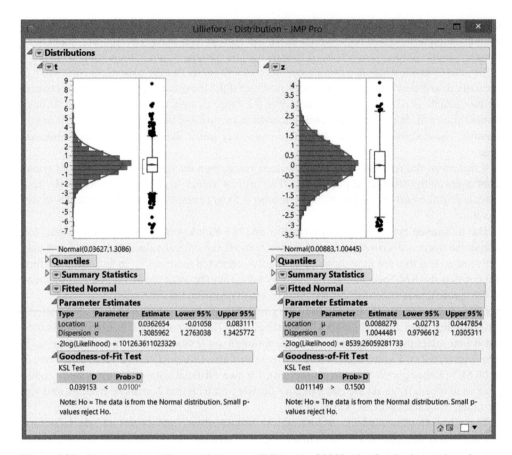

Figure 6.23 The JMP output for a Lilliefors test (KSL test) of 3000 (simulated) observations from a t-distribution with five degrees of freedom (left), and a standard normal distribution (right).

JMP does not use the same test for all distributions: for some distributions it uses the Kolmogorov–Smirnov test, while the Cramér – von Mises test is used for other distributions. In some cases, JMP uses the Kolmogorov–Smirnov test even for discrete probability distributions.

6.3 Discussion

The Shapiro–Wilk test in combination with the p-values calculated using the methods of Royston, is, according to many researchers, the most powerful normality test. Unfortunately, even the power of the Shapiro–Wilk test is disappointingly low for small samples. In other words, the Shapiro–Wilk test will rarely reject the null hypothesis, even if that hypothesis is blatantly wrong. As a consequence, it makes no sense to perform normality tests for small samples.

Many researchers still perform normality tests using small samples, to "prove" that their data points come from a normally distributed population. However, the low power of the normality tests means that this method has absolutely no corroborating value. Therefore, a normality test cannot be used as a justification for the use of z-tests and t-tests for a small number of observations. For small samples, therefore, the use of nonparametric tests is recommended.

A good process to determine which approach to use for testing hypotheses is as follows:

- If $n < 15$, use nonparametric hypothesis tests, unless the probability distribution or probability density of the underlying population is known with certainty.
- If $15 \leq n < 30$, use the Shapiro–Wilk test to determine whether the data comes from a normally distributed population. If the null hypothesis of normality cannot be rejected, you can use a t-test for the population mean or a χ^2-test for the population variance, or one of the related parametric tests for the comparison of two or more population means or population variances.
- If $n \geq 30$, then, due to the central limit theorem, the t-test can be used for testing a population mean. The χ^2-test for a population variance still necessitates a test for normality, since it requires the individual observations to be normally distributed, so that the test statistic is χ^2-distributed.

Finally, it is always useful to look at histograms, box plots, and quantile diagrams of every variable under investigation. These figures help us to identify outliers (extremely large or extremely small observations). If outliers are present, they sometimes influence the sample mean and the sample variance very strongly. A nonparametric hypothesis test is therefore preferable in cases where the data contains one or more outliers.

Part Three

Two Populations

7

Independent Versus Paired Samples

Her paper began by demolishing, somehow, the assumption that beards were more 'natural' or easier to maintain than clean-shavenness – she actually published statistics from Gillette's research department comparing the amount of time that bearded and beardless men spent in the bathroom each day, proving that the difference was not statistically significant.

(from *Cryptonomicon*, Neal Stephenson, p. 94)

In this part, we study hypothesis tests for comparing two populations. The goal of hypothesis tests for two populations is to discover similarities or differences between two populations or processes, and to measure them.

The most general hypothesis for two populations examines the equality or inequality of their two probability distributions or densities. If F_1 and F_2 are the cumulative distribution functions of the two populations, then the null hypothesis and alternative hypothesis of this general test can be written as

$$H_0 : F_1 = F_2$$

and

$$H_a : F_1 \neq F_2.$$

Acceptance of the null hypothesis of this test implies that the means, medians, variances, and any other populations statistics of both populations are not statistically different. In practice, this general hypothesis test for two distribution functions is rarely used. Researchers typically have more concrete questions, and therefore use more specific hypothesis tests. In this part, we discuss hypothesis tests for comparing means, variances, proportions, and medians of two populations instead of their distributions.

There are two possible scenarios in the investigation of two populations. It is important for a researcher to recognize which scenario applies to his work. In the first scenario, the data is collected using two **independent samples** (sometimes, one speaks of nonpaired observations). The second scenario involves so-called dependent samples or **paired observations**.

Statistics with JMP: Hypothesis Tests, ANOVA and Regression, First Edition. Peter Goos and David Meintrup.
© 2016 John Wiley & Sons, Ltd. Published 2016 by John Wiley & Sons, Ltd.
Companion Website: http://www.wiley.com/go/goosandmeintrup/JMP

We illustrate the distinction by means of two examples. Example 7.0.1 involves independent samples, whereas Example 7.0.2 illustrates a scenario with paired observations.

Example 7.0.1   *To test the wear of a new type of shoe sole, a number of subjects are recruited. Half of the subjects receive a pair of shoes with the new soles. The other half of the subjects receive a pair of shoes with conventional soles. After the test period, all shoes are collected and the wear of the soles of both types is quantified and compared.*

Example 7.0.2   *To test the wear of a new type of shoe sole, a number of subjects are recruited. All subjects receive one shoe with the new sole, and one shoe with the conventional sole. Each subject receives one right shoe and one left shoe. After the test period, all shoes are collected and the wear of the soles of both types is quantified and compared.*

The advantage of giving each subject two different shoes, as in Example 7.0.2, is that we know that they are used to the same extent. The wear of the two soles can therefore be compared directly. For such an experiment, the term **paired observations** is used, because this approach leads to pairs of observations: each observation from one data set (the sample data on the new soles) is inseparably linked to one observation from the second data set (the sample data on the old soles).

If the setup of Example 7.0.1 is used, then any difference in wear between each pair of soles can be due to a difference in wear resistance, but also to a difference in the use of the soles. In fact, there is no guarantee that the new and the old soles will have been used exactly to the same extent during the trial period. Maybe, one subject (wearing one kind of shoe) will have used the shoes more often than another (wearing another kind of shoe). Also, one subject might have a greater weight than another, so that the soles of that subject suffer more and exhibit more wear. In short, when using independent samples, any differences in wear between the two types of soles are also affected by factors other than the type of sole.

It is extremely important that a researcher recognizes whether his or her observations are paired. This is because statistical hypothesis tests for paired data have a larger power than those for independent observations. The following example, which is described in the book *Statistics for Experiments: Design, Innovation, and Discovery* by G.E.P. Box, J.S. Hunter and W.G. Hunter, illustrates, by means of a few simple graphs, why statistical techniques based on paired observations are more powerful than techniques based on independent observations.

Example 7.0.3   *Table 7.1 shows the data from an experiment with 10 subjects, each testing two types of shoes by wearing a shoe sole with Material A on one foot and a shoe sole with Material B on the other. In total, there are 20 measurements of the wear of shoes.*

Figure 7.1 provides a first graphical representation of the 20 measurements of wear. In the figure, the 10 measurements for Material A are indicated with white circles, while the 10 measurements for Material B are indicated by black circles. On the vertical axis of the figure, the wear is displayed. If we compare the locations of the white and black circles, we do not see a systematic difference between the two materials. Based on this figure, there is no evidence that one material is more wear resistant than the other.

A weakness of Figure 7.1 is that it does not consider the paired nature of the observations. Figure 7.2 corrects this by putting the 10 subjects on the horizontal axis. This figure allows us to compare the wear of Materials A and B for each subject separately. For eight of the

Table 7.1   The data from an experiment with paired observations on the wear resistance of shoe soles. The numbers in columns "Material A" and "Material B" are measurements for the amount of wear.

Person	Material A	Material B	Difference	Foot with Material A	Foot with Material B
1	13.2	14.0	−0.8	Left	Right
2	8.2	8.8	−0.6	Left	Right
3	10.9	11.2	−0.3	Right	Left
4	14.3	14.2	0.1	Left	Right
5	10.7	11.8	−1.1	Right	Left
6	6.6	6.4	0.2	Left	Right
7	9.5	9.8	−0.3	Left	Right
8	10.8	11.3	−0.5	Left	Right
9	8.8	9.3	−0.5	Right	Left
10	13.3	13.6	−0.3	Left	Right

Source: Box et al., 2005. Reproduced with permission of Wiley.

10 *subjects, the black circle appears higher than the white circle, indicating that Material B shows more wear than Material A for eight of the 10 subjects. For the two remaining subjects, subjects 4 and 6, Material A wears off more than Material B. However, Figure 7.2 clearly shows that the difference in favor of Material B for these two subjects is minimal. In short, Figure 7.2 graphically provides strong evidence for the fact that Material A is more wear resistant.*

Figure 7.2 not only shows that Material A wears off to a lesser extent, but also that some subjects' shoes exhibit more wear than other subjects' shoes. The reason for this is that some

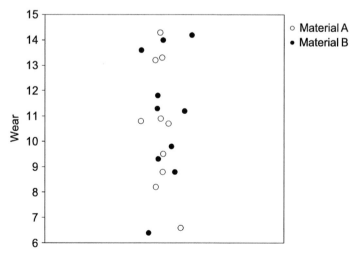

Figure 7.1   A first graphical representation of the 20 measurements of the wear of shoe soles in Example 7.0.3.

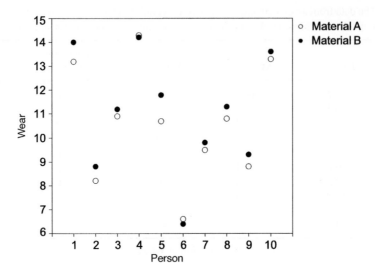

Figure 7.2 A second graphical representation of the 20 measurements of the wear of shoe soles in Example 7.0.3.

subjects use their shoes more intensively than others and/or have a greater weight. From Figure 7.2, it is clear that both shoes of subjects 1, 4, and 10 suffer most, while the two shoes of subject 6 suffer least. This difference in the testing conditions from subject to subject is the cause of Figure 7.1's failure to show a difference in performance between the two types of soles.

The following chapters describe hypothesis tests for two independent samples and for paired observations. In Chapters 8 and 9, we discuss tests for independent samples, while Chapters 10 and 11 deal with tests for paired samples.

8

Hypothesis Tests for the Means, Proportions, or Variances of Two Independent Samples

Intense. Intense. Right? But we did it. Our overall average is at 93, which is normally not so good, but okay considering the new systems and increased flow. Congratulations.

(from The Circle, Dave Eggers, p. 149)

The goal of hypothesis tests for two populations is to discover and measure similarities or differences between two populations or processes. We have already mentioned at the beginning of Chapter 7 that the most general hypothesis for two populations examines the equality or inequality of their two probability distributions or densities. If F_1 and F_2 are the cumulative distribution functions of the two populations, then the null hypothesis and alternative hypothesis of this general test can be written as

$$H_0 : F_1 = F_2$$

and

$$H_a : F_1 \neq F_2.$$

Researchers, however, typically use more specific hypothesis tests, because they are specifically interested in comparing means, locations, proportions, or variances. This chapter, which deals with parametric tests, therefore covers hypothesis tests for the comparison of the means, variances, and proportions of two populations.

8.1 Tests for Two Population Means for Independent Samples

When comparing two population means, researchers are generally interested in whether or not the means are equal. The null hypothesis tested is then

$$H_0 : \mu_1 = \mu_2,$$

Statistics with JMP: Hypothesis Tests, ANOVA and Regression, First Edition. Peter Goos and David Meintrup.
© 2016 John Wiley & Sons, Ltd. Published 2016 by John Wiley & Sons, Ltd.
Companion Website: http://www.wiley.com/go/goosandmeintrup/JMP

where μ_1 represents the population mean of the first population under investigation, and μ_2 the population mean of the second population under investigation. A right-tailed alternative hypothesis is

$$H_a : \mu_1 > \mu_2,$$

while a left-tailed and a two-tailed alternative hypothesis can be written as

$$H_a : \mu_1 < \mu_2$$

and

$$H_a : \mu_1 \neq \mu_2,$$

respectively. To test these hypotheses, the sample means $\overline{X}_1$ of a first sample (with data from the first population) and $\overline{X}_2$ of a second sample (with data from the second population) are compared with each other. The number of observations is denoted by n_1 for the first sample and by n_2 for the second sample. These numbers of observations do not need to be the same. The population variances for the two populations under study are indicated by σ_1^2 and σ_2^2.

8.1.1   The Starting Point

First, it is important to realize that the hypotheses tested can be rewritten as

$$H_0 : \mu_1 - \mu_2 = 0,$$
$$H_a : \mu_1 - \mu_2 > 0,$$
$$H_a : \mu_1 - \mu_2 < 0,$$

and

$$H_a : \mu_1 - \mu_2 \neq 0.$$

In order to decide whether the above null hypothesis should be rejected, the difference $\mu_1 - \mu_2$ has to be estimated and the resulting estimate should be compared to zero. Since, independently of the probability distribution or probability density of the populations under study, $\overline{X}_1$ is an unbiased estimator of μ_1 and $\overline{X}_2$ is an unbiased estimator of μ_2, the difference $\overline{X}_1 - \overline{X}_2$ is an unbiased estimator of $\mu_1 - \mu_2$. Indeed, because $E(\overline{X}_1) = \mu_1$ and $E(\overline{X}_2) = \mu_2$, we know that $E(\overline{X}_1 - \overline{X}_2) = \mu_1 - \mu_2$: the expected value of a difference of random variables is equal to the difference of their expected values (see the book *Statistics with JMP: Graphs, Descriptive Statistics and Probability*).

As it is also possible to determine the variance and the probability density of $\overline{X}_1 - \overline{X}_2$, the decision rules for hypothesis tests for two population means are based on the difference between the two sample means.

The variance of the estimator $\overline{X}_1 - \overline{X}_2$ is

$$\text{var}(\overline{X}_1 - \overline{X}_2) = \frac{\sigma_1^2}{n_1} + \frac{\sigma_2^2}{n_2},$$

because $\mathrm{var}(\overline{X}_1) = \sigma_1^2/n_1$ and $\mathrm{var}(\overline{X}_2) = \sigma_2^2/n_2$, and because the two samples are independent. The independence of the samples ensures that $\overline{X}_1$ and $\overline{X}_2$ are independent random variables. Since the variance of a difference of independent random variables is equal to the sum of their variances, we know that the variance $\mathrm{var}(\overline{X}_1 - \overline{X}_2)$ is equal to $\mathrm{var}(\overline{X}_1) + \mathrm{var}(\overline{X}_2)$ (see *Statistics with JMP: Graphs, Descriptive Statistics and Probability*).

If the populations under investigation are normally distributed, then, of course, the sample means $\overline{X}_1$ and $\overline{X}_2$ are normally distributed as well. If the populations under study are not normally distributed, but n_1 and n_2 are at least 30, then $\overline{X}_1$ and $\overline{X}_2$ are approximately normally distributed. In both cases, we can assume that $\overline{X}_1 - \overline{X}_2$ is normally distributed, with expected value $\mu_1 - \mu_2$ and variance

$$\frac{\sigma_1^2}{n_1} + \frac{\sigma_2^2}{n_2}.$$

This follows from the fact that a difference between normally distributed random variables is also normally distributed (see *Statistics with JMP: Graphs, Descriptive Statistics and Probability*).

If the populations under study are known to have a distribution other than the normal and n_1 and/or n_2 are smaller than 30, then it is no longer safe to assume that $\overline{X}_1 - \overline{X}_2$ is approximately normally distributed. It is then better to use one of the nonparametric hypothesis tests from Chapter 9.

When the probability distributions or densities of the populations studied are unknown (which is often the case), and at least one of the two samples contains fewer than 15 observations, we should again not rely on the normal distribution for $\overline{X}_1 - \overline{X}_2$. In that case, it is also better to use one of the nonparametric hypothesis tests from Chapter 9 (see also Section 6.3). If the smaller sample contains between 15 and 30 observations, the normality of the data in that sample should be tested. If the second sample also contains between 15 and 30 observations, the same should be done for the second sample. If the normality test(s) suggest(s) that the small sample(s) actually do(es) originate from a normally distributed population, it is appropriate to assume that $\overline{X}_1 - \overline{X}_2$ is normally distributed. If not, then one of the nonparametric hypothesis tests from Chapter 9 should be used.

In the next few sections, we study various scenarios in which we can rely on the normal distribution of $\overline{X}_1$ and $\overline{X}_2$ when testing two population means. First, we look at the scenario in which the population variances σ_1^2 and σ_2^2 are known. Next, we study two situations in which the population variances σ_1^2 and σ_2^2 are unknown.

8.1.2 Known Variances σ_1^2 and σ_2^2

A realistic scenario in which the population variances σ_1^2 and σ_2^2 can be considered known is when a lot of historical data is available from the populations under study. In that case, it is sometimes better to determine σ_1^2 and σ_2^2 based on this historical data than to estimate these variances from a new, small, sample. If the variances σ_1^2 and σ_2^2 express the precision of machines, calibrated measurement equipment, or standardized measurement procedures, then it can also happen that σ_1^2 and σ_2^2 are known. This is because manufacturers of machinery and developers of measurement equipment often conduct many tests to get a good picture of the variability in the performance of the machinery and the equipment. In most studies, however, the population variances have be estimated from a new data sample, due to the lack of reliable historical data or information concerning performance variability in the machinery and equipment utilized.

In this section, we assume that σ_1^2 and σ_2^2 are known and that both $\overline{X}_1$ and $\overline{X}_2$ are (at least approximately) normally distributed. There are two possible reasons why $\overline{X}_1$ and $\overline{X}_2$ may be normally distributed: (i) the populations under investigation are themselves normally distributed; and/or (ii) each of the data samples from the two populations contain at least 30 observations. In any case, if $\overline{X}_1$ and $\overline{X}_2$ are at least approximately normally distributed, then $\overline{X}_1 - \overline{X}_2$ is also at least approximately normally distributed, namely with expected value $\mu_1 - \mu_2$ and variance

$$\frac{\sigma_1^2}{n_1} + \frac{\sigma_2^2}{n_2}.$$

Consequently,

$$\frac{\overline{X}_1 - \overline{X}_2 - (\mu_1 - \mu_2)}{\sqrt{\dfrac{\sigma_1^2}{n_1} + \dfrac{\sigma_2^2}{n_2}}}$$

is at least approximately standard normally distributed.

If the null hypothesis $H_0 : \mu_1 - \mu_2 = 0$ is correct, then we can say that

$$\frac{\overline{X}_1 - \overline{X}_2}{\sqrt{\dfrac{\sigma_1^2}{n_1} + \dfrac{\sigma_2^2}{n_2}}} \tag{8.1}$$

is also at least approximately standard normally distributed. Due to this result, the test statistic used for two population means with known variances is

$$\frac{\overline{X}_1 - \overline{X}_2}{\sqrt{\dfrac{\sigma_1^2}{n_1} + \dfrac{\sigma_2^2}{n_2}}}.$$

Based on this test statistic, decision rules for the various hypothesis tests for two population means can be developed.

The Right-Tailed Test

For a right-tailed hypothesis test, the null hypothesis $H_0 : \mu_1 - \mu_2 = 0$ is rejected (and the alternative hypothesis $H_a : \mu_1 - \mu_2 > 0$ accepted) if the first computed sample mean $\bar{x}_1$ is considerably larger than the second computed sample mean $\bar{x}_2$. In other words, the null hypothesis is rejected if $\bar{x}_1 - \bar{x}_2$ is considerably larger than zero. In order to decide whether $\bar{x}_1 - \bar{x}_2$ is sufficiently larger than zero, we need to determine a critical value c, as in Section 3.2.1. This time, the critical value is based on the normal distribution of the difference in sample means, instead of the normal distribution of a single sample mean.

Following the same kind of reasoning as for the right-tailed test in Section 3.2.1, we obtain the following critical value:

$$c = z_\alpha \sqrt{\frac{\sigma_1^2}{n_1} + \frac{\sigma_2^2}{n_2}}.$$

This leads to the following decision rule for the right-tailed test for the difference in population means $\mu_1 - \mu_2$:

- Reject the null hypothesis H_0 and accept the alternative hypothesis H_a if

$$\bar{x}_1 - \bar{x}_2 > z_\alpha \sqrt{\frac{\sigma_1^2}{n_1} + \frac{\sigma_2^2}{n_2}}.$$

- Accept the null hypothesis if

$$\bar{x}_1 - \bar{x}_2 \leq z_\alpha \sqrt{\frac{\sigma_1^2}{n_1} + \frac{\sigma_2^2}{n_2}}.$$

In practice, one generally works with the test statistic

$$Z = \frac{\bar{X}_1 - \bar{X}_2}{\sqrt{\frac{\sigma_1^2}{n_1} + \frac{\sigma_2^2}{n_2}}}$$

rather than with the difference in sample means $\bar{X}_1 - \bar{X}_2$. If the value computed for the test statistic based on the observed sample data,

$$z = \frac{\bar{x}_1 - \bar{x}_2}{\sqrt{\frac{\sigma_1^2}{n_1} + \frac{\sigma_2^2}{n_2}}},$$

is larger than the critical value z_α, then the null hypothesis is rejected. Otherwise, it is accepted. This decision rule, based on the critical value z_α for the test statistic, always leads to the very same results as the decision rule involving the critical value c for the difference in means.

The p-value of the right-tailed test is

$$p = P\left(Z > \frac{\bar{x}_1 - \bar{x}_2}{\sqrt{\dfrac{\sigma_1^2}{n_1} + \dfrac{\sigma_2^2}{n_2}}} \right) = P(Z > z).$$

The p-value is the probability that a standard normally distributed random variable Z exceeds the computed test statistic z. The null hypothesis is rejected if the p-value is smaller than the significance level α.

The Left-Tailed Test

For a left-tailed hypothesis test, the null hypothesis $H_0 : \mu_1 - \mu_2 = 0$ is rejected (and the alternative hypothesis $H_a : \mu_1 - \mu_2 < 0$ accepted) if $\bar{x}_1 - \bar{x}_2$ is strongly negative; that is, if that difference is smaller than the critical value

$$c = -z_\alpha \sqrt{\frac{\sigma_1^2}{n_1} + \frac{\sigma_2^2}{n_2}}.$$

Otherwise, the null hypothesis is accepted. Alternatively, the null hypothesis is rejected if the computed test statistic

$$z = \frac{\bar{x}_1 - \bar{x}_2}{\sqrt{\dfrac{\sigma_1^2}{n_1} + \dfrac{\sigma_2^2}{n_2}}}$$

is smaller than the critical value $-z_\alpha$.

The p-value of the left-tailed test is

$$p = P\left(Z < \frac{\bar{x}_1 - \bar{x}_2}{\sqrt{\dfrac{\sigma_1^2}{n_1} + \dfrac{\sigma_2^2}{n_2}}} \right) = P(Z < z).$$

The null hypothesis is rejected if this p-value is smaller than the significance level α.

The Two-Tailed Test

For a two-tailed hypothesis test, the null hypothesis $H_0 : \mu_1 - \mu_2 = 0$ is rejected (and the alternative hypothesis $H_a : \mu_1 - \mu_2 \neq 0$ accepted) if $\bar{x}_1 - \bar{x}_2$ is either strongly negative or strongly positive. To decide whether $\bar{x}_1 - \bar{x}_2$ is sufficiently negative, a left critical value c_L is needed. To decide whether $\bar{x}_1 - \bar{x}_2$ is sufficiently positive, a right critical value c_R is needed. These two critical values can be obtained using the same kind of reasoning as in Section 3.2.3. They can be written as

$$c_L = -z_{\alpha/2} \sqrt{\frac{\sigma_1^2}{n_1} + \frac{\sigma_2^2}{n_2}}$$

and

$$c_R = z_{\alpha/2} \sqrt{\frac{\sigma_1^2}{n_1} + \frac{\sigma_2^2}{n_2}}.$$

If the computed difference in sample means, $\bar{x}_1 - \bar{x}_2$, lies between these two values, then the null hypothesis is accepted. If $\bar{x}_1 - \bar{x}_2$ is smaller than c_L or larger than c_R, then the null hypothesis is rejected in favor of the alternative hypothesis.

Another way of putting this is that we reject the null hypothesis if the absolute value of $\bar{x}_1 - \bar{x}_2$ is larger than c_R. As usual, we can also express this decision rule in terms of the test statistic: we reject the null hypothesis if the absolute value of the computed test statistic

$$|z| = \left| \frac{\bar{x}_1 - \bar{x}_2}{\sqrt{\frac{\sigma_1^2}{n_1} + \frac{\sigma_2^2}{n_2}}} \right|$$

is larger than the critical value $z_{\alpha/2}$.

The p-value of the two-tailed test is

$$p = 2 P \left(Z > \left| \frac{\bar{x}_1 - \bar{x}_2}{\sqrt{\frac{\sigma_1^2}{n_1} + \frac{\sigma_2^2}{n_2}}} \right| \right) = 2 P(Z > |z|).$$

The p-value is thus twice the probability that a standard normally distributed random variable takes values larger than the absolute value of the computed test statistic. The null hypothesis is rejected if the p-value is smaller than the chosen significance level α.

A Generalized Hypothesis Test

In some cases, a researcher is interested in testing whether a difference between two population means is equal to a certain value Δ_0. In that case, the null hypothesis is

$$H_0 : \mu_1 - \mu_2 = \Delta_0,$$

and the possible alternative hypotheses may be written as

$$H_a : \mu_1 - \mu_2 > \Delta_0,$$
$$H_a : \mu_1 - \mu_2 < \Delta_0$$

and

$$H_a : \mu_1 - \mu_2 \neq \Delta_0.$$

If the null hypothesis is true, then the expected value of $\overline{X}_1 - \overline{X}_2$ equals Δ_0. If the populations under study are normally distributed and/or both samples are large enough, then $\overline{X}_1 - \overline{X}_2$ is (at least approximately) normally distributed with expected value Δ_0 and variance

$$\frac{\sigma_1^2}{n_1} + \frac{\sigma_2^2}{n_2}.$$

The test statistic then becomes

$$Z = \frac{\overline{X}_1 - \overline{X}_2 - \Delta_0}{\sqrt{\dfrac{\sigma_1^2}{n_1} + \dfrac{\sigma_2^2}{n_2}}}.$$

This statistic is standard normally distributed if the null hypothesis is correct.

Apart from the way in which the test statistic is calculated, nothing changes for the generalized hypothesis test. In particular, the decision rules based on the comparison of the test statistic with z_α, $-z_\alpha$ and $z_{\alpha/2}$ and on the p-value remain valid. However, the critical values c, c_L and c_R do change. For example, for the right-tailed test, the critical value c changes to

$$c = \Delta_0 + z_\alpha \sqrt{\frac{\sigma_1^2}{n_1} + \frac{\sigma_2^2}{n_2}}.$$

for the generalized test.

8.1.3 Unknown Variances σ_1^2 and σ_2^2

If the population variances σ_1^2 and σ_2^2 are unknown, then the test statistics

$$Z = \frac{\overline{X}_1 - \overline{X}_2}{\sqrt{\dfrac{\sigma_1^2}{n_1} + \dfrac{\sigma_2^2}{n_2}}}$$

and

$$Z = \frac{\overline{X}_1 - \overline{X}_2 - \Delta_0}{\sqrt{\dfrac{\sigma_1^2}{n_1} + \dfrac{\sigma_2^2}{n_2}}}$$

from the previous pages cannot be used. The variances σ_1^2 and σ_2^2 then have to be estimated from the sample data.

The most intuitive thing to do to this end is to estimate σ_1^2 using the sample variance S_1^2 of the first sample (from the first population), and σ_2^2 using the sample variance S_2^2 of the second sample (from the second population). As a matter of fact, sample variances are unbiased estimators of population variances. In the expressions for the test statistics, we can then replace σ_1^2 and σ_2^2 by S_1^2 and S_2^2.

However, this approach is not optimal if the population variances σ_1^2 and σ_2^2 are equal. Moreover, in the event that σ_1^2 and σ_2^2 are unequal, the resulting test statistics

$$\frac{\overline{X}_1 - \overline{X}_2}{\sqrt{\dfrac{S_1^2}{n_1} + \dfrac{S_2^2}{n_2}}}$$

and

$$\frac{\overline{X}_1 - \overline{X}_2 - \Delta_0}{\sqrt{\dfrac{S_1^2}{n_1} + \dfrac{S_2^2}{n_2}}}$$

have a probability density that we have not yet encountered.

We first study the scenario in which $\sigma_1^2 = \sigma_2^2$, and then look at the scenario in which the population variances σ_1^2 and σ_2^2 are unequal. In both cases, we first focus on the null hypothesis

$$H_0 : \mu_1 - \mu_2 = 0$$

and the corresponding right-tailed, left-tailed, and two-tailed alternative hypotheses.

To decide which scenario is relevant for a particular data analysis, it is usually necessary to conduct a hypothesis test to check whether both population variances can be considered to be equal. The required hypothesis test for two population variances is discussed in Section 8.3.2.

Equal but Unknown Variances

If the hypothesis test for comparing the two variances σ_1^2 and σ_2^2 indicates that the two variances are equal, we can replace σ_1^2 and σ_2^2 by a single symbol, σ^2. We call this variance the common variance because it is common to both populations under study. If $\overline{X}_1$ and $\overline{X}_2$ are at least approximately normally distributed, then $\overline{X}_1 - \overline{X}_2$ is also at least approximately normally distributed, namely with expected value $\mu_1 - \mu_2$ and variance

$$\frac{\sigma_1^2}{n_1} + \frac{\sigma_2^2}{n_2} = \frac{\sigma^2}{n_1} + \frac{\sigma^2}{n_2} = \sigma^2 \left(\frac{1}{n_1} + \frac{1}{n_2} \right).$$

This implies that

$$\frac{\overline{X}_1 - \overline{X}_2 - (\mu_1 - \mu_2)}{\sigma \sqrt{\dfrac{1}{n_1} + \dfrac{1}{n_2}}}$$

is standard normally distributed. If the null hypothesis $H_0 : \mu_1 - \mu_2 = 0$ is correct, then

$$\frac{\overline{X}_1 - \overline{X}_2}{\sigma \sqrt{\dfrac{1}{n_1} + \dfrac{1}{n_2}}} \tag{8.2}$$

is also standard normally distributed.

However, this result cannot be used immediately because the common variance σ^2 is unknown and has to be estimated. As σ^2 is the variance of both the first and the second population under study, we can estimate it using the data from the first sample as well as using the data from the second sample. However, we cannot simply merge the data from both samples and compute a single sample variance from the pooled data, because the population mean μ_1 may differ from the population mean μ_2.

The best method to estimate σ^2 is to compute the sample variances s_1^2 and s_2^2 separately from the first and the second samples. Then, we can combine these two estimates for σ^2. The best possible combination of the two sample variances s_1^2 and s_2^2 is the weighted mean

$$s_p^2 = \frac{(n_1 - 1)s_1^2 + (n_2 - 1)s_2^2}{n_1 + n_2 - 2}.$$

In this weighted mean, the weight $(n_1 - 1)/(n_1 + n_2 - 2)$ is assigned to the first sample variance, s_1^2, and the weight $(n_2 - 1)/(n_1 + n_2 - 2)$ is assigned to the second sample variance, s_2^2. In this way, the sample variance that is based on the largest number of observations has the largest weight. This makes sense because the precision of an estimator increases with the number of observations. This also applies to sample variances.

It can now be shown that the estimator

$$S_p^2 = \frac{(n_1 - 1)S_1^2 + (n_2 - 1)S_2^2}{n_1 + n_2 - 2}$$

is an unbiased estimator for σ^2, regardless of whether or not the null hypothesis is correct. Moreover, it can be shown that

$$\frac{(n_1 + n_2 - 2)S_p^2}{\sigma^2}$$

is χ^2-distributed with $n_1 + n_2 - 2$ degrees of freedom if the populations under study are normally distributed.

The estimator S_p^2 is called a **pooled estimator**. The adjective "pooled" means that the information of the two samples is combined to form a good estimator for the common variance σ^2.

Substituting σ by S_p in Equation (8.2) leads to the test statistic

$$T = \frac{\overline{X}_1 - \overline{X}_2}{S_p \sqrt{\dfrac{1}{n_1} + \dfrac{1}{n_2}}}.$$

This statistic is t-distributed with $n_1 + n_2 - 2$ degrees of freedom for samples from normally distributed populations and large samples from nonnormally distributed populations.

The Right-Tailed Test

For a right-tailed hypothesis test, the null hypothesis $H_0 : \mu_1 - \mu_2 = 0$ is rejected (and the alternative hypothesis $H_a : \mu_1 - \mu_2 > 0$ accepted) if the computed test statistic

$$t = \frac{\overline{x}_1 - \overline{x}_2}{s_p \sqrt{\dfrac{1}{n_1} + \dfrac{1}{n_2}}}$$

is larger than the critical value $t_{\alpha;n_1+n_2-2}$. The p-value of the right-tailed hypothesis test is

$$p = P\left(t_{n_1+n_2-2} > \frac{\overline{x}_1 - \overline{x}_2}{s_p \sqrt{\dfrac{1}{n_1} + \dfrac{1}{n_2}}} \right) = P(t_{n_1+n_2-2} > t).$$

The p-value is thus the probability that a t-distributed random variable with $n_1 + n_2 - 2$ degrees of freedom exceeds the computed test statistic. The null hypothesis is rejected if the p-value is smaller than the significance level α.

The Left-Tailed Test

For a left-tailed hypothesis test, the null hypothesis $H_0 : \mu_1 - \mu_2 = 0$ is rejected (and the alternative hypothesis $H_a : \mu_1 - \mu_2 < 0$ accepted) if the computed test statistic

$$t = \frac{\overline{x}_1 - \overline{x}_2}{s_p \sqrt{\dfrac{1}{n_1} + \dfrac{1}{n_2}}}$$

is smaller than the critical value $-t_{\alpha;n_1+n_2-2}$. The p-value of the left-tailed test is

$$p = P\left(t_{n_1+n_2-2} < \frac{\bar{x}_1 - \bar{x}_2}{s_p\sqrt{\dfrac{1}{n_1} + \dfrac{1}{n_2}}}\right) = P(t_{n_1+n_2-2} < t).$$

The p-value is thus the probability that a t-distributed random variable with $n_1 + n_2 - 2$ degrees of freedom is smaller than the computed test statistic. The null hypothesis is rejected if the p-value is smaller than the significance level α.

The Two-Tailed Test

For a two-tailed hypothesis test, the null hypothesis $H_0 : \mu_1 - \mu_2 = 0$ is rejected (and the alternative hypothesis $H_a : \mu_1 - \mu_2 \neq 0$ accepted) if the absolute value of the computed test statistic

$$|t| = \left|\frac{\bar{x}_1 - \bar{x}_2}{s_p\sqrt{\dfrac{1}{n_1} + \dfrac{1}{n_2}}}\right|$$

is larger than the critical value $t_{\alpha/2;n_1+n_2-2}$. The p-value of the test is

$$p = 2\,P\left(t_{n_1+n_2-2} > \left|\frac{\bar{x}_1 - \bar{x}_2}{s_p\sqrt{\dfrac{1}{n_1} + \dfrac{1}{n_2}}}\right|\right) = 2\,P(t_{n_1+n_2-2} > |t|).$$

The p-value is thus twice the probability that a t-distributed random variable with $n_1 + n_2 - 2$ degrees of freedom will take a value larger than the absolute value of the computed test statistic. The null hypothesis is rejected if the p-value is smaller than the significance level α.

A Generalized Hypothesis Test

If one is interested in testing whether the difference between two population means is equal to a given number Δ_0, the test statistic is

$$T = \frac{\bar{X}_1 - \bar{X}_2 - \Delta_0}{s_p\sqrt{\dfrac{1}{n_1} + \dfrac{1}{n_2}}}.$$

Apart from the computation of the test statistic, nothing changes in the execution of the test. In particular, the decision rules listed above remain valid.

Example 8.1.1 *Two machines are supposed to produce identical steel cables. A sample of $n_1 = 20$ observations from the first machine has a mean thickness of $\bar{x}_1 = 50.9033$ μm, and a sample variance of $s_1^2 = 0.1848$ μm^2. A sample of $n_2 = 20$ observations from the second machine results in a mean of $\bar{x}_2 = 50.2525$ μm with sample variance $s_2^2 = 0.2837$ μm^2. We test whether the mean thicknesses of the cables originating from the two machines are different. The appropriate hypotheses to test are*

$$H_0 : \mu_1 = \mu_2$$

and

$$H_a : \mu_1 \neq \mu_2.$$

Because the sample variances are not too different from each other (they differ by a factor of less than 2), we assume for convenience that σ_1^2 and σ_2^2 are equal. We also assume that the data is normally distributed.
  The pooled variance is

$$s_p^2 = \frac{(20 - 1)(0.1848) + (20 - 1)(0.2837)}{20 + 20 - 2} = 0.2342,$$

so that $s_p = \sqrt{0.2342} = 0.4840$. The computed test statistic is

$$t = \frac{50.9033 - 50.2525}{0.4840\sqrt{\frac{1}{20} + \frac{1}{20}}} = 4.2521,$$

so that the p-value is

$$p = 2\, P(t_{38} > 4.2521) = 0.000133.$$

This p-value can be calculated directly in JMP using the formula "$2 * (1 - t$ Distribution$(4.2521, 38))$".

The scripts "Hypothesis Test for Two Means Using Summary Statistics" and "Hypothesis Test for Two Means Using Data" are available in JMP to perform hypothesis tests for two population means. Figure 8.1 illustrates the use of the former script for Example 8.1.1. The script provides a rounded p-value of 0.0001.
  Note also that you can specify in the script whether you want a test based on the normal distribution (when σ^2 is known) or based on the t-distribution (when σ^2 is not known). If

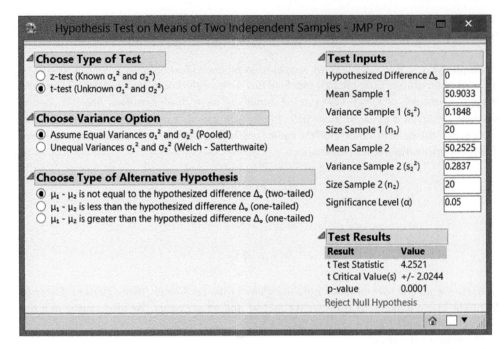

Figure 8.1   The use of the JMP script "Hypothesis Test for Two Means Using Summary Statistics" for Example 8.1.1.

you opt for a t-test, you can choose between a test that assumes equal variances and one that assumes unequal variances.

Unknown and Unequal Variances

The inequality of the variances of the two populations under study complicates the test for the equality of two population means considerably. As before, the starting point for the hypothesis test in the event of unequal variances is the random variable

$$\frac{\overline{X}_1 - \overline{X}_2}{\sqrt{\dfrac{\sigma_1^2}{n_1} + \dfrac{\sigma_2^2}{n_2}}},$$

which, under the null hypothesis $H_0 : \mu_1 - \mu_2 = 0$, has a standard normal distribution, assuming that the populations are normally distributed and/or the data samples are large enough to rely on the central limit theorem.

If the variances σ_1^2 and σ_2^2 are not equal, then the sample variance S_1^2 of the first sample is used as an estimator for σ_1^2, and the sample variance S_2^2 of the second sample is used as estimator for σ_2^2. Both sample variances are unbiased estimators of the corresponding population variances.

Replacing σ_1^2 and σ_2^2 by the estimators S_1^2 and S_2^2, respectively, in the above expression yields

$$\frac{\overline{X}_1 - \overline{X}_2}{\sqrt{\dfrac{S_1^2}{n_1} + \dfrac{S_2^2}{n_2}}},$$

which is used as test statistic. The probability density of this random variable is called the Behrens–Fisher density. This is a complex density. Ideally, we would determine critical values and p-values accordingly, but this is a difficult task.

To circumvent this problem, the Behrens–Fisher probability density is approximated by a t-distribution with v degrees of freedom. With regard to determining the number of degrees of freedom v of the required t-distribution, several formulas can be found in the statistical literature. One approach, named after the researchers Welch and Satterthwaite, involves setting v equal to

$$\frac{\left(\dfrac{s_1^2}{n_1} + \dfrac{s_2^2}{n_2}\right)^2}{\dfrac{(s_1^2/n_1)^2}{n_1 - 1} + \dfrac{(s_2^2/n_2)^2}{n_2 - 1}}. \tag{8.3}$$

In most cases, the resulting number of degrees of freedom v is not an integer. While, so far, we have only encountered t-distributions with integer numbers of degrees of freedom, the t-distribution also exists when the number of degrees of freedom is not an integer. Critical values and p-values based t-distributions with noninteger numbers of degrees of freedom can be calculated easily using JMP. If you only have access to a table of percentiles or quantiles for t-distributions with integer numbers of degrees of freedom (such as the one in Appendix D), you have to round v to the nearest integer.

Apart from the way in which the number of degrees of freedom is determined and the way in which the test statistic is computed, performing the hypothesis test for two population means assuming unequal variances is completely analogous to performing the test assuming equal variances.

The Right-Tailed Test

For a right-tailed hypothesis test, the null hypothesis $H_0 : \mu_1 - \mu_2 = 0$ is rejected (and the alternative hypothesis $H_a : \mu_1 - \mu_2 > 0$ accepted) if the computed test statistic

$$t = \frac{\overline{x}_1 - \overline{x}_2}{\sqrt{\dfrac{s_1^2}{n_1} + \dfrac{s_2^2}{n_2}}}$$

is larger than the critical value $t_{\alpha;v}$.

The *p*-value of the right-tailed test is

$$
p = P\left(t_\nu > \dfrac{\bar{x}_1 - \bar{x}_2}{\sqrt{\dfrac{s_1^2}{n_1} + \dfrac{s_2^2}{n_2}}}\right) = P(t_\nu > t).
$$

The *p*-value is thus the probability that a *t*-distributed random variable with ν degrees of freedom exceeds the computed test statistic. The null hypothesis is rejected if the *p*-value is smaller than the significance level α.

The Left-Tailed Test

For a left-tailed hypothesis test, the null hypothesis $H_0 : \mu_1 - \mu_2 = 0$ is rejected (and the alternative hypothesis $H_a : \mu_1 - \mu_2 < 0$ accepted) if the computed test statistic

$$
t = \dfrac{\bar{x}_1 - \bar{x}_2}{\sqrt{\dfrac{s_1^2}{n_1} + \dfrac{s_2^2}{n_2}}}
$$

is smaller than the critical value $-t_{\alpha;\nu}$.

The *p*-value for the left-tailed test is

$$
p = P\left(t_\nu < \dfrac{\bar{x}_1 - \bar{x}_2}{\sqrt{\dfrac{s_1^2}{n_1} + \dfrac{s_2^2}{n_2}}}\right) = P(t_\nu < t).
$$

The *p*-value is thus the probability that a *t*-distributed random variable with ν degrees of freedom is smaller than the computed test statistic. The null hypothesis is rejected if the *p*-value is smaller than the significance level α.

The Two-Tailed Test

For a two-tailed hypothesis test, the null hypothesis $H_0 : \mu_1 - \mu_2 = 0$ is rejected (and the alternative hypothesis $H_a : \mu_1 - \mu_2 \neq 0$ accepted) if the absolute value of the computed test statistic

$$
t = \dfrac{\bar{x}_1 - \bar{x}_2}{\sqrt{\dfrac{s_1^2}{n_1} + \dfrac{s_2^2}{n_2}}}
$$

is larger than the critical value $t_{\alpha/2;\nu}$.

The p-value of the two-tailed test is

$$p = 2\,P\left(t_\nu > \left|\frac{\bar{x}_1 - \bar{x}_2}{\sqrt{\dfrac{s_1^2}{n_1} + \dfrac{s_2^2}{n_2}}}\right|\right) = 2\,P(t_\nu > |t|).$$

Hence, the p-value is twice the probability that a t-distributed random variable with ν degrees of freedom is larger than the absolute value of the computed test statistic. The null hypothesis is rejected if the p-value is smaller than the significance level α.

A Generalized Hypothesis Test

In some cases, a researcher wants to test whether the difference between two population means is equal to a certain number Δ_0. The test statistic is then

$$T = \frac{\bar{X}_1 - \bar{X}_2 - \Delta_0}{\sqrt{\dfrac{s_1^2}{n_1} + \dfrac{s_2^2}{n_2}}}.$$

Apart from the way in which the test statistic is computed, the procedures for conducting right-tailed, left-tailed, and two-tailed hypothesis tests do not change. In particular, the decision rules outlined above remain valid.

Example 8.1.2 *In Example 8.1.1, we performed a test for two population means assuming that the population variances σ_1^2 and σ_2^2 were equal, and using the pooled sample variance. However, if we do not want to make the assumption that σ_1^2 and σ_2^2 are equal, then the approach for unequal variances has to be used. The computed test statistic is then*

$$t = \frac{\bar{x}_1 - \bar{x}_2}{\sqrt{\dfrac{s_1^2}{n_1} + \dfrac{s_2^2}{n_2}}} = \frac{50.9033 - 50.2525}{\sqrt{\dfrac{0.1848}{20} + \dfrac{0.2837}{20}}} = 4.2521.$$

The required number of degrees of freedom for the (approximate) t-distribution is

$$\nu = \frac{\left(\dfrac{s_1^2}{n_1} + \dfrac{s_2^2}{n_2}\right)^2}{\dfrac{(s_1^2/n_1)^2}{n_1 - 1} + \dfrac{(s_2^2/n_2)^2}{n_2 - 1}} = \frac{\left(\dfrac{0.1848}{20} + \dfrac{0.2837}{20}\right)^2}{\dfrac{(0.1848/20)^2}{20 - 1} + \dfrac{(0.2837/20)^2}{20 - 1}} = 36.3789,$$

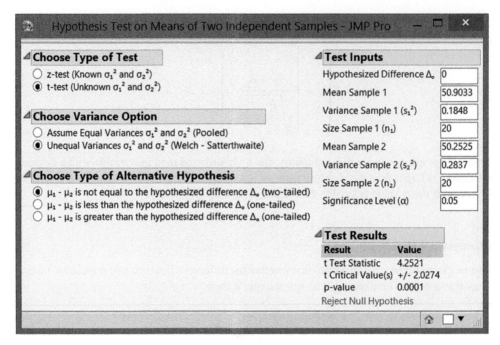

Figure 8.2   The use of the JMP script "Hypothesis Test for Two Means Using Summary Statistics" for Example 8.1.2.

so that the p-value is

$$p = 2\,P(t_{36.3789} > 4.2521) = 0.000141.$$

*This p-value can be calculated in JMP using the formula "2 * (1 − t* `Distribution`*(4.2521, 36.3789))", and, of course, leads to a rejection of the null hypothesis. Figure 8.2 shows the output of the JMP script "Hypothesis Test for Two Means Using Summary Statistics" if unequal variances* σ_1^2 *and* σ_2^2 *are assumed. The critical values* ± 2.0274 *indicated in the output can be calculated using the formulas "*`t Quantile`*(0.025, 36.3789)" and "*`t Quantile`*(0.975, 36.3789)".*

8.1.4   Confidence Intervals for a Difference in Population Means

In Section 3.2.3, we recommended that whenever reporting the results of a statistical analysis, one should not only mention the hypotheses tested and the associated *p*-values, but also the confidence intervals for the population parameters under study. The reason is that a confidence interval provides a clear picture of the uncertainty regarding the population parameters.

A confidence interval can also be computed for the difference of two population means, μ_1 and μ_2. The exact method for calculating this interval depends on whether or not the variances σ_1^2 and σ_2^2 of the populations under study are known, and, when they are unknown, whether the variances can be assumed to be equal. In any case, the computation of a confidence interval for $\mu_1 - \mu_2$ is interesting because it expresses the uncertainty concerning the estimate for $\mu_1 - \mu_2$.

Known σ_1^2 and σ_2^2

We first assume that the populations under study are normally distributed, or that the samples from both populations are large. If the variances σ_1^2 and σ_2^2 are known, then

$$\frac{\overline{X}_1 - \overline{X}_2 - (\mu_1 - \mu_2)}{\sqrt{\dfrac{\sigma_1^2}{n_1} + \dfrac{\sigma_2^2}{n_2}}}$$

is standard normally distributed, because $\overline{X}_1 - \overline{X}_2$ is an unbiased estimator of $\mu_1 - \mu_2$ with standard deviation

$$\sqrt{\frac{\sigma_1^2}{n_1} + \frac{\sigma_2^2}{n_2}}.$$

This allows us to derive the following confidence interval for the difference $\mu_1 - \mu_2$ of two population means:

$$\left[\overline{X}_1 - \overline{X}_2 - z_{\alpha/2} \sqrt{\frac{\sigma_1^2}{n_1} + \frac{\sigma_2^2}{n_2}} \, , \, \overline{X}_1 - \overline{X}_2 + z_{\alpha/2} \sqrt{\frac{\sigma_1^2}{n_1} + \frac{\sigma_2^2}{n_2}} \right].$$

If this interval contains the value zero, then we say that the difference in the population means is not significantly different from zero, or that the two population means are not significantly different from each other. In the opposite case, we say that the difference in the population means is significantly different from zero, or that the two population means differ significantly.

If the confidence interval contains the value Δ_0, then we say that the difference in the population means is not significantly different from Δ_0. If the confidence interval does not contain the value Δ_0, then the difference in the population means is significantly different from Δ_0.

So, here again, we observe a correspondence between a confidence interval and a two-tailed hypothesis test.

Unknown but Equal σ_1^2 and σ_2^2

If the variances σ_1^2 and σ_2^2 are unknown but can be assumed to be equal, then they can both be estimated by the pooled sample variance S_p^2. The fraction

$$\frac{\overline{X}_1 - \overline{X}_2 - (\mu_1 - \mu_2)}{S_p \sqrt{\dfrac{1}{n_1} + \dfrac{1}{n_2}}}$$

is then t-distributed with $n_1 + n_2 - 2$ degrees of freedom (see Section 8.1.3). This allows us to derive the following confidence interval for the difference in the population means, $\mu_1 - \mu_2$:

$$\left[\overline{X}_1 - \overline{X}_2 - t_{\alpha/2;n_1+n_2-2} \, S_p \sqrt{\frac{1}{n_1} + \frac{1}{n_2}} \,, \; \overline{X}_1 - \overline{X}_2 + t_{\alpha/2;n_1+n_2-2} \, S_p \sqrt{\frac{1}{n_1} + \frac{1}{n_2}} \right].$$

Unknown and Unequal σ_1^2 and σ_2^2

If the variances σ_1^2 and σ_2^2 are unknown and cannot be assumed to be equal, then σ_1^2 and σ_2^2 must be estimated separately. The fraction

$$\frac{\overline{X}_1 - \overline{X}_2 - (\mu_1 - \mu_2)}{\sqrt{\dfrac{S_1^2}{n_1} + \dfrac{S_2^2}{n_2}}}$$

has a Behrens–Fisher probability density function. That distribution can be approximated by a t-distribution with v degrees of freedom, where v is calculated using Equation (8.3).

Using the t-distribution with v degrees of freedom, we can derive the following confidence interval for a difference $\mu_1 - \mu_2$ in population means:

$$\left[\overline{X}_1 - \overline{X}_2 - t_{\alpha/2;v} \sqrt{\frac{S_1^2}{n_1} + \frac{S_2^2}{n_2}} \,, \; \overline{X}_1 - \overline{X}_2 + t_{\alpha/2;v} \sqrt{\frac{S_1^2}{n_1} + \frac{S_2^2}{n_2}} \right].$$

8.2 A Hypothesis Test for Two Population Proportions

Sometimes a researcher wants to compare the proportions of "success" in two populations. Typically, the first population proportion is represented by π_1, whereas the second population proportion is denoted by π_2. The null hypothesis of a test for two population proportions is

$$H_0 : \pi_1 = \pi_2 \quad \text{or} \quad H_0 : \pi_1 - \pi_2 = 0.$$

The right-tailed alternative hypothesis is

$$H_a : \pi_1 > \pi_2 \quad \text{or} \quad H_a : \pi_1 - \pi_2 > 0,$$

and the left-tailed and two-tailed alternative hypotheses are

$$H_a : \pi_1 < \pi_2 \quad \text{or} \quad H_a : \pi_1 - \pi_2 < 0$$

and

$$H_a : \pi_1 \neq \pi_2 \quad \text{or} \quad H_a : \pi_1 - \pi_2 \neq 0,$$

respectively.

The details of the hypothesis tests for two population proportions are similar to those for two population means with known σ, because the tests for two proportions are also based on the normal distribution (see Section 8.1.2).

8.2.1 The Starting Point

In practice, the population proportions are estimated by means of the sample proportions $\hat{P}_1$ and $\hat{P}_2$, based on samples of n_1 and n_2 observations from the two populations under study. If the samples are sufficiently large[1], then these sample proportions are approximately normally distributed:

$$\hat{P}_1 \sim N\left(\pi_1, \frac{\pi_1(1-\pi_1)}{n_1}\right)$$

and

$$\hat{P}_2 \sim N\left(\pi_2, \frac{\pi_2(1-\pi_2)}{n_2}\right).$$

If the two samples are independent, then the two estimators $\hat{P}_1$ and $\hat{P}_2$ are also independent. In that case, the difference between $\hat{P}_1$ and $\hat{P}_2$ is normally distributed, with expected value $\pi_1 - \pi_2$ and variance

$$\frac{\pi_1(1-\pi_1)}{n_1} + \frac{\pi_2(1-\pi_2)}{n_2}.$$

This can be written as

$$\hat{P}_1 - \hat{P}_2 \sim N\left(\pi_1 - \pi_2, \frac{\pi_1(1-\pi_1)}{n_1} + \frac{\pi_2(1-\pi_2)}{n_2}\right).$$

If the null hypothesis is correct, then we can introduce a common symbol for π_1 and π_2, namely $\pi_1 = \pi_2 = \pi$. Then, we have that

$$\hat{P}_1 - \hat{P}_2 \sim N\left(0, \frac{\pi(1-\pi)}{n_1} + \frac{\pi(1-\pi)}{n_2}\right).$$

This result is the starting point of the hypothesis test. To proceed, we need to estimate the common population proportion π. This is done with a "pooled" estimator:

$$\overline{P} = \frac{n_1\hat{P}_1 + n_2\hat{P}_2}{n_1 + n_2}.$$

This estimator is equal to the overall proportion of successes in the two samples.

---

[1] The samples are sufficiently large if their sizes satisfy the conditions $n_1\hat{p}_1 > 5$, $n_1(1-\hat{p}_1) > 5$, $n_2\hat{p}_2 > 5$, and $n_2(1-\hat{p}_2) > 5$.

This leads us to the following test statistic:

$$Z = \frac{\hat{P}_1 - \hat{P}_2}{\sqrt{\overline{P}(1-\overline{P})\left(\frac{1}{n_1}+\frac{1}{n_2}\right)}}.$$

This test statistic is approximately standard normally distributed, if the null hypothesis is correct.

8.2.2   The Right-Tailed Test

The null hypothesis is rejected if the computed test statistic exceeds the critical value z_α, or if the p-value

$$p = P\left(Z > \frac{\hat{p}_1 - \hat{p}_2}{\sqrt{\overline{p}(1-\overline{p})\left(\frac{1}{n_1}+\frac{1}{n_2}\right)}}\right) = P(Z > z)$$

is smaller than the significance level α.

8.2.3   The Left-Tailed Test

The null hypothesis is rejected if the computed test statistic is smaller than the critical value $-z_\alpha$, or if the p-value

$$p = P\left(Z < \frac{\hat{p}_1 - \hat{p}_2}{\sqrt{\overline{p}(1-\overline{p})\left(\frac{1}{n_1}+\frac{1}{n_2}\right)}}\right) = P(Z < z)$$

is smaller than the significance level α.

8.2.4   The Two-Tailed Test

The null hypothesis is rejected if the computed test statistic is smaller than the left critical value $-z_{\alpha/2}$ or larger than the right critical value $z_{\alpha/2}$. In other words, the null hypothesis is rejected if the absolute value of the computed test statistic exceeds $z_{\alpha/2}$. Finally, we can also say that we reject the null hypothesis if the p-value

$$p = 2\,P\left(Z > \left|\frac{\hat{p}_1 - \hat{p}_2}{\sqrt{\overline{p}(1-\overline{p})\left(\frac{1}{n_1}+\frac{1}{n_2}\right)}}\right|\right) = 2\,P(Z > |z|)$$

is smaller than the significance level α.

Example 8.2.1 *Based on a sample, the proportion of First Class flight passengers arriving late (less than one hour before the scheduled departure time) at the check-in is estimated to be $\hat{p}_1 = 0.78$. A similar study for Economy passengers yields an estimate of $\hat{p}_2 = 0.52$. Both estimates are sample proportions based on 50 observations.*

A two-tailed hypothesis test (significance level 5%) for the population proportions of passengers arriving late requires the computation of the following test statistic:

$$z = \frac{\hat{p}_1 - \hat{p}_2}{\sqrt{\overline{p}(1-\overline{p})\left(\frac{1}{n_1} + \frac{1}{n_2}\right)}} = \frac{0.78 - 0.52}{\sqrt{(0.65)(1-0.65)\left(\frac{1}{50} + \frac{1}{50}\right)}} = 2.7255.$$

The critical values for the two-tailed test are $-z_{\alpha/2} = -z_{0.025} = -1.96$ and $z_{\alpha/2} = z_{0.025} = 1.96$. As a result, the computed test statistic z clearly lies outside the interval determined by the critical values, so that the null hypothesis can be rejected. The p-value is

$$p = 2\,P(Z > 2.7255) = 2(0.0032) = 0.0064,$$

which is clearly smaller than 5%. The p-value can be calculated using either the function "2 * (1 − Normal Distribution(2.7255))" or the function "2 * (1 − Normal Distribution(2.7255, 0, 1))". Due to the small p-value, the null hypothesis can be rejected, and we can say that the proportion of late passengers is significantly different between the First Class and the Economy passengers.

8.2.5 Generalized Hypothesis Tests

In some cases, a researcher is more interested in knowing whether the difference between two population proportions is equal to a given value Δ_0. In that case, the null hypothesis is

$$H_0 : \pi_1 - \pi_2 = \Delta_0,$$

and the possible alternative hypotheses can be written as

$$H_a : \pi_1 - \pi_2 > \Delta_0,$$
$$H_a : \pi_1 - \pi_2 < \Delta_0$$

and

$$H_a : \pi_1 - \pi_2 \neq \Delta_0.$$

If the null hypothesis is correct for such a generalized hypothesis test, then

$$\hat{P}_1 - \hat{P}_2 \sim N\left(\Delta_0, \frac{\pi_1(1-\pi_1)}{n_1} + \frac{\pi_2(1-\pi_2)}{n_2}\right).$$

Since π_1 and π_2 are different (even if the null hypothesis is true), we cannot use the pooled sample proportion $\overline{P}$. The population proportions π_1 and π_2 therefore have to be estimated

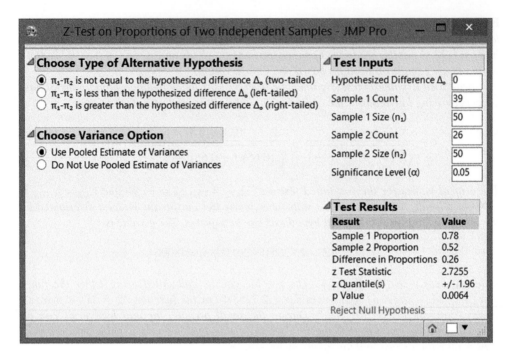

Figure 8.3   The use of the JMP script "Hypothesis Test for Two Proportions Using Summary Statistics" for Example 8.2.1.

separately by means of the sample proportions $\hat{P}_1$ and $\hat{P}_2$, based on the first and second samples. The test statistic is therefore

$$Z = \frac{\hat{P}_1 - \hat{P}_2 - \Delta_0}{\sqrt{\dfrac{\hat{P}_1(1-\hat{P}_1)}{n_1} + \dfrac{\hat{P}_2(1-\hat{P}_2)}{n_2}}}.$$

This test statistic is approximately standard normally distributed if the samples are large enough.

In JMP, the scripts "Hypothesis Test for Two Proportions Using Summary Statistics" and "Hypothesis Test for Two Proportions Using Data" are available to perform hypothesis tests for two population proportions. Figure 8.3 illustrates the use of the former script for Example 8.2.1, where $\Delta_0 = 0$. The script returns a p-value of 0.0064 when using the pooled sample proportion.

8.2.6   The Confidence Interval for a Difference in Population Proportions

Assuming that the samples from the two populations under study are large enough,

$$\hat{P}_1 - \hat{P}_2 \sim N\left(\pi_1 - \pi_2, \frac{\pi_1(1-\pi_1)}{n_1} + \frac{\pi_2(1-\pi_2)}{n_2} \right),$$

so that a confidence interval with confidence level $1 - \alpha$ is given by

$$\left[\hat{P}_1 - \hat{P}_2 - z_{\alpha/2} \sqrt{\frac{\hat{P}_1(1-\hat{P}_1)}{n_1} + \frac{\hat{P}_2(1-\hat{P}_2)}{n_2}}, \hat{P}_1 - \hat{P}_2 + z_{\alpha/2} \sqrt{\frac{\hat{P}_1(1-\hat{P}_1)}{n_1} + \frac{\hat{P}_2(1-\hat{P}_2)}{n_2}} \right].$$

Note that for this confidence interval, we do not assume that the two population proportions π_1 and π_2 are equal. Therefore, the pooled sample proportion is not used. Instead, π_1 and π_2 are estimated separately using the sample proportions $\hat{P}_1$ and $\hat{P}_2$.

8.3 A Hypothesis Test for Two Population Variances

One can also perform a hypothesis test for the comparison of two variances. In some cases, this is necessary because of a genuine interest in variances. In other cases, a hypothesis test for two variances is an intermediate step before testing two population means. Indeed, in the case of unknown variances σ_1^2 and σ_2^2, there are two types of hypothesis tests for comparing two population means, namely one where the variances σ_1^2 and σ_2^2 are assumed to be equal, and one in which the variances σ_1^2 and σ_2^2 are believed to be different. In order to decide which test to use for comparing two population means, one can perform a formal hypothesis test for the variances σ_1^2 and σ_2^2.

The classic hypothesis test for two variances, which is discussed in this section, is only valid for normally distributed populations. It has already been mentioned that it is impossible to verify whether a small number of sample data originates from a normally distributed population. Therefore, it is not advisable to rely on the hypothesis test described here if $n < 15$. For larger samples ($n \geq 15$), one should perform a normality test before using the hypothesis test described here (see Section 6.2).

The null hypothesis tested when comparing two variances is

$$H_0 : \sigma_1^2 = \sigma_2^2,$$

while the possible alternative hypotheses may be written as

$$H_a : \sigma_1^2 > \sigma_2^2,$$
$$H_a : \sigma_1^2 < \sigma_2^2,$$

and

$$H_a : \sigma_1^2 \neq \sigma_2^2.$$

An alternative way of writing these hypotheses is

$$H_0 : \sigma_1^2/\sigma_2^2 = 1,$$
$$H_a : \sigma_1^2/\sigma_2^2 > 1,$$
$$H_a : \sigma_1^2/\sigma_2^2 < 1,$$

and

$$H_a : \sigma_1^2 / \sigma_2^2 \neq 1.$$

The most traditional hypothesis test for two population variances is based on the sample variances. The starting point for that test is the fact that, for normally distributed populations,

$$\frac{(n_1 - 1)S_1^2}{\sigma_1^2} \sim \chi_{n_1-1}^2$$

and

$$\frac{(n_2 - 1)S_2^2}{\sigma_2^2} \sim \chi_{n_2-1}^2.$$

This relationship between sample variances and the χ^2-distribution was discussed in detail in Section 1.7.4. Next, the hypothesis test uses the result that the quotient of two independent χ^2-distributed random variables, both divided by their number of degrees of freedom, is F-distributed. Hence,

$$\frac{\dfrac{(n_1 - 1)S_1^2}{\sigma_1^2}}{n_1 - 1} \Bigg/ \frac{\dfrac{(n_2 - 1)S_2^2}{\sigma_2^2}}{n_2 - 1} = \frac{\dfrac{S_1^2}{\sigma_1^2}}{\dfrac{S_2^2}{\sigma_2^2}} \sim F_{n_1-1;n_2-1}.$$

This F-distribution is characterized by two parameters. The first parameter here is $n_1 - 1$ (the number of observations in the first sample minus 1). That parameter value is the numerator degrees of freedom. The second parameter is $n_2 - 1$ (the number of observations in the second sample minus 1). That parameter value is the denominator degrees of freedom.

8.3.1 Fisher's F-Distribution

The F-distribution is named after the British statistician Sir Ronald Fisher, who lived at the end of the nineteenth and the beginning of the twentieth century. Fisher found much of his inspiration in the agricultural experiments that he conducted at the Rothamsted Experimental Station in England. The F-distribution has two parameters, v_1 and v_2, which must both be positive. They are called numerator degrees of freedom and denominator degrees of freedom, respectively. The probability density of an F- distributed random variable X is given by

$$f_X(x; v_1, v_2) = \frac{\Gamma\left(\frac{v_1 + v_2}{2}\right) v_1^{v_1/2} v_2^{v_2/2}}{\Gamma\left(\frac{v_1}{2}\right)\Gamma\left(\frac{v_2}{2}\right)} \frac{x^{v_1/2-1}}{(v_2 + v_1 x)^{(v_1+v_2)/2}}, \text{ for } x \geq 0.$$

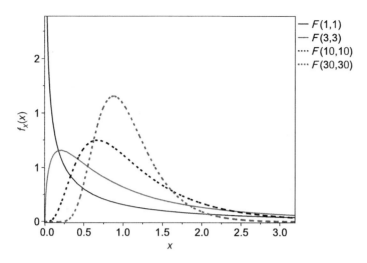

Figure 8.4 A graphical representation of four F-distributions involving different numbers of degrees of freedom v_1 and v_2.

Some F-distributions are shown in Figure 8.4. The figure shows that all F-distributions are skewed to the right and that F-distributed random variables can only take positive values. The expected value of an F-distributed random variable only depends on v_2 and equals

$$\frac{v_2}{v_2 - 2}$$

if $v_2 > 2$. The variance is equal to

$$\frac{2v_2^2(v_1 + v_2 - 2)}{v_1(v_2 - 2)^2(v_2 - 4)}$$

if $v_2 > 4$.

In JMP, function values of the F-distribution can be determined using the formula "F Density (x, v_1, v_2)". For the cumulative distribution function, the formula "F Distribution (x, v_1, v_2)" can be used. Finally, percentiles or quantiles can be determined using the formula "F Quantile (p, v_1, v_2)".

As mentioned above, a quotient of two independent χ^2-distributed random variables, both divided by their degrees of freedom, is F-distributed. If X_1 is a χ^2-distributed random variable with v_1 degrees of freedom, and X_2 is a χ^2-distributed random variable with v_2 degrees of freedom, then

$$\frac{\dfrac{X_1}{v_1}}{\dfrac{X_2}{v_2}}$$

is an F-distributed random variable with v_1 numerator degrees of freedom and v_2 denominator degrees of freedom. Note that the random variables X_1 and X_2 should be independent.

Moreover, the square of a t-distributed random variable with v degrees of freedom is also F-distributed. The degrees of freedom of this F-distribution are v for the denominator and 1 for the numerator. We know that a t-distributed random variable is equal to the quotient of a standard normally distributed random variable and the square root of a χ^2-distributed random variable divided by its degrees of freedom (where the random variables in the numerator and denominator of the quotient must be independent). If we denote the standard normally distributed random variable by Z, the χ^2-distributed random variable by X, and the degrees of freedom by v, then

$$\frac{Z}{\sqrt{\frac{X}{v}}}$$

is t-distributed with v degrees of freedom. The square of this is

$$\left(\frac{Z}{\sqrt{\frac{X}{v}}}\right)^2 = \frac{Z^2}{\frac{X}{v}} = \frac{\frac{Z^2}{1}}{\frac{X}{v}}.$$

The numerator of this expression contains the square of a standard normally distributed random variable, Z^2. This square is χ^2-distributed with one degree of freedom, and is divided by 1. The denominator of this expression contains a χ^2-distributed random variable with v degrees of freedom, divided by v. Consequently, the square of a t-distributed random variable is the quotient of two χ^2-distributed random variables, both divided by their number of degrees of freedom. As a result, the square of a t-distributed random variable is F-distributed. The number of degrees of freedom for the numerator is 1, while the number of degrees of freedom for the denominator is v.

The inverse of an F-distributed random variable with v_1 numerator degrees of freedom and v_2 denominator degrees of freedom is also an F-distributed random variable, but now with v_2 numerator degrees of freedom and v_1 denominator degrees of freedom.

8.3.2   The F-Test for the Comparison of Two Population Variances

The hypothesis test for the comparison of two population variances makes use of the F-distribution. If the null hypothesis of this test is true, then $\sigma_1^2 = \sigma_2^2 = \sigma^2$, and

$$\frac{\frac{(n_1 - 1)S_1^2}{\sigma_1^2}}{n_1 - 1} = \frac{\frac{S_1^2}{\sigma_1^2}}{\frac{S_2^2}{\sigma_2^2}} = \frac{\frac{S_1^2}{\sigma^2}}{\frac{S_2^2}{\sigma^2}} = \frac{S_1^2}{S_2^2}$$

is F-distributed with $n_1 - 1$ numerator degrees of freedom and $n_2 - 1$ denominator degrees of freedom. The fraction

$$F = \frac{S_1^2}{S_2^2}$$

serves as the test statistic. The exact decision rules based on this test statistic depend on the nature of the alternative hypothesis.

The Right-Tailed Test

In a right-tailed test, the null hypothesis $H_0 : \sigma_1^2/\sigma_2^2 = 1$ is rejected in favor of the alternative hypothesis $H_a : \sigma_1^2/\sigma_2^2 > 1$ if the computed sample variance of the first sample, s_1^2, is considerably larger than the computed sample variance of the second sample, s_2^2. In that case, the computed test statistic $f = s_1^2/s_2^2$ is considerably larger than 1. The (right) critical value required to determine whether the computed test statistic is large enough for the null hypothesis to be rejected is $F_{\alpha;n_1-1;n_2-1}$. In other words, we reject the null hypothesis if

$$f = s_1^2/s_2^2 > F_{\alpha;n_1-1;n_2-1}.$$

Alternatively, we can compare the p-value

$$p = P\left(F_{n_1-1;n_2-1} > \frac{s_1^2}{s_2^2}\right) = P(F_{n_1-1;n_2-1} > f)$$

to the significance level α selected and reject the null hypothesis if the p-value is smaller than α. The p-value will be small if s_1^2 is substantially larger than s_2^2; that is, if s_1^2/s_2^2 is substantially larger than 1.

The critical values $F_{\alpha;n_1-1;n_2-1}$ can be found in Appendix G, for the significance levels $0.01, 0.05$, and 0.10. The formula "F Quantile (p, ν_1, ν_2)" allows us to determine critical values for other values of the significance level and for numbers of degrees of freedom that are not listed in the table in Appendix G.

The Left-Tailed Test

In a left-tailed test, the null hypothesis is rejected in favor of the alternative hypothesis $H_a : \sigma_1^2/\sigma_2^2 < 1$ if the computed test statistic $f = s_1^2/s_2^2$ is much smaller than 1; that is, if the computed test statistic $f = s_1^2/s_2^2$ is smaller than the (left) critical value $F_{1-\alpha;n_1-1;n_2-1}$. Alternatively, we can compare the p-value

$$p = P\left(F_{n_1-1;n_2-1} < \frac{s_1^2}{s_2^2}\right) = P(F_{n_1-1;n_2-1} < f)$$

to the significance level α and reject the null hypothesis if the p-value is smaller than α.

The table in Appendix G only contains the right critical values $F_{0.10;v_1;v_2}$, $F_{0.05;v_1;v_2}$, and $F_{0.01;v_1;v_2}$ needed for the right-tailed test, but not the left critical values $F_{0.90;v_1;v_2}$, $F_{0.95;v_1;v_2}$, and $F_{0.99;v_1;v_2}$ needed for the left-tailed test. This means that, at first sight, the table cannot be used for left-tailed tests. However, a simple solution to this problem is to ensure that sample 1 is always the data set with the larger sample variance and to use a right-tailed instead of a left-tailed hypothesis test. This approach leads to the same test results, as the inverse of an F-distributed random variable with v_1 and v_2 degrees of freedom is also F-distributed, but with v_2 and v_1 degrees of freedom, and

$$P(F_{v_1;v_2} > f) = P(F_{v_2;v_1} < 1/f).$$

Of course, using JMP also avoids this problem, since JMP allows us to determine the critical values $F_{0.90;v_1;v_2}$, $F_{0.95;v_1;v_2}$, and $F_{0.99;v_1;v_2}$.

The Two-Tailed Test

In a two-tailed test, the null hypothesis is rejected in favor of the alternative hypothesis $H_a : \sigma_1^2/\sigma_2^2 \neq 1$ if the computed test statistic $f = s_1^2/s_2^2$ is considerably smaller or considerably larger than 1. To decide whether the computed test statistic is small enough for the null hypothesis to be rejected, we need a left critical value. In contrast, to decide whether the computed test statistic is large enough for the null hypothesis to be rejected, we need a right critical value. We obtain both critical values from the F-distribution. The left critical value is $F_{1-\alpha/2;n_1-1;n_2-1}$, while the right critical value is equal to $F_{\alpha/2;n_1-1;n_2-1}$. If either

$$f = s_1^2/s_2^2 < F_{1-\alpha/2;n_1-1;n_2-1}$$

or

$$f = s_1^2/s_2^2 > F_{\alpha/2;n_1-1;n_2-1},$$

we reject the null hypothesis and accept the alternative hypothesis. If $f = s_1^2/s_2^2$ lies between $F_{1-\alpha/2;n_1-1;n_2-1}$ and $F_{\alpha/2;n_1-1;n_2-1}$, we accept the null hypothesis.

Again, a researcher who does not have access to a software package such as JMP has to circumvent the limitations of the table in Appendix G by ensuring that the first sample has the larger sample variance. In that case, $F_{\alpha/2;n_1-1;n_2-1}$ is the only value that matters. This is because, if $s_1^2 > s_2^2$, then $f = s_1^2/s_2^2$ is larger than 1, and the test statistic is certainly also larger than $F_{1-\alpha/2;n_1-1;n_2-1}$. The only remaining comparison to be done is then to verify whether $f = s_1^2/s_2^2$ is larger than $F_{\alpha/2;n_1-1;n_2-1}$. If that is the case, we can reject the null hypothesis.

An alternative approach is to compare the p-value with the selected significance level α and to reject the null hypothesis if it is smaller than α. In the computation of the p-value, one should distinguish between two cases. If $s_1^2 \geq s_2^2$, then the p-value is

$$p = 2 P\left(F_{n_1-1;n_2-1} > \frac{s_1^2}{s_2^2} \right) = 2 P(F_{n_1-1;n_2-1} > f).$$

If not, then the p-value is

$$p = 2\,P\left(F_{n_1-1;n_2-1} < \frac{s_1^2}{s_2^2}\right) = 2\,P(F_{n_1-1;n_2-1} < f).$$

Example 8.3.1   *In Example 8.1.1, a test was performed to compare two population means, assuming equal population variances. Now, we conduct a two-tailed test at a 5% significance level to verify whether the assumption of equal variances was justified. To be able to do so, we need to assume that the data is normally distributed. The two samples of size 20 from the two populations under study resulted in sample variances of $s_1^2 = 0.1848$ and $s_2^2 = 0.2837$. The test statistic is therefore*

$$f = \frac{s_1^2}{s_2^2} = \frac{0.1848}{0.2837} = 0.6514.$$

The left and right critical values required for the two-tailed test can be calculated using the JMP formulas "F Quantile(0.025, 19, 19)" and "F Quantile(0.975, 19, 19)". They amount to 0.3958 and 2.5265, respectively. The computed test statistic lies between these two values, so that the null hypothesis of equal variances should be accepted. The p-value is

$$p = 2\,P(F_{n_1-1;n_2-1} < f) = 2\,P(F_{19;19} < 0.6514) = 0.3583,$$

*as $s_1^2 < s_2^2$. This p-value can be computed in JMP using the formula "2 * F Distribution(0.6514, 19, 19)".*

   Note that we would obtain exactly the same p-value if we assigned the number 1 to the second sample and the number 2 to the first sample. In this case, s_1^2 would become 0.2837 and s_2^2 would become 0.1848, so that the computed test statistic would be

$$f = \frac{0.2837}{0.1848} = \frac{1}{0.6514} = 1.5352.$$

The corresponding p-value would be calculated as

$$p = 2\,P(F_{n_1-1;n_2-1} > f) = 2\,P(F_{19;19} > 1.5352) = 0.3583,$$

which is the same value we obtained initially. In any case, the conclusion from this large p-value is that there is no statistical evidence for the null hypothesis to be rejected. This justifies the use of the pooled sample variance for the hypothesis test for the two population means in Example 8.1.1.

A Generalized Hypothesis Test

In some cases, it is necessary to use a generalized hypothesis test. In the generalized test, the question is whether, for instance, one variance is twice or three times or ω_0 times as large as the other. To test whether the population variance σ_1^2 equals ω_0 times the population variance σ_2^2, the null hypothesis is

$$H_0 : \sigma_1^2 = \omega_0\sigma_2^2,$$

while the possible alternative hypotheses can be written as

$$H_a : \sigma_1^2 > \omega_0\sigma_2^2,$$
$$H_a : \sigma_1^2 < \omega_0\sigma_2^2,$$

and

$$H_a : \sigma_1^2 \neq \omega_0\sigma_2^2.$$

An alternative way of writing these hypotheses is

$$H_0 : \sigma_1^2/\sigma_2^2 = \omega_0,$$
$$H_a : \sigma_1^2/\sigma_2^2 > \omega_0,$$
$$H_a : \sigma_1^2/\sigma_2^2 < \omega_0,$$

and

$$H_a : \sigma_1^2/\sigma_2^2 \neq \omega_0.$$

If the null hypothesis is correct, we have that

$$\frac{\dfrac{(n_1 - 1)S_1^2}{\sigma_1^2}}{n_1 - 1} \bigg/ \frac{\dfrac{(n_2 - 1)S_2^2}{\sigma_2^2}}{n_2 - 1} = \frac{\dfrac{(n_1 - 1)S_1^2}{\omega_0\sigma_2^2}}{n_1 - 1} \bigg/ \frac{\dfrac{(n_2 - 1)S_2^2}{\sigma_2^2}}{n_2 - 1} = \frac{S_1^2}{\omega_0 S_2^2} \sim F_{n_1 - 1; n_2 - 1}.$$

Hence, the computed test statistic for the generalized hypothesis test for two population variances is

$$f = \frac{s_1^2}{\omega_0 s_2^2}.$$

The determination of the critical values and p-values for the right-tailed, left-tailed, and two-tailed tests can be done in the same way as when testing the equality of two variances.

To perform tests for two population variances in JMP, the scripts "Hypothesis Test for Two Variances Using Summary Statistics" and "Hypothesis Test for Two Variances Using Data" are available. If desired, standard deviations can be used in these scripts instead of variances.

8.3.3 The Confidence Interval for a Quotient of Two Population Variances

Starting from the fact that

$$
\frac{\dfrac{(n_1 - 1)S_1^2}{\sigma_1^2}}{n_1 - 1} \bigg/ \frac{\dfrac{(n_2 - 1)S_2^2}{\sigma_2^2}}{n_2 - 1} = \frac{\dfrac{S_1^2}{\sigma_1^2}}{\dfrac{S_2^2}{\sigma_2^2}} \sim F_{n_1 - 1; n_2 - 1},
$$

it is not difficult to show that

$$
\left[\frac{1}{F_{\alpha/2; n_1 - 1; n_2 - 1}} \frac{S_1^2}{S_2^2}, \; \frac{1}{F_{1 - \alpha/2; n_1 - 1; n_2 - 1}} \frac{S_1^2}{S_2^2} \right]
$$

is a confidence interval for σ_1^2/σ_2^2 with confidence level $1 - \alpha$. By taking the square root of the lower limit and the upper limit of the confidence interval, we obtain the following confidence interval for the quotient of the standard deviations, σ_1/σ_2:

$$
\left[\frac{1}{\sqrt{F_{\alpha/2; n_1 - 1; n_2 - 1}}} \frac{S_1}{S_2}, \; \frac{1}{\sqrt{F_{1 - \alpha/2; n_1 - 1; n_2 - 1}}} \frac{S_1}{S_2} \right].
$$

8.4 Hypothesis Tests for Two Independent Samples in JMP

The hypothesis tests described in this chapter, as well as the confidence intervals, can be computed in JMP. To this end, we need to use the "Fit Y by X" option in the "Analyze" menu. When doing so, it is important that the data is in the correct format: one column in the data table must contain the values of the variable under study, while another column should indicate to which sample each observation belongs. The latter column should be declared as a nominal variable in JMP. In the event of a t-test for the comparison of two population means or an F-test for the comparison of two variances, the first column must contain a quantitative variable. For a z-test comparing two population proportions, the first column has to be a nominal variable.

8.4.1 Two Population Means

We illustrate the use of JMP to test two population means and variances using data on 30 Japanese and 49 American cars. The summary statistics for the Japanese and American cars

Summary Statistics		Summary Statistics	
Mean	2881	Mean	3187.6531
Std Dev	623.06999	Std Dev	471.32802
Std Err Mean	113.7565	Std Err Mean	67.332575
Upper 95% Mean	3113.6582	Upper 95% Mean	3323.0343
Lower 95% Mean	2648.3418	Lower 95% Mean	3052.2718
N	30	N	49

| (a) Japan | (b) USA |

Figure 8.5 Summary statistics for the weight of 30 Japanese and 49 American cars.

are shown in Figure 8.5. These statistics show that, on average, American cars are heavier than Japanese ones. A 95% confidence interval for the weight of Japanese cars extends from 2648.34 to 3113.66 pounds, while a 95% confidence interval for the weight of American cars ranges from 3052.27 to 3323.03 pounds.

Now, suppose that we want to use JMP to determine whether the mean weight of Japanese cars differs significantly from that of American cars. Figure 8.6 shows what the required data table should look like in JMP. It is important that the variable "Weight" is a quantitative variable, while the variable "Country" is a nominal variable.

Because, alphabetically, "Japan" is ranked before "USA", JMP automatically assumes that the first population of cars under study is from Japan, while the second population of cars under investigation is from the USA. Next, JMP automatically calculates the difference

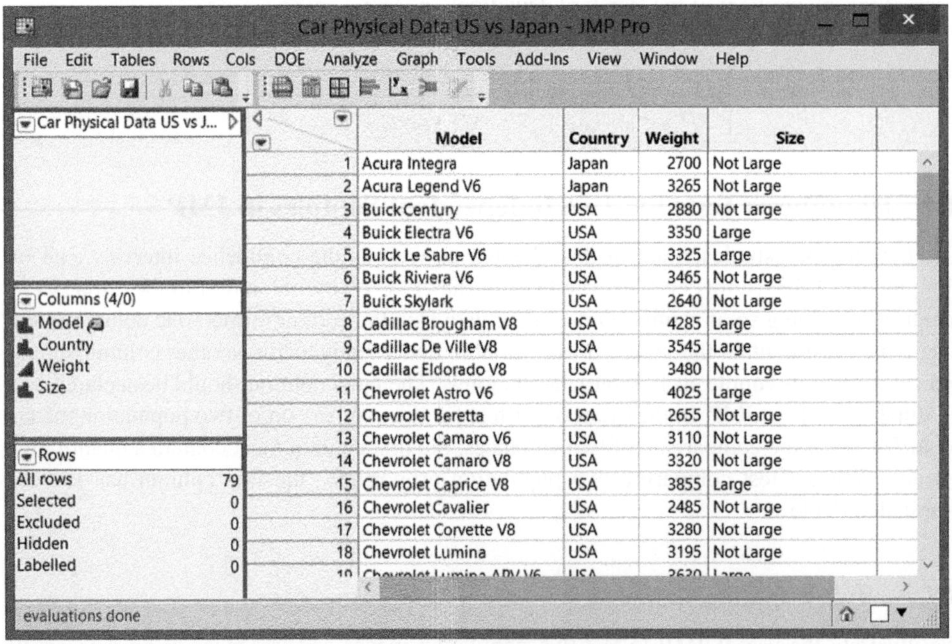

Figure 8.6 The JMP data table for a test comparing the mean weights of Japanese and American cars.

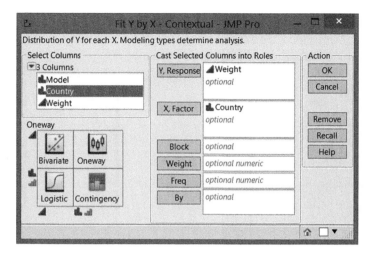

Figure 8.7 The dialog window of the "Fit Y by X" option in the "Analyze" menu for the analysis of the variable "Weight".

$\mu_2 - \mu_1 = \mu_{\text{USA}} - \mu_{\text{Japan}}$ instead of the difference $\mu_1 - \mu_2 = \mu_{\text{Japan}} - \mu_{\text{USA}}$. Therefore, all of the test statistics in JMP are also based on the difference $\mu_2 - \mu_1$, instead of the difference $\mu_1 - \mu_2$. JMP thus works in a slightly different way than described earlier in this chapter. Provided that the JMP output is interpreted correctly, this alternative approach does not have any impact on the conclusions. A correct interpretation is especially important when performing the one-sided hypothesis tests and when the confidence intervals in the output are studied.

To conduct a test for two population means, you need to choose the option "Fit Y by X" in the "Analyze" menu. The corresponding dialog window is shown in Figure 8.7. In this dialog window, the variable "Weight" has been entered in the field "Y, Response", and the variable "Country" has been entered in the field "X, Factor". If you click the button "OK" in the dialog window, you obtain the output shown in Figure 8.8. This output contains a plot that graphically compares the weights of the Japanese and the American cars.

To obtain the results of a formal hypothesis test for the mean weights of the cars, you have to click the hotspot (red triangle) next to the term "Oneway Analysis of Weight by Country". As can be seen in Figure 8.9, a menu containing, among other things, the options "Means/Anova/Pooled t", "Means and Std Dev", "t Test", and "Unequal Variances" will appear.

Equal Variances

The option "Means/Anova/Pooled t" performs a *t*-test for equal variances σ_1^2 and σ_2^2. This test uses the pooled sample variance S_p^2. The corresponding output is shown in Figure 8.10. The output contains *p*-values for a two-tailed test, a right-tailed test, and a left-tailed test. To differentiate between these *p*-values, JMP uses the labels "Prob > |t|", "Prob > t", and "Prob < t", respectively. JMP also reports the test statistic's value of 2.479151 next to the

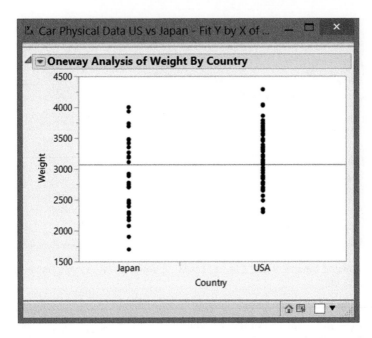

Figure 8.8 The initial output of the "`Fit Y by X`" option in the "`Analyze`" menu for the analysis of the variable "Weight".

term "`t Ratio`", and 77 degrees of freedom next to the abbreviation "`DF`". The degrees of freedom can be calculated as $n_1 + n_2 - 2 = 30 + 49 - 2$. The test statistic can be found as follows:

$$\frac{\bar{x}_2 - \bar{x}_1}{s_p\sqrt{\dfrac{1}{n_1} + \dfrac{1}{n_2}}} = \frac{\text{Difference}}{\text{Std Err Dif}} = \frac{306.653}{123.693} = 2.479151.$$

The abbreviation "`Std Err Dif`" stands for **standard error of difference**, which is the standard error

$$s_p\sqrt{\frac{1}{n_1} + \frac{1}{n_2}}$$

of the difference in sample means $\bar{x}_2 - \bar{x}_1$. The output also contains the lower and upper limits of the confidence interval for $\mu_2 - \mu_1$, next to the labels "`Lower CL Dif`" and "`Upper CL Dif`". It is not difficult to reconstruct these limits based on the rest of the output.

The p-value for the two-tailed test (alternative hypothesis $\mu_{\text{USA}} \neq \mu_{\text{Japan}}$ or $\mu_{\text{USA}} - \mu_{\text{Japan}} \neq 0$) in Figure 8.10 is 0.0154. The p-value for the one-sided test with alternative hypothesis $\mu_{\text{USA}} > \mu_{\text{Japan}}$ or $\mu_{\text{USA}} - \mu_{\text{Japan}} > 0$ is 0.0077. Finally, the p-value for the one-sided test with alternative hypothesis $\mu_{\text{USA}} < \mu_{\text{Japan}}$ or $\mu_{\text{USA}} - \mu_{\text{Japan}} < 0$, is 0.9923. This p-value is particularly large because the sample data contradicts the alternative hypothesis that $\mu_{\text{USA}} <$

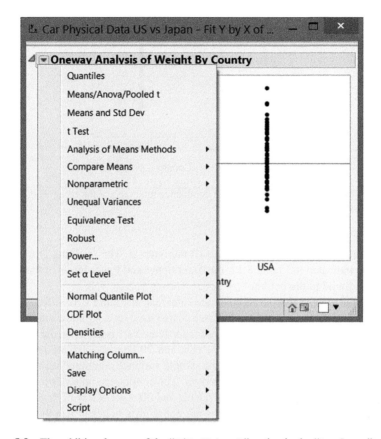

Figure 8.9 The additional menu of the "Fit Y by X" option in the "Analyze" menu.

μ_{Japan}. Indeed, the sample mean for the United States (USA) is larger than the one for Japan. The null hypothesis has to be rejected in two of the three tests.

The confidence interval for $\mu_{\text{USA}} - \mu_{\text{Japan}}$ is also entirely positive. This confirms that the mean weights of Japanese and American cars differ significantly from each other.

Choosing the option "Means/Anova/Pooled t" changes the graph in Figure 8.9: for both the mean weight of Japanese cars and the mean weight of American cars, the 95% confidence intervals are added to the figure in the form of a diamond. The bottom and the top of the

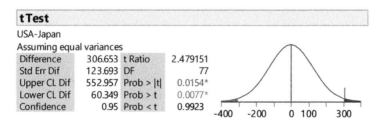

Figure 8.10 The output of the option "Means/Anova/Pooled t".

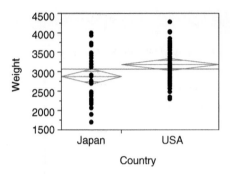

Figure 8.11 A graphical representation of the confidence intervals for the mean weights of Japanese and American cars.

diamond indicate the lower and upper limits of the interval. The resulting graph is shown in Figure 8.11. As indicated in Figure 8.12, the exact upper and lower limits of these confidence intervals can be found in the output.

It is interesting to note that the confidence intervals in Figure 8.12 differ from those in Figure 8.5. The confidence intervals in Figure 8.5 are based on a separate analyses of the data for Japanese and American cars. The interval for Japanese cars is based on the sample variance computed using the weights of 30 Japanese cars and 29 degrees of freedom. Similarly, the interval for the American cars is based on the sample variance computed using the weights of the 49 American cars and 48 degrees of freedom. The confidence intervals in Figure 8.12 are both calculated using the pooled sample variance, and $30 + 49 - 2 = 77$ degrees of freedom. If the variances of both studied populations are equal, it is better to use the pooled sample variance. If not, then it is better to use the individual confidence intervals.

Unequal Variances

The option "t Test" in Figure 8.9 performs a t-test assuming unequal variances σ_1^2 and σ_2^2. The corresponding output is shown in Figure 8.13. The appearance of this output is similar to that in Figure 8.10. However, the test statistic, the degrees of freedom, and the p-values are different from those in Figure 8.10.

The p-value for a two-tailed test (alternative hypothesis $\mu_{USA} \neq \mu_{Japan}$ or $\mu_{USA} - \mu_{Japan} \neq 0$) in Figure 8.13 is 0.0245. The p-value for the one-sided test with alternative hypothesis

Means for Oneway Anova

Level	Number	Mean	Std Error	Lower 95%	Upper 95%
Japan	30	2881.00	97.416	2687.0	3075.0
USA	49	3187.65	76.224	3035.9	3339.4

Std Error uses a pooled estimate of error variance

Figure 8.12 A table-based representation of the confidence intervals for the mean weights of Japanese and American cars.

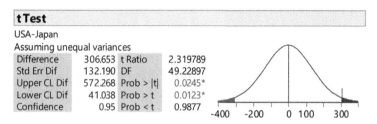

Figure 8.13   The output of the option "t Test".

$\mu_{\text{USA}} > \mu_{\text{Japan}}$ or $\mu_{\text{USA}} - \mu_{\text{Japan}} > 0$ is 0.0123. Finally, the p-value for the one-sided test with alternative hypothesis $\mu_{\text{USA}} < \mu_{\text{Japan}}$ or $\mu_{\text{USA}} - \mu_{\text{Japan}} < 0$, is 0.9877. The null hypothesis would be rejected in two of the three tests. The confidence interval for $\mu_{\text{USA}} - \mu_{\text{Japan}}$ is also entirely positive. This confirms that the mean weights of Japanese and American cars differ significantly.

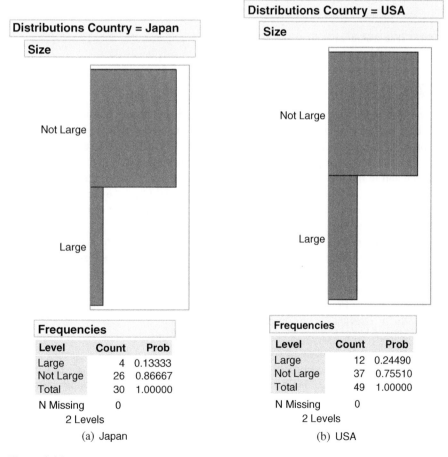

Figure 8.14   An overview of the numbers of large and nonlarge Japanese and American cars.

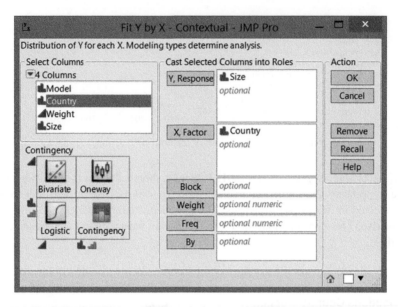

Figure 8.15 The dialog window of the "Fit Y by X" option in the "Analyze" menu for the analysis of the variable "Size".

8.4.2 Two Population Proportions

Another variable in the data table on Japanese and American cars is a nominal variable called "Size", which is a binary variable indicating whether or not a car is large. Figure 8.14 shows that the proportion of large cars in Japan is equal to 0.1333, while it is equal to 0.2449 in the USA. Via the "Fit Y by X" option, we can now test whether the proportions of large cars in Japan and in the USA are significantly different. The required input for this option is shown in Figure 8.15. Note that the variable "Size" in the field "Y, Response" is a nominal variable with only two possible values. This is necessary so that JMP can perform a test for the comparison of two proportions.

The initial output of the "Fit Y by X" option is shown in Figure 8.16 and includes a mosaic plot and a cross-tabulation (for a detailed discussion of these, see *Statistics with JMP: Graphs, Descriptive Statistics and Probability*). A new menu can be obtained via the hotspot (red triangle) next to the term "Contingency Analysis of Size by Country". This menu is shown in Figure 8.17. In the new menu, the option "Two Sample Test for Proportions" is relevant for our purposes.

The option "Two Sample Test for Proportions" initially leads to the output in Figure 8.18, involving a confidence interval for the difference $\pi_1 - \pi_2 = \pi_{\text{Japan}} - \pi_{\text{USA}}$, where $\pi_1 = \pi_{\text{Japan}}$ represents the proportion of large cars in Japan and $\pi_2 = \pi_{\text{USA}}$ is the proportion of large cars in the USA. JMP indicates this clearly in the output, using the notation[2]

---

[2] The vertical line in the expressions "P(Large | Japan)" and "P(Large | USA)" has the same meaning as the vertical line in a conditional probability. The term "P(Large | Japan)" should therefore be read as the proportion of large cars "given that" the country under investigation is Japan, or the probability that a car is large, "given that" the car was made in Japan. Conditional probabilities are discussed in detail in *Statistics with JMP: Graphs, Descriptive Statistics and Probability*.

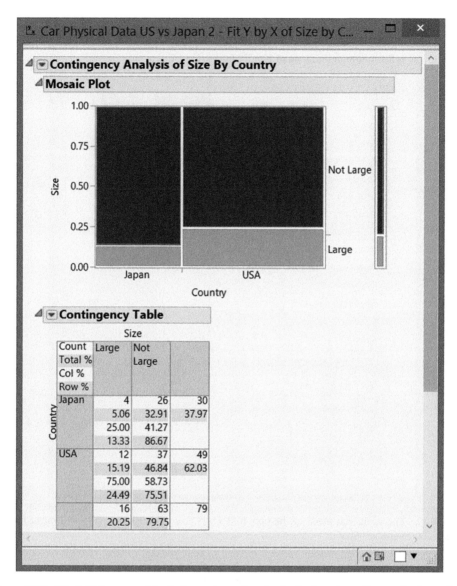

Figure 8.16 The initial output of the "`Fit Y by X`" option in the "`Analyze`" menu for the analysis of the variable "Size".

"P(Large|Japan)" and "P(Large|USA)". However, JMP also provides the opportunity to study the proportion of nonlarge cars. To take advantage of that opportunity, indicate at the bottom of the output that the category of cars you are interested in is "Not Large" (see Figure 8.19). If you opt for the outcome category "Not Large", you obtain the alternative output shown in Figure 8.20.

The outputs in Figures 8.18 and 8.20 include the p-values for two-tailed, right-tailed, and left-tailed hypothesis tests. None of the p-values is smaller than 5%. We can therefore conclude

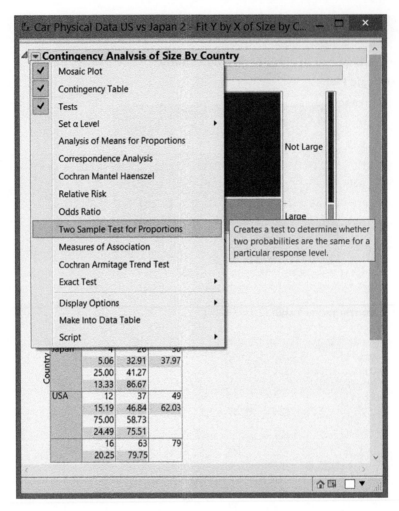

Figure 8.17 The additional menu of the option "Fit Y by X" in the "Analyze" menu for the analysis of the variable "Size".

Two Sample Test for Proportions

Description	Proportion Difference	Lower 95%	Upper 95%
P(Large|Japan)-P(Large|USA)	-0.11156	-0.27224	0.074934

Adjusted Wald Test	Prob
P(Large|Japan)-P(Large|USA) ≥ 0	0.8673
P(Large|Japan)-P(Large|USA) ≤ 0	0.1327
P(Large|Japan)-P(Large|USA) = 0	0.2653

Figure 8.18 The output of the option "Two Sample Test for Proportions" in the analysis of the variable "Size".

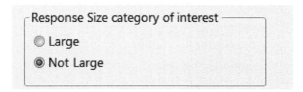

Figure 8.19   The choice between an analysis of the proportion of large cars or of nonlarge cars.

that the proportion of large cars in Japan does not differ significantly from the proportion of large cars in the USA.

The confidence intervals and the p-values that JMP reports are calculated in a different way than is explained in Section 8.2. JMP uses the so-called adjusted Wald procedure, which is not discussed in detail here.

The results of Fisher's exact test for two population proportions are automatically listed in JMP's standard output. The corresponding output is shown in Figure 8.21. Fisher's test does not use approximations in the calculation of p-values – therefore, this test is called exact. The p-values in Fisher's exact test are computed using the hypergeometric distribution (see *Statistics with JMP: Graphs, Descriptive Statistics and Probability*). The use of the hypergeometric distribution in the context of Fisher's exact test is beyond the scope of this book.

8.4.3   Two Population Variances

If we use the option "Unequal variances" in Figure 8.9, we can test whether the variances of the two populations under study are equal. For the data on the weight of Japanese and American cars, this generates the output shown in Figure 8.22.

The result of the F-test from Section 8.3.2 appears next to the label "F Test 2-sided" in the output. The p-value for the F-test is 0.0849. This p-value was computed from an F-distribution with 29 numerator degrees of freedom and 48 denominator degrees of freedom. The test statistic is 1.7475, which is the quotient of the sample variances of the weights of the Japanese and American cars, $(623.0700)^2 = 388216.2069$ and $(471.33)^2 = 222150.1063$. The p-value for the F-test is larger than the usual significance level of 5%, so that we would accept the null hypothesis of equal variances based on this test.

Two Sample Test for Proportions

Description	Proportion Difference	Lower 95%	Upper 95%
P(Not Large|Japan)-P(Not Large|USA)	0.111565	-0.07493	0.272238

Adjusted Wald Test	Prob
P(Not Large|Japan)-P(Not Large|USA) ≥ 0	0.1327
P(Not Large|Japan)-P(Not Large|USA) ≤ 0	0.8673
P(Not Large|Japan)-P(Not Large|USA) = 0	0.2653

Figure 8.20   The alternative output of the option "Two Sample Test for Proportions" in the analysis of the variable "Size".

Tests

N	DF	-LogLike	RSquare (U)
79	1	0.75088697	0.0189

Test	ChiSquare	Prob>ChiSq
Likelihood Ratio	1.502	0.2204
Pearson	1.434	0.2311

Fisher's Exact Test

	Prob	Alternative Hypothesis
Left	0.1826	Prob(Size=Not Large) is greater for Country=Japan than USA
Right	0.9346	Prob(Size=Not Large) is greater for Country=USA than Japan
2-Tail	0.2649	Prob(Size=Not Large) is different across Country

Figure 8.21   The JMP output for Fisher's exact test in the analysis of the variable "Size".

The F-test for two variances is only useful for normally distributed populations. There are, however, alternative tests for the comparison of two or more variances that also work if the studied populations are not normally distributed. Some of these tests are performed by JMP when the option "Unequal Variances" is chosen. The corresponding p-values are also shown in Figure 8.22. It is remarkable that the O'Brien, Brown–Forsythe, and Levene tests all yield p-values that are smaller than the significance level. Based on these three tests, we would

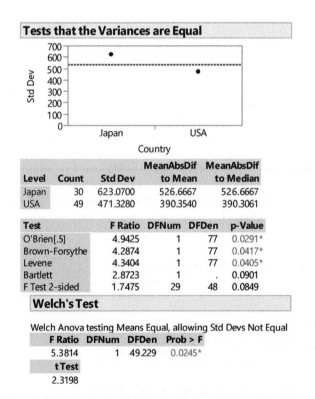

Tests that the Variances are Equal

Level	Count	Std Dev	MeanAbsDif to Mean	MeanAbsDif to Median
Japan	30	623.0700	526.6667	526.6667
USA	49	471.3280	390.3540	390.3061

Test	F Ratio	DFNum	DFDen	p-Value
O'Brien[.5]	4.9425	1	77	0.0291*
Brown-Forsythe	4.2874	1	77	0.0417*
Levene	4.3404	1	77	0.0405*
Bartlett	2.8723	1	.	0.0901
F Test 2-sided	1.7475	29	48	0.0849

Welch's Test

Welch Anova testing Means Equal, allowing Std Devs Not Equal

F Ratio	DFNum	DFDen	Prob > F
5.3814	1	49.229	0.0245*

t Test
2.3198

Figure 8.22   The output of the option "Unequal Variances".

have to reject the null hypothesis and assume that the variances are unequal. The Bartlett test produces a p-value that is larger than the significance level of 5%.

Given that there is evidence that the variances of the weights of Japanese and American cars might not be equal, it is better to use a t-test for unequal variances when comparing the population means.

We will study the O'Brien, Brown–Forsythe, Levene, and Bartlett tests in detail in Chapter 14.

9

A Nonparametric Hypothesis Test for the Medians of Two Independent Samples

If the convoys always avoid the U-boats, if the air forces always go to the German convoys, then it is clear to the Germans – I'm speaking of a very bright sort of German here, a German of the professor type – that there is no randomness here. This German can find correlations. He can also see we know more than we should.

(from *Cryptonomicon*, Neal Stephenson, pp. 152–153)

The *t*-tests and *z*-tests for two population means from the previous chapter can only be used if the populations under study are normally distributed, or if the two samples from the studied populations contain at least 30 observations. If there is no certainty about the normal distribution of the populations and at least one of the samples is small, then it is appropriate to apply a nonparametric test for the medians or the locations of the two populations, instead of a test for the population means. The recommended test in that case is the Wilcoxon rank-sum test.

If the variables under study are ordinal rather than quantitative, the Wilcoxon rank-sum test can still be used. For the rank-sum test, it is not necessary that the populations under study have a symmetric probability distribution or probability density.

The Wilcoxon rank-sum test is also called the Mann–Whitney *U*-test. Wilcoxon, on the one hand, and Mann and Whitney, on the other, independently developed the same test, in 1945 and 1947, respectively. A similarity between the rank-sum test and the signed-rank test is that both tests use ranks. However, in the rank-sum test no signs are assigned to the ranks. Also, the rank-sum test is used for two independent samples, while the signed-rank test is either used to test one population (see Chapter 5) or to test paired observations (see Chapter 11).

Unlike the *t*-test and the *z*-test, the Wilcoxon rank-sum test is insensitive to outliers in the data. The rank-sum test is slightly less powerful than the *t*-test and *z*-test if the populations under study are normally distributed. However, the rank-sum test is much better than the *t*-test or the *z*-test in certain other cases.

Statistics with JMP: Hypothesis Tests, ANOVA and Regression, First Edition. Peter Goos and David Meintrup.
© 2016 John Wiley & Sons, Ltd. Published 2016 by John Wiley & Sons, Ltd.
Companion Website: http://www.wiley.com/go/goosandmeintrup/JMP

9.1   The Hypotheses Tested

The rank-sum test is designed to determine whether the medians of two populations are equal. The test's null hypothesis is

$$H_0 : \mathrm{Me}_1 = \mathrm{Me}_2,$$

where Me_1 and Me_2 represent the medians of the first population and the second population, respectively. The corresponding potential alternative hypotheses are

$$H_a : \mathrm{Me}_1 > \mathrm{Me}_2,$$
$$H_a : \mathrm{Me}_1 < \mathrm{Me}_2,$$

and

$$H_a : \mathrm{Me}_1 \neq \mathrm{Me}_2.$$

Some researchers say that the rank-sum test is about the locations of two populations. The locations of two populations are equal if their probability distributions or densities overlap completely. The locations are unequal if the probability distribution or density of one population lies to the right or to the left of the other population's probability distribution or density. In terms of the locations of the populations under study, the null hypothesis is

$$H_0 : \text{Location 1} = \text{Location 2}.$$

The corresponding right-tailed alternative hypothesis is

$$H_a : \text{Location 1} > \text{Location 2},$$

while the left-tailed alternative hypothesis is

$$H_a : \text{Location 1} < \text{Location 2}.$$

For a two-tailed test, the alternative hypothesis is

$$H_a : \text{Location 1} \neq \text{Location 2}.$$

9.1.1   The Procedure

The rank-sum test is based on two independent samples, which do not need to have the same size. We call the number of observations in the first sample (from the first population) n_1 and the number of observations in the second sample (from the second population) n_2. The total number of observations is $n = n_1 + n_2$.

The test procedure is as follows:

(1) Assign the number "1" to the smaller sample and the number "2" to the larger sample. If both samples have the same size, the assignment does not matter.

(2) Join the two samples and order the sample observations from small to large.
(3) Assign a rank number to each observation: 1 for the smallest observation in the pooled data set, 2 for the next smallest observation, and so on. (If the same observation occurs several times, use an average rank. Suppose that the value in sixth place occurs four times. Then, the rank to be used four times is $(6 + 7 + 8 + 9)/4 = 7.5$. For the next observation, the rank 10 is used.)
(4) Sum all ranks that belong to the first sample. Call the resulting sum t_1.
(5) Sum all ranks that belong to the second sample. Call the resulting sum t_2.
(6) Verify that

$$t_1 + t_2 = \frac{n(n + 1)}{2} = \frac{(n_1 + n_2)(n_1 + n_2 + 1)}{2}.$$

In the above procedure, we denote the rank sums t_1 and t_2 using lowercase letters. This is because we have assumed that the data has been collected. Where no data are available yet, the observations themselves, their ranks, and the rank sums are still random variables. Therefore, as long as no data are available, we denote the rank sums by T_1 and T_2, using uppercase letters.

9.1.2 The Starting Point

If the null hypothesis is correct, large ranks and small ranks will be spread more or less evenly across the two samples. If both samples have the same size ($n_1 = n_2$) and the null hypothesis is correct, then T_1 and T_2 will take about the same value. If the two samples do not have the same size (where $n_1 < n_2$) and the null hypothesis is correct, T_1 will take a value close to n_1/n_2 times the value taken by T_2.

This can be explained as follows. The rank of an individual observation is a random variable. If the null hypothesis is correct, then, for each observation (regardless of whether it is an observation from the first sample or an observation from the second sample), it is equally likely to be assigned rank 1 as it is to get rank 2, rank 3, ... The rank of an individual observation thus has a uniform probability distribution: each rank has probability $1/n = 1/(n_1 + n_2)$. The expected value of an individual observation's rank number is therefore

$$\frac{n + 1}{2}.$$

The sum T_1 of all n_1 ranks for the first sample then has the expected value

$$E(T_1) = \frac{n_1(n + 1)}{2} = \frac{n_1(n_1 + n_2 + 1)}{2},$$

while the sum T_2 of all n_2 ranks for the second sample has the expected value

$$E(T_2) = \frac{n_2(n + 1)}{2} = \frac{n_2(n_1 + n_2 + 1)}{2}.$$

These calculations imply that, if the null hypothesis is true,

$$\frac{n+1}{2} = \frac{E(T_1)}{n_1} = \frac{E(T_2)}{n_2},$$

so that

$$E(T_1) = \frac{n_1}{n_2} E(T_2).$$

This equation states that if the null hypothesis is correct, the computed value of T_1 will be close to the computed value of $n_1 T_2 / n_2$.

If, however, the computed value of T_1 is much larger than the computed value of $n_1 T_2 / n_2$, then this suggests that the probability distribution or density of the first population lies to the right of the probability distribution or density of the second population. In that case, the median of the first population, Me_1, is greater than the median of the second population, Me_2.

If the computed value of T_1 is much smaller than the computed value of $n_1 T_2 / n_2$, then this suggests that the probability distribution or density of the first population lies to the left of the probability distribution or density of the second population. In that case, the median of the first population, Me_1, is smaller than the median of the second population, Me_2.

The expected values $E(T_1)$ and $E(T_2)$ are not affected by the presence of ties, but the variances of T_1 and T_2 are. If there are no ties, then

$$\text{var}(T_1) = \text{var}(T_2) = \frac{n_1 n_2 (n+1)}{12}.$$

The proof of this result, which is beyond the scope of this book, is based on sampling theory for finite populations. If ties do occur, then

$$\text{var}(T_1) = \text{var}(T_2) = \frac{n_1 n_2}{12} \left(n + 1 - \frac{\sum_{i=1}^{g} (u_i - 1) u_i (u_i + 1)}{n(n-1)} \right),$$

where g represents the number of groups of ties, and u_i is the number of observations in the ith group of ties. In the presence of ties, the variances of T_1 and T_2 thus are slightly lower than without ties.

To complete the Wilcoxon rank-sum test, we need to find critical values and p-values. As with the signed-rank test in Section 5.2, there is an exact method for determining the p-values. For large samples, however, an approximation method has to be used to find the p-values.

9.2   Exact p-Values in the Absence of Ties

For the determination of exact p-values, all possible scenarios for the assignment of ranks have to be enumerated. We illustrate the enumeration procedure for the computation of exact p-values using an example where $n_1 = 2$ and $n_2 = 4$. In other words, we assume that two observations are made from the first population and four observations are taken from the second population. The total number of observations is therefore $n = n_1 + n_2 = 6$. If there are no ties, we need to assign the integer ranks $1, 2, \ldots, 6$ to the six observations.

Table 9.1 Fifteen possible scenarios for the assignment of ranks to a first sample with two observations and a second sample with four observations when implementing Wilcoxon's rank-sum test without ties.

Scenario	Sample 1			Sample 2			t_1	t_2	Probability
1	1	2	3	4	5	6	3	18	$\frac{1}{15}$
2	1	3	2	4	5	6	4	17	$\frac{1}{15}$
3	1	4	2	3	5	6	5	16	$\frac{1}{15}$
4	1	5	2	3	4	6	6	15	$\frac{1}{15}$
5	1	6	2	3	4	5	7	14	$\frac{1}{15}$
6	2	3	1	4	5	6	5	16	$\frac{1}{15}$
7	2	4	1	3	5	6	6	15	$\frac{1}{15}$
8	2	5	1	3	4	6	7	14	$\frac{1}{15}$
9	2	6	1	3	4	5	8	13	$\frac{1}{15}$
10	3	4	1	2	5	6	7	14	$\frac{1}{15}$
11	3	5	1	2	4	6	8	13	$\frac{1}{15}$
12	3	6	1	2	4	5	9	12	$\frac{1}{15}$
13	4	5	1	2	3	6	9	12	$\frac{1}{15}$
14	4	6	1	2	3	5	10	11	$\frac{1}{15}$
15	5	6	1	2	3	4	11	10	$\frac{1}{15}$

If the null hypothesis is correct, then any assignment of these ranks to the six observations is equally likely. One possibility is that the two observations from the first sample will be assigned the ranks 1 and 2, but it is equally possible that they will get ranks 3 and 4, or ranks 2 and 6. In total, there are

$$\binom{n}{n_1} = \frac{n!}{n_1! \, (n - n_1)!} = \frac{n!}{n_1! \, n_2!} = \frac{6!}{2! \, 4!} = 15$$

possible scenarios. All these scenarios are listed in Table 9.1.

The table shows that the possible values for T_1 are the integers 3 to 11, while the possible values for T_2 are the integers 10 to 18. The sum of the ranks in the second sample can take larger values, because the second sample contains more observations. Therefore, for the calculation of the value of T_2, a larger number of ranks has to be summed. Table 9.1 also shows that not all values of T_1 and T_2 are equally likely. For example, the value 7 for T_1 is three times as likely as the value 3.

Starting from Table 9.1, we can derive the probability distribution for T_1 in Table 9.2. Based on that probability distribution, it is not difficult to verify that

$$E(T_1) = \frac{n_1(n + 1)}{2} = \frac{2(6 + 1)}{2} = 7$$

Table 9.2 The probability distribution of T_1 in the rank-sum test for a first sample with two observations and a second sample with four observations without ties.

t_1	Number of scenarios	$P(T_1 = t_1)$		$P(T_1 \leq t_1)$	$P(T_1 \geq t_1)$
3	1	$\frac{1}{15}$	0.06667	0.06667	1.00000
4	1	$\frac{1}{15}$	0.06667	0.13334	0.93333
5	2	$\frac{2}{15}$	0.13333	0.26667	0.86666
6	2	$\frac{2}{15}$	0.13333	0.40000	0.73333
7	3	$\frac{3}{15}$	0.20000	0.60000	0.60000
8	2	$\frac{2}{15}$	0.13333	0.73333	0.40000
9	2	$\frac{2}{15}$	0.13333	0.86666	0.26667
10	1	$\frac{1}{15}$	0.06667	0.93333	0.13334
11	1	$\frac{1}{15}$	0.06667	1.00000	0.06667

and that

$$\text{var}(T_1) = \frac{n_1 n_2 (n+1)}{12} = \frac{2 \times 4 \times (6+1)}{12} = 4.6667.$$

The probability distribution for T_1 in Table 9.2 allows us to determine p-values for the rank-sum test. More specifically, the probabilities $P(T_1 \leq t_1)$ and $P(T_1 \geq t_1)$ are used as p-values.

A detailed table with exceedance probabilities $P(T_1 \geq t_1)$ for various sample sizes n_1 and n_2 is given in Appendix H. Because the probability distribution of T_1 is symmetric around the expected value $E(T_1) = n_1(n+1)/2$, a probability of the type $P(T_1 \leq t_1^*)$ can always be found as

$$P(T_1 \leq t_1^*) = P(T_1 \geq n_1(n+1) - t_1^*).$$

For example, using Table 9.2, we can verify that

$$P(T_1 \leq 5) = P(T_1 \geq 2(6+1) - 5) = P(T_1 \geq 9) = 0.26667.$$

9.2.1 The Right-Tailed Test

When performing a right-tailed test, we reject the null hypothesis if T_1 takes a large value t_1; that is, a value that is unlikely when the null hypothesis is correct. The right-tailed or exceedance probability,

$$p = P(T_1 \geq t_1),$$

quantifies the improbability of the computed t_1 value. If the computed value t_1 is large, the exceedance probability will be small, suggesting that the null hypothesis might be incorrect. We reject the null hypothesis if the exceedance probability, which acts as p-value, is smaller than the significance level α chosen.

9.2.2   The Left-Tailed Test

When performing a left-tailed test, we reject the null hypothesis if T_1 takes a small value t_1; that is, a value that is unlikely if the null hypothesis is correct. The left-tailed probability

$$p = P(T_1 \le t_1)$$

quantifies the improbability of the t_1 value computed. If the computed value t_1 is small, then the left-tailed probability will also be small. This again suggests that the null hypothesis might be incorrect. We reject the null hypothesis if the left-tailed probability, which acts as p-value, is smaller than the significance level α.

Note that the left-tailed probability $P(T_1 \le t_1)$ can be derived from the table in Appendix H by exploiting the fact that

$$P(T_1 \le t_1^*) = P\left(T_1 \ge n_1(n+1) - t_1^*\right).$$

This approach is illustrated below by means of a small example.

Example 9.2.1   *A journalist asks several economists, $n_1 = 6$ government economists and $n_2 = 7$ academics, for a prediction of the price inflation rate over the next four years. The reason for the study is that the journalist has the impression that government economists are overly optimistic concerning the inflation rate, in order not to embarrass the government. In other words, the journalist is convinced that government economists systematically underestimate the inflation rate. The predictions made by the 13 interviewees are shown in Table 9.3.*

The journalist wants to show that the predictions of the government economists are smaller than those of the academics, at a significance level of $\alpha = 0.05$. The hypothesis test is therefore a left-tailed one. The tested hypotheses can be written as

$$H_0 : \text{The predictions of government economists and academics}$$
$$\text{are equal}$$

Table 9.3   Predictions of the inflation rate by government economists and academics for Example 9.2.1, along with the corresponding ranks.

Government economists		Academics	
Prediction	Rank	Prediction	Rank
3.1	4	4.4	6
4.8	7	5.8	9
2.3	2	3.9	5
5.6	8	8.7	11
0.0	1	6.3	10
2.9	3	10.5	12
		10.8	13

versus

$$H_a : \text{ The predictions of government economists are lower}$$
$$\text{than those of academics.}$$

In technical jargon, the hypotheses tested are

$$H_0 : Location \; 1 \; (government) = Location \; 2 \; (academics)$$

versus

$$H_a : Location \; 1 < Location \; 2.$$

Ordering the $n = n_1 + n_2 = 6 + 7 = 13$ observations from small to large leads to the ranks in the second and fourth columns of Table 9.3. Summing all ranks that belong to sample 1 yields $t_1 = 25$. For sample 2, the sum of all ranks is $t_2 = 66$. The sum of t_1 and t_2 is thus $n(n + 1)/2 = 13(13 + 1)/2 = 91$.

Since the journalist conducts a left-tailed hypothesis test, the p-value is

$$P(T_1 \leq 25) = P(T_1 \geq n_1(n + 1) - 25) = P(T_1 \geq 6(13 + 1) - 25)$$
$$= P(T_1 \geq 59) = 0.007.$$

This p-value can be found in the table in Appendix H for $n_1 = 6$ and $n_2 = 7$. As the p-value is smaller than 0.05, we can conclude that the predictions of government economists are significantly lower than those of academics.

The left-tailed test can also be carried out in JMP. To this end, we first have to enter the data in the correct format. Figure 9.1 shows how this should be done. It is important that the variable "Prediction" is a quantitative variable, while the variable "Type of Economist" is a nominal variable.

Next, we have to use the option "`Fit Y by X`" in the "`Analyze`" menu. The required input in the dialog window of the option "`Fit Y by X`" is shown in Figure 9.2. This input initially results in the graphical output shown in Figure 9.3. To obtain exact p-values for the rank-sum test, click the hotspot (red triangle) next to the word "`Oneway Analysis of Predic-tion by Type of Economist`" and choose "`Nonparametric`", "`Exact Test`", and "`Wilcoxon Exact Test`". The successive choices are shown in Figure 9.4.

The final output is shown in Figure 9.5. The output consists of four parts. The first section contains summary information involving, among other things, the values n_1 and n_2, and the computed values t_1 and t_2. The values n_1 and n_2 appear in the column labeled "`Count`", while the values t_1 and t_2 appear in the column "`Score Sum`". Note that JMP first shows the results for academics, because the word "academic" alphabetically comes before the word "government". The next part of the output shows the result of an approximate z-test, as explained in Section 9.4. The third part of the output contains a χ^2-test, which is not addressed in this book. Finally, the last part of the output provides the exact p-values for one-sided and two-sided rank-sum tests. The p-value for the one-sided test is labeled "`Prob≤S`", while the p-value for the two-sided test is labeled "`Prob≥|S-Mean|`".

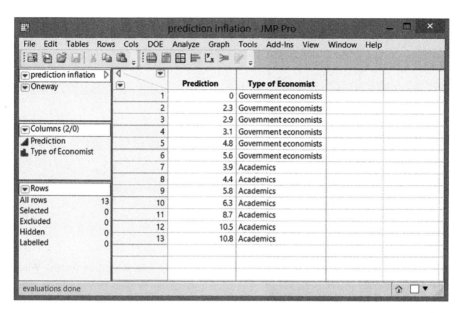

Figure 9.1   The JMP data table with the predictions of government economists and academics for the Wilcoxon rank-sum test in Example 9.2.1.

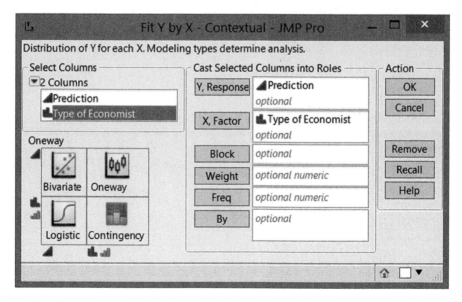

Figure 9.2   The required input in the dialog window of the option "Fit Y by X" in JMP for the Wilcoxon rank-sum test in Example 9.2.1.

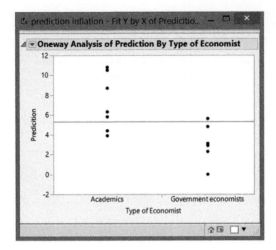

Figure 9.3 A plot of the predictions of government economists and academics for the Wilcoxon rank-sum test in Example 9.2.1.

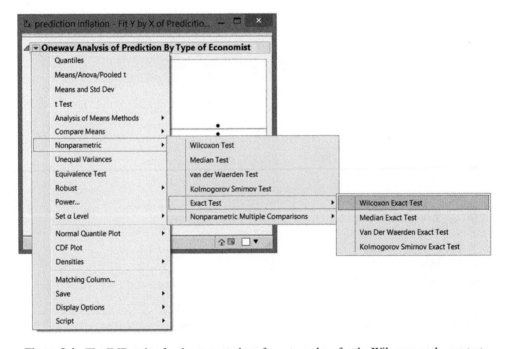

Figure 9.4 The JMP option for the computation of exact p-values for the Wilcoxon rank-sum test.

Wilcoxon / Kruskal-Wallis Tests (Rank Sums)					
Level	Count	Score Sum	Expected Score	Score Mean	(Mean-Mean0)/Std0
Academics	7	66.000	49.000	9.42857	2.357
Government economists	6	25.000	42.000	4.16667	-2.357

2-Sample Test, Normal Approximation		
S	Z	Prob>|Z|
25	-2.35714	0.0184*

1-way Test, ChiSquare Approximation		
ChiSquare	DF	Prob>ChiSq
5.8980	1	0.0152*

2-Sample: Exact Test		
S	Prob≤S	Prob≥|S-Mean|
25	0.0070*	0.0140*

Figure 9.5 The output of the option "`Wilcoxon Exact Test`" in JMP for the Wilcoxon rank-sum test in Example 9.2.1.

It is interesting to note that the exact p-value of 0.007 in the final part of the output differs from the p-value of 0.0184 in the second part, which is based on an approximation using the normal distribution.

9.2.3   The Two-Tailed Test

When performing a two-tailed test, we reject the null hypothesis if T_1 takes a small value t_1, or if T_1 takes a large value t_1. The computation of the p-value for a two-sided test is, as usual, slightly more complicated than for a one-sided test. For the two-tailed rank-sum test, the p-value is

$$p = 2\,P(T_1 \geq t_1)$$

if $t_1 \geq n_1(n + 1)/2$, and

$$p = 2\,P(T_1 \leq t_1)$$

if $t_1 \leq n_1(n + 1)/2$. In exceptional cases[1], this method leads to a p-value larger than 1. If this happens, we simply report the value 1 as the p-value.

We reject the null hypothesis if the p-value is smaller than the significance level α. In the opposite case, we cannot reject the null hypothesis.

9.3   Exact p-Values in the Presence of Ties

The table in Appendix H is only valid if there are no ties. If ties occur, it is necessary to reenumerate all possible scenarios, taking the ties into account. Then, a conditional probability

---

[1] These situations occur when the computed sum t_1 is exactly equal to the expected value $E(T_1) = n_1(n + 1)/2$.

Table 9.4   A modified data table with predictions of government
economists and academics, involving a tie.

Government economists		Academics	
Prediction	Rank	Prediction	Rank
3.1	2.5	4.4	4
5.9	6	5.8	5
2.3	1	3.1	2.5

distribution can be constructed for the sum of the ranks in the first sample, T_1. This probability
distribution is called conditional because it is only valid for the particular pattern of ties at hand.

We illustrate the construction of the conditional probability distribution by means of a small
example, where $n_1 = n_2 = 3$. Consequently, $n = n_1 + n_2 = 6$. The data for the small example
is shown in Table 9.4, along with the ranks. The smallest observation receives rank 1, the
three but smallest observation receives rank 4, the one but largest observation receives rank
5, and the largest observation receives rank 6. The two remaining observations are equal, and
should therefore be assigned the same rank. Normally, these two observations would occupy
the positions 2 and 3 in the ranking. Since the mean of 2 and 3 is equal to 2.5, the average rank
2.5 is assigned to both observations.

We now have to enumerate all possible assignments of the six ranks 1, 2.5, 2.5, 4, 5, and
6 to the two samples. When doing so, it is important to make a distinction between the two
occurrences of the average rank 2.5. We do so by denoting the first occurrence of the average
rank 2.5 as 2.5^a and the second occurrence by 2.5^b.

If the null hypothesis is correct, then any assignment of the six ranks to the two samples is
equally likely. In total, there are

$$\binom{n}{n_1} = \frac{n!}{n_1!\,(n - n_1)!} = \frac{n!}{n_1!\,n_2!} = \frac{6!}{3!\,3!} = 20$$

possible scenarios. All these scenarios are listed in Table 9.5.

The table shows that there are 12 possible values for T_1 and 12 possible values for T_2. This
time, not all of the possible values are integers. As the two samples have the same size, the
sums of the ranks in the first and in the second sample can take exactly the same values.

Based on Table 9.5, the conditional probability distribution for T_1 in Table 9.6 can be
constructed. The probability distribution of T_2 is identical to that of T_1. Table 9.6 shows that
not all values of T_1 (and T_2) are equally likely. For example, the value of 8.5 for T_1 (and T_2)
is twice as likely as the value 9.

Based on the conditional probability distribution in Table 9.6, it is not difficult to verify that

$$E(T_1) = \frac{n_1(n + 1)}{2} = \frac{3(6 + 1)}{2} = 10.5,$$

Table 9.5 Twenty possible scenarios for the assignment of the ranks 1, 2.5^a, 2.5^b, 4, 5, and 6 to a first sample with three observations and a second sample with three observations when performing the Wilcoxon rank-sum test, where 2.5^a denotes the first and 2.5^b the second occurrence of the average rank 2.5.

Scenario	Ranks						t_1	t_2	Probability
	Sample 1			Sample 2					
1	1	2.5^a	2.5^b	4	5	6	6	15	1/20
2	1	2.5^a	4	2.5^b	5	6	7.5	13.5	1/20
3	1	2.5^a	5	2.5^b	4	6	8.5	12.5	1/20
4	1	2.5^a	6	2.5^b	4	5	9.5	11.5	1/20
5	1	2.5^b	4	2.5^a	5	6	7.5	13.5	1/20
6	1	2.5^b	5	2.5^a	4	6	8.5	12.5	1/20
7	1	2.5^b	6	2.5^a	4	5	9.5	11.5	1/20
8	1	4	5	2.5^a	2.5^b	6	10	11	1/20
9	1	4	6	2.5^a	2.5^b	5	11	10	1/20
10	1	5	6	2.5^a	2.5^b	4	12	9	1/20
11	2.5^a	2.5^b	4	1	5	6	9	12	1/20
12	2.5^a	2.5^b	5	1	4	6	10	11	1/20
13	2.5^a	2.5^b	6	1	4	5	11	10	1/20
14	2.5^a	4	5	1	2.5^b	6	11.5	9.5	1/20
15	2.5^a	4	6	1	2.5^b	5	12.5	8.5	1/20
16	2.5^a	5	6	1	2.5^b	4	13.5	7.5	1/20
17	2.5^b	4	5	1	2.5^a	6	11.5	9.5	1/20
18	2.5^b	4	6	1	2.5^a	5	12.5	8.5	1/20
19	2.5^b	5	6	1	2.5^a	4	13.5	7.5	1/20
20	4	5	6	1	2.5^a	2.5^b	15	6	1/20

Table 9.6 The conditional probability distribution of T_1 for a rank-sum test with a first sample of three observations and a second sample of three observations involving a tie.

t_1	Number of scenarios	$P(T_1 = t_1)$	$P(T_1 \leq t_1)$	$P(T_1 \geq t_1)$
6	1	0.05	0.05	1.00
7.5	2	0.10	0.15	0.95
8.5	2	0.10	0.25	0.85
9	1	0.05	0.30	0.75
9.5	2	0.10	0.40	0.70
10	2	0.10	0.50	0.60
11	2	0.10	0.60	0.50
11.5	2	0.10	0.70	0.40
12	1	0.05	0.75	0.30
12.5	2	0.10	0.85	0.25
13.5	2	0.10	0.95	0.15
15	1	0.05	1.00	0.05

and

$$\mathrm{var}(T_1) = \frac{n_1 n_2}{12}\left(n + 1 - \frac{\sum_{i=1}^{g}(u_i - 1)u_i(u_i + 1)}{n(n-1)}\right),$$

$$= \frac{3 \times 3}{12}\left(6 + 1 - \frac{(2-1)\times 2 \times (2+1)}{6(6-1)}\right),$$

$$= 5.1.$$

In fact, there is one group involving two tied observations, so that $g = 1$ and $u_1 = 2$.

Table 9.6 can be used to determine probabilities of the types $P(T_1 \leq t_1)$ and $P(T_1 \geq t_1)$ that are needed to determine p-values for one-sided or two-sided rank-sum tests. The precise procedure for the determination of the p-values is similar to that described in Section 9.2 in the absence of ties.

Example 9.3.1   *Performing a left-tailed rank-sum test for the data in Table 9.4 results in a t_1 value of 9.5 and a t_2 value of 11.5. The sum of these two values is $n(n + 1)/2 = 6(6 + 1)/2 = 21$. The p-value of the left-tailed test is*

$$P(T_1 \leq 9.5) = P(T_1 \geq n_1(n+1) - 9.5) = P(T_1 \geq 3(6+1) - 9.5)$$

$$= P(T_1 \geq 11.5) = 0.40.$$

Since this p-value is larger than the journalist's significance level of 0.05, the data in Table 9.4 do not allow the null hypothesis to be rejected. The prediction of the government economists is not significantly smaller than that of the academics in the modified example involving a tie. The JMP output for the test is shown in Figure 9.6.

Wilcoxon / Kruskal-Wallis Tests (Rank Sums)

Level	Count	Score Sum	Expected Score	Score Mean	(Mean-Mean0)/Std0
Academics	3	11.500	10.500	3.83333	0.221
Government economists	3	9.500	10.500	3.16667	-0.221

2-Sample Test, Normal Approximation

S	Z	Prob>|Z|
9.5	-0.22140	0.8248

1-way Test, ChiSquare Approximation

ChiSquare	DF	Prob>ChiSq
0.1961	1	0.6579

Small sample sizes. Refer to statistical tables for tests, rather than large-sample approximations.

2-Sample: Exact Test

S	Prob≤S	Prob≥|S-Mean|
9.5	0.4000	0.8000

Figure 9.6   The output of the option "Wilcoxon Exact Test" in JMP for the Wilcoxon rank-sum test in Example 9.3.1.

9.4 Approximate p-Values

If the two samples contain more than 10 observations, another approach for determining p-values can be used. It turns out that, if the null hypothesis is true, T_1 is approximately normally distributed with expected value

$$E(T_1) = n_1(n+1)/2 = n_1(n_1 + n_2 + 1)/2$$

and variance

$$\operatorname{var}(T_1) = n_1 n_2(n+1)/12 = n_1 n_2(n_1 + n_2 + 1)/12. \tag{9.1}$$

This variance is only valid if there are no ties. If there are g groups of $u_1, u_2, \ldots, u_g$ tied observations, then the variance decreases to

$$
\begin{aligned}
\operatorname{var}(T_1) &= \frac{n_1 n_2}{12}\left(n+1 - \frac{\sum_{i=1}^{g}(u_i-1)u_i(u_i+1)}{n(n-1)}\right), \\
&= \frac{n_1 n_2}{12}\left(n_1 + n_2 + 1 - \frac{\sum_{i=1}^{g}(u_i-1)u_i(u_i+1)}{n(n-1)}\right).
\end{aligned}
\tag{9.2}
$$

Note that Equation (9.2) reduces to Equation (9.1) if there are no ties. Indeed, if there are no ties, $g = 0$.

As a result of the approximate normal distribution of T_1, the fraction

$$
\frac{T_1 - n_1(n_1 + n_2 + 1)/2}{\sqrt{\dfrac{n_1 n_2}{12}\left(n_1 + n_2 + 1 - \dfrac{\sum_{i=1}^{g}(u_i-1)u_i(u_i+1)}{n(n-1)}\right)}}
\tag{9.3}
$$

is approximately standard normally distributed. This fraction forms the basis for calculating approximate p-values for the Wilcoxon rank-sum test. It is, however, recommended to use a continuity correction in the numerator. The exact expression for the correction depends on whether a right-tailed, a left-tailed, or a two-tailed test is performed.

9.4.1 The Right-Tailed Test

The test statistic for an approximate right-tailed rank-sum test is

$$
Z = \frac{T_1 - n_1(n_1 + n_2 + 1)/2 - 1/2}{\sqrt{\dfrac{n_1 n_2}{12}\left(n_1 + n_2 + 1 - \dfrac{\sum_{i=1}^{g}(u_i-1)u_i(u_i+1)}{n(n-1)}\right)}}.
\tag{9.4}
$$

Note that this expression has a numerator that is $1/2$ smaller than Equation (9.3). This is due to the so-called continuity correction. The computed value of the test statistic Z is compared with the critical value z_α. If the computed value for the test statistic exceeds the critical value z_α, we reject the null hypothesis. Otherwise, we accept the null hypothesis.

The justification for this procedure is as follows. If the alternative hypothesis in a right-tailed rank-sum test is true, then T_1 takes a value that is large compared to the expected value $E(T_1) = n_1(n_1 + n_2 + 1)/2$. In that case, the numerator of the test statistic in Equation (9.4) is a large positive value, as is the complete test statistic. The critical value z_α allows us to judge whether the test statistic is sufficiently large for the null hypothesis to be rejected.

The approximate p-value for the right-tailed rank-sum test is

$$p = P\left(Z > \frac{t_1 - n_1(n_1 + n_2 + 1)/2 - 1/2}{\sqrt{\frac{n_1 n_2}{12}\left(n_1 + n_2 + 1 - \frac{\sum_{i=1}^{g}(u_i - 1)u_i(u_i + 1)}{n(n-1)}\right)}}\right).$$

If the p-value is smaller than the significance level α, then we reject the null hypothesis in favor of the alternative hypothesis. Otherwise, we accept the null hypothesis.

9.4.2 The Left-Tailed Test

The test statistic for a left-tailed rank-sum test is

$$Z = \frac{T_1 - n_1(n_1 + n_2 + 1)/2 + 1/2}{\sqrt{\frac{n_1 n_2}{12}\left(n_1 + n_2 + 1 - \frac{\sum_{i=1}^{g}(u_i - 1)u_i(u_i + 1)}{n(n-1)}\right)}}. \tag{9.5}$$

Note that this expression has a numerator that is $1/2$ larger than Equation (9.3). This is again due to the so-called continuity correction. The computed value of the test statistic Z is compared with the critical value $-z_\alpha$. If the computed value for the test statistic is smaller than the critical value $-z_\alpha$, we reject the null hypothesis. Otherwise, we accept the null hypothesis.

Indeed, if the alternative hypothesis of a left-tailed rank-sum test is true, then T_1 takes a value that is small compared to the expected value $E(T_1) = n_1(n_1 + n_2 + 1)/2$. In that case, the numerator of the test statistic in Equation (9.5) takes a strong negative value, as does the complete test statistic. The critical value $-z_\alpha$ allows us to judge whether the test statistic takes a sufficiently large negative value for the null hypothesis to be rejected.

The p-value for the left-tailed rank-sum test is

$$p = P\left(Z < \frac{t_1 - n_1(n_1 + n_2 + 1)/2 + 1/2}{\sqrt{\frac{n_1 n_2}{12}\left(n_1 + n_2 + 1 - \frac{\sum_{i=1}^{g}(u_i - 1)u_i(u_i + 1)}{n(n-1)}\right)}}\right).$$

If the p-value is smaller than the significance level α, then we reject the null hypothesis and accept the alternative hypothesis. If the p-value is larger than α, then we cannot reject the null hypothesis.

9.4.3   The Two-Tailed Test

The test statistic for a two-tailed rank-sum test is

$$Z = \frac{T_1 - n_1(n_1 + n_2 + 1)/2 + 1/2}{\sqrt{\dfrac{n_1 n_2}{12}\left(n_1 + n_2 + 1 - \dfrac{\sum_{i=1}^{g}(u_i - 1)u_i(u_i + 1)}{n(n - 1)}\right)}} \tag{9.6}$$

if the computed value for T_1 is smaller than $E(T_1) = n_1(n_1 + n_2 + 1)/2$, and

$$Z = \frac{T_1 - n_1(n_1 + n_2 + 1)/2 - 1/2}{\sqrt{\dfrac{n_1 n_2}{12}\left(n_1 + n_2 + 1 - \dfrac{\sum_{i=1}^{g}(u_i - 1)u_i(u_i + 1)}{n(n - 1)}\right)}} \tag{9.7}$$

if the computed value for T_1 is larger than $E(T_1) = n_1(n_1 + n_2 + 1)/2$.

The computed value of the test statistic Z is compared with the critical value $-z_{\alpha/2}$ (if $t_1 < n_1(n_1 + n_2 + 1)/2$) or with the critical value $z_{\alpha/2}$ (if $t_1 > n_1(n_1 + n_2 + 1)/2$). If the computed value for the test statistic is sufficiently negative or sufficiently positive, then we reject the null hypothesis. In the opposite case, we accept the null hypothesis.

The p-value for the two-tailed rank-sum test is

$$p = 2P\left(Z < \frac{t_1 - n_1(n_1 + n_2 + 1)/2 + 1/2}{\sqrt{\dfrac{n_1 n_2}{12}\left(n_1 + n_2 + 1 - \dfrac{\sum_{i=1}^{g}(u_i - 1)u_i(u_i + 1)}{n(n - 1)}\right)}}\right)$$

if the computed value for T_1 is smaller than $E(T_1) = n_1(n_1 + n_2 + 1)/2$, and

$$p = 2P\left(Z > \frac{t_1 - n_1(n_1 + n_2 + 1)/2 - 1/2}{\sqrt{\dfrac{n_1 n_2}{12}\left(n_1 + n_2 + 1 - \dfrac{\sum_{i=1}^{g}(u_i - 1)u_i(u_i + 1)}{n(n - 1)}\right)}}\right)$$

if the computed value for T_1 is greater than $E(T_1) = n_1(n_1 + n_2 + 1)/2$.

Again, we reject the null hypothesis if the *p*-value is smaller than the significance level α. If the *p*-value is larger than α, we cannot reject the null hypothesis and so we accept it.

Example 9.4.1   *Nowadays, university professors are evaluated on the basis of three criteria: teaching, research, and administration. For the evaluation of a professor's research performance, the number of publications in internationally renowned scientific journals is important, as well as the number of times that his articles are referred to by other authors. The latter figure is the number of citations.*

For a fair comparison of the number of citations of a professor in marketing and a professor in statistics, it is useful to compare how often scientific articles in the field of marketing are cited with how often scientific articles in the field of statistics are cited. The following table shows the numbers of citations for 12 marketing articles and 14 statistics articles:

Marketing	2, 3, 6, 6, 6, 9, 9, 10, 14, 15, 16, 72
Statistics	6, 8, 9, 10, 16, 18, 20, 22, 22, 23, 23, 26, 28, 59

The interest in this example lies in figuring out whether the citation patterns in marketing and in statistics are different. Therefore, a two-tailed test is appropriate. The data for marketing and statistics are counts, which are often Poisson distributed. Since the two data sets are small, it is unwise to conduct a test for two population means. Instead, the rank-sum test for the medians or locations of two independent samples is suitable. The hypotheses tested are the following:

$$H_0 : Location\ 1\ (Marketing) = Location\ 2\ (Statistics)$$

and

$$H_a : Location\ 1\ (Marketing) \neq Location\ 2\ (Statistics).$$

For the calculation of the rank sums t_1 and t_2, the data has to be ordered[2] and rank numbers have to be assigned to all observations. This was done in Table 9.7.

The table contains two columns with ranks, because there are many ties. The first set of ranks (in the penultimate column of the table) contains an initial rough ranking, not yet using average ranks to correct for ties. The second set of ranks in the last column of the table is adjusted for ties and involves average ranks. In the two samples, there is a total of four observations with the value 6. In the rough ranking from small to large, these four observations occupy positions 3, 4, 5, and 6. The average of these four ranks is $(3 + 4 + 5 + 6)/4 = 4.5$. This average rank is eventually used for the four observations with value 6.

In total, the original ranking contains six groups of ties, so that $g = 6$. The first group of ties contains the four observations with value 6, so that $u_1 = 4$. The second group of ties contains three observations with value 9, so that $u_2 = 3$. The remaining groups of ties each contain two observations (with values 10, 16, 22, and 23), so that u_3, u_4, u_5, and u_6 are equal to 2.

---

[2] Ranking data in JMP requires only a small effort: right-click on the header of the column you want to use for sorting and select the option "Sort". Alternatively, you can use the option "Sort" in the "Tables" menu.

Table 9.7   The detailed calculation of the ranks required for the rank-sum test in Example 9.4.1. Sample 1 refers to articles in marketing, whereas sample 2 refers to articles in statistics.

Sample	Observation	Rank[a]	Rank[b]
1	2	1	1
1	3	2	2
1	6	3	4.5
1	6	4	4.5
1	6	5	4.5
2	6	6	4.5
2	8	7	7
1	9	8	9
1	9	9	9
2	9	10	9
1	10	11	11.5
2	10	12	11.5
1	14	13	13
1	15	14	14
1	16	15	15.5
2	16	16	15.5
2	18	17	17
2	20	18	18
2	22	19	19.5
2	22	20	19.5
2	23	21	21.5
2	23	22	21.5
2	26	23	23
2	28	24	24
2	59	25	25
1	72	26	26

[a]Not adjusted for ties.
[b]Adjusted for ties.

Based on Table 9.7, it can be verified that $t_1 = 114.5$ *and* $t_2 = 236.5$, *and that* $t_1 + t_2 = n(n + 1)/2 = 26(26 + 1)/2 = 351$. *Since* t_1 *is smaller than* $E(T_1) = n_1(n_1 + n_2 + 1)/2 = 12(12 + 14 + 1)/2 = 162$, *the computed value of the test statistic is*

$$
z = \frac{t_1 - n_1(n_1 + n_2 + 1)/2 + 1/2}{\sqrt{\dfrac{n_1 n_2}{12}\left(n_1 + n_2 + 1 - \dfrac{\sum_{i=1}^{g}(u_i - 1)u_i(u_i + 1)}{n(n - 1)}\right)}},
$$

$$
= \frac{114.5 - 12(12 + 14 + 1)/2 + 1/2}{\sqrt{\dfrac{12 \times 14}{12}\left(12 + 14 + 1 - \dfrac{\sum_{i=1}^{6}(u_i - 1)u_i(u_i + 1)}{26 \times (26 - 1)}\right)}},
$$

$$= \frac{-47}{\sqrt{14 \left(27 - \dfrac{108}{650}\right)}},$$

$$= \frac{-47}{\sqrt{14 \times 26.83384615}},$$

$$= -2.42489.$$

The approximate p-value of the two-tailed test is

$$p = 2 \, P(Z < -2.42489) = 0.0153.$$

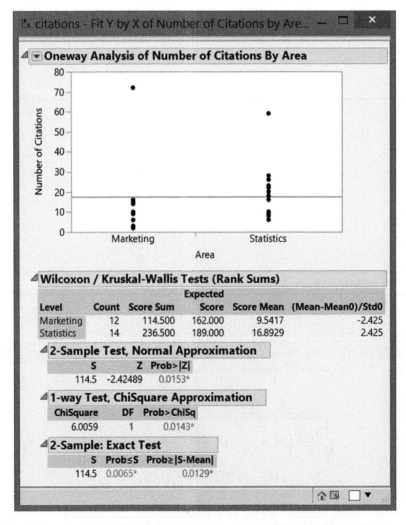

Figure 9.7   The output of the option "Wilcoxon Exact Test" in JMP for the Wilcoxon rank-sum test of Example 9.4.1.

This p-value is smaller than the usual significance level of 5%, so that the null hypothesis can be rejected. We therefore conclude that the numbers of citations in marketing journals and in statistical journals are significantly different. This implies that the research performances of marketing professors and statistics professors cannot be compared by merely counting their numbers of citations.

The JMP output for this example is shown in Figure 9.7. This output not only contains the approximate p-value based on the standard normal distribution, but also the exact p-value. The approximate p-value of 0.0153 is reported in the output part called "2-Sample Test, Normal Approximation". Note that here we perform a two-tailed test and the corresponding approximate p-value is labeled "Prob>|Z|" in the JMP output. The exact p-value can be found in the final part of the output, called "2-Sample: Exact Test", next to the label "Prob=|S-Mean|". It is equal to 0.0129.

10

Hypothesis Tests for the Means of Two Paired Samples

Therese had collected more than thirty tapes of data for me on barrels per month by country. I had to format these and load my own data by keypunch to produce trends on production, consumption and distribution. Then, I had to write the programs that would analyze it – all before this conference took place.

(from *The Eight*, Katherine Neville, p. 336)

If it is possible to identify pairs of observations in the two studied samples, then a more powerful approach for the comparison of two population means can be used. Directly comparing the observed results within each pair typically eliminates a substantial portion of the variability in the data, and thus leads to better results.

This has already been illustrated in Example 7.0.3, where the wear resistance of two types of shoe soles was investigated. In that example, we directly compared the two observations for each subject. This ensured that any variability in the wear measurements due to differences between subjects did not affect the study. This had a positive impact on the precision with which we could compare the population means.

The hypothesis test for the comparison of two population means for paired observations requires quantitative data, and either normally distributed populations or large samples.

10.1 The Hypotheses Tested

As in the test for two population means for independent samples, the tested null hypothesis is

$$H_0 : \mu_1 = \mu_2.$$

A right-tailed alternative hypothesis can again be written as

$$H_a : \mu_1 > \mu_2,$$

Statistics with JMP: Hypothesis Tests, ANOVA and Regression, First Edition. Peter Goos and David Meintrup.
© 2016 John Wiley & Sons, Ltd. Published 2016 by John Wiley & Sons, Ltd.
Companion Website: http://www.wiley.com/go/goosandmeintrup/JMP

while the left-tailed and two-tailed alternative hypotheses are

$$H_a : \mu_1 < \mu_2$$

and

$$H_a : \mu_1 \neq \mu_2.$$

These hypotheses can be rewritten as

$$H_0 : \mu_1 - \mu_2 = 0,$$
$$H_a : \mu_1 - \mu_2 > 0,$$
$$H_a : \mu_1 - \mu_2 < 0,$$

and

$$H_a : \mu_1 - \mu_2 \neq 0.$$

For paired observations, the difference $\mu_1 - \mu_2$ is generally replaced by the Greek letter δ. The null and alternative hypotheses then become

$$H_0 : \delta = 0,$$
$$H_a : \delta > 0,$$
$$H_a : \delta < 0,$$

and

$$H_a : \delta \neq 0.$$

This notation is the preferred one for paired data, because the hypothesis test requires the calculation of the differences between the two observations within each pair. The value δ is the expected value of this variable.

10.2 The Procedure

Example 7.0.3 has already made clear that the two observations within each pair have to be compared directly. If the observations are quantitative measurements, we can compute the difference between the two observations for each pair. By calculating this difference, we reduce the original two samples to a single sample.

10.2.1 The Starting Point

The start of the test procedure is shown schematically in Table 10.1. In the table, capital letters are used to emphasize that, as long as we have not collected any data yet, each individual observation is a random variable. Once we have collected the data, we can, of course, replace the uppercase letters by lowercase letters.

Table 10.1 A schematic representation of the data in the case of paired samples, along with the calculation of the pairwise differences $D_1, D_2, \ldots, D_n$.

Observation	Sample 1	Sample 2	Difference
1	X_{11}	X_{21}	$D_1 = X_{11} - X_{21}$
2	X_{12}	X_{22}	$D_2 = X_{12} - X_{22}$
3	X_{13}	X_{23}	$D_3 = X_{13} - X_{23}$
$\vdots$	$\vdots$	$\vdots$	$\vdots$
n	X_{1n}	X_{2n}	$D_n = X_{1n} - X_{2n}$

The table shows that, for paired observations, the numbers of observations in both samples are identical. We call this number of observations n. The ith observation in the first sample (from the first population) is called X_{1i}, while the ith observation in the second sample (from the second population) is called X_{2i}. For paired samples, these two observations, X_{1i} and X_{2i}, are inseparably linked to each other and need to be compared directly. This direct comparison is made by calculating the difference

$$D_i = X_{1i} - X_{2i}$$

for each of the n pairs of observations. In this way, we obtain a single sample of n differences $D_1, D_2, \ldots, D_n$.

The sample mean of the differences in this new sample is

$$\overline{D} = \frac{1}{n} \sum_{i=1}^{n} D_i,$$

while the sample variance of the new sample of differences is

$$S_D^2 = \frac{1}{n-1} \sum_{i=1}^{n} (D_i - \overline{D})^2.$$

Note that the sample mean $\overline{D}$ can also be calculated as

$$\overline{D} = \frac{1}{n} \sum_{i=1}^{n} (X_{1i} - X_{2i}),$$

$$= \frac{1}{n} \sum_{i=1}^{n} X_{1i} - \frac{1}{n} \sum_{i=1}^{n} X_{2i},$$

$$= \overline{X}_1 - \overline{X}_2,$$

and that

$$E(\overline{D}) = E(\overline{X}_1 - \overline{X}_2) = E(\overline{X}_1) - E(\overline{X}_2) = \mu_1 - \mu_2 = \delta.$$

The population variance of the differences $D_1, D_2, \ldots, D_n$ is denoted by σ_D^2. As the sample variance is an unbiased estimator of the population variance,

$$E\left(S_D^2\right) = \sigma_D^2.$$

In addition,

$$\mathrm{var}(\overline{D}) = \frac{\sigma_D^2}{n}.$$

If the differences D_i are normally distributed, then the sample mean $\overline{D}$ is automatically also normally distributed. The differences are normally distributed if the two populations under investigation are both normally distributed, because the difference of two normally distributed random variables is again normally distributed (see the book *Statistics with JMP: Graphs, Descriptive Statistics and Probability*). If the populations under investigation are not normally distributed, then the differences $D_i = X_{1i} - X_{2i}$ are also not normally distributed. The sample mean $\overline{D}$ is then approximately normally distributed, provided that n, the size of the two samples, is at least 30.

If the differences D_i are not normally distributed and both samples contain fewer than 30 observations, one can only use the nonparametric tests described in Chapter 11, namely the sign test and Wilcoxon's signed-rank test.

Whenever $\overline{D}$ is at least approximately normally distributed, we can perform the hypothesis test for two means for paired samples. In that case,

$$\frac{\overline{D} - \delta}{\sigma_D / \sqrt{n}}$$

is standard normally distributed. This result can be used immediately for a hypothesis test when the population standard deviation σ_D is known. If σ_D is not known, then this population standard deviation can be estimated using the sample standard deviation S_D. In order to perform the hypothesis test for two means in that scenario, we need to make use of the fact that

$$\frac{\overline{D} - \delta}{S_D / \sqrt{n}}$$

is t-distributed with $n - 1$ degrees of freedom.

The exact procedure for the hypothesis test with known σ_D is similar to the one for a hypothesis test for one population mean and known σ in Chapter 3. The exact procedure for the hypothesis test with unknown σ_D is similar to the one for a hypothesis test for one population mean and unknown σ in Section 4.1.

10.2.2   Known σ_D

For known σ_D, the test statistic is

$$Z = \frac{\overline{D}}{\sigma_D/\sqrt{n}}.$$

This test statistic is standard normally distributed if the null hypothesis, which states that $\delta = 0$, is correct.

The Right-Tailed Test

In a right-tailed test for two population means, the alternative hypothesis ($H_a : \mu_1 > \mu_2$, which is the same as $H_a : \delta > 0$) is accepted if the differences D_i are predominantly positive, so that the sample mean $\overline{D}$ takes a strongly positive value. The test statistic computed from the data,

$$z = \frac{\overline{d}}{\sigma_D/\sqrt{n}},$$

then is a large positive number.

To determine whether the computed test statistic z is sufficiently positive for the null hypothesis to be rejected in favor of the alternative hypothesis, we need to compare it with the critical value z_α. If $z > z_\alpha$, we reject the null hypothesis and accept the alternative hypothesis. In the opposite case, the computed test statistic is not sufficiently positive, and we can only accept the null hypothesis. An alternative is to compare the p-value

$$p = P(Z > z)$$

with the selected significance level α. If the p-value is smaller than α, we opt for the alternative hypothesis. If the p-value is greater than or equal to α, then we accept the null hypothesis.

When we choose the alternative hypothesis, we conclude that δ is significantly larger than zero or, alternatively, that μ_1 is significantly larger than μ_2.

The Left-Tailed Test

In a left-tailed test for two population means, the alternative hypothesis ($H_a : \mu_1 < \mu_2$, or $H_a : \delta < 0$) is accepted if the differences D_i are predominantly negative, so that the sample mean $\overline{D}$ takes a strongly negative value. The computed test statistic z then is a large negative number. To determine whether this number is sufficiently negative for the null hypothesis to be rejected in favor of the alternative hypothesis, we need to compare it with the critical value $-z_\alpha$. If $z < -z_\alpha$, then we reject the null hypothesis and choose the alternative hypothesis. In the opposite case, we accept the null hypothesis. An alternative is to compute the p-value

$$p = P(Z < z).$$

If this p-value is smaller than α, we opt for the alternative hypothesis. If the p-value is greater than or equal to α, then we choose the null hypothesis. When we choose the alternative hypothesis, we conclude that δ is significantly smaller than zero or, alternatively, that μ_1 is significantly smaller than μ_2.

The Two-Tailed Test

In a two-tailed test for two population means, the alternative hypothesis ($H_a : \mu_1 \neq \mu_2$, or $H_a : \delta \neq 0$) is accepted if the differences D_i are either predominantly negative or predominantly positive. In other words, the alternative hypothesis is accepted if the sample mean $\overline{D}$ takes either a strongly negative or a strongly positive value. The computed test statistic z is then either a large negative or a large positive number. In order to determine whether the test statistic z is sufficiently negative or sufficiently positive for the null hypothesis to be rejected in favor of the alternative hypothesis, we compare it to the critical values $-z_{\alpha/2}$ and $z_{\alpha/2}$. If $z < -z_{\alpha/2}$ or $z > z_{\alpha/2}$ (thus, if $|z| > z_{\alpha/2}$), then we accept the alternative hypothesis. Otherwise, we choose the null hypothesis. An alternative is to compute the p-value

$$p = 2\, P(Z > |z|).$$

If this p-value is smaller than α, then we accept the alternative hypothesis. If the p-value is greater than or equal to α, then we accept the null hypothesis. If we reject the null hypothesis and choose the alternative hypothesis, we conclude that δ is significantly different from zero, or that μ_1 is significantly different from μ_2.

10.2.3   Unknown σ_D

If σ_D is not known, we use the test statistic

$$t_{n-1} = \frac{\overline{D}}{S_D/\sqrt{n}}.$$

This test statistic is t-distributed with $n - 1$ degrees of freedom if the null hypothesis that $\delta = 0$ is correct.

The Right-Tailed Test

For a right-tailed test for two population means, the alternative hypothesis ($H_a : \mu_1 > \mu_2$, or $H_a : \delta > 0$) is accepted if the differences are predominantly positive, so that the sample mean $\overline{D}$ takes a strongly positive value, and the computed test statistic

$$t = \frac{\overline{d}}{s_D/\sqrt{n}}$$

is a large positive number. To determine whether the computed test statistic t is sufficiently positive for the null hypothesis to be rejected in favor of the alternative hypothesis, we need to compare it with the critical value $t_{\alpha;n-1}$. If $t > t_{\alpha;n-1}$, then we reject the null hypothesis and choose the alternative hypothesis. If $t \leq t_{\alpha;n-1}$, the computed test statistic is not sufficiently positive and we have to accept the null hypothesis. An alternative is to compare the p-value

$$p = P(t_{n-1} > t)$$

with the selected significance level α. If the p-value is smaller than α, we opt for the alternative hypothesis. If the p-value is greater than or equal to α, then we accept the null hypothesis.

The Left-Tailed Test

In a left-tailed test for two population means, the alternative hypothesis ($H_a : \mu_1 < \mu_2$, or $H_a : \delta < 0$) is accepted if the differences are predominantly negative, so that the sample mean $\overline{D}$ takes a strongly negative value, and the computed test statistic t is also a large negative number. To determine whether this value is sufficiently negative for the null hypothesis to be rejected in favor of the alternative hypothesis, we need to compare it with the critical value $-t_{\alpha;n-1}$. If $t < -t_{\alpha;n-1}$, then we reject the null hypothesis and choose the alternative hypothesis. In the opposite case, we accept the null hypothesis. An alternative is to compute the p-value

$$p = P(t_{n-1} < t).$$

If this p-value is smaller than α, we opt for the alternative hypothesis. Otherwise, we accept the null hypothesis.

The Two-Tailed Test

In a two-tailed test for two population means, the alternative hypothesis ($H_a : \mu_1 \neq \mu_2$, or $H_a : \delta \neq 0$) is accepted if the differences are either predominantly negative or predominantly positive. In other words, the alternative hypothesis is accepted if the sample mean $\overline{D}$ takes either a strongly negative or a strongly positive value. The computed test statistic t is then either a large negative or a large positive number. To determine whether the computed test statistic is sufficiently negative or sufficiently positive for the null hypothesis to be rejected in favor of the alternative hypothesis, we compare it with the critical values $-t_{\alpha/2;n-1}$ and $t_{\alpha/2;n-1}$. If $t < -t_{\alpha/2;n-1}$ or $t > t_{\alpha/2;n-1}$ (thus, if $|t| > t_{\alpha/2;n-1}$), then we accept the alternative hypothesis. If $-t_{\alpha/2;n-1} \leq t \leq t_{\alpha/2;n-1}$ (and thus, if $|t| \leq t_{\alpha/2;n-1}$), we accept the null hypothesis. An alternative is to compute the p-value

$$p = 2\, P(t_{n-1} > |t|).$$

If this p-value is smaller than α, then we accept the alternative hypothesis. If the p-value is greater than or equal to α, then we accept the null hypothesis. If we reject the null hypothesis

Table 10.2 The logarithms of the numbers of bacteria in 12 milk samples before and after heating, and the differences required for the t-test for paired samples.

Milk sample	x_{1i} (before)	x_{2i} (after)	d_i (before $-$ after)
1	6.98	6.95	0.03
2	7.08	6.94	0.14
3	8.34	7.17	1.17
4	5.30	5.15	0.15
5	6.26	6.28	-0.02
6	6.77	6.81	-0.04
7	7.03	6.59	0.44
8	5.56	5.34	0.22
9	5.97	5.98	-0.01
10	6.64	6.51	0.13
11	7.03	6.84	0.19
12	7.69	6.99	0.70

and choose the alternative hypothesis, we conclude that δ is significantly different from zero, or that μ_1 is significantly different from μ_2.

10.3   Examples

In this section, we study two practical problems. First, we perform a test for the comparison of the number of bacteria in milk samples. Next, we apply the t-test for paired samples to the data in Example 7.0.3, concerning two types of shoes. In both cases, not only do we perform the test for paired samples, but we also perform a test that incorrectly assumes independent samples.

Example 10.3.1   *To investigate whether the heating of milk is effective in reducing the number of bacteria, a microscope is used to count the bacteria in 12 milk samples. This happens both before and after the heating of the milk. It is clear that this leads to 12 pairs of sample observations that can be compared with each other directly.*

Because the distribution of the logarithm of the number of bacteria is often closer to the normal distribution than the number of bacteria itself, we carry out the hypothesis test using the logarithm[1] of the number of bacteria. The log-transformed data and the pairwise differences are displayed in Table 10.2. We assume, for convenience, that the logarithm of the number of bacteria is indeed normally distributed, so that we can use the t-test.

---

[1] Such a transformation is often used to convert nonnormal data into data that could have been generated by a normal distribution. This has already been illustrated in Section 6.2.

As we want to show that the heating results in a decrease in the number of bacteria, the tested hypotheses are

$$H_0 : \mu_1 = \mu_2$$

and

$$H_a : \mu_1 > \mu_2.$$

In these hypotheses, μ_1 represents the (logarithmized) mean number of bacteria before the heating, while μ_2 is the (logarithmized) mean number of bacteria after the heat treatment. An alternative notation for these hypotheses is

$$H_0 : \delta = 0$$

and

$$H_a : \delta > 0.$$

In this notation, δ is the mean difference between the log-transformed numbers of bacteria before and after treatment.

The sample mean $\overline{d}$ of all observed differences $d_1, d_2, \dots, d_n$ in Table 10.2 is equal to 0.2583. The sample variance s_D^2 of the 12 differences is 0.1271, while the sample standard deviation is equal to 0.3565. The computed test statistic is therefore

$$t = \frac{0.2583}{0.3565/\sqrt{12}} = 2.5101.$$

The p-value of the right-tailed test is

$$p = P(t_{n-1} > 2.5101) = P(t_{11} > 2.5101) = 0.0145.$$

The p-value is smaller than the usual significance level of 5%. As a result, we can reject the null hypothesis at a confidence level of 95%. In JMP, the p-value can be computed using the formula "$1 - $ t Distribution(2.5101, 11)". We reach the same conclusion if we compare the computed test statistic with $t_{\alpha,n-1} = t_{0.05;11} = 1.7959$. This quantile can be found in Appendix D, or calculated using the JMP formula "t Quantile(0.95, 11)".

In any case, we can conclude that the heating has a beneficial effect on the number of bacteria present in the milk.

The use of the test for paired samples allows us to take more powerful decisions. We can demonstrate this by testing the same hypotheses as in the preceding example, but now ignoring the paired nature of the samples. As a result, we pretend that the two samples of measurements before and after the heating are independent of each other.

Example 10.3.2   *Consider again the data in Table 10.2. It is not difficult to verify using JMP that $\bar{x}_1 = 6.7208$, $\bar{x}_2 = 6.4625$, $s_1^2 = 0.7363$ and $s_2^2 = 0.4341$. Assuming normally distributed data, independent samples, and unequal variances, we obtain the computed test statistic*

$$t = \frac{\bar{x}_1 - \bar{x}_2}{\sqrt{\dfrac{s_1^2}{n_1} + \dfrac{s_2^2}{n_2}}} = \frac{6.7208 - 6.4625}{\sqrt{\dfrac{0.7363}{12} + \dfrac{0.4341}{12}}} = 0.8272.$$

To compute the corresponding p-value, we first have to determine the number of degrees of freedom ν for the required t-distribution (see Section 8.1.3), based on Equation (8.3):

$$\nu = \frac{\left(\dfrac{s_1^2}{n_1} + \dfrac{s_2^2}{n_2}\right)^2}{\dfrac{\left(s_1^2/n_1\right)^2}{n_1 - 1} + \dfrac{\left(s_2^2/n_2\right)^2}{n_2 - 1}} = \frac{\left(\dfrac{0.7363}{12} + \dfrac{0.4341}{12}\right)^2}{\dfrac{(0.7363/12)^2}{12 - 1} + \dfrac{(0.4341/12)^2}{12 - 1}} = 20.6247.$$

The p-value for the right-tailed hypothesis test can now be computed as

$$p = P(t_\nu > 0.8272) = P(t_{20.6247} > 0.8272) = 0.2088.$$

This is a large p-value, indicating that we cannot reject the null hypothesis. As a result, when we ignore the paired nature of the samples and pretend that the two samples are independent, we conclude that the mean numbers of bacteria before and after the heating of the milk are not significantly different. This contradicts the conclusion in Example 10.3.1, where the paired nature of the samples was recognized and exploited.

This shows that the hypothesis test based on independent samples has a lower power (and, therefore, it is less likely to reject the null hypothesis) than the test based on paired observations.

It is not difficult to implement the *t*-test for paired observations in JMP. To this end, you need to enter the data from the two samples in two separate columns in a data table, as in Figure 10.1. In the "Analyze" menu, you then have to select the option "Matched Pairs" (see Figure 10.2). This produces the dialog window in Figure 10.3. Finally, you need to enter the variable "After", followed by the variable "Before" in the field "Y, Paired Response", and click on "OK".

The resulting output is shown in Figure 10.4. The output shows the mean values of all observations in the two samples, as well as the mean value of the differences, namely 0.25833. In addition, the computed value of the test statistic (2.5101) and the number of degrees of freedom (11) are displayed. JMP again provides three *p*-values: one for a right-tailed test ("Prob > t"), one for a left-tailed test ("Prob < t"), and one for a two-tailed test ("Prob > |t|"). The relevant *p*-value for Example 10.3.1 is the one for the right-tailed hypothesis test, 0.0145.

An important part of the output is the plot, which clearly indicates that the differences are positive for most observations. Each point in the graph corresponds to one of the pairs. A

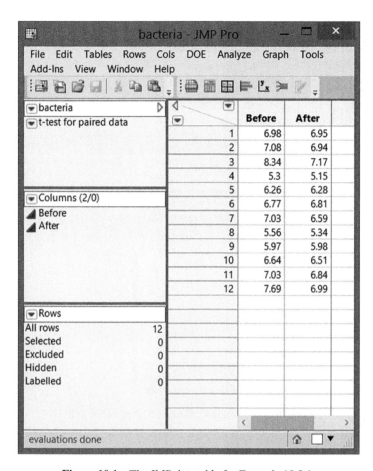

Figure 10.1   The JMP data table for Example 10.3.1.

point that is above zero has a positive value for the difference variable. A point below zero has a negative value for the difference variable. The leftmost point on the graph corresponds to the milk sample that contains the smallest average number of bacteria. This is the fourth milk sample, which had a value of 5.3 before and 5.15 after the heating. The average of these two values is 5.225, which is the smallest average of all of the milk samples in the data table in Figure 10.1.

By means of a solid horizontal line, the graph in Figure 10.4 also shows the mean of the difference variable over the 12 milk samples, 0.25833, and a 95% confidence interval for the difference in population means $\delta = \mu_1 - \mu_2$. This interval has a lower limit of 0.0318 and an upper limit of 0.4849. The interval does not contain the value zero, so that we can conclude that μ_1 (the mean number of bacteria before heating) is significantly different from μ_2 (the mean number of bacteria after heating). The calculation of the confidence interval for $\delta = \mu_1 - \mu_2$ is discussed in Section 10.6.

The output also shows the Pearson correlation coefficient between the observations in the first sample and the observations in the second sample. This correlation is 0.9227. The importance of this strong positive correlation is discussed in Section 10.4.

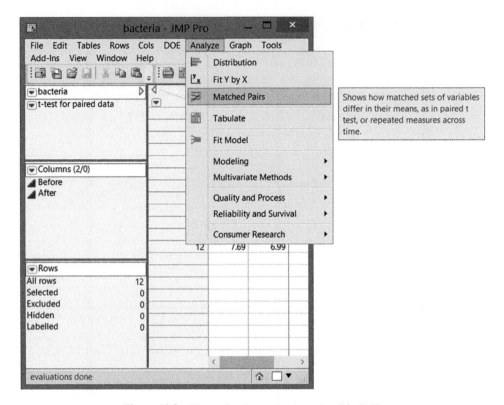

Figure 10.2 The option "Matched Pairs" in JMP.

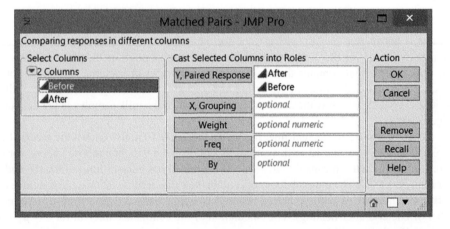

Figure 10.3 The dialog window for the *t*-test for paired samples in JMP for Example 10.3.1.

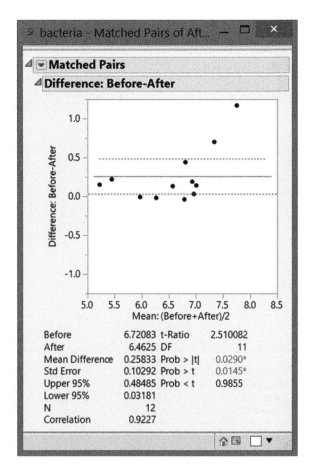

Figure 10.4   The JMP output for the *t*-test for paired samples for Example 10.3.1.

Finally, it is worth pointing out that, by default, JMP uses a significance level of 5% and thus a confidence level of 95%. You can always set the significance level to another value by clicking on the hotspot (red triangle) next to the word "Matched Pairs". This is illustrated in Figure 10.5.

The next example involves an experiment with two types of shoe sole. The graphical analysis in Example 7.0.3 has already revealed that it is crucial to take the paired nature of the observations into account in order to detect a difference in the wear of the two types of shoe.

Example 10.3.3   *Example 7.0.3 presents data from an experiment with paired observations, comparing two types of shoe sole made of Materials A and B. The JMP analysis using the option "Matched Pairs" yields the output shown in Figure 10.6. The output shows a confidence interval that is completely negative for the difference "Material A − Material B". This indicates that the wear of Material B is significantly greater than that of Material A. The p-value of the two-tailed t-test is 0.0085, which also indicates a significant difference in wear.*

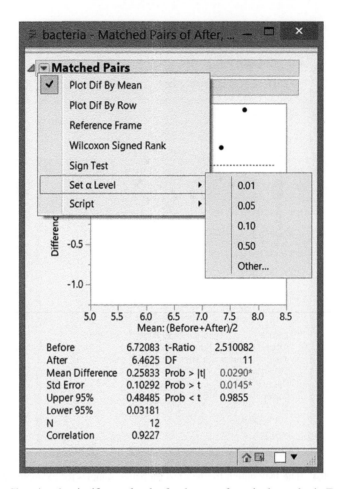

Figure 10.5   Changing the significance level α for the t-test for paired samples in Example 10.3.1.

The mean wear for Material A is 10.63, *while it is* 11.04 *for Material B. This results in a sample mean of* $10.63 - 11.04 = -0.41$ *for the difference variable D. The difference variable has a sample variance* s_D^2 *of* 0.1499 *and a sample standard deviation* s_D *of* 0.3872. *The standard error of the difference variable is* $s_D/\sqrt{n} = 0.3872/\sqrt{12} = 0.12243$. *The 95% confidence interval for* $\delta = \mu_1 - \mu_2$ *has lower and upper limits of* -0.6870 *and* -0.1330, *respectively.*

The computed test statistic for the paired t-test is

$$t = \frac{\overline{d}}{s_D/\sqrt{n}} = \frac{-0.41}{0.3872/\sqrt{12}} = \frac{-0.41}{0.12243} = -3.3489.$$

Therefore, the p-value for the two-tailed t-test is

$$p = 2\,P(t_9 > |-3.3489|) = 2 \times 0.00427 = 0.00854.$$

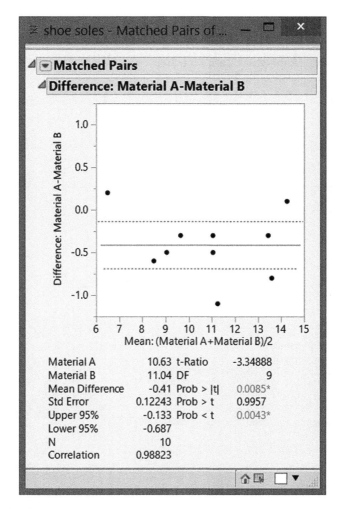

Figure 10.6 The JMP output for the *t*-test for paired samples for Examples 7.0.3 and 10.3.3.

If we were accidentally to implement the t-test for independent samples, we would not find a significant difference between the wear of the two types of shoes. This follows from the JMP output in Figure 10.7 and illustrates the importance of a correct data analysis; that is, one using the t-test for paired samples.

Note that we have assumed in the previous examples that the observations have come from normally distributed populations. We did not formally test this at any time. Given the small sample size, testing normality would make little sense here anyway. The examples in this chapter should therefore be viewed as illustrations of how the *t*-test for paired observations works, and of the extent to which this test differs from the *t*-test for independent samples. Because of the limited sample size in this chapter's examples, we recommend using the nonparametric hypothesis tests that we discuss in the next chapter.

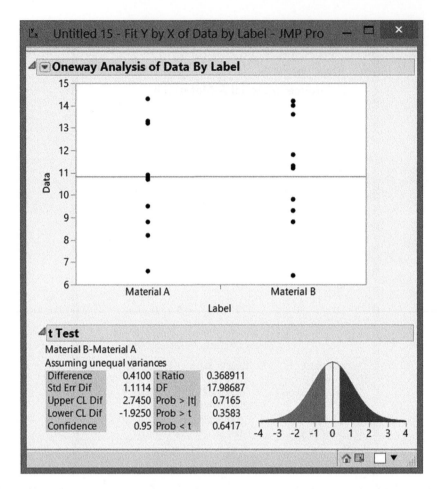

Figure 10.7   The JMP output for the *t*-test for independent samples for Examples 7.0.3 and 10.3.3.

10.4   The Technical Background

The reason for the differing results between the hypothesis test for independent samples, on the one hand, and the hypothesis test for paired samples that uses the differences $D_1, D_2, \ldots, D_n$, on the other, is that for paired samples, a large value in the first sample will often correspond to a large value in the second sample, and a small value in the first sample will often correspond to a small value in the second sample. In Example 10.3.1, a large (small) number of bacteria in the milk sample before the heating usually corresponds to a fairly large (small) number of bacteria after the heating. A similar observation can be made in Example 10.3.3. As a consequence, the values in the two paired samples are positively correlated, and thus they have a positive covariance σ_{12}.

The impact of the positive correlation and covariance between the observations from the first sample, $X_{11}, X_{12}, \ldots, X_{1n}$, and the observations from the second sample, $X_{21}, X_{22}, \ldots, X_{2n}$, can be quantified. To this end, suppose that the population variances of the observations in the first and second samples are equal to σ_1^2 and σ_2^2, and that the population covariance between the observations from the first and second samples is equal to σ_{12}. Then, the variance of a difference variable D_i is

$$
\begin{aligned}
\sigma_D^2 &= \text{var}(D_i), \\
&= \text{var}(X_{1i} - X_{2i}), \\
&= \text{var}(X_{1i}) + \text{var}(X_{2i}) - 2\text{cov}(X_{1i}, X_{2i}), \\
&= \sigma_1^2 + \sigma_2^2 - 2\sigma_{12}.
\end{aligned}
$$

This shows that the variance of the difference variable D_i is smaller than the sum of the variances of the two populations. A smaller variance for a variable under investigation leads to narrower confidence intervals and to hypothesis tests with a greater power. This is exactly what happens when using the difference variables in the case of paired samples.

10.5   Generalized Hypothesis Tests

In some cases, one does not want to test whether μ_1 and μ_2 are the same. Instead, one might be interested in finding out whether the difference between the two population means is equal to a certain predetermined value Δ_0. The corresponding hypotheses are

$$
\begin{aligned}
H_0 &: \mu_1 - \mu_2 = \Delta_0, \\
H_a &: \mu_1 - \mu_2 > \Delta_0, \\
H_a &: \mu_1 - \mu_2 < \Delta_0,
\end{aligned}
$$

and

$$
H_a : \mu_1 - \mu_2 \neq \Delta_0.
$$

For paired samples, these hypotheses are usually rewritten as

$$
\begin{aligned}
H_0 &: \delta = \Delta_0, \\
H_a &: \delta > \Delta_0, \\
H_a &: \delta < \Delta_0,
\end{aligned}
$$

and

$$
H_a : \delta \neq \Delta_0.
$$

If $\overline{D}$ is at least approximately normally distributed, we know that

$$
\frac{\overline{D} - \delta}{\sigma_D / \sqrt{n}}
$$

is standard normally distributed. This result can be used immediately for a hypothesis test in which the population standard deviation σ_D is known. If σ_D is not known, then this population standard deviation should be estimated using the sample standard deviation S_D. In that case, for the hypothesis test, we build on the fact that

$$\frac{\overline{D} - \delta}{S_D / \sqrt{n}}$$

is t-distributed with $n - 1$ degrees of freedom.

If the null hypothesis is true, then $\delta = \Delta_0$, and

$$Z = \frac{\overline{D} - \Delta_0}{\sigma_D / \sqrt{n}}$$

and

$$T = \frac{\overline{D} - \Delta_0}{S_D / \sqrt{n}}$$

are standard normally distributed and t-distributed with $n - 1$ degrees of freedom, respectively. The latter two expressions are the test statistics in the generalized test for two population means for paired samples.

The procedures for right- and left-tailed tests and for two-tailed tests are completely analogous to those in Section 10.2.2 for known σ_D, and Section 10.2.3 for unknown σ_D.

JMP does not have an explicit way of performing a generalized t-test for paired observations. However, it is possible to use a confidence interval to carry out the two-tailed generalized t-test for paired samples implicitly.

10.6   A Confidence Interval for a Difference of Two Population Means

It is recommended not only to perform a hypothesis test, but also to determine a confidence interval for $\mu_1 - \mu_2$. The precise computation of the confidence interval depends on whether or not σ_D is known.

If a confidence interval does not contain the value zero, this indicates that the population means μ_1 and μ_2 are significantly different. If the confidence interval does not contain the value Δ_0, then the difference between μ_1 and μ_2 is significantly different from Δ_0.

10.6.1   Known σ_D

The confidence interval for known σ_D is

$$\left[\overline{D} - z_{\alpha/2} \frac{\sigma_D}{\sqrt{n}}, \overline{D} + z_{\alpha/2} \frac{\sigma_D}{\sqrt{n}} \right],$$

which is the same as

$$\left[\overline{X}_1 - \overline{X}_2 - z_{\alpha/2} \frac{\sigma_D}{\sqrt{n}}, \overline{X}_1 - \overline{X}_2 + z_{\alpha/2} \frac{\sigma_D}{\sqrt{n}} \right].$$

10.6.2   Unknown σ_D

The confidence interval for unknown σ_D is

$$\left[\overline{D} - t_{\alpha/2;n-1} \frac{S_D}{\sqrt{n}}, \overline{D} + t_{\alpha/2;n-1} \frac{S_D}{\sqrt{n}} \right],$$

or

$$\left[\overline{X}_1 - \overline{X}_2 - t_{\alpha/2;n-1} \frac{S_D}{\sqrt{n}}, \overline{X}_1 - \overline{X}_2 + t_{\alpha/2;n-1} \frac{S_D}{\sqrt{n}} \right].$$

Example 10.6.1   *In Example 10.3.1, where $n = 12$, $\overline{d} = 0.2583$, and $s_D = 0.3565$, the 95% confidence interval for $\delta = \mu_1 - \mu_2$ can be computed as*

$$\left[0.2583 - t_{0.025;11} \frac{0.3565}{\sqrt{12}}, 0.25833 + t_{0.025;11} \frac{0.3565}{\sqrt{12}} \right].$$

This can be reduced to

$$[0.0318, 0.4849],$$

since $t_{0.025;11} = 2.2010$.

Example 10.6.2   *In Example 10.3.3, where $n = 10$, $\overline{d} = -0.41$, and $s_D = 0.3872$, the 95% confidence interval for $\delta = \mu_1 - \mu_2$ can be computed as*

$$\left[-0.41 - t_{0.025;9} \frac{0.3872}{\sqrt{10}}, -0.41 + t_{0.025;9} \frac{0.3872}{\sqrt{10}} \right].$$

This can be reduced to

$$[-0.687, -0.133],$$

since $t_{0.025;9} = 2.2622$.

11

Two Nonparametric Hypothesis Tests for Paired Samples

Nothing seemed to fit, yet there were plenty of clues to indicate everything was somehow related. I knew the random probability of so many coincidences was zero.

(from *The Eight*, Katherine Neville, p. 120)

The *t*-test and the *z*-test for paired samples, which were described in detail in Chapter 10, are only useful when the populations under investigation are normally distributed, or when the sample size *n* is sufficiently large. Because collecting a large number of observations is time-consuming and expensive, many researchers have to work with small numbers of observations. For small samples, it is impossible to determine whether the populations under study are normally distributed. This issue was discussed in Section 6.3.

In such cases, researchers have to rely on nonparametric hypothesis tests. For paired samples, two nonparametric hypothesis tests exist, namely the sign test and the signed-rank test. These tests have already been used in Chapter 5 for the location or the median of a single population. The only new aspect in this chapter is that the sign test and the signed-rank test are now used for comparing the locations of two populations with paired samples. This new application of the sign test and signed-rank test requires us to compute a difference variable. This difference variable has already played an important role in the *t*-test and the *z*-test in Chapter 10. Its calculation is shown schematically in Table 11.1.

In this chapter, we initially focus on the sign test. This test can be used for ordinal data and (discrete and continuous) quantitative data. Next, we study the signed-rank test, which can only be used for continuous quantitative data if the probability density function of the difference variable is symmetrical. For both the sign test and the signed-rank test for paired samples, the tested hypotheses are

$$H_0 : \text{Location 1} = \text{Location 2},$$

$$H_a : \text{Location 1} > \text{Location 2}$$

Statistics with JMP: Hypothesis Tests, ANOVA and Regression, First Edition. Peter Goos and David Meintrup.
© 2016 John Wiley & Sons, Ltd. Published 2016 by John Wiley & Sons, Ltd.
Companion Website: http://www.wiley.com/go/goosandmeintrup/JMP

Table 11.1   A schematic representation of the data in the case of paired samples, along with the calculation of the pairwise differences $D_1, D_2, \ldots, D_n$.

Observation	Sample 1	Sample 2	Difference
1	X_{11}	X_{21}	$D_1 = X_{11} - X_{21}$
2	X_{12}	X_{22}	$D_2 = X_{12} - X_{22}$
3	X_{13}	X_{23}	$D_3 = X_{13} - X_{23}$
$\vdots$	$\vdots$	$\vdots$	$\vdots$
n	X_{1n}	X_{2n}	$D_n = X_{1n} - X_{2n}$

for a right-tailed test,

$$H_a : \text{Location 1} < \text{Location 2}$$

for a left-tailed test, and

$$H_a : \text{Location 1} \neq \text{Location 2}$$

for a two-tailed test.

We illustrate the test procedure by revisiting the examples of Chapter 10 on the number of bacteria in milk, and the wear of two types of shoe sole. In addition, we look at an example from economics on the Maastricht Treaty.

11.1   The Sign Test

The sign test can be used not only for quantitative data, but also for ordinal data. The reason for this is that the sign test only uses the sign of the computed differences, and not their absolute value. In order to understand the procedure of the sign test, a sound understanding of the median is crucial.

11.1.1   The Hypotheses Tested

The sign test for paired samples is a hypothesis test for the population median of the difference variable D. If we call this median Me_D, then the tested null hypothesis is

$$H_0 : \text{Me}_D = 0.$$

The alternative hypothesis for a right-tailed test is

$$H_a : \text{Me}_D > 0.$$

For a left-tailed hypothesis test, it is

$$H_a : \text{Me}_D < 0,$$

while it is

$$H_a : \text{Me}_D \neq 0$$

for a two-tailed hypothesis test. The null hypothesis literally says that the median of the difference variable is zero. In other words, the null hypothesis states that the difference variable $D_i = X_{1i} - X_{2i}$ takes positive values in 50% of the cases, and takes negative values in the other 50%. Alternatively, the null hypothesis says that an observation X_{1i} from the first population is larger than the corresponding observation X_{2i} from the second population in 50% of the cases, and smaller in the other 50%.

If the null hypothesis is true, then the probability that the first observation in a pair, X_{1i}, is larger than the second observation, X_{2i}, equals the probability that the second observation, X_{2i}, is larger than the first observation, X_{1i}. Of course, these probabilities are then equal to $1/2$. The alternative hypothesis in a right-tailed test states that the observations from the first population are larger than the corresponding observations from the second population in more than half of the cases. A left-tailed alternative hypothesis states that the observations from the first population are greater than the corresponding observations from the second population in less than half of the cases. Alternative formulations for the hypotheses tested in the sign test for paired samples are therefore as follows:

H_0 : The probability that an observation from the first population is larger than an observation from the second population is $1/2$,

H_a : The probability that an observation from the first population is larger than an observation from the second population is greater than $1/2$,

H_a : The probability that an observation from the first population is larger than an observation from the second population is less than $1/2$

and

H_a : The probability that an observation from the first population is larger than an observation from the second population is different from $1/2$.

If we have n pairs of observations in our data and the null hypothesis is correct, then the number of pairs for which the difference $D_i = X_{1i} - X_{2i}$ is positive is binomially distributed, with parameters n and $1/2$. This fact forms the basis of the computation of the p-value of the sign test.

As explained in Section 5.1, the sign test uses a decision rule based only on the p-value. A correct, exact p-value can be calculated using the binomial distribution. Therefore, the sign test (both for one population and for two populations with paired samples) looks quite similar to the binomial test that we met in Section 4.2.2. Whenever calculating an exact p-value based on the binomial distribution is a problem, then that p-value can be approximated using the normal probability density. However, for this approximation to work well, the sample size n should be sufficiently large.

11.1.2 Practical Implementation

A first step in the implementation of the sign test is to calculate the value of the difference variable for each of the n pairs of observations. When some of these differences are equal to zero, the corresponding observations need to be left out of the data. In that case, the value of n has to be reduced by the number of omitted zero differences.

The computed test statistic for the sign test for paired samples is denoted by s. The definition of s and the computation of the p-value depend on the type of hypothesis test that is performed:

- For a right-tailed sign test, s is the number of pairs for which x_{1i} is larger than x_{2i} (which is equal to the number of strictly positive values of the difference variable).
- For a left-tailed sign test, s is the number of pairs for which x_{1i} is smaller than x_{2i} (which is equal to the number of strictly negative values of the difference variable).
- For a two-tailed test, s is the larger of the following two numbers:
  - the number of pairs for which x_{1i} is larger than x_{2i} (or the number of strictly positive values of the difference variable), and
  - the number of pairs for which x_{1i} is smaller than x_{2i} (or the number of strictly negative values of the difference variable).

If the null hypothesis is correct, then the test statistic, which we denote by S as long as we have not collected any data, is a binomial random variable with n (the number of pairs in the data set minus the number of times the difference variable is zero) as its first parameter, and success probability $\pi = 0.5$ as its second parameter. The expected value of S is $n/2$, while its variance is equal to $n/4$, as shown in Section 5.1. The exact p-value for the sign test is calculated in the following way:

- For a right- or left-tailed sign test, the exact p-value is $p = P(S \geq s)$.
- For a two-tailed sign test, the exact p-value is $p = 2\,P(S \geq s)$.

When the exact p-value cannot be determined, one can calculate an approximate p-value. This approach relies on the fact that the binomial distribution of S can be approximated by the normal distribution, with expected value $n/2$ and variance $n/4$ if $n > 10$ (for a more detailed explanation, see Section 5.1). The computation of the p-value uses a continuity correction:

- For a right- or left-tailed sign test, the approximate p-value is

$$p = P\left(Z > \frac{s - \frac{1}{2} - \frac{n}{2}}{\frac{\sqrt{n}}{2}}\right).$$

- For a two-tailed sign test, the approximate p-value is

$$p = 2\,P\left(Z > \frac{s - \frac{1}{2} - \frac{n}{2}}{\frac{\sqrt{n}}{2}}\right).$$

Due to the different definitions of the test statistic in the left-tailed, right-tailed, and two-tailed hypothesis tests, the same kind of exceedance probability has to be calculated to determine the p-value in each case. It is not difficult to see that a large value for the computed test statistic s, which provides strong evidence against the null hypothesis, implies a small p-value. If the p-value is sufficiently small (smaller than the selected significance level α), we reject the null hypothesis and accept the alternative hypothesis.

Example 11.1.1 *In Example 10.3.1, we performed a t-test for the comparison of the number of bacteria in milk before and after heating it. As the number of observations in this example is extremely small, it is not possible to test whether the populations under investigation and the difference variable are normally distributed. It is therefore recommended to use a nonparametric test instead of a t-test. For the sign test, we need to associate a sign with each observed value of the difference variable. This is illustrated in Table 11.2. The hypotheses tested in this example are as follows:*

H_0 : *The probability that the number of bacteria is greater before than after the heating is equal to $1/2$*

and

H_a : *The probability that the number of bacteria is greater before than after the heating is greater than $1/2$.*

Thus, the alternative hypothesis is right-tailed.

Table 11.2 contains nine plus signs and three minus signs in the last column. Therefore, for nine pairs of observations, the value of x_{1i} is larger than the corresponding value of x_{2i}, while, for the three other pairs, the value of x_{2i} is larger than the corresponding value of x_{1i}.

Table 11.2 The logarithms of the numbers of bacteria in milk before and after heating it, the difference variable needed for the sign test, and the sign associated with each observation of the difference variable.

Milk sample i	x_{1i} (before)	x_{2i} (after)	Difference d_i (before − after)	Sign
1	6.98	6.95	0.03	+
2	7.08	6.94	0.14	+
3	8.34	7.17	1.17	+
4	5.30	5.15	0.15	+
5	6.26	6.28	−0.02	−
6	6.77	6.81	−0.04	−
7	7.03	6.59	0.44	+
8	5.56	5.34	0.22	+
9	5.97	5.98	−0.01	−
10	6.64	6.51	0.13	+
11	7.03	6.84	0.19	+
12	7.69	6.99	0.70	+

Because the alternative hypothesis is right-tailed, the computed test statistic s is equal to 9, the number of plus signs in the last column of Table 11.2. Therefore, the exact p-value is

$$p = P(S \geq 9) = 0.0730.$$

This probability can be computed in JMP using the command "1-Binomial Distribution(0.5, 12, 8)". It can also be found in the table for n = 12 in Appendix A. If we use a significance level of 5%, the sign test does not allow us to conclude that the number of bacteria is significantly reduced: the p-value is larger than the significance level chosen.

Based on the t-test in the previous chapter, we concluded that there was a significant decrease in the number of bacteria. However, to use the t-test, we had to assume that the logarithms of the numbers of bacteria were normally distributed, and there is no way to verify the normality of a data set with as few as 10 observations.

Therefore, in this example, we observe a contradiction between the results of the t-test and the sign test. However, it is important to point out that although the sign test has a smaller power (especially for small numbers of observations, as in this example), it still yields a fairly small p-value (0.0730). This relatively small p-value suggests that we need to look at the null hypothesis with at least some skepticism and that the alternative hypothesis might well be correct.

Example 11.1.2   *Example 7.0.3 presents data from an experiment with 10 paired observations comparing two types of shoe sole, made of Materials A and B. In Example 10.3.3, we performed a two-tailed t-test for the same data. Now, we conduct a two-tailed sign test. The first step is to compute the difference variable for each pair of observations, and to determine the sign of each difference. The results are shown in Table 11.3.*

In eight cases, the difference variable is negative, and in the two other cases, it takes a positive value. Because we perform a two-tailed test in this example, our computed test

Table 11.3   Measurements of the amount of wear for Materials A and B in Example 10.3.3, the difference variable needed for the sign test, and the sign associated with each observation of the difference variable.

Person	Material A	Material B	Difference	Sign
1	13.2	14.0	−0.8	−
2	8.2	8.8	−0.6	−
3	10.9	11.2	−0.3	−
4	14.3	14.2	0.1	+
5	10.7	11.8	−1.1	−
6	6.6	6.4	0.2	+
7	9.5	9.8	−0.3	−
8	10.8	11.3	−0.5	−
9	8.8	9.3	−0.5	−
10	13.3	13.6	−0.3	−

statistic s equals 8. This is because, in a two-tailed sign test, s is the greater of the number of positive values and the number of negative values. The exact p-value is

$$p = 2\,P(S \geq 8) = 2 \times 0.0547 = 0.1094.$$

Hence, based on the sign test and using a significance level of 5%, we cannot conclude that there is a significant difference in wear between the two materials for the shoe soles. Here, again, the p-value can be computed using JMP. However, the probability $P(S \geq 8) = 0.0547$ can also be read from the table for $n = 10$ in Appendix A.

The analysis using the t-test in Example 10.3.3 led to a rejection of the null hypothesis. In other words, the t-test suggested that there is a significant difference in wear. As demonstrated in Example 11.1.1, this kind of contradictory result is not exceptional, especially for small samples. We discuss this in more detail in Section 11.3.

Example 11.1.3   *In December 1991, the Maastricht Treaty on the European Economic and Monetary Union (later simply referred to as the European Union, or EU) defined standards for the government deficits of the then 15 Member States, their inflation rates, the mutual exchange rates, and the interest rates. An official of the European Commission, responsible for monitoring the macroeconomic performance of the member states, wants to investigate whether the treaty indeed had an attenuating effect on inflation in the Member States. She collects the inflation rates of the 15 countries of the EU (more specifically, the mean inflation rates over the periods 1985–1991 and 1992–1998). The $n = 15$ observations are shown in Table 11.4.*

The official wants to test whether there was a significant decline in inflation, with a minimum of assumptions regarding the data. The sign test is ideal for this purpose, because it does

Table 11.4   The inflation rates (expressed in %) of the 15 Member States of the EU before and after the Maastricht Treaty, together with the difference variable and the sign of the difference variable.

Country	Inflation before	Inflation after	Difference	Sign
Belgium	3.54	2.33	1.21	+
Denmark	3.73	2.10	1.63	+
Germany	2.60	2.41	0.19	+
Greece	17.13	9.84	7.29	+
Spain	7.40	3.97	3.43	+
France	3.74	1.60	2.14	+
Ireland	3.33	1.93	1.40	+
Italy	7.33	3.93	3.40	+
Luxembourg	2.26	2.41	−0.16	−
The Netherlands	1.23	1.94	−0.71	−
Austria	2.76	2.43	0.33	+
Portugal	14.40	4.97	9.43	+
Finland	5.14	1.61	3.53	+
Sweden	7.03	2.00	5.03	+
United Kingdom	5.76	2.81	2.94	+

not require any assumption concerning the normality of the data or the symmetry of their distribution. The tested hypotheses are as follows:

$$H_0 : \text{The Maastricht Treaty had no effect on inflation}$$

and

$$H_a : \text{The Maastricht Treaty led to a reduction of inflation.}$$

If we define the difference variable D as the difference in inflation before and after the treaty, then the hypothesis test is right-tailed. The tested hypotheses can then be formulated in the following way:

$$H_0 : Me_D = 0$$

and

$$H_a : Me_D > 0.$$

To perform the test, we first need to determine the values of the difference variable and the signs of these values. The results are shown in the last two columns of Table 11.4. In total, there are 13 plus signs and only two minus signs. Because we are performing a right-tailed test, the number of plus signs is the value of the test statistic, so s = 13. Therefore, the p-value of the test is

$$p = P(S \geq 13) = 0.0037.$$

The p-value is very small; hence we reject the null hypothesis. As the sign test does not make any assumptions on the distribution of the data, this is a very strong indication that the Maastricht Treaty has had the desired effect on inflation.

11.1.3   JMP

To perform the sign test for paired observations in JMP, first enter the data as shown in Figure 10.1 for Example 10.3.1. Next, in the menu "Analyze", select the option "Matched Pairs" (see Figure 10.2). This produces the dialog window shown in Figure 10.3. In that window, first enter the variable "After", followed by the variable "Before" in the field "Y, Paired Response", and click "OK". JMP then automatically performs a paired *t*-test.

To conduct the sign test on top of the paired *t*-test, click on the hotspot (red triangle) next to "Matched Pairs". In the resulting menu, shown in Figure 11.1, choose the option "Sign Test". This results in an additional piece of output[1], with exact *p*-values corresponding

---

[1] Note that JMP uses a certain value *M* as test statistic. In the output in Figure 11.2, this test statistic takes the value 3. In general, in JMP, *M* equals the difference between the number of plus signs and $n/2$. In Example 11.1.1, this means that $M = 9 - 12/2 = 3$.

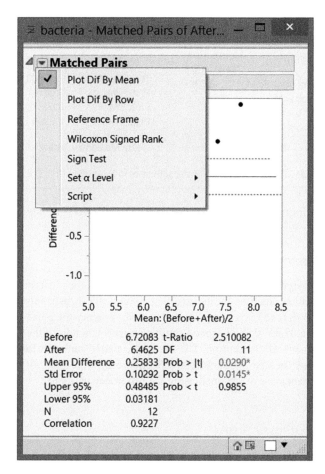

Figure 11.1 The extra JMP menu for tests with paired observations.

to a two-tailed alternative hypothesis (denoted by "Prob ≥ |M|"), a right-tailed alternative hypothesis (denoted by "Prob ≥ M"), and a left-tailed alternative hypothesis (denoted by "Prob ≤ M"). In Example 11.1.1, the relevant p-value is the one corresponding to a right-tailed test. As a result, the relevant p-value in the JMP output in Figure 11.2 is 0.0730.

Sign Test	
	Before-After
Test Statistic M	3.000
Prob ≥ |M|	0.1460
Prob ≥ M	0.0730
Prob ≤ M	0.9807

Figure 11.2 The JMP output for the sign test in Example 11.1.1.

Sign Test	
	Material A- Material B
Test Statistic M	−3.000
Prob ≥ |M|	0.1094
Prob ≥ M	0.9893
Prob ≤ M	0.0547

(a) Shoe soles

Sign Test	
	Before- After
Test Statistic M	5.500
Prob ≥ |M|	0.0074*
Prob ≥ M	0.0037*
Prob ≤ M	0.9995

(b) Maastricht Treaty

Figure 11.3   The JMP outputs for the sign tests in Examples 11.1.2 and 11.1.3.

The JMP outputs for the sign tests in Examples 11.1.2 and 11.1.3 are shown in Figure 11.3.

11.2   The Wilcoxon Signed-Rank Test

A weakness of the sign test is that its power is low when the difference variable has a symmetrical distribution, such as the normal distribution, the t-distribution, or the uniform distribution. If the difference is a normally distributed variable, the t-test from the previous chapter is appropriate. The Wilcoxon signed-rank test is recommended if the difference variable has a different symmetrical distribution. The use of the signed-rank test for paired samples is very similar to its use for one population (see Section 5.2). The main difference is that, for paired samples, the ranks are determined using the difference variable.

11.2.1   The Hypotheses Tested

As in the sign test, the tested null hypothesis in the signed-rank test is

$$H_0 : \mathrm{Me}_D = 0.$$

The alternative hypothesis for a right-tailed test is

$$H_a : \mathrm{Me}_D > 0.$$

For a left-tailed test, the alternative hypothesis is

$$H_a : \mathrm{Me}_D < 0,$$

while for a two-tailed test, it is

$$H_a : \mathrm{Me}_D \neq 0.$$

The hypotheses deal with the median of the difference variable D. The use of the signed-rank test assumes that the difference variable has a symmetrical distribution. Whenever a random variable has a symmetrical distribution, the median is identical to the population mean. The hypotheses listed above can therefore be rewritten as

$$H_0 : \mathrm{Me}_D = \mu_D = 0,$$

$$H_a : \mathrm{Me}_D = \mu_D > 0,$$

$$H_a : \mathrm{Me}_D = \mu_D < 0,$$

and

$$H_a : \mathrm{Me}_D = \mu_D \neq 0,$$

when the assumption that the difference variable has a symmetrical distribution is fulfilled.

Finally, as the mean of a difference of two random variables is identical to the difference of the means of these two variables, the tested hypotheses can also be written as

$$H_0 : \mu_1 = \mu_2,$$

$$H_a : \mu_1 > \mu_2,$$

$$H_a : \mu_1 < \mu_2,$$

and

$$H_a : \mu_1 \neq \mu_2,$$

if the difference variable has a symmetrical distribution.

11.2.2   Practical Implementation

To perform the signed-rank test for paired observations, the following steps need to be taken:

(1) Calculate, for each pair of observations, the difference $d_i = x_{1i} - x_{2i}$.
(2) Typically, some of the differences $d_1, d_2, \ldots, d_n$ are positive, whereas other differences are negative. If certain differences are zero, then delete these differences and the corresponding observations from the data, and continue with the smaller number of observations[2]. The available number of observations, n, decreases correspondingly.
(3) Determine the absolute value of all remaining differences, $|d_1|, |d_2|, \ldots, |d_n|$.
(4) Rank the absolute values $|d_1|, |d_2|, \ldots, |d_n|$ from small to large and, based on the resulting ranking, assign ranks from 1 to n to all observations.
(5) If there are ties, determine an average rank for the corresponding observations. Denote by r_1 the final rank of the first observation, by r_2 the final rank of the second observation, $\ldots$, and by r_n the final rank of the last observation.

---

[2] Alternatively, one could rank the zeros along with the other differences, and then drop the ranks of the zeros. This so-called Pratt method is implemented in JMP 12.

(6) Assign a plus sign to all ranks r_i corresponding to a positive difference $d_i = x_{1i} - x_{2i}$, and assign a minus sign to all ranks r_i corresponding to a negative difference $d_i = x_{1i} - x_{2i}$. Denote the resulting signed ranks by s_i.

(7) Calculate the sum of all positive signed ranks s_i and call this sum t_+.

(8) Calculate the sum of the absolute values of all negative signed ranks s_i and call this sum t_-.

(9) Check that the sum of t_+ and t_- is equal to $n(n + 1)/2$. As the sum of the natural numbers from 1 to n is equal to $n(n + 1)/2$, the sum of all ranks $r_1, r_2, \ldots, r_n$, and thus the sum of t_+ and t_-, must also be $n(n + 1)/2$.

Because the signed-rank test uses the magnitude of the differences between the values of x_{1i} and x_{2i}, this test is not suitable for ordinal data. In fact, differences cannot be calculated meaningfully for ordinal variables.

Ideally, exact p-values are used in the signed-rank test. The calculation of exact p-values for the signed-rank test is discussed in detail in Sections 5.2.3 and 5.2.4. Appendix E contains a table with exact p-values for the signed-rank test statistic values when there are no ties. JMP is able to calculate exact p-values in the presence of ties as well as in the absence of ties. The precise procedure is explained below for the three types of alternative hypotheses. We denote the test statistic for the signed-rank test by T_+ as long as no data has been collected, and by t_+ as soon as the data has been acquired and the test statistic has been computed based on the data.

The Right-Tailed Test

If the alternative hypothesis that $\mathrm{Me}_D = \mu_D > 0$ is correct, then there are many positive differences in the sample, and t_+ will be large. To determine whether t_+ is large enough for the null hypothesis to be rejected in favor of the alternative hypothesis, we have to quantify how likely the observed value of t_+ is under the assumption that the null hypothesis is correct. To this end, we determine the probability $P(T_+ \geq t_+)$. This probability is our p-value for the right-tailed test. If the p-value is smaller than the selected significance level α, we reject the null hypothesis.

The Left-Tailed Test

If the alternative hypothesis that $\mathrm{Me}_D = \mu_D < 0$ is correct, then there are many negative differences in the sample, and t_+ will be small. To determine whether t_+ is small enough for the null hypothesis to be rejected in favor of the alternative hypothesis, we again have to quantify how likely the observed value of t_+ is under the assumption that the null hypothesis is correct. To this end, we calculate the probability $P(T_+ \leq t_+)$, which acts as the p-value for the left-tailed test. If the p-value is smaller than the significance level α chosen, we reject the null hypothesis.

The Two-Tailed Test

If the alternative hypothesis that $\mathrm{Me}_D = \mu_D \neq 0$ is correct, then there are two possibilities. One possibility is that $\mathrm{Me}_D = \mu_D < 0$. In this case, t_+ will be small. The second possibility is that $Me_D = \mu_D > 0$. In that case, t_+ will be large. In order to find out whether the value of t_+ is sufficiently large or small for the null hypothesis to be rejected, we need to determine the corresponding p-value. How this should be done depends on the exact value of t_+:

$$
p = \begin{cases}
2\,P(T_+ \leq t_+), & \text{if } t_+ \leq \dfrac{n(n+1)}{4}, \\[2ex]
2\,P(T_+ \geq t_+), & \text{if } t_+ \geq \dfrac{n(n+1)}{4}.
\end{cases}
$$

Again, we reject the null hypothesis if the p-value is smaller than the chosen significance level.

Example 11.2.1   *Example 7.0.3 presents data from an experiment with 10 paired observations comparing two types of shoe sole made of Materials A and B. In Example 10.3.3, we performed a two-tailed t-test for this data set, whereas in Example 11.1.2, we carried out a two-tailed sign test. Now, we conduct a two-tailed signed-rank test.*

As a first step in the execution of the signed-rank test, we compute all values of the difference variable, and check the symmetry of the distribution of these values. The values for the difference variable are calculated in the fourth column of Table 11.5, which also shows the original data. In principle, there are formal hypothesis tests for the symmetry of a distribution around an unknown median. However, such a test is beyond the scope of this book. Therefore, we restrict ourselves to a graphical assessment of the symmetry of the probability distribution or density. Via the "Distribution" platform in the "Analyze" menu in JMP, we can quickly generate a histogram and a box plot of the difference variable. These two graphs are shown in Figure 11.4. The JMP output of the "Distribution" platform also shows that the sample median of the difference variable is -0.4, while the sample mean is -0.41. The sample standard deviation of the difference variable is 0.39.

Although the histogram is not perfectly symmetrical, the box plot in Figure 11.4 does look very symmetrical. Moreover, the sample mean and the sample median are very close to each other (especially if we take the standard deviation of 0.39 into account), which is typical for

Table 11.5   Measurements of the amount of wear for Materials A and B in Example 11.2.1, the required difference variable for the signed-rank test, and the final signed ranks s_i for each pair of observations.

Person i	Material A x_{1i}	Material B x_{2i}	Difference d_i	| Difference | $\|d_i\|$	Signed rank s_i
1	13.2	14.0	−0.8	0.8	−9
2	8.2	8.8	−0.6	0.6	−8
3	10.9	11.2	−0.3	0.3	−4
4	14.3	14.2	0.1	0.1	1
5	10.7	11.8	−1.1	1.1	−10
6	6.6	6.4	0.2	0.2	2
7	9.5	9.8	−0.3	0.3	−4
8	10.8	11.3	−0.5	0.5	−6.5
9	8.8	9.3	−0.5	0.5	−6.5
10	13.3	13.6	−0.3	0.3	−4

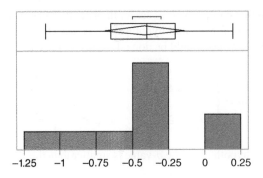

Figure 11.4   A histogram and a box plot to check the symmetry of the distribution of the difference variable in Example 11.2.1.

a symmetrical distribution. Fisher's skewness coefficient[3] is equal to −0.08, which is close to zero and thus also suggests that the distribution is symmetrical. As a result, it seems safe to use the signed-rank test in this example.

The next step in the execution of the signed-rank test is to determine the rank for each pair of observations. Table 11.5 illustrates how this is done.

In total, there are eight negative and two positive signed ranks. Since there are three pairs with the value −0.3 for the difference variable, an average (signed) rank is used for these three pairs, namely −4. In addition, there are two pairs of observations with a value of −0.5 for the difference variable. The average (signed) rank here is −6.5. The sum of all positive ranks amounts to $t_+ = 3$, while the sum of the absolute values of all negative ranks is $t_- = 52$. The sum of t_+ and t_- is $n(n + 1)/2 = 55$.

As there are ties in the ranking of the values of the difference variable, Appendix E cannot be used to find the exact p-value. If we use JMP, however, we obtain

$$p = 2\, P(T_+ \leq 3) = 0.0078$$

as the exact p-value. Based on this p-value, we can conclude that there is a significant difference in the wear between the two materials. This decision is consistent with the result of the t-test in Example 10.3.3, but it contradicts the conclusion obtained from the sign test in Example 11.1.2. The JMP output for the signed-rank test is shown in Figure 11.5.

Wilcoxon Signed Rank	
	Material A- Material B
Test Statistic S	−24.500
Prob > |S|	0.0078*
Prob > S	0.9961
Prob < S	0.0039*

Figure 11.5   The JMP output for the signed-rank test in Example 11.2.1.

---

[3] Positive values for Fisher's skewness coefficient indicate a distribution that is skewed to the right, while negative values point to a left-skewed distribution. A value close to zero indicates symmetry. For more details, see the book *Statistics with JMP: Graphs, Descriptive Statistics and Probability.*

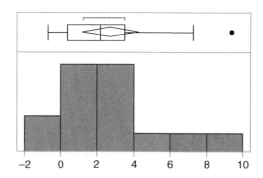

Figure 11.6 A histogram and a box plot to check the symmetry of the distribution of the difference variable in Example 11.2.2.

Example 11.2.2 *In Example 11.1.3, we performed a one-sided sign test to assess whether inflation rates had dropped significantly in the EU as a result of the Maastricht Treaty. Before we apply the signed-rank test to this example, we should study the distribution of the difference variable. A histogram and a box plot of the values of the difference variable are shown in Figure 11.6. Both graphs indicate a right-skewed distribution of the difference variable. The sample median of the difference variable is 2.14, while the sample mean is 2.74. The sample standard deviation of the difference variable is 2.81. Therefore, the sample mean exceeds the sample median by almost a quarter of the standard deviation. This suggests a right-skewed distribution, as does Fisher's skewness coefficient, which is 1.13.*

It is clear that the symmetry condition is most likely not satisfied for the difference variable in this example. Therefore, the signed-rank test should not be used to investigate whether the inflation rates have dropped.

Another consideration is that it is not very useful to apply the signed-rank test, given that the sign test in Example 11.1.3 has already allowed us to reject the null hypothesis that the median of the difference variable is zero, and to conclude that inflation has decreased significantly. If the sign test (characterized by a low power) allows us to reject this null hypothesis, the signed-rank test (characterized by a larger power) has no added value.

Example 11.2.3 *In Example 10.3.1, we performed a one-sided t-test for the comparison of the number of bacteria in milk before and after heating it. In Example 11.1.1, we carried out a one-sided sign test for the same data. Before conducting the signed-rank test, we check whether the difference variable has a symmetrical distribution by means of the histogram and the box plot in Figure 11.7. Both graphs suggest a right-skewed distribution, as does Fisher's skewness coefficient, which is equal to 1.84.*

As in Example 11.2.2, it does not seem appropriate to perform the Wilcoxon signed-rank test here. If we proceed using JMP anyway, we obtain the output shown in Figure 11.8. The relevant p-value in this output is 0.0046, since a right-tailed test is needed here. The p-value is

$$p = P(T_+ \geq 71) = 0.0046,$$

as in this example the sum of the positive ranks t_+ is equal to 71 (see Table 11.6 for all positive and negative ranks). The p-value, rounded to three decimal places, can be read off the table in Appendix E, because in this example there are no ties.

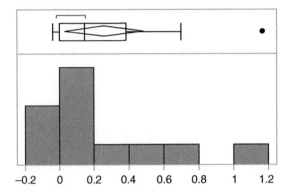

Figure 11.7 A histogram and a box plot to check the symmetry of the distribution of the difference variable in Example 11.2.3.

Wilcoxon Signed Rank	
	Before-After
Test Statistic S	32.000
Prob > |S|	0.0093*
Prob > S	0.0046*
Prob < S	0.9954

Figure 11.8 The JMP output for the signed-rank test in Example 11.2.3.

Table 11.6 The logarithms of the numbers of bacteria in milk before and after heating it, the difference variable required for the signed-rank test, and the rank for each pair of observations.

Milk sample i	x_{1i} (before)	x_{2i} (after)	Difference d_i (before − after)	| Difference | | d_i |	Signed rank s_i
1	6.98	6.95	0.03	0.03	3
2	7.08	6.94	0.14	0.14	6
3	8.34	7.17	1.17	1.17	12
4	5.30	5.15	0.15	0.15	7
5	6.26	6.28	−0.02	0.02	−2
6	6.77	6.81	−0.04	0.04	−4
7	7.03	6.59	0.44	0.44	10
8	5.56	5.34	0.22	0.22	9
9	5.97	5.98	−0.01	0.01	−1
10	6.64	6.51	0.13	0.13	5
11	7.03	6.84	0.19	0.19	8
12	7.69	6.99	0.70	0.70	11

Based on the signed-rank test, we would reject the null hypothesis that there is no reduction of the number of bacteria in favor of the alternative hypothesis that there is a decrease. The signed-rank test would lead to the same conclusion as the t-test in Chapter 10, but to a different conclusion from the more valid sign test.

11.2.3   Approximate p-Values

If one cannot use statistical software to calculate exact p-values, one has to rely on tables such as the one in Appendix E to obtain exact p-values. However, this table is only perfectly valid when there are no ties. Moreover, it only contains p-values for sample sizes up to 20. Therefore, when no software is available, it is possible that an approximate p-value will have to be computed. As explained in Sections 5.2.5 and 5.2.6, there are two approaches to determining approximate p-values for the signed-rank test. The first approach is based on the standard normal distribution, while the second one is based on the t-distribution.

The first method for calculating approximate p-values uses the fact that both the sum of the positive ranks, T_+, and the sum of the absolute values of the negative ranks, T_-, are approximately normally distributed. This follows from the fact that T_+ and T_- are both sums of random variables, so that the central limit theorem ensures that they are approximately normally distributed, if the sample size is sufficiently large. If the null hypothesis is correct, T_+ and T_- are therefore approximately normally distributed with expected value

$$E(T_+) = E(T_-) = n(n+1)/4$$

and variance

$$\mathrm{var}(T_+) = \mathrm{var}(T_-) = n(n+1)(2n+1)/24,$$

if there are no ties. The decision rules, the continuity correction, and a correction for ties are shown in detail in Section 5.2.5.

The second method for calculating an approximate p-value uses the test statistic

$$t = \frac{t_+ - \dfrac{n(n+1)}{4}}{\sqrt{\dfrac{n^2(n+1)(2n+1)}{24(n-1)} - \dfrac{t_+ - \dfrac{n(n+1)}{4}}{n-1}}}, \tag{11.1}$$

which is approximately t-distributed with $n-1$ degrees of freedom. The corresponding decision rules are shown in detail in Section 5.2.6.

11.2.4   JMP

In order to perform a signed-rank test in JMP, enter the data as shown in Figure 10.1 for Example 10.3.1, and choose the option "Matched Pairs" in the "Analyze" menu (see Figure 10.2). This produces the dialog window shown in Figure 10.3. After identifying the paired variables to compare, click on "OK", and JMP will automatically perform a t-test.

To obtain the results for the signed-rank test, click on the hotspot (red triangle) next to "Matched Pairs". In the menu shown in Figure 11.1, choose the option "Wilcoxon Signed Rank". This yields an additional piece of output[4], as in Figures 11.5 and 11.8. This additional piece of output contains exact p-values corresponding to a two-tailed alternative hypothesis (denoted by "Prob > |S|"), a right-tailed alternative hypothesis (denoted by "Prob > S"), and a left-tailed alternative hypothesis (denoted by "Prob < S").

11.3   Contradictory Results

In this and the previous chapter, and more specifically in Examples 10.3.1, 11.1.1, and 11.2.3, we carried out three different hypothesis tests to investigate whether heating milk reduces the number of bacteria. This led to contradictory results: the sign test did not allow us to reject the null hypothesis, while the t-test and the signed-rank test did allow us to do so.

The sign test will, as in Examples 10.3.1, 11.1.1, and 11.2.3, usually lead to the largest p-value. It is therefore the least likely to lead to a rejection of the null hypothesis. The reason for this is that the absolute values of the difference variable are not taken into account (only the sign is used) in the implementation of the test. In other words, the sign test does not use all the information that is contained in the data. Although this seems unattractive, for small samples with nonsymmetrical distributions of the difference variable, the sign test is the only allowable hypothesis test.

The t-test makes direct use of the values of the difference variable and thus makes maximum use of the information that is contained in the data. The problem with the t-test is that it is only valid when the difference variable is normally distributed, or when there is enough data to be able to rely on the central limit theorem. The signed-rank test also uses the values of the difference variable, but indirectly, namely via the ranks. Indeed, large positive or negative values for the difference variable will result in large positive or large negative ranks. The problem with the signed-rank test is that it is only valid when the difference variable has a symmetrical distribution.

If the t-test is justified, it should be used because that test uses the data optimally and has the largest power. The next best choice is the signed-rank test, which neglects part of the information in the data by using ranks, and therefore has a slightly smaller power. The last choice, but the only acceptable one for a nonsymmetrical distribution of the difference variable, is the sign test.

For the data set on bacteria in milk, the sign test is the only test that can be used without any doubt, because it seems that the difference variable has a right-skewed distribution. This test provides a p-value of around 7%. If a significance level of 5% were to be used, this p-value would lead to the conclusion that the number of bacteria after heating was not significantly lower than before heating. However, such a small p-value, resulting from a test with a low power, may be interpreted as an indication that the null hypothesis could possibly be false, and as an incentive to collect additional data. In fact, some additional observations will allow a more definitive choice between the null and alternative hypotheses.

---

[4] Note that JMP uses a certain value S as a test statistic. In the outputs in Figures 11.5 and 11.8, this test statistic takes the values -24.5 and 32, respectively. In general, in JMP, S is equal to $(t_+ - t_-)/2$. For Example 11.2.1, this expression yields the value $(3 - 52)/2 = -24.5$, while, for Example 11.2.3, it yields $(71 - 7)/2 = 32$.

Part Four

More Than Two Populations

12

Hypothesis Tests for More Than Two Population Means: One-Way Analysis of Variance

Before he bombarded the boys and that single stray girl with discrete stochasts, probability space, and Bayes's theorem, he challenged them to relate the most spectacular piece of coincidence they'd ever experienced. Let's have it, your most outrageous twist of fate, your spookiest fluke. As a teaser he made a probability analysis on the blackboard of the most remarkable of the students' anecdotes.

(from *Bonita Avenue*, Peter Buwalda, p. 37)

In Chapter 8, we learnt how to test whether the means μ_1 and μ_2 of two populations are equal using two independent samples. In this chapter, we discuss a more general test for the comparison of more than two population means. We restrict our attention to independent samples from normally distributed populations with equal variances. For the Welch test, which copes with unequal variances, we refer to Section 14.6. For a hypothesis test for dependent samples, we refer to more advanced literature.

Example 12.0.1 *Plastic foil is produced on three different extruders. In order to determine whether the mean thickness for each of the three extruders is the same, measurements are made at six randomly chosen positions of the foil produced by each machine. The quality department wants to verify whether the three machines produce foil with the same mean thickness. To this end, they test the hypotheses*

$$H_0 : \mu_1 = \mu_2 = \mu_3$$

and

$$H_a : \text{at least one population mean } \mu_i \text{ differs from the others.}$$

Statistics with JMP: Hypothesis Tests, ANOVA and Regression, First Edition. Peter Goos and David Meintrup.
© 2016 John Wiley & Sons, Ltd. Published 2016 by John Wiley & Sons, Ltd.
Companion Website: http://www.wiley.com/go/goosandmeintrup/JMP

Table 12.1   The data for the three extruders from the factory in France.

	\multicolumn{6}{c}{Measurement (μm)}					
	1	2	3	4	5	6
Extruder 1	39	45	43	53	49	56
Extruder 2	56	48	54	58	60	63
Extruder 3	38	37	40	55	50	50

Table 12.2   The data for the three extruders from the factory in Germany.

	\multicolumn{6}{c}{Measurement (μm)}					
	1	2	3	4	5	6
Extruder 1	35	41	39	57	53	60
Extruder 2	50	42	48	64	66	69
Extruder 3	34	33	36	59	54	54

Two data sets for this problem can be found in Tables 12.1 and 12.2. The data in Table 12.1 is from a production site in France, while the figures in Table 12.2 come from a factory in Germany. The two data sets are represented graphically in Figures 12.1 and 12.2.

12.1   One-Way Analysis of Variance

In this chapter, we generalize the hypothesis test for the null hypothesis

$$H_0 : \mu_1 = \mu_2$$

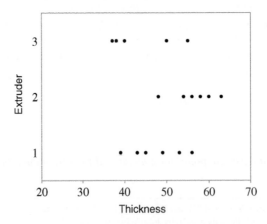

Figure 12.1   A plot of the data in Table 12.1, from the factory in France.

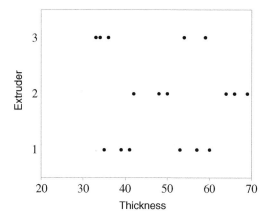

Figure 12.2 A plot of the data in Table 12.2, from the factory in Germany.

and the two-tailed alternative hypothesis

$$H_a : \mu_1 \neq \mu_2$$

to situations involving more than two populations. If g populations are studied, then the hypotheses tested are

$$H_0 : \mu_1 = \mu_2 = \cdots = \mu_g$$

and

$$H_a : \text{ at least one population mean } \mu_i \text{ differs from the others.}$$

This hypothesis test is commonly referred to as **one-way analysis of variance** or **one-way ANOVA**.

In general, a one-way analysis of variance can be used to test the equality of g population means $\mu_1, \mu_2, \ldots, \mu_g$. To this end, a data sample is collected for each of the g populations studied. If the sample from the first population contains n_1 observations, the sample from the second population contains n_2 observations, $\ldots$, and the sample from the gth population contains n_g observations, then the data can be represented in the following way:

Sample 1	Sample 2	Sample 3	$\ldots$	Sample g
X_{11}	X_{21}	X_{31}		X_{g1}
X_{12}	X_{22}	X_{32}		X_{g2}
X_{13}	X_{23}	X_{33}		X_{g3}
$\vdots$	X_{24}	X_{34}		X_{g4}
X_{1n_1}	$\vdots$	X_{35}		$\vdots$
	X_{2n_2}	$\vdots$		X_{gn_g}
		X_{3n_3}		

In this table, X_{ij} represents the jth observation in the ith sample. Note that it is not required that all samples have the same size. In total, there are

$$n = n_1 + n_2 + n_3 + \cdots + n_g = \sum_{i=1}^{g} n_i$$

observations.

Since the hypothesis test concerns the means of g populations, it is based on the sample means of the g samples from these populations. The sample mean of the first sample is denoted by $\overline{X}_1$ and computed as

$$\overline{X}_1 = \frac{1}{n_1}(X_{11} + X_{12} + \cdots + X_{1n_1}) = \frac{1}{n_1} \sum_{j=1}^{n_1} X_{1j}.$$

The sample means of the samples $2, \ldots, g$ are denoted by $\overline{X}_2, \ldots, \overline{X}_g$, respectively, and computed as

$$\overline{X}_2 = \frac{1}{n_2}(X_{21} + X_{22} + \cdots + X_{2n_2}) = \frac{1}{n_2} \sum_{j=1}^{n_2} X_{2j},$$

$$\vdots$$

$$\overline{X}_g = \frac{1}{n_g}(X_{g1} + X_{g2} + \cdots + X_{gn_g}) = \frac{1}{n_g} \sum_{j=1}^{n_g} X_{gj}.$$

Example 12.1.1   *For the French data in Table 12.1 and Figure 12.1, the means of the three samples are $\bar{x}_1 = 47.5$, $\bar{x}_2 = 56.5$, and $\bar{x}_3 = 45$. The three computed sample means are thus quite different, but the question is whether the differences are large enough for us to decide that they are statistically significant.*

For the German data in Table 12.2 and Figure 12.2, the means of the three samples are also $\bar{x}_1 = 47.5$, $\bar{x}_2 = 56.5$, and $\bar{x}_3 = 45$. So, for the German data, the three computed sample means are also quite different. Again, the question is whether the differences are large enough for us to decide that they are statistically significant.

Although the three sample means for the two studied data sets are the same, there is a fundamental difference between the two data sets. The difference is clearly visible when comparing Figures 12.1 and 12.2. The points in the latter figure have a larger spread than those in the former. This means that the thickness of the plastic foil produced in the French factory has less variability than the thickness of the foil produced in the German factory. A statistical consequence is that the sample means computed from the French data are more precise estimates of the corresponding population means than those calculated from the German data. Indeed, the variance of a sample mean decreases if the variance of the studied population decreases. This follows from Theorem 1.5.2.

Because of the higher precision of $\bar{x}_1$, $\bar{x}_2$, and $\bar{x}_3$ in the French data, we can conclude that the data from France contains stronger evidence that the population means μ_1, μ_2, and μ_3 differ than the data from Germany.

We also see that in order to make a decision on whether or not to reject the null hypothesis, we have to weigh the differences in $\bar{x}_1$, $\bar{x}_2$, and $\bar{x}_3$ against the variances of the populations under investigation.

To decide whether the differences between the computed sample means $\bar{x}_1, \bar{x}_2, \ldots, \bar{x}_g$ are large enough to be considered statistically significant, we have to check whether the differences are large compared to the variances of the populations under study. We can do so by comparing two statistics: one statistic that quantifies the magnitude of the differences between the populations (i.e., how different the sample means $\bar{x}_1, \bar{x}_2, \ldots, \bar{x}_g$ are), and one statistic that quantifies the variability within the different populations. We obtain these two statistics by splitting the total variability in the entire data set.

The measure used for the total variability in the entire data set is based on the deviation of the individual observations from the overall sample mean, $\overline{X}$, which we define as

$$
\begin{aligned}
\overline{X} &= \frac{1}{n}\,(X_{11} + X_{12} + \cdots + X_{1n_1} \\
&\quad + X_{21} + X_{22} + \cdots + X_{2n_2} \\
&\quad + \ldots \\
&\quad + X_{g1} + X_{g2} + \cdots + X_{gn_g}), \\
&= \frac{1}{n}\sum_{i=1}^{g}\sum_{j=1}^{n_i} X_{ij}.
\end{aligned}
$$

For each observation X_{ij}, we can compute a total deviation

$$
X_{ij} - \overline{X}.
$$

We can rewrite this total deviation as

$$
\begin{aligned}
X_{ij} - \overline{X} &= X_{ij} - \overline{X} + \overline{X}_i - \overline{X}_i, \\
&= (X_{ij} - \overline{X}_i) + (\overline{X}_i - \overline{X}).
\end{aligned}
$$

Consequently, the total deviation of the jth observation in population i can be split into

(1) the deviation of observation X_{ij} from the ith sample mean $\overline{X}_i$, and
(2) the deviation of the ith sample mean $\overline{X}_i$ from the global mean $\overline{X}$.

The first type of deviation provides an indication of the variability within population i, while the second type of deviation gives an indication of the difference in mean between population i and the other populations.

This split of the total deviation can be made for each observation in each sample. Now, if we square all these total deviations and sum all the resulting squares, we obtain

$$\sum_{i=1}^{g}\sum_{j=1}^{n_i}(X_{ij}-\overline{X})^2 = \sum_{i=1}^{g}\sum_{j=1}^{n_i}\left((X_{ij}-\overline{X}_i)+(\overline{X}_i-\overline{X})\right)^2,$$

$$= \sum_{i=1}^{g}\sum_{j=1}^{n_i}\left((X_{ij}-\overline{X}_i)^2+(\overline{X}_i-\overline{X})^2+2(X_{ij}-\overline{X}_i)(\overline{X}_i-\overline{X})\right),$$

$$= \sum_{i=1}^{g}\sum_{j=1}^{n_i}(X_{ij}-\overline{X}_i)^2+\sum_{i=1}^{g}\sum_{j=1}^{n_i}(\overline{X}_i-\overline{X})^2+2\sum_{i=1}^{g}\sum_{j=1}^{n_i}(X_{ij}-\overline{X}_i)(\overline{X}_i-\overline{X}),$$

$$= \sum_{i=1}^{g}\sum_{j=1}^{n_i}(X_{ij}-\overline{X}_i)^2+\sum_{i=1}^{g}\sum_{j=1}^{n_i}(\overline{X}_i-\overline{X})^2+2\sum_{i=1}^{g}(\overline{X}_i-\overline{X})\left(\sum_{j=1}^{n_i}(X_{ij}-\overline{X}_i)\right).$$

For each sample i, we have that

$$\sum_{j=1}^{n_i}(X_{ij}-\overline{X}_i)=0,$$

which follows from the definition of the sample mean[1]. Therefore, the sum of the squared total deviations can be simplified to

$$\sum_{i=1}^{g}\sum_{j=1}^{n_i}(X_{ij}-\overline{X})^2 = \sum_{i=1}^{g}\sum_{j=1}^{n_i}(X_{ij}-\overline{X}_i)^2+\sum_{i=1}^{g}\sum_{j=1}^{n_i}(\overline{X}_i-\overline{X})^2,$$

and

$$\sum_{i=1}^{g}\sum_{j=1}^{n_i}(X_{ij}-\overline{X})^2 = \sum_{i=1}^{g}\sum_{j=1}^{n_i}(X_{ij}-\overline{X}_i)^2+\sum_{i=1}^{g}n_i(\overline{X}_i-\overline{X})^2. \qquad (12.1)$$

---

[1] The ith sample mean is

$$\overline{X}_i = \frac{1}{n_i}\sum_{j=1}^{n_i}X_{ij}.$$

Hence,

$$\sum_{j=1}^{n_i}X_{ij}=n_i\overline{X}_i=\sum_{j=1}^{n_i}\overline{X}_i.$$

From the equality of the first and last terms in this equation, it follows that

$$\sum_{j=1}^{n_i}(X_{ij}-\overline{X}_i)=0.$$

Each of the three terms in this equation is a sum of squares. The term on the left-hand side of the equation is called the **total sum of squares**. Numerous abbreviations for the total sum of squares can be found, both in the literature and in software packages. In this book, we use the abbreviation SSTO for the total sum of squares:

$$\text{SSTO} = \sum_{i=1}^{g} \sum_{j=1}^{n_i} (X_{ij} - \overline{X})^2.$$

The SSTO quantifies the total variation in the data.

The first term on the right-hand side of Equation (12.1) is called the **within sum of squares** and is abbreviated as SSW:

$$\text{SSW} = \sum_{i=1}^{g} \sum_{j=1}^{n_i} (X_{ij} - \overline{X}_i)^2. \tag{12.2}$$

It is a measure of the variation of the individual observations around their sample mean. The second term on the right-hand side of Equation (12.1) is called the **between sum of squares** and is abbreviated as SSB:

$$\text{SSB} = \sum_{i=1}^{g} n_i (\overline{X}_i - \overline{X})^2. \tag{12.3}$$

The between sum of squares is a measure of the difference between the sample means. Using the newly introduced notation, Equation (12.1) can be rewritten as

$$\text{SSTO} = \text{SSW} + \text{SSB}.$$

If all populations had the same population mean, the sample means $\overline{X}_1, \overline{X}_2, \dots, \overline{X}_g$ would be close to each other and SSB would be small.

We illustrate the computation of the sums of squares SSTO, SSW, and SSB using the data in Tables 12.1 and 12.2. In the illustration, we use lowercase instead of uppercase letters to clearly distinguish between the values computed from data that has been collected already (ssto, ssw, and ssb) and the corresponding random variables (SSTO, SSW, and SSB).

Example 12.1.2 *For the French data in Table 12.1, we can verify that* ssto $= 1066$, *ssw* $= 627$, *and* ssb $= 439$. *To find these values, we first have to determine the overall sample mean:*

$$\overline{x} = \frac{1}{18}(39 + 45 + 43 + \cdots + 55 + 50 + 50) = 49.667.$$

The three sums of squares are then

$$\text{ssto} = (39 - 49.667)^2 + (45 - 49.667)^2 + \cdots + (50 - 49.667)^2 + (50 - 49.667)^2,$$
$$= 1066,$$

$$\begin{aligned}
\text{ssw} = & (39 - 47.5)^2 + (45 - 47.5)^2 + \cdots + (49 - 47.5)^2 + (56 - 47.5)^2 \\
& + (56 - 56.5)^2 + (48 - 56.5)^2 + \cdots + (60 - 56.5)^2 + (63 - 56.5)^2 \\
& + (38 - 45)^2 + (37 - 45)^2 + \cdots + (50 - 45)^2 + (50 - 45)^2, \\
= & \ 627,
\end{aligned}$$

and

$$\text{ssb} = n_1(47.5 - 49.6667)^2 + n_2(56.5 - 49.6667)^2 + n_3(45 - 49.6667)^2 = 439,$$

since $n_1 = n_2 = n_3 = 6$. For the German data in Table 12.2, we obtain ssto = 2318, ssw = 1879, *and* ssb = 439.

Note that the two data sets yield the same between sum of squares, namely 439. However, there is an important difference between the two data sets: in the French data set, the between sum of squares constitutes a large portion of the total sum of squares (1066), while that is not the case in the German data set (where the total sum of squares is 2318). In other words, in the French data set, a relatively large proportion of the variability is due to the difference between the machines. Therefore, the first data set contains a stronger indication of a difference in the means of the machines than the second data set. This will also be reflected in the formal hypothesis test, which compares the value of the between sum of squares, ssb, *with the within sum of squares,* ssw.

The split of the total variability, measured by the sum of squares SSTO, in the two other sums of squares, SSW and SSB, explains why the name "analysis of variance" is used for the hypothesis test described in this chapter: one literally analyzes the variability in the data.

12.2 The Test

12.2.1 Variance Within and Between Groups

Because the three sums of squares, SSTO, SSW, and SSB, are not based on the same numbers of summed squares, they cannot be compared directly. Before comparing them, one should take into account the degrees of freedom of each sum of squares. The total sum of squares SSTO has $n - 1$ degrees of freedom. The within sum of squares SSW has $n - g$ degrees of freedom, and the between sum of squares SSB has $g - 1$ degrees of freedom.

These degrees of freedom are used to convert the within sum of squares and the between sum of squares into a mean sums of squares:

$$\text{MSW} = \frac{\text{SSW}}{n - g}$$

and

$$\text{MSB} = \frac{\text{SSB}}{g - 1}.$$

These two quantities, which can be compared directly, are called the **within-group variance** and the **between-group variance**[2], respectively. A relatively large value for MSB compared to MSW will indicate a difference between the g studied populations.

We now establish a hypothesis test that allows us to judge whether MSB is large enough relative to MSW to conclude that the populations have different population means. In order to do so, we need to make two important assumptions:

(1) The g populations under study are normally distributed.
(2) The g populations under study have the same variance: $\sigma_1^2 = \sigma_2^2 = \cdots = \sigma_g^2 = \sigma^2$.

The second assumption allows us to prove two theorems, which are useful for the construction of the hypothesis test.

Theorem 12.2.1 *The expected value of the within-group variance MSW is*

$$E(\text{MSW}) = \sigma^2$$

for g populations with the same variance σ^2.

Proof. First, we have

$$\text{MSW} = \frac{\text{SSW}}{n - g},$$

$$= \frac{1}{n - g} \sum_{i=1}^{g} \sum_{j=1}^{n_i} (X_{ij} - \overline{X}_i)^2,$$

$$= \frac{1}{n - g} \sum_{i=1}^{g} (n_i - 1) \frac{\sum_{j}^{n_i} (X_{ij} - \overline{X}_i)^2}{n_i - 1}.$$

The fraction in the latter expression is simply the sample variance S_i^2 of the observations in sample i:

$$S_i^2 = \frac{\sum_{j=1}^{n_i} (X_{ij} - \overline{X}_i)^2}{n_i - 1}.$$

---

[2] In some textbooks, the within-group variance MSW is denoted by MSE, while the between-group variance MSB is denoted by MSTr. The letter "E" in the abbreviation MSE stands for the word "Error", while the letters "Tr" in MSTr are for "Treatment". Likewise, the sums of squares SSW and SSB are sometimes denoted by SSE and SSTr, respectively. In textbooks that focus on regression techniques, the abbreviations SSR and MSR are used as alternatives for SSTr and MSTr, respectively. The last letter in SSR and MSR then refers to the word regression. Regression techniques are discussed in Chapters 18 and 19.

Consequently,

$$\text{MSW} = \frac{1}{n-g} \sum_{i=1}^{g} (n_i - 1)S_i^2.$$

Using Theorem 1.7.1, which states that a sample variance is an unbiased estimator of the corresponding population variance, the expected value of the within-group variance MSW can be written as

$$E(\text{MSW}) = \frac{1}{n-g} \sum_{i=1}^{g} (n_i - 1)E\left(S_i^2\right),$$

$$= \frac{1}{n-g} \sum_{i=1}^{g} (n_i - 1)\sigma_i^2,$$

where σ_i^2 represents the population variance of the ith population under investigation.

If we now assume that $\sigma_1^2 = \sigma_2^2 = \cdots = \sigma_g^2 = \sigma^2$, then we can simplify this expression to

$$E(\text{MSW}) = \frac{1}{n-g} \sum_{i=1}^{g} (n_i - 1)\sigma^2,$$

$$= \frac{(n-g)\sigma^2}{n-g},$$

$$= \sigma^2.$$

The within-group variance MSW is thus an unbiased estimator of the common population variance $\sigma_1^2 = \sigma_2^2 = \cdots = \sigma_g^2 = \sigma^2$. It acts as a pooled estimator of the variance σ^2, where the data from all g samples is combined to obtain a single estimator. ∎

The next theorem implies that, in general, the between-group variance MSB is a biased estimator of the common population variance σ^2. In this theorem, μ represents the weighted average of all population means $\mu_1, \mu_2, \ldots, \mu_g$:

$$\mu = \frac{n_1\mu_1 + n_2\mu_2 + \cdots + n_g\mu_g}{n} = \frac{1}{n} \sum_{i=1}^{g} n_i\mu_i.$$

If the null hypothesis is true, then

$$\mu = \mu_1 = \mu_2 = \cdots = \mu_g.$$

However, if the alternative hypothesis is true, then, in general,

$$\mu \neq \mu_i.$$

Theorem 12.2.2   *The expected value of the between-group variance* MSB *is*

$$E(\text{MSB}) = \sigma^2 + \frac{\sum_{i=1}^{g} n_i(\mu_i - \mu)^2}{g - 1}$$

for g populations with the same variance σ^2.

   The proof of this statement is much more complicated than the proof of Theorem 12.2.1. For this reason, we omit it here. The key point of the theorem is that the expected value of the between-group variance MSB depends on the unknown population means $\mu_1, \mu_2, \ldots, \mu_g$. When these means are not all the same, then, inevitably,

$$\mu \neq \mu_i$$

for some populations. In that case,

$$\mu_i - \mu \neq 0$$

and

$$(\mu_i - \mu)^2 > 0.$$

As a result,

$$\sum_{i=1}^{g} n_i(\mu_i - \mu)^2 > 0$$

and

$$E(\text{MSB}) > \sigma^2.$$

This implies that the between-group variance MSB overestimates σ^2 when not all of the population means are equal; in other words, when the alternative hypothesis is correct.
   Conversely, when the null hypothesis is correct, then all population means μ_i are equal to μ, and, for each population,

$$\mu = \mu_i,$$
$$\mu_i - \mu = 0,$$

and

$$(\mu_i - \mu)^2 = 0.$$

As a consequence,

$$\sum_{i=1}^{g} n_i(\mu_i - \mu)^2 = 0$$

and, finally,

$$E(\text{MSB}) = \sigma^2.$$

This means that the between-group variance is an unbiased estimator of σ^2 when the null hypothesis is true.

12.2.2   The Test Statistic

From the previous subsection, we know the following:

(1) If the null hypothesis is true, then we have two unbiased estimators of σ^2, namely the within-group variance MSW and the between-group variance MSB. If the null hypothesis is true, then the computed values for the within-group variance and the between-group variance will usually lie close to σ^2, and thus also close to each other. In that case, the quotient msb/msw is close to 1.
(2) If the null hypothesis is not true, then only the within-group variance MSW is an unbiased estimator of σ^2. The between-group variance then overestimates σ^2. In other words, if the null hypothesis is not true, then the computed value for the between-group variance will typically be larger than σ^2, and thus will usually also be larger than the computed value for the within-group variance. In that case, the quotient msb/msw is larger than 1.

To decide whether we should reject the null hypothesis in the analysis of variance, we compare the computed values for the within-group variance and the between-group variance. If both values are almost identical (i.e., if their quotient is close to 1), this suggests that the null hypothesis might be true. If the between-group variance is substantially larger than the within-group variance (i.e., if the quotient msb/msw is substantially larger than 1), this suggests that the alternative hypothesis is likely to be true.

To cast this test procedure into a formal framework, we need to define a test statistic. This test statistic is the ratio

$$F = \frac{\text{MSB}}{\text{MSW}}. \tag{12.4}$$

This quotient is F-distributed with $g - 1$ degrees of freedom for the numerator and $n - g$ degrees of freedom for the denominator, assuming that the g populations under study are all normally distributed with variance σ^2. This follows directly from the fact that both

$$\frac{\text{SSW}}{\sigma^2} = \frac{(n - g)\,\text{MSW}}{\sigma^2}$$

and

$$\frac{\text{SSB}}{\sigma^2} = \frac{(g-1)\,\text{MSB}}{\sigma^2}$$

are independent χ^2-distributed random variables with $n - g$ and $g - 1$ degrees of freedom, respectively, when the null hypothesis that $\mu_1 = \mu_2 = \cdots = \mu_g$ is correct and when the g populations under study are normally distributed with variance σ^2. In Section 8.3, we have learnt that a ratio of two independent χ^2-distributed random variables (each divided by its degrees of freedom) is F-distributed.

If the null hypothesis is wrong, then the numerator of the test statistic is not χ^2-distributed and the test statistic in Equation (12.4) is not F-distributed.

Example 12.2.3   *The F-distribution of the test statistic in Equation (12.4) under the null hypothesis is illustrated in Figure 12.3. The illustration shows six normally distributed populations 1, 2, 3, 4, 5, and 6, all having an expected value of 20 and a variance of 9, so that $\mu_1 = \mu_2 = \cdots = \mu_6 = 20$, $\sigma^2 = 9$ and the null hypothesis is correct. This can be seen in the panel at the top left of the figure, where the expected value for each population is displayed by means of a horizontal line at the value 20. For this scenario, five observations were simulated for each population 1000 times (thus 1000 times $(6 \times 5) = 30$ observations were simulated in total). An example of one simulated data set is shown by means of the dots in the top left panel.*

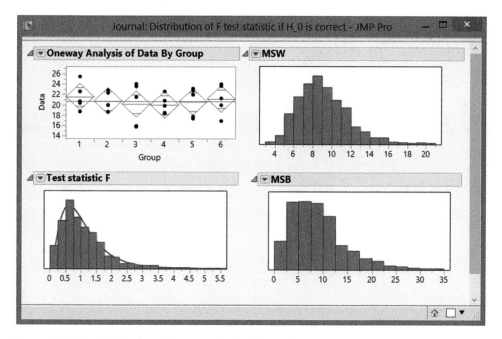

Figure 12.3   The distribution of the test statistic F in Equation (12.4) if the null hypothesis $\mu_1 = \mu_2 = \cdots = \mu_6$ is correct.

The six sample means associated with these dots are indicated by means of the centerlines in the six diamonds in that panel.

For each of the 1000 simulated data sets, the within-group variance, the between-group variance, and the test statistic were computed. These 1000 test statistics were plotted in a histogram. This histogram is shown in the bottom left panel of the figure. The histogram has exactly the same shape as an F-distribution with $g - 1 = 6 - 1 = 5$ degrees of freedom for the numerator and $n - g = 30 - 6 = 24$ degrees of freedom for the denominator. This F-distribution is also displayed in the bottom left panel in Figure 12.3, by means of the solid curve. The histogram in the bottom left panel of the figure shows that the values of the test statistic are usually in the vicinity of 1, which indicates that the within-group variance and the between-group variance are generally close to each other.

The two right-hand panels in the figure contain the histograms of the within-group and between-group variances, MSW and MSB. Both the within-group variance MSW and the between-group variance MSB usually take values in the vicinity of 9, the value of σ^2. This illustrates that when the null hypothesis is true, the within-group variance and the between-group variance are unbiased estimators of σ^2.

Figure 12.4 illustrates the consequences of an incorrect null hypothesis. This figure has been created for a situation where $\mu_1 = 25$, $\mu_4 = 15$, $\mu_2 = \mu_3 = \mu_5 = \mu_6 = 20$, and $\sigma^2 = 9$. For this figure, 1000 data sets of 30 observations (5 for each population) were simulated as well, and the corresponding 1000 test statistic values were computed. A histogram of these 1000 computed test statistics is shown in the bottom left panel in Figure 12.4. The computed test statistics are now almost always considerably larger than 1. The histogram of these test

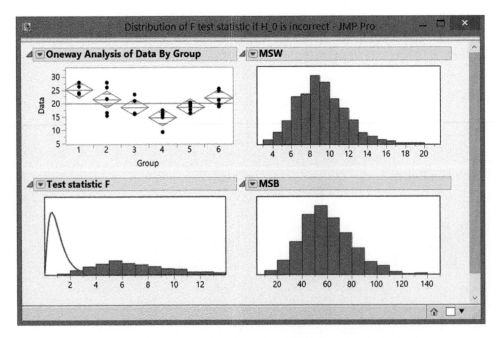

Figure 12.4 The distribution of the test statistic F in Equation (12.4) if the null hypothesis $\mu_1 = \mu_2 = \cdots = \mu_6$ is not correct.

statistics has a completely different form than the one shown in Figure 12.3 and does not correspond at all to the F-distribution with $g - 1 = 6 - 1 = 5$ degrees of freedom for the numerator and $n - g = 30 - 6 = 24$ degrees of freedom for the denominator (shown by the solid curve).

The two right-hand panels of the figure contain the histograms of the within-group variance MSW and the between-group variance MSB. The within-group variance MSW is again an unbiased estimator of σ^2: it generally takes values in the vicinity of 9. The between-group variance MSB takes much larger values, in the vicinity of its expected value

$$E(\text{MSB}) = \sigma^2 + \frac{\sum_{i=1}^{g} n_i(\mu_i - \mu)^2}{g - 1}$$

$$= 9 + \frac{\begin{array}{c} 5(25 - 20)^2 + 5(20 - 20)^2 + 5(20 - 20)^2 + \\ 5(15 - 20)^2 + 5(20 - 20)^2 + 5(20 - 20)^2 \end{array}}{6 - 1}$$

$$= 59,$$

as $g = 6$, $\sigma^2 = 9$, $\mu_1 = 25$, $\mu_4 = 15$, $\mu_2 = \mu_3 = \mu_5 = \mu_6 = 20$, $n_1 = \cdots = n_6 = 5$ and

$$\mu = \frac{n_1\mu_1 + n_2\mu_2 + \cdots + n_g\mu_g}{n},$$

$$= \frac{5 \times 25 + 5 \times 20 + 5 \times 20 + 5 \times 15 + 5 \times 20 + 5 \times 20}{30},$$

$$= 20.$$

This illustrates that the between-group variance can be much larger than the within-group variance if the null hypothesis is incorrect.

12.2.3 The Decision Rule and the p-Value

Large values for the F test statistic point in the direction of the alternative hypothesis. Therefore, in an analysis of variance, we will reject the null hypothesis if the computed test statistic is large – more specifically, if it exceeds a certain critical value. An appropriate critical value is based on the significance level α chosen and the F-distribution of the test statistic under the null hypothesis. We denote that critical value by $F_{\alpha;g-1;n-g}$.

In a one-way analysis of variance, we reject the null hypothesis if the computed test statistic

$$f = \frac{\dfrac{\text{ssb}}{g - 1}}{\dfrac{\text{ssw}}{n - g}} = \frac{\text{msb}}{\text{msw}}$$

is larger than the critical value $F_{\alpha;g-1;n-g}$. In the opposite case, we cannot reject the null hypothesis.

The p-value in an analysis of variance should be small when the computed test statistic f is large and thus points in the direction of the alternative hypothesis. Therefore, we define the p-value of the hypothesis test to be

$$p = P\left(F_{g-1;n-g} > \frac{\frac{\text{ssb}}{g-1}}{\frac{\text{ssw}}{n-g}}\right) = P\left(F_{g-1;n-g} > \frac{\text{msb}}{\text{msw}}\right) = P(F_{g-1;n-g} > f).$$

The p-value in a one-way analysis of variance is the probability that an F-distributed random variable with $g-1$ and $n-g$ degrees of freedom for the numerator and denominator exceeds the computed test statistic f. The null hypothesis is rejected if the p-value is smaller than the significance level α chosen.

Example 12.2.4 *For the data from the factory in France in Table 12.1, the test statistic is $f = \text{msb}/\text{msw} = 219.5/41.8 = 5.251$. This value is larger than the critical value $F_{0.05;2;15} = 3.682$. Therefore, based on this data, we reject the null hypothesis that the three extruders produce the same mean thickness. The p-value is*

$$p = P(F_{2;15} > 5.251) = 0.0187,$$

which is smaller than the usual significance level of 5%. So, the p-value also suggests that we should reject the null hypothesis. We conclude that the three extruders in the French factory produce significantly different mean thicknesses.

*The critical value $F_{0.05;2;15}$ can be read off the table in Appendix G or computed in JMP using the command "*F Quantile(0.95, 2, 15)*". The p-value can be computed in JMP using the command "*1 - F Distribution(5.251, 2, 15)*".*

Example 12.2.5 *For the data from the German factory in Table 12.2, the test statistic is $f = \text{msb}/\text{msw} = 219.5/125.267 = 1.752$. This value is smaller than the critical value $F_{0.05;2;15} = 3.682$. Therefore, based on the German data, we cannot reject the null hypothesis that the three extruders produce the same mean thickness. The p-value is*

$$p = P(F_{2;15} > 1.752) = 0.2071,$$

which is larger than the usual significance level of 5%. The p-value thus also indicates that we cannot reject the null hypothesis. We conclude that the three extruders in the German factory do not produce significantly different mean thicknesses.

*The p-value for the German data can be computed in JMP using the command "*1 - F Distribution(1.752, 2, 15)*".*

12.2.4   The ANOVA Table

All information related to the hypothesis test in the context of a one-way analysis of variance is usually neatly summarized in a table, called the analysis of variance table, or ANOVA table:

Source of variation	Sum of squares	Degrees of freedom	Mean sum of squares	F test statistic	p-value
Between	SSB	$g - 1$	MSB	F	p
Within	SSW	$n - g$	MSW		
Total	SSTO	$n - 1$			

Example 12.2.6   *The ANOVA tables corresponding to the data in Tables 12.1 and 12.2 are shown in Tables 12.3 and 12.4. As can be seen in these tables, the French data set results in a small p-value (0.0187), while the German data set produces a large p-value (0.2071). Based on these p-values, we reject the null hypothesis for the French data set, while we accept it for the German data set. We reach the same conclusion by comparing the computed test statistics 5.251 and 1.752 with the critical value $F_{0.05;2;15} = 3.682$.*

At this point, it is important to stress that our acceptance of the null hypothesis (or our failure to reject the null hypothesis) for the German factory does not mean that we have proven that the mean thicknesses for the three extruders are identical. Failing to reject the hypothesis that $\mu_1 = \mu_2 = \mu_3$ is not the same as proving that it is true. A detailed discussion of this topic can be found in Chapter 16.

Our inability to reject the null hypothesis for the German data is due to its larger variability. Therefore, the quality control managers in the German factory should perhaps start worrying about excessive variability in the thicknesses produced, and not just about the possible differences in means.

Table 12.3   The ANOVA table for the French data in Table 12.1.

Source of variation	Sum of squares	Degrees of freedom	Mean sum of squares	F test statistic	p-value
Between	439	2	219.500	5.251	0.0187
Within	627	15	41.800		
Total	1066	17			

Table 12.4   The ANOVA table for the German data in Table 12.2.

Source of variation	Sum of squares	Degrees of freedom	Mean sum of squares	F test statistic	p-value
Between	439.00	2	219.500	1.752	0.2071
Within	1879.00	15	125.267		
Total	2318.00	17			

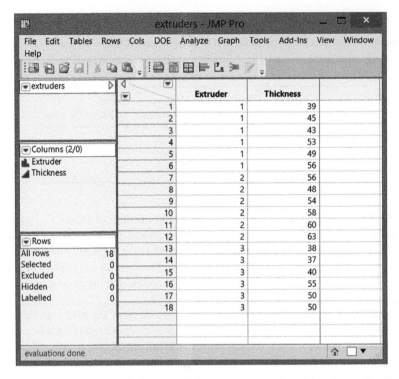

Figure 12.5 JMP data table suitable for an analysis of variance, with the French data from Table 12.1.

12.3   One-Way Analysis of Variance in JMP

It is not difficult to perform a one-way analysis of variance in JMP. It is, however, important that the data is entered in an appropriate fashion: one column should contain the quantitative variable that we wish to study, while another column should indicate from which population each observation originates (i.e., to which sample each observation belongs). Therefore, this second column must contain a nominal variable. A suitable JMP data table is shown in Figure 12.5. In this table, the first column contains the nominal variable "Extruder", while the second column contains the quantitative variable "Thickness".

After entering the data, we need to choose the option "Fit Y by X" in the "Analyze" menu (see Figure 12.6). In the resulting dialog window, we have to enter the quantitative variable "Thickness" in the field "Y, Response" and the nominal variable "Extruder" in the field "X, Factor" (see Figure 12.7). Clicking "OK" then produces the initial output shown in Figure 12.8.

The initial output generated by JMP is purely graphical. To perform the analysis of variance, click on the hotspot (red triangle) next to the term "Oneway Analysis of Thickness By Extruder", and choose "Means/Anova", as shown in Figure 12.9.

For the French data in Table 12.1, the option "Means/Anova" leads to the output in Figure 12.10. The ANOVA table in this output is identical to the one in Table 12.3. For the German data in Table 12.2, the option "Means/Anova" leads to the output in Figure 12.11. The ANOVA table in this output is identical to the one in Table 12.4.

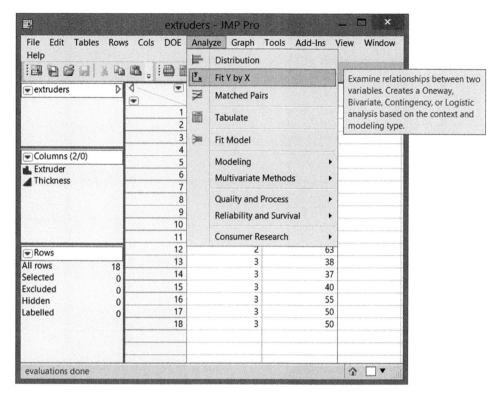

Figure 12.6   The "Fit Y by X" option in the "Analyze" menu.

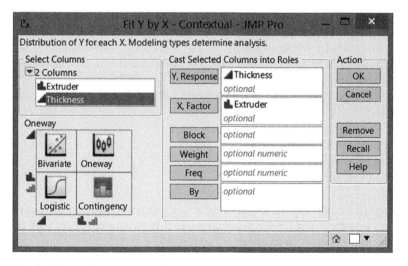

Figure 12.7   The dialog window of the option "Fit Y by X" in the "Analyze" menu for a one-way analysis of variance.

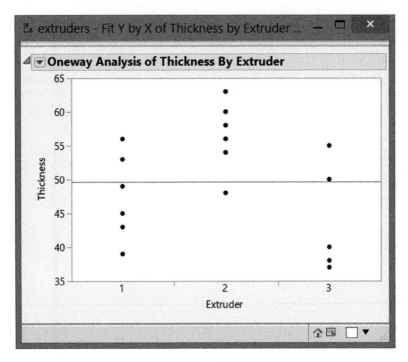

Figure 12.8 The initial output of the "Fit Y by X" option in the "Analyze" menu for the French data from Table 12.1.

12.4 Pairwise Comparisons

Table 12.3 shows that there is a significant difference between the mean thicknesses of the plastic foil produced by the three extruders in France. In that case, it is natural to wonder whether the first extruder produces a different mean thickness than extruder 2, whether the first extruder yields a different mean thickness than extruder 3, or whether the second extruder has a different mean thickness than extruder 3.

To answer these questions, we need to perform the following three pairwise hypothesis tests simultaneously:

- $H_0 : \mu_1 = \mu_2$ versus $H_a : \mu_1 \neq \mu_2$;
- $H_0 : \mu_1 = \mu_3$ versus $H_a : \mu_1 \neq \mu_3$;
- $H_0 : \mu_2 = \mu_3$ versus $H_a : \mu_2 \neq \mu_3$.

When looking at these null hypotheses and alternative hypotheses, it may seem like a reasonable idea to use the hypothesis test for two population means from Section 8.1 for each of the three pairwise tests. This is not recommended for two reasons:

(1) First, it is advisable to estimate σ^2 based on all available data; that is, based on all three samples from the three extruders at the same time. The ideal estimator for σ^2 is the within-group variance MSW. This estimator is based on $n - g$ degrees of freedom. The test from

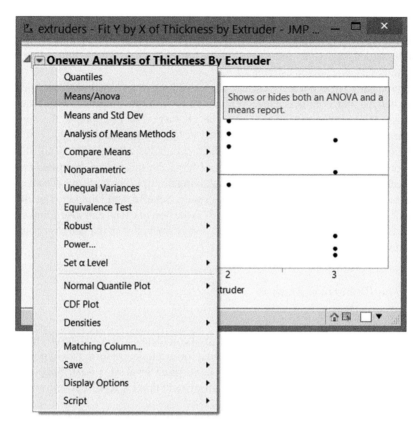

Figure 12.9   Selection of the option "Means/Anova" to perform a one-way analysis of variance.

Analysis of Variance

Source	DF	Sum of Squares	Mean Square	F Ratio	Prob > F
Extruder	2	439.0000	219.500	5.2512	0.0187*
Error	15	627.0000	41.800		
C. Total	17	1066.0000			

Figure 12.10   The ANOVA table produced by the option "Means/Anova" for the French data in Table 12.1.

Analysis of Variance

Source	DF	Sum of Squares	Mean Square	F Ratio	Prob > F
Extruder	2	439.0000	219.500	1.7523	0.2071
Error	15	1879.0000	125.267		
C. Total	17	2318.0000			

Figure 12.11   The ANOVA table produced by the option "Means/Anova" for the German data in Table 12.2.

Section 8.1 would lead to an estimate of σ^2 based on only two samples, with a smaller number of degrees of freedom. The resulting hypothesis test would have a smaller power and thus would lead to an increased risk of a type II error. The within-group variance MSW thus acts as a pooled estimator for σ^2.

(2) If the three pairwise hypothesis tests are performed separately, each with confidence level $1 - \alpha$, then the joint confidence level is less than $1 - \alpha$. If the three tests were independent, then their joint confidence level[3] would be $(1 - \alpha)^3$. For example, if the significance level chosen is 5%, then the confidence level is $1 - \alpha = 0.95$ and $(1 - \alpha)^3 = 0.8574$. As 0.8574 is smaller than 0.95, the entire statistical analysis, consisting of all three tests, does not have the desired confidence level of 95%. For this reason, modified procedures have been developed for hypothesis tests involving multiple comparisons. The best-known procedures are the methods of Bonferroni, Tukey, Scheffé, and Dunnett. The Bonferroni method[4] is the easiest, while Tukey's method[5] is the best one if all pairwise comparisons are performed. Dunnett's method is used if one is only interested in the pairwise comparisons with one specific population mean.

12.4.1   The Bonferroni Method

The Starting Point

To gain insight into joint significance levels and joint confidence levels, we start from a situation where we want to make two statistical statements; for example, perform two hypothesis tests, with a significance level α (and thus with a confidence level of $1 - \alpha$). Suppose that U_1 represents the event that the first statement is wrong, and that U_2 represents the event that the second statement is wrong.

The significance level of the complete statistical analysis, consisting of the two statements, is the probability that at least one of the statements is wrong:

$$
\begin{aligned}
P(U_1 \text{ or } U_2) &= P(U_1 \cup U_2), \\
&= P(U_1) + P(U_2) - P(U_1 \cap U_2), \\
&= \alpha + \alpha - P(U_1 \cap U_2), \\
&= 2\alpha - P(U_1 \cap U_2).
\end{aligned}
$$

---

[3] Suppose that the probabilities for the occurrence of the independent events A, B, and C are all three equal to $1 - \alpha$. The joint probability $P(A \cap B \cap C)$ is, due to the independence of the events, $P(A)P(B)P(C) = (1 - \alpha)(1 - \alpha)(1 - \alpha) = (1 - \alpha)^3$. The confidence level of a hypothesis test is the probability of a correct decision, if the null hypothesis is true. The joint confidence level of three hypothesis tests is the probability that these three tests simultaneously lead to a correct decision if the null hypotheses of all three tests are true.

[4] Carlo Emilio Bonferroni (1892–1960) was an Italian mathematician who made some contributions to probability theory, with important applications in statistics.

[5] John Tukey (1915–2000) was an American mathematician who is considered the inventor of the box plot. He was involved in the development of numerous statistical methods and also invented a fast way to calculate Fourier transforms.

The complement of this probability is the probability that both statements are correct at the same time. This is the joint confidence level:

$$P(\overline{U}_1 \text{ and } \overline{U}_2) = 1 - P(U_1 \text{ or } U_2),$$
$$= 1 - \left[2\alpha - P(U_1 \cap U_2)\right],$$
$$= 1 - 2\alpha + P(U_1 \cap U_2).$$

Since any probability is greater than or equal to zero, we have that $P(U_1 \cap U_2) \geq 0$. As a result, the joint confidence level of the two statements is greater than or equal to $1 - 2\alpha$:

$$P(\overline{U}_1 \text{ and } \overline{U}_2) \geq 1 - 2\alpha.$$

The practical relevance of this result is that two statistical statements, each of which separately has a confidence level of $1 - \alpha$, have a joint confidence level of at least $1 - 2\alpha$. Thus, if we want a joint confidence level of 95%, then we have to use a confidence level of 97.5% for each separate statement. In other words, we have to decrease the significance level to 2.5% for each individual statement, so that the joint significance level does not exceed 5%.

If we now have a total of m different statements and we want an overall confidence level of $1 - \alpha$, then we should use a significance level of α/m for each separate statement. This follows from the fact that

$$P(\overline{U}_1 \text{ and } \overline{U}_2 \text{ and } \ldots \text{ and } \overline{U}_m) \geq 1 - m\alpha$$

for m statistical statements with an individual significance level of α.

Pairwise Hypothesis Tests

The test statistic for the pairwise comparison of two means in the context of an analysis of variance is

$$c_{ij} = \frac{\bar{x}_i - \bar{x}_j}{\sqrt{\text{msw}\left(\frac{1}{n_i} + \frac{1}{n_j}\right)}}.$$

This test statistic is t-distributed with $n - g$ degrees of freedom if $\mu_i = \mu_j$. It is denoted by the letter "c" because the test seeks to contrast populations i and j with each other. The "c" is short for the word "contrast".

If this test statistic is strongly positive or strongly negative, this indicates a significant difference between the ith population mean μ_i and the jth population mean μ_j. The critical values to decide whether the null hypothesis that $\mu_i = \mu_j$ should be rejected are $-t_{\alpha/(2m);n-g}$ and $t_{\alpha/(2m);n-g}$. Here, m represents the total number of pairwise comparisons. In the expression for the quantiles needed, $-t_{\alpha/(2m);n-g}$ and $t_{\alpha/(2m);n-g}$, the value α is divided by m due to the Bonferroni method, and by 2 because the hypothesis test is two-tailed. In general, $m = g(g-1)/2$ whenever the interest is in all pairwise comparisons.

Using the Bonferroni method, we work with a significance level of $\alpha^* = \alpha/m$ for each individual pairwise comparison. Therefore, we reject the null hypothesis that $\mu_i = \mu_j$ if

$$2\,P\left(t_{n-g} > \left|\frac{\bar{x}_i - \bar{x}_j}{\sqrt{\text{msw}\left(\frac{1}{n_i} + \frac{1}{n_j}\right)}}\right|\right), \tag{12.5}$$

which is essentially the "usual" p-value for a two-tailed test for two population means using independent samples, is smaller than α/m. So, to make the decision whether or not to reject the null hypothesis, we compare the expression in Equation (12.5) to α/m. This is equivalent to comparing m times the expression in Equation (12.5) to α. Therefore, the Bonferroni-adjusted p-value is defined as

$$p = 2m\,P\left(t_{n-g} > \left|\frac{\bar{x}_i - \bar{x}_j}{\sqrt{\text{msw}\left(\frac{1}{n_i} + \frac{1}{n_j}\right)}}\right|\right)$$

and the null hypothesis is rejected if $p < \alpha$. A disadvantage of this procedure is that the resulting p-value no longer represents a probability, so that, in some cases, it may take values larger than 1. This is illustrated in Example 12.4.1.

Confidence Intervals

Confidence intervals for m pairwise comparisons of population means can be computed using the expression

$$(\bar{x}_i - \bar{x}_j) \pm t_{\alpha/(2m);n-g}\sqrt{\text{msw}\left(\frac{1}{n_i} + \frac{1}{n_j}\right)},$$

if the Bonferroni method is applied. In this expression, $\bar{x}_i$ and $\bar{x}_j$ again represent the means of the samples from populations i and j, $t_{\alpha/(2m);n-g}$ is the required quantile of the t-distribution with $n - g$ degrees of freedom, and n_i and n_j are the numbers of observations in samples i and j. This expression is similar to Equation (8.1.4).

Example 12.4.1   *In the examples concerning the three extruders in the French factory, $g = 3$. As a result, three pairwise comparisons can be performed. Therefore, m is equal to 3, and, according to the Bonferroni method, we need to work with a significance level of $\alpha/3$ for each individual test. Assuming that a significance level of 5% is desired, the required quantiles of the t-distribution with $n - g = 18 - 3 = 15$ degrees of freedom are*

$\pm t_{0.05/(2\times3);15} = t_{0.05/6;15} = \pm 2.6937$. *The values ± 2.6937 can be found using the JMP commands "*t Quantile *(0.05/6, 15)" and "*t Quantile*(1 − 0.05/6, 15)".*

As an illustration, we first investigate whether there is a significant difference between the mean thicknesses produced by extruders 1 and 2 for the French data in Table 12.1. For these two extruders, the difference[6] in the sample means is $\bar{x}_2 - \bar{x}_1 = 56.5 - 47.5 = 9$. The within-group variance msw *is equal to 41.8. The sample sizes for the two extruders are $n_1 = 6$ and $n_2 = 6$. Therefore, the test statistic is*

$$c_{21} = \frac{56.5 - 47.5}{\sqrt{41.8 \left(\frac{1}{6} + \frac{1}{6}\right)}} = 2.411.$$

This value can now be compared with the critical values $\pm t_{0.05/6;15} = \pm 2.6937$. The test statistic lies between these two critical values, so that we conclude that μ_1 and μ_2 are not significantly different. The p-value of this hypothesis test is

$$p = 2 \times 3 \, P\left(t_{15} > 2.411\right) = 0.0876$$

*and can be computed in JMP using the formula "6 * (1 − *t Distribution *(2.411, 15))".*

The Bonferroni confidence interval for the difference $\mu_2 - \mu_1$ can be computed as

$$(56.5 - 47.5) \pm t_{0.05/6,15} \sqrt{41.8 \left(\frac{1}{6} + \frac{1}{6}\right)} = (56.5 - 47.5) \pm 2.6937 \times 3.7327,$$

which gives a lower limit of −1.0549 and an upper limit of 19.0549. This interval contains the value zero, so that, again, we conclude that μ_1 and μ_2 do not differ significantly.

In the same way, we can compute p-values and confidence intervals for the comparison of μ_1 with μ_3 and for the comparison of μ_2 with μ_3, using the Bonferroni method. The confidence interval for $\mu_1 - \mu_3$ is $[-7.5549, 12.5549]$, while it is equal to $[1.4451, 21.5549]$ for $\mu_2 - \mu_3$. The former interval contains the value zero, so μ_1 and μ_3 are not significantly different. The latter interval has positive lower and upper limits. Therefore, it does not contain the value zero, which leads us to conclude that μ_2 and μ_3 are significantly different. The p-values corresponding to the two intervals are 1.5396 and 0.0228. The p-value greater than 1 is due to the fact that the exceedance probability value is multiplied by $2 \times 3 = 6$ when applying the Bonferroni method. Statistical software reports a p-value of 1 if the original calculation produces a value larger than 1.

We cannot easily verify these calculations in JMP, because the Bonferroni method is not available in JMP. The reason for this is that Tukey's method is better than the Bonferroni method when it comes to making all pairwise comparisons.

---

[6] Note that we subtract the smaller sample mean from the larger one, because this is the approach taken by JMP when performing pairwise comparisons.

12.4.2 Tukey's Method

If we want to perform hypothesis tests for all possible pairwise comparisons, or if we need all pairwise confidence intervals, then Tukey's method is better than the Bonferroni method. This means that Tukey's method results in narrower confidence intervals and hypothesis tests with a larger power. The disadvantage of Tukey's method[7] is that it is more complicated, because it makes use of a new probability distribution, namely the so-called Studentized range distribution or Q-distribution.

The Studentized Range Distribution

The **Studentized range** distribution deals with the range of observations from a normally distributed population. Suppose that we collect g observations $Y_1, Y_2, \ldots, Y_g$ from a normally distributed population with variance σ^2, and that S^2 is an unbiased estimator of σ^2 based on v degrees of freedom. Then, the random variable

$$Q_{g;v} = \frac{\max(Y_1, Y_2, \ldots, Y_g) - \min(Y_1, Y_2, \ldots, Y_g)}{S}$$

has the Studentized range distribution.

Appendix I contains a table with useful percentiles or quantiles $Q_{\alpha;g;v}$ from the Studentized range distribution. These quantiles can be computed in JMP using the command "`Tukey HSD Quantile(1 - α, g, v) * Sqrt(2)`". The abbreviation HSD in this command is short for "honest significant difference".

Pairwise Hypothesis Tests

The test statistic c_{ij} for the pairwise comparison of two population means μ_i and μ_j in the context of an analysis of variance using Tukey's method is

$$c_{ij} = \sqrt{2} \times \frac{\bar{x}_i - \bar{x}_j}{\sqrt{\mathrm{msw}\left(\frac{1}{n_i} + \frac{1}{n_j}\right)}}.$$

If this test statistic is strongly positive or strongly negative, this indicates a significant difference between the ith population mean μ_i and the jth population mean μ_j. The critical values to decide whether the null hypothesis that $\mu_i = \mu_j$ should be rejected are $-Q_{\alpha;g;n-g}$ and $Q_{\alpha;g;n-g}$.

---

[7] Tukey's method is sometimes also called the Tukey–Kramer method, because, in 1956, Clyde Kramer generalized the original method for equal sample sizes to scenarios with unequal sample sizes.

The *p*-value is now computed using the Studentized range distribution:

$$p = P\left(Q_{g;n-g} > \left|\sqrt{2} \times \frac{\bar{x}_i - \bar{x}_j}{\sqrt{\mathrm{msw}\left(\frac{1}{n_i} + \frac{1}{n_j}\right)}}\right|\right) = P\left(Q_{g;n-g} > |c_{ij}|\right).$$

This *p*-value can be computed in JMP using the command "Tukey HSD P Value $(c_{ij}/\sqrt{2}$, g, $n-g)$".

Confidence Intervals

Confidence intervals for *m* pairwise comparisons of population means can be computed from the expression

$$(\bar{x}_i - \bar{x}_j) \pm \frac{1}{\sqrt{2}} Q_{\alpha;g;n-g} \sqrt{\mathrm{msw}\left(\frac{1}{n_i} + \frac{1}{n_j}\right)},$$

when Tukey's method is used. This expression differs from the one in Section 12.4.1 because the estimated standard deviation of $\bar{x}_i - \bar{x}_j$ is multiplied by

$$\frac{1}{\sqrt{2}} Q_{\alpha;g;n-g}$$

instead of

$$t_{\alpha/(2m);n-g}.$$

It is useful to observe that

$$\frac{1}{\sqrt{2}} Q_{\alpha;g;n-g} < t_{\alpha/(2m);n-g},$$

which implies that the confidence intervals obtained by applying Tukey's method are narrower than those obtained by applying the Bonferroni method.

Example 12.4.2   *We now use Tukey's method to investigate whether there is a significant difference between the mean thicknesses produced by extruders 1 and 2 for the data in Table 12.1. For these two extruders, the difference in the sample means is $\bar{x}_2 - \bar{x}_1 = 56.5 - 47.5 = 9$. The within-group variance is* msw $= 41.8$. *The sample sizes for the two extruders* *are $n_1 = 6$ and $n_2 = 6$. Therefore, the test statistic is*

$$c_{21} = \sqrt{2} \times \frac{56.5 - 47.5}{\sqrt{41.8 \left(\frac{1}{6} + \frac{1}{6} \right)}} = \sqrt{2} \times 2.411 = 3.410.$$

This value can be compared with the critical values $\pm Q_{0.05;3;15} = \pm 3.673$. The computed test statistic lies between these two critical values, so we conclude that μ_1 and μ_2 are not significantly different. The critical value $Q_{0.05;3;15}$ can be computed in JMP using the command "`Tukey HSD Quantile`(0.95, 3, 15) * `Sqrt(2)`". *The p-value of the hypothesis test comparing μ_1 and μ_2 can be calculated in JMP using the formula* "`Tukey HSD P Value`(2.411, 3, 15)". *This yields a p-value of 0.0708.*

　The confidence interval for the difference $\mu_2 - \mu_1$ can be computed as

$$(56.5 - 47.5) \pm \frac{1}{\sqrt{2}} Q_{0.05;3;15} \sqrt{41.8 \left(\frac{1}{6} + \frac{1}{6} \right)}$$

$$= (56.5 - 47.5) \pm \frac{1}{\sqrt{2}} \times 3.673 \times 3.7327,$$

which gives a lower limit of -0.69568 and an upper limit of 18.69568. This interval contains the value zero, so, again, we have to conclude that μ_1 and μ_2 are not significantly different.

　In contrast with the Bonferroni method, Tukey's method is implemented in JMP. The main portion of the JMP output delivered by the option "`All Pairs, Tukey HSD`" *is shown in Figure 12.12. The output includes both confidence intervals and p-values for all pairwise comparisons. There are three pairwise comparisons, which are ranked by JMP in ascending order of the p-values. The most significant result (with the smallest p-value) is shown first, while the least significant result (with the largest p-value) is shown last.*

　The JMP output clearly shows that, of the three pairwise comparisons, only one produces a difference that is significant at the 5% level. More specifically, only the difference between μ_2 and μ_3 is significant. The point estimate for the difference $\mu_2 - \mu_3$ is $\bar{x}_2 - \bar{x}_3 = 11.5$. The Tukey confidence interval has a lower limit of 1.80432 and an upper limit of 21.19568. This interval does not contain the value zero, which confirms the significant difference between μ_2 and μ_3.

　The population means μ_1 and μ_2 are not significantly different (the p-value is 0.0708), nor are the population means μ_1 and μ_3 (the p-value is 0.7842). The confidence intervals for both $\mu_2 - \mu_1$ and $\mu_1 - \mu_3$ do contain the value zero.

Ordered Differences Report						
Level	**- Level**	**Difference**	**Std Err Dif**	**Lower CL**	**Upper CL**	**p-Value**
2	3	11.50000	3.732738	1.80432	21.19568	0.0196*
2	1	9.00000	3.732738	-0.69568	18.69568	0.0708
1	3	2.50000	3.732738	-7.19568	12.19568	0.7842

Figure 12.12 The JMP output for all pairwise comparisons of two population means with Tukey's method, applied to the French data in Table 12.1.

12.4.3 Dunnett's Method

In a one-way analysis of variance, researchers often want to compare some new "treatments" with a standard treatment or a placebo. For example, in marketing studies, researchers may wish to compare the mean sales for a number of new types of packaging for a product to the mean sales of the current packaging. In medical experiments, the results of new drugs are compared to those of an existing drug or a placebo. In agricultural experiments, the yield of crops obtained using new fertilizers is compared to the yield of the same crop with a conventional fertilizer. In the statistical jargon, the standard treatment is often referred to as the "control treatment". In situations involving such a control treatment, the researchers are initially often mainly interested in a pairwise comparison of the control treatment with all other treatments.

In this section, we assume that the first of all treatments is the control treatment.

The Required Probability Density

Dunnett's method is based on the joint probability density of all weighted differences of sample means

$$\frac{|\overline{X}_i - \overline{X}_1|}{\sqrt{\text{MSW}\left(\frac{1}{n_i} + \frac{1}{n_1}\right)}},$$

where the numerator is the difference between the sample mean for treatment i and that for the control treatment, and the denominator is the estimated standard deviation of this difference. The probability density is a multivariate t-distribution.

Appendices J and K contain tables with useful percentiles or quantiles $d_{\alpha;g-1;n-g}$ for Dunnett's method. These values are only valid when the number of observations is identical for all treatments; that is, when $n_1 = n_2 = \cdots = n_g$. The quantiles in Appendices J and K can be computed in JMP using the commands "Dunnett Quantile($1 - \alpha$, $g - 1$, $n - g$, $\{1/\text{sqrt}(2), \ldots, 1/\text{sqrt}(2)\}$)" and "Dunnett Quantile($1 - 2\alpha$, $g - 1$, $n - g$, $\{1/\text{sqrt}(2), \ldots, 1/\text{sqrt}(2)\}$)", respectively. The number of times that the value $1/\text{sqrt}(2)$ must be specified is $g - 1$. When the sample sizes $n_1, n_2, \ldots, n_g$ are unequal, the values "$1/\text{sqrt}(2)$" have to be replaced by

$$\sqrt{\frac{n_j}{n_j + n_1}},$$

where n_j is the sample size for treatment j and n_1 is the sample size for the control treatment.

Pairwise Hypothesis Tests

The test statistic c_{i1} for the pairwise comparison of the population mean μ_i of treatment i and the population mean μ_1 of the control treatment for Dunnett's method is

$$c_{i1} = \frac{\overline{x}_i - \overline{x}_1}{\sqrt{\text{msw}\left(\frac{1}{n_i} + \frac{1}{n_1}\right)}}.$$

The corresponding p-value can be calculated using the multivariate t-distribution, as described by Dunnett in 1955 and 1964. If $n_1 = n_2 = \cdots = n_g$, the p-value can be computed in JMP using the command "Dunnett P Value(c_{i1}, $g-1$, $n-g$, {1/sqrt(2), ..., 1/sqrt(2)})". Again, the number of times the value 1/sqrt(2) has to be specified is $g-1$. When the sample sizes are unequal, the values 1/sqrt(2) have to be modified as described above.

Instead of using p-values, one can also compare the test statistic c_{i1} with the quantiles or percentiles in Appendices J and K. The quantiles in Appendix J are the critical values for a two-tailed test, while the quantiles in Appendix K are the critical values for a right-tailed test. Appendix K is therefore useful if one wants to test whether treatment i yields a larger population mean than the control treatment. This is indeed a common scenario, as one often wants to demonstrate that a new treatment is better than a standard or placebo treatment. In such a scenario, a one-tailed hypothesis test is indeed appropriate.

Confidence Intervals

Confidence intervals for $g-1$ pairwise comparisons between, on the one hand, the population mean of the control treatment, and, on the other, the population means of the other treatments can be computed using the expression

$$(\bar{x}_i - \bar{x}_1) \pm d_{\alpha,g-1,n-g} \sqrt{\mathrm{msw}\left(\frac{1}{n_i} + \frac{1}{n_1}\right)},$$

when Dunnett's method with equal sample sizes ($n_1 = n_2 = \cdots = n_g$) is used. This expression differs from the one in Section 12.4.1 because the standard deviation of $\bar{x}_i - \bar{x}_1$ is multiplied by $d_{\alpha;g-1;n-g}$ instead of $t_{\alpha/(2m);n-g}$. For unequal sample sizes, the value $d_{\alpha;g-1;n-g}$ has to be adjusted.

Example 12.4.3   *Tomato plants are prone to infections, which can have a negative impact on the harvest. However, some fungi in the soil can help the tomato plants to overcome certain infections or to slow down the development of certain diseases. For example, some researchers have investigated to what extent certain fungi from the genus* Trichoderma *can slow down the growth of the fungus* Botrytis cinerea *(which causes botrytis bunch rot, or gray mold). Figure 12.13 shows a graphical summary of the results of an agricultural study on the impact of three* Trichoderma *strains and one control treatment (no* Trichoderma *strain) on the fungus* Botrytis cinerea. *The measurements reported in Figure 12.13 are diameters (expressed in millimeters) of a lesion (injury) of 200 different tomato plants. Samples of 50 observations were taken from the three* Trichoderma *strains and the control treatment. The mean diameters of the four samples are $\bar{x}_1 = 8.10$ mm (control treatment), $\bar{x}_2 = 6.61$ mm (strain 140103), $\bar{x}_3 = 7.41$ mm (strain 190127), and $\bar{x}_4 = 5.70$ mm (strain BFA).*

The F-test in the ANOVA table in Figure 12.14 produces a very small p-value. We can therefore conclude that there are significant differences in lesion diameter between the four groups of observations. The researchers were interested in differences in the lesion diameter between, on the one hand, the tomato plants treated with one of the Trichoderma *strains and,*

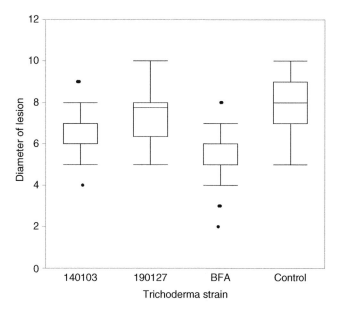

Figure 12.13 Box plots for the 200 measurements from the agricultural experiment in Example 12.4.3.

on the other, the tomato plants that were not treated (i.e., the tomato plants that underwent the control treatment). Therefore, Dunnett's method was the most appropriate.

Figure 12.15 shows the complete output produced by the JMP option "With Control, Dunnett's". The p-values for the three Trichoderma strains are all smaller than 5%, so there are significant differences in lesion diameter between plants treated with a Trichoderma strain and untreated plants. In the output, the three differences are ranked from least significant to most significant. The strain called "BFA" thus leads to the largest difference in lesion diameter.

Note that JMP performs Dunnett's two-tailed test. There is no option to select the one-sided version.

Also, JMP does not report confidence intervals for the three pairwise comparisons. With the formula in Section 12.4.3, however, we can compute confidence intervals. For the difference between the mean lesion diameter of the "BFA" strain and that of the control treatment, $\mu_4 - \mu_1$, we obtain, for example,

$$5.70 - 8.10 \pm d_{0.05;3;196} \sqrt{1.4948 \left(\frac{1}{50} + \frac{1}{50} \right)}.$$

Analysis of Variance

Source	DF	Sum of Squares	Mean Square	F Ratio	Prob > F
Trichoderma strain	3	160.60500	53.5350	35.8130	<.0001*
Error	196	292.99000	1.4948		
C. Total	199	453.59500			

Figure 12.14 The ANOVA table for the agricultural experiment in Example 12.4.3.

Comparisons with a control using Dunnett's Method

Control Group = Control

Confidence Quantile

|d|	Alpha
2.36734	0.05

Difference Matrix

Dif=Mean[i]-Mean[j]

	Control	190127	140103	BFA
Control	0.0000	0.6900	1.4900	2.4000
190127	-0.6900	0.0000	0.8000	1.7100
140103	-1.4900	-0.8000	0.0000	0.9100
BFA	-2.4000	-1.7100	-0.9100	0.0000

LSD Threshold Matrix

Level	Abs(Dif)-LSD	p-Value
Control	-0.58	1.0000
190127	0.111	0.0146*
140103	0.911	<.0001*
BFA	1.821	<.0001*

Positive values show pairs of means that are significantly different.

Figure 12.15   The JMP output with the results of the (two-sided) Dunnett test for Example 12.4.3.

Note that we can find the computed value for the within-group variance msw, *namely* 1.4948, *in the ANOVA table in Figure 12.14. The required quantile* $d_{0.05;3;196}$ *cannot be found in the table in Appendix J for a two-sided test and a significance level of 5%, because the appendix does not contain quantiles for* $n - g = 196$. *However, we can deduce from this table that the precise value of* $d_{0.05;3;196}$ *lies between* 2.385 *(the value for* $n - g = 100$*) and* 2.349 *(the value for* $n - g \to +\infty$*). By means of the JMP command* "Dunnett Quantile (0.95, 3, 196, {1/sqrt(2), 1/sqrt(2), 1/sqrt(2)})", *we find the exact value,* 2.367. *This value is also reported in the JMP output in Figure 12.15, under the title* "Confidence Quantile". *This value allows us to finalize the calculation of the confidence interval for* $\mu_4 - \mu_1$:

$$-2.4 \pm 2.36734\sqrt{0.059792},$$

or

$$[-2.98, -1.82].$$

In a similar fashion, we can verify that the 95% confidence intervals for $\mu_2 - \mu_1$ *and* $\mu_3 - \mu_1$ *are given by*

$$[-2.06, -0.90]$$

and

$$[-1.24, -0.08],$$

respectively. None of the three intervals contains the value zero, which confirms the results of the two-tailed hypothesis tests.

12.5   The Relation Between a One-Way Analysis of Variance and a *t*-Test for Two Population Means

It is useful to note that the hypothesis test for the comparison of g population means introduced in this chapter can also be used for a two-tailed hypothesis test for just two population means. The comparison of two population means through the use of independent samples has already been done in Chapter 8, using a *t*-test.

In fact, the two-sided *t*-test from Chapter 8 for equal variances yields exactly the same p-value as the *F*-test in a one-way analysis of variance. Moreover, the computed test statistic in a one-way analysis of variance is always the square of the computed test statistic of a two-sided *t*-test for the comparison of two means. It can also be shown that the square of a *t*-distributed random variable with, for example, v degrees of freedom, is an *F*-distributed random variable with one degree of freedom for the numerator and v degrees of freedom for the denominator.

It is a useful exercise to verify all this, based on an example. For instance, you can perform a one-way analysis of variance for the data from extruders 1 and 2 in Table 12.1. Then, you can compare the p-value of this analysis with the p-value that you obtain from a two-tailed *t*-test for two population means with equal variances.

12.6   Power

In Section 4.4, we used the noncentral *t*-distribution for determining the power and the probability of a type II error of a *t*-test for a single population mean and a given number of observations. We also learnt how to determine the number of observations so that the *t*-test has a prespecified power. In this section, we use the noncentral *F*-distribution for the computation of the power of a one-way analysis of variance and the required number of observations to achieve a certain desired power.

12.6.1   The Noncentral F-Distribution

In Section 8.3, we introduced the *F*-distribution and observed that the *F*-distribution has two parameters, v_1 and v_2. We called these parameters the degrees of freedom for the numerator and the degrees of freedom for the denominator. The reason for this is that an *F*-distributed random variable is defined as the quotient of two independent χ^2-distributed random variables, each divided by their number of degrees of freedom v_1 and v_2.

In the statistical literature, however, there is a distinction between the central *F*-distribution and the noncentral *F*-distribution. So far in this book, we have only used the central *F*-distribution for determining critical values and computing p-values. For simplicity, we have omitted the adjective "central".

For the computation of the power of a one-way analysis of variance, it is, however, necessary to use both the usual central *F*-distribution and the noncentral *F*-distribution. A noncentral *F*-distribution has one additional parameter. This additional parameter is called the noncentrality parameter ϕ. It has to be greater than or equal to zero. A central *F*-distribution is a special case of a noncentral *F*-distribution, with the noncentrality parameter being equal to zero.

Figure 12.16 shows one central *F*-distribution and two noncentral *F*-distributions. All three distributions have five numerator degrees of freedom and five denominator degrees of

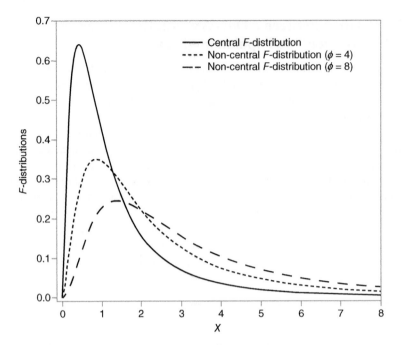

Figure 12.16 The central and two noncentral F-distributions with five degrees of freedom for both the numerator and the denominator.

freedom, so that $v_1 = v_2 = 5$. The figure shows that the F-distribution shifts to the right if the noncentrality parameter is increased.

JMP allows us to calculate probabilities for random variables with a noncentral F-distribution with parameters v_1, v_2, and ϕ. To do so, the required command is "F Distribution(f, v_1, v_2, ϕ)". We used the command "F Distribution" earlier in this book, but without explicitly mentioning the noncentrality parameter ϕ. JMP then assumes that the noncentrality parameter is zero, and uses the usual central F-distribution. In the same way, if we want to determine percentiles or quantiles for random variables with a noncentral F-distribution, or if we want to create a graphical representation of a noncentral F-distribution, we can add ϕ as an additional parameter to the commands "F Quantile" and "F Density".

12.6.2 The Noncentral F-Distribution and Analysis of Variance

The test statistic in an analysis of variance, namely the quotient

$$F = \frac{\text{MSB}}{\text{MSW}},$$

has a central F-distribution with $g - 1$ degrees of freedom for the numerator and $n - g$ degrees of freedom for the denominator, if the null hypothesis is true (and assuming that the studied populations are normally distributed with equal variance σ^2).

However, if the null hypothesis is false, then the test statistic no longer has a central F-distribution. Instead, the test statistic then has a noncentral F-distribution with $g - 1$ degrees of freedom for the numerator, $n - g$ degrees of freedom for the denominator, and noncentrality parameter

$$\phi = \frac{1}{\sigma^2} \sum_{i=1}^{g} n_i(\mu_i - \mu)^2. \tag{12.6}$$

Here,

$$\mu = \frac{1}{n} \sum_{i=1}^{g} n_i \mu_i.$$

The larger the differences between the population means, the larger is the noncentrality parameter ϕ. In other words, the less the null hypothesis is true, the larger is the noncentrality parameter.

The noncentrality parameter also depends on the sample sizes $n_1, n_2, \ldots, n_g$ and the variance σ^2. The larger the sample sizes, the larger is the noncentrality parameter. The larger the variance σ^2, the smaller is the noncentrality parameter.

Example 12.6.1 *Figure 12.4 in Example 12.2.3 was constructed for a scenario in which six populations are studied (so, $g = 6$), and $\mu_1 = 25$, $\mu_4 = 15$, $\mu_2 = \mu_3 = \mu_5 = \mu_6 = 20$, and $\sigma^2 = 9$. Five observations were collected from each population, so that $n_1 = n_2 = \cdots = n_6 = 5$ and $n = 30$. As a result,*

$$\mu = \frac{1}{30}(5 \times 25 + 5 \times 20 + 5 \times 20 + 5 \times 15 + 5 \times 20 + 5 \times 20) = \frac{600}{30} = 20.$$

Therefore, the noncentrality parameter ϕ for this scenario is

$$\phi = \frac{1}{9}[5(25 - 20)^2 + 5(20 - 20)^2 + 5(20 - 20)^2 + 5(15 - 20)^2$$
$$+ 5(20 - 20)^2 + 5(20 - 20)^2],$$
$$= \frac{250}{9} \approx 27.7778.$$

As a conclusion, the test statistic $F = \text{MSB}/\text{MSW}$ in this example has a noncentral F-distribution with $g - 1 = 5$ degrees of freedom for the numerator, $n - g = 24$ degrees of freedom for the denominator, and noncentrality parameter $\phi \approx 27.7778$.

12.6.3 *The Power and the Probability of a Type II Error*

To determine the power of the analysis of variance, it is important to know the critical value $F_{\alpha;g-1;n-g}$. This critical value delimits the acceptance region and the rejection region of the null hypothesis. If the test statistic f takes a value smaller than $F_{\alpha;g-1;n-g}$, then the null hypothesis is accepted. If the test statistic f takes a value larger than $F_{\alpha;g-1;n-g}$, then the null hypothesis is rejected.

To compute the power, we need to start from a scenario in which the null hypothesis is false. We have to choose different values for the population means $\mu_1, \mu_2, \ldots, \mu_g$, and pick the sample sizes $n_1, n_2, \ldots, n_g$, and the variance σ^2. Once we have chosen these values, we can determine the noncentrality parameter ϕ of the required noncentral F-distribution using Equation (12.6).

The power is then the probability that a random variable with a noncentral F-distribution with $g - 1$ degrees of freedom for the numerator, $n - g$ degrees of freedom for the denominator, and noncentrality parameter ϕ takes a larger value than the critical value $F_{\alpha;g-1;n-g}$. The complement of this probability is the probability of a type II error.

The full JMP command to compute the power is "1-F Distribution(F Quantile(1−α, g−1, n−g), g−1, n−g, φ)". In this command, the first argument, "F Quantile(1 − α, g − 1, n − g)", produces the critical value $F_{\alpha;g-1;n-g}$. The command required to calculate the probability of a type II error is "F Distribution(F Quantile (1 − α, g − 1, n − g), g − 1, n − g, φ)". Alternatively, one can use the 'Distribution Calculator'.

Example 12.6.2 *In Example 12.6.1, we found that the noncentrality parameter ϕ for the scenario in Figure 12.4 (where $\mu_1 = 25$, $\mu_4 = 15$, $\mu_2 = \mu_3 = \mu_5 = \mu_6 = 20$, $\sigma^2 = 9$, $n_1 = n_2 = $*

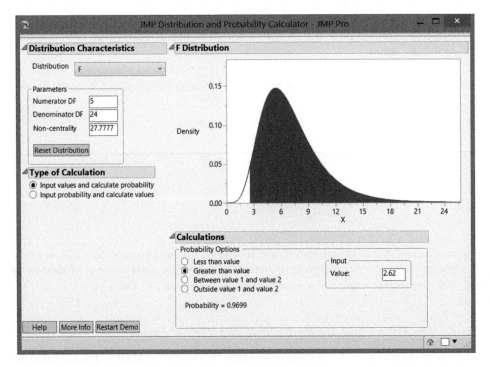

Figure 12.17 The use of the JMP script "Distribution Calculator" for computing the power for Example 12.2.3, where $\mu_1 = 25$, $\mu_4 = 15$, $\mu_2 = \mu_3 = \mu_5 = \mu_6 = 20$, and $\sigma^2 = 9$.

$\cdots = n_6 = 5$, and $n = 30$) is equal to 27.7778. Therefore, in that example, the test statistic $F = MSB/MSW$ has a noncentral F-distribution with five degrees of freedom for the numerator, 24 degrees of freedom for the denominator, and noncentrality parameter $\phi \approx 27.7778$.

Figure 12.17 shows how the JMP script called "Distribution Calculator" can be used to calculate the power for this problem. At the top left of the screen, the three parameters of the noncentral F-distribution were entered. At the bottom right, an exceedance probability of 0.9699 was obtained for the critical value $F_{0.05;5;24} = 2.62$. That probability is the power of the analysis of variance test here. The complement of this probability, 0.0301, is the probability of a type II error.

The critical value $F_{0.05;5;24} = 2.62$ can also be calculated using the "Distribution Calculator". To this end, the noncentrality parameter should be set to zero, because the critical value is determined based on the ordinary central F-distribution. An alternative is to make use of the command "`F Quantile(0.95, 5, 24)`". The power can then be calculated using the command "`1 - F Distribution(2.62, 5, 24, 27.7778)`" or "`1 - F Distribution(F Quantile(0.95, 5, 24), 5, 24, 27.7778)`". The probability of a type II error can be computed using the command "`F Distribution(2.62, 5, 24, 27.7778)`" or "`F Distribution(F Quantile(0.95, 5, 24), 5, 24, 27.7778)`".

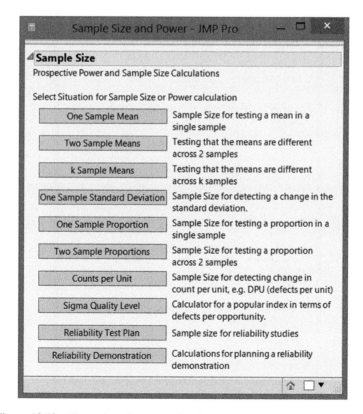

Figure 12.18 The options for computing the power in the "DOE" menu in JMP.

12.6.4 *Determining the Sample Size and Power in JMP*

JMP allows us to compute the power for various scenarios and numbers of observations $n_1, n_2, \ldots, n_g$ in the "DOE" menu. Alternatively, we can also determine the number of observations required to achieve a certain power.

To this end, we need to select the option "Sample Size and Power" in the "DOE" menu. In the resulting list of options (see Figure 12.18), we need to pick "k Sample Means", since we want to compare a number of sample means using analysis of variance. We then obtain the dialog window shown in Figure 12.19, where we can enter all relevant information.

To illustrate the calculation of a power, we use the scenario with six populations shown in Figure 12.4, where $\mu_1 = 25$, $\mu_4 = 15$, $\mu_2 = \mu_3 = \mu_5 = \mu_6 = 20$, and $\sigma^2 = 9$. This scenario, involving five observations for each population and a total of 30 observations, was discussed in Example 12.2.3. The six population means, the value for σ (namely 3), and the value for n (namely 30) have to be entered into the dialog window in Figure 12.19, as shown in Figure 12.20. The value for σ has to be entered in the field called "Std Dev", while the

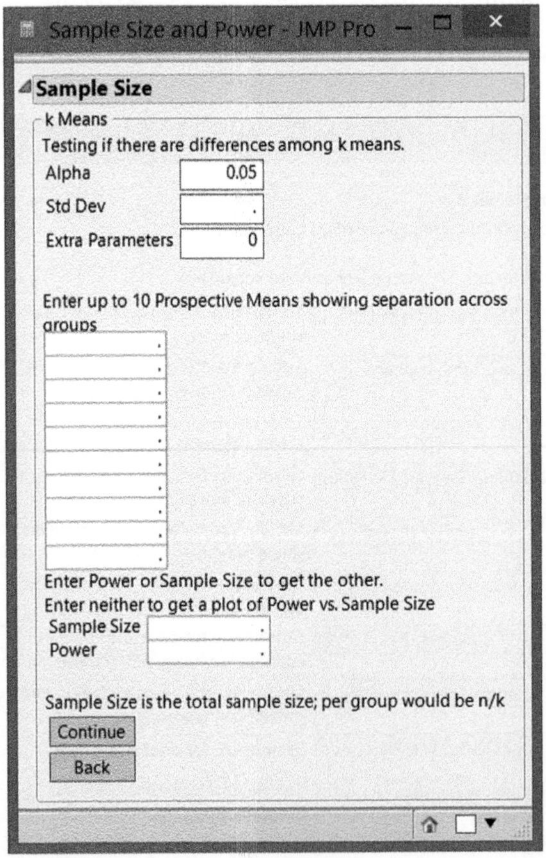

Figure 12.19 The dialog window for computing the power of a one-way analysis of variance in the "DOE" menu in JMP.

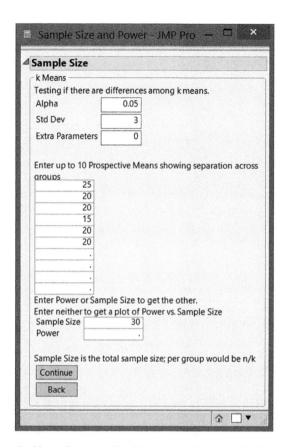

Figure 12.20   The required input for computing the power in Example 12.2.3, where $\mu_1 = 25$, $\mu_4 = 15$, $\mu_2 = \mu_3 = \mu_5 = \mu_6 = 20$, $n = 30$, and $\sigma^2 = 9$.

value for n has to be entered in the field called "Sample Size". JMP assumes that the 30 observations are evenly distributed across all samples. The value for "Extra Parameters" should remain zero. If we click on "Continue", JMP reports a power of 0.9699. This is shown in Figure 12.21.

Now, suppose that we are not satisfied with this power and that we want to determine how many observations are required to obtain a power of 0.9999. To this end, in the dialog window shown in Figure 12.20, we need to remove the value 30 in the field called "Sample Size", enter the value 0.9999 in the field called "Power", and click the button "Continue". As indicated in Figure 12.22, the required total sample size is 54. In that case, the six samples in the example should include 9 observations instead of 5.

If we leave both fields "Sample Size" and "Power" blank before clicking "Continue", JMP produces the graph shown in Figure 12.23. The graph shows how the power of the hypothesis test in an analysis of variance increases with the total number of observations.

The graph in Figure 12.23 was constructed for $\sigma^2 = 9$, or $\sigma = 3$. Since we rarely know the exact value of σ or σ^2, we should examine the impact of σ on the power. We expect that a smaller value of σ will lead to a larger power. In fact, a smaller variability in the measurements

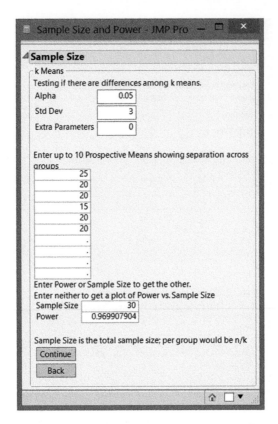

Figure 12.21 Computing the power for Example 12.2.3, where $\mu_1 = 25$, $\mu_4 = 15$, $\mu_2 = \mu_3 = \mu_5 = \mu_6 = 20$, $n = 30$, and $\sigma^2 = 9$.

should lead to better decision-making; that is, to fewer type II errors and a larger power. Similarly, a larger value for σ will lead to a smaller power.

Figure 12.24 shows three curves. The leftmost curve shows the relationship between the power and the total number of observations n for Example 12.2.3 when $\sigma = 2$. The middle curve shows the relationship between the power and the total number of observations for $\sigma = 3$ (this curve is identical to the one in Figure 12.23). Finally, the rightmost curve shows the relationship between the power and the total number of observations for $\sigma = 4$.

All of the above calculations were made for a situation in which $\mu_1 = 25$, $\mu_4 = 15$, $\mu_2 = \mu_3 = \mu_5 = \mu_6 = 20$. In this case, two of the six populations under investigation have a clearly different mean. Figure 12.25 demonstrates how sensitive the power of the hypothesis test in analysis of variance is to the values of μ_1 and μ_4. The leftmost curve shows the relation between the power and the number of observations when $\mu_1 = 29$ and $\mu_4 = 11$. The next curve was created for a situation in which $\mu_1 = 27$ and $\mu_4 = 13$. The middle curve applies when $\mu_1 = 25$ and $\mu_4 = 15$ (the middle curve is therefore identical to the one in Figure 12.23). The fourth curve was constructed for $\mu_1 = 23$ and $\mu_4 = 17$. The rightmost or lowest curve corresponds to the values 21 and 19 for μ_1 and μ_4.

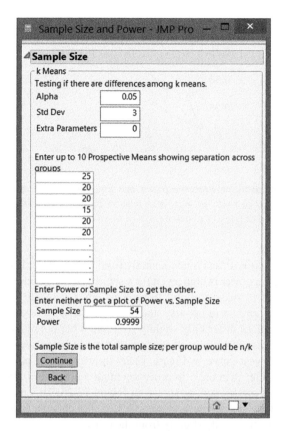

Figure 12.22 The computation of the number of observations required for a power of 0.9999 in Example 12.2.3, where $\mu_1 = 25$, $\mu_4 = 15$, $\mu_2 = \mu_3 = \mu_5 = \mu_6 = 20$, and $\sigma^2 = 9$.

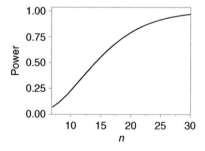

Figure 12.23 The relationship between the power and the total number of observations in Example 12.2.3, where $\mu_1 = 25$, $\mu_4 = 15$, $\mu_2 = \mu_3 = \mu_5 = \mu_6 = 20$, and $\sigma^2 = 9$.

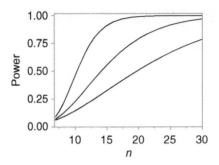

Figure 12.24 The relationship between the power and the total number of observations in Example 12.2.3, where $\mu_1 = 25$, $\mu_4 = 15$, $\mu_2 = \mu_3 = \mu_5 = \mu_6 = 20$. The leftmost curve is for $\sigma = 2$, the middle curve for $\sigma = 3$, and the rightmost curve for $\sigma = 4$.

The first of these scenarios differs fundamentally from the one described by the null hypothesis. For this scenario, the power is high, even for a small number of observations. This means that it is relatively easy to reject the null hypothesis when it is completely wrong. In the latter scenario, where $\mu_1 = 21$ and $\mu_4 = 19$, the null hypothesis is very close to the truth: the population means 21 and 19 differ only slightly from the other population means, which are all equal to 20. In that case, the power of the hypothesis test is low, even for a large number of observations. The lowest curve in Figure 12.25 shows that the power for a total of 40 observations is still smaller than 0.25. In order to obtain a power of 0.95 in this scenario, as many as 540 data points are needed, as shown in Figure 12.26. This means that we need 90 observations in each of the six samples in Example 12.2.3 to obtain a power of 0.95 when two of the population means differ only slightly from the others.

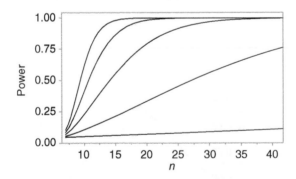

Figure 12.25 The relationship between the power and the total number of observations in Example 12.2.3 for different values of μ_1 and μ_4. The leftmost curve is for the largest value of μ_1 and the smallest value of μ_4 (and corresponds to a scenario that is radically different from the one in the null hypothesis). The rightmost or lowest curve is for the smallest value of μ_1 and the largest value of μ_4 (and corresponds to a scenario that is only a little different from the one in the null hypothesis).

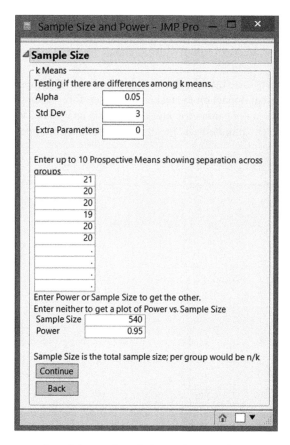

Figure 12.26 The computation of the number of observations required for a power of 0.95 in Example 12.2.3, if $\mu_1 = 21$, $\mu_4 = 19$, $\mu_2 = \mu_3 = \mu_5 = \mu_6 = 20$, and $\sigma^2 = 9$.

12.7   Analysis of Variance for Nonnormal Distributions and Unequal Variances

In a one-way analysis of variance, a nonnormal distribution of the studied populations is not a big problem as long as the deviation from normality is not extreme. Even probability densities that are not symmetrical have no drastic impact on the validity of the decisions that can be drawn from the analysis of variance. The estimators of the population means remain unbiased. The confidence level and the power of the F-test are not affected substantially. One therefore says that the F-test in an analysis of variance is robust against deviations from normality.

If the variances of the populations under study are unequal, the F-test still works properly provided that the sample sizes $n_1, n_2, \ldots, n_g$ are approximately equal. Therefore, it is recommended to try to collect an equal number of observations for each population studied. Where the variances are unequal, the Welch method, discussed in Section 14.6, can be used. How to test whether g population variances are equal is explained in Chapter 14.

There is another reason to try to conduct an analysis of variance with equal numbers of observations in all samples. For a given total number of observations, the pairwise comparisons of all population means are performed with maximal precision when the variances of all populations are equal.

For small samples, there is no meaningful test of normality, and outliers and deviations from normality may have a great impact on the test results. Therefore, when the samples are small, it is recommended to use a nonparametric alternative to an analysis of variance. Probably the best-known nonparametric alternative is the Kruskal–Wallis test, which is discussed in the next chapter.

13

Nonparametric Alternatives to an Analysis of Variance

Once we get full participation from all schools and districts, we'll be able to keep daily ratings, with every test, every pop quiz incorporated instantly. And of course these can be broken up between public and private, regional, and the rankings can be merged, weighted and analyzed to see trends among various other factors – socioeconomic, race, ethnicity, everything.

(from *The Circle*, Dave Eggers, pp. 343–344)

In the previous chapter, we discussed one-way analysis of variance, or ANOVA. One-way analysis of variance is suitable for comparing population means of independent samples when the populations under study are all normally distributed and have the same variance. In this chapter, we study nonparametric alternatives to one-way analysis of variance. These alternatives can be used when the populations studied are not normally distributed. The best-known alternative is the Kruskal–Wallis test, which generalizes the Wilcoxon rank-sum test (also known as the Mann–Whitney U-test) from Chapter 9 for two populations to more than two populations. In addition to the Kruskal–Wallis test, we also consider the van der Waerden test and the median test.

A prerequisite for the use of the Kruskal–Wallis test is that the only difference between the populations under study is their position or location. This means that the probability densities or distributions of the studied populations should have the same shape. The only difference allowed between the probability densities or distributions is that they are shifted relative to each other.

When a total of g populations are studied, then the hypotheses tested by the Kruskal–Wallis test and the other alternatives to one-way analysis of variance are

$$H_0 : \text{Location } 1 = \text{Location } 2 = \cdots = \text{Location } g$$

and

$$H_a : \text{At least one location differs from the others.}$$

Statistics with JMP: Hypothesis Tests, ANOVA and Regression, First Edition. Peter Goos and David Meintrup.
© 2016 John Wiley & Sons, Ltd. Published 2016 by John Wiley & Sons, Ltd.
Companion Website: http://www.wiley.com/go/goosandmeintrup/JMP

To test these hypotheses, sample data is collected for each of the g populations under study. We denote the number of observations from the first population by n_1, the number of observations from the second population by $n_2, \ldots$, and the number of observations from the gth population by n_g. The data can then be represented in the following way:

Sample 1	Sample 2	Sample 3	...	Sample g
X_{11}	X_{21}	X_{31}		X_{g1}
X_{12}	X_{22}	X_{32}		X_{g2}
X_{13}	X_{23}	X_{33}		X_{g3}
$\vdots$	X_{24}	X_{34}		X_{g4}
X_{1n_1}	$\vdots$	X_{35}		$\vdots$
	X_{2n_2}	$\vdots$		X_{gn_g}
		X_{3n_3}		

In this table, X_{ij} is the jth observation in the ith sample. Note that it is not required that all samples have the same size. In total, there are

$$n = n_1 + n_2 + n_3 + \cdots + n_g = \sum_{i=1}^{g} n_i$$

observations.

For the analysis of variance, the test statistic was computed based on all observations X_{ij}, the g sample means $\overline{X}_i$, and the overall mean $\overline{X}$. This is no longer the case for the nonparametric alternatives. The Kruskal–Wallis test uses ranks. The van der Waerden test uses ranks as well, but, in addition, it transforms the ranks into so-called van der Waerden scores. The median test, finally, is similar to the sign test and turns the individual observations into zeros or ones.

13.1 The Kruskal–Wallis Test

13.1.1 Computing the Test Statistic

Like the signed-rank test and the Wilcoxon rank-sum test, the Kruskal–Wallis test is based on ranks. A test statistic is generated from the ranks and a corresponding p-value is computed. Again, exact p-values as well as approximate p-values can be computed. The steps required to assign the ranks and compute the test statistic are as follows:

(1) Merge the observations X_{ij} from all g samples.
(2) Rank all observations in the resulting data set from small to large.
(3) Assign ranks from 1 to n to the ordered observations. In the event of ties, use average ranks. Denote the final rank of observation X_{ij} by R_{ij}.
(4) For each sample i, sum all final ranks $R_{i1}, R_{i2}, \ldots, R_{in_i}$ and call this sum R_i:

$$R_i = \sum_{j=1}^{n_i} R_{ij}.$$

(5) For each sample i, compute the mean of all ranks:

$$\overline{R}_i = \frac{1}{n_i} \sum_{j=1}^{n_i} R_{ij} = \frac{R_i}{n_i}.$$

(6) Compute the test statistic H based on one of the following formulas:

$$H = \frac{12}{n(n+1)} \sum_{i=1}^{g} n_i \left(\overline{R}_i - \frac{n+1}{2}\right)^2, \tag{13.1}$$

$$= \left(\frac{12}{n(n+1)} \sum_{i=1}^{g} \frac{R_i^2}{n_i}\right) - 3(n+1). \tag{13.2}$$

This test statistic is always positive.

(7) If ties occurred when ranking the data, adjust the test statistic H by dividing it by the correction factor

$$c = 1 - \frac{\sum_{i=1}^{v} \left(t_i^3 - t_i\right)}{n^3 - n},$$

where v represents the number of groups of observations with the same rank, and t_i is the number of observations in group i. In the presence of ties, the corrected test statistic is

$$H_c = \frac{H}{c} = \frac{H}{1 - \dfrac{\sum\limits_{i=1}^{v} \left(t_i^3 - t_i\right)}{n^3 - n}}.$$

When there are ties, the corrected test statistic H_c is larger than the original test statistic H (since the correction factor c always takes values between 0 and 1). The larger the number of ties, the greater is the difference between H and H_c. Note that the correction factor c and the corrected test statistic H_c can also be computed when there are no ties. In that case, it is as if there are $v = n$ groups, each containing a single observation, so that $t_1 = t_2 = \cdots = t_v = 1$. The correction factor c is then equal to 1, so that $H_c = H$ in the absence of ties. We can therefore always use H_c, regardless of whether or not there are ties.

13.1.2   The Behavior of the Test Statistic

In the Kruskal–Wallis test, ranks from 1 to n are assigned to all n observations. The sum of all ranks is therefore $n(n + 1)/2$. Since there are n observations, the mean of all ranks is $[n(n + 1)/2]/n = (n + 1)/2$.

This average rank appears explicitly in the definition of the statistic H in Equation (13.1). More specifically, in this expression, the difference

$$\bar{R}_i - \frac{n+1}{2}$$

is calculated for each sample i. In other words, the test statistic H compares all values $\bar{R}_1, \bar{R}_2, \ldots, \bar{R}_g$ with the average rank $(n+1)/2$. When all values $\bar{R}_1, \bar{R}_2, \ldots, \bar{R}_g$ lie close to $(n+1)/2$, then H (and also H_c) is a small positive number. If some of the values $\bar{R}_1, \bar{R}_2, \ldots, \bar{R}_g$ differ substantially from $(n+1)/2$, then H (and also H_c) takes a large positive value.

Now, suppose that the null hypothesis, that the locations of all populations studied are equal, is true. In that case, large and small observations will be spread nicely over the g different samples. In other words, none of the samples will systematically contain larger or smaller observations than any other sample. Each individual observation is then equally likely to be given rank 1, rank 2, rank 3, ... The rank that is assigned to an individual observation is then completely random. More specifically, the rank number of an individual observation then follows a discrete uniform distribution[1] with probability $1/n$ for each of the possible ranks $1, 2, \ldots, n$.

Continuing this line of reasoning, the expected rank of an individual observation is equal to the expected value of a uniformly distributed discrete random variable, $(n+1)/2$. If we assume for simplicity that the ranks are assigned independently to the individual observations, then the expected value of R_i is

$$E(R_i) = E\left(\sum_{j=1}^{n_i} R_{ij}\right) = \sum_{j=1}^{n_i} E(R_{ij}) = \frac{n_i(n+1)}{2},$$

and the expected value of $\bar{R}_i$ is

$$E(\bar{R}_i) = E\left(\frac{R_i}{n_i}\right) = \frac{1}{n_i} E(R_i) = \frac{n_i(n+1)}{2n_i} = \frac{n+1}{2}.$$

As a consequence, if the null hypothesis is correct, we expect that all computed values of $\bar{R}_1, \bar{R}_2, \ldots, \bar{R}_g$ will be around $(n+1)/2$, so that the final value of the test statistic H (and thus H_c) will be a small positive number.

If the null hypothesis is false, some samples will contain systematically smaller or larger observations than other samples. If sample i systematically contains small observations, the ranks for this sample, namely $R_{i1}, R_{i2}, \ldots, R_{in_i}$, will all be fairly small. Obviously, their sum,

---

[1] The discrete uniform probability distribution is discussed in the book *Statistics with JMP: Graphs, Descriptive Statistics and Probability*. A uniformly distributed discrete random variable with the integers $1, 2, \ldots, n$ as possible outcomes has expected value $(n+1)/2$.

R_i, will also be small, as well as their mean $\overline{R}_i$. In other words, if sample i systematically contains small observations, then $\overline{R}_i$ will be considerably smaller than $(n + 1)/2$,

$$\overline{R}_i - \frac{n+1}{2}$$

will be strongly negative, and

$$\left(\overline{R}_i - \frac{n+1}{2} \right)^2$$

will take a large positive value. For a sample i that systematically contains large observations, the observations will all have fairly large ranks $R_{i1}, R_{i2}, \ldots, R_{in_i}$, so that the sum of all ranks of that sample, R_i, and the mean value, $\overline{R}_i$, will be large. As a result, $\overline{R}_i$ will be considerably larger than $(n + 1)/2$,

$$\overline{R}_i - \frac{n+1}{2}$$

will be strongly positive, and

$$\left(\overline{R}_i - \frac{n+1}{2} \right)^2$$

will again take a large positive value. It is clear that, whenever the null hypothesis is incorrect, the test statistic H (and thus also the corrected test statistic H_c) will take a large value.

The above reasoning allows us to conclude that a large value for H or H_c points in the direction of the alternative hypothesis, whereas a small value points in the direction of the null hypothesis. To decide whether the test statistic in a particular application is large enough for the null hypothesis to be rejected, we need to compute a p-value. Large values for H or H_c will lead to small p-values, and thus to a rejection of the null hypothesis. Small values of H or H_c will lead to large p-values, and thus will not allow the null hypothesis to be rejected.

Example 13.1.1 *Table 13.1 contains the life span (expressed in number of days) for* 15 *different lamps. Three different brands were studied. Five lamps from each brand were used in the study. The question is whether there is a significant difference in the life span between the three brands.*

Table 13.1 The life span (in days) of five lamps for three different brands.

Observation	Brand 1	Brand 2	Brand 3
1	952	1334	987
2	1084	1126	913
3	836	1324	1079
4	898	1538	857
5	1003	988	1045

Table 13.2   The assignment of ranks to the life span measurements in Example 13.1.1.

Life span	Rank	Brand
836	1	Brand 1
857	2	Brand 3
898	3	Brand 1
913	4	Brand 3
952	5	Brand 1
987	6	Brand 3
988	7	Brand 2
1003	8	Brand 1
1045	9	Brand 3
1079	10	Brand 3
1084	11	Brand 1
1126	12	Brand 2
1324	13	Brand 2
1334	14	Brand 2
1538	15	Brand 2

Given the small sample sizes for the three brands and the fact that life spans are generally not normally distributed, it is not appropriate to perform a one-way analysis of variance here. A good alternative is to conduct a Kruskal–Wallis test. The first step in the test procedure is to order the 15 observations and to assign ranks. Table 13.2 shows how this is done. A comparison of Tables 13.1 and 13.2 tells us that the first observation in the first sample, $x_{11} = 952$, is given the rank $r_{11} = 5$, the second observation in the first sample, $x_{12} = 1084$, is given the rank $r_{12} = 11$, and so on.

The sum of the ranks for all observations in the first sample, r_1, is 28. The sum of the ranks of the observations in the second sample, r_2, is 61, and the sum of the ranks in the third sample, r_3, is 31. To check the correctness of our calculations, we can verify that $r_1 + r_2 + r_3 = n(n+1)/2 = 15(15+1)/2 = 120$. The mean ranks for the three samples, $\bar{r}_1$, $\bar{r}_2$, and $\bar{r}_3$, are 5.6, 12.2, and 6.2, respectively.

The computed test statistic is

$$h = \frac{12}{15(15+1)} \left[5 \left(5.6 - \frac{15+1}{2} \right)^2 + 5 \left(12.2 - \frac{15+1}{2} \right)^2 + 5 \left(6.2 - \frac{15+1}{2} \right)^2 \right],$$

$$= \frac{1}{20} \left(5 \times (-2.4)^2 + 5 \times (4.2)^2 + 5 \times (-1.8)^2 \right),$$

$$= 6.66.$$

Two methods exist for determining the corresponding p-value. The first provides an exact p-value, while the second yields an approximate p-value. Of course, ideally, exact p-values are used.

13.1.3 Exact p-Values

To calculate exact p-values, we have to know the exact probability distribution of H or H_c, assuming that the null hypothesis is correct. To determine this probability distribution, it is necessary to enumerate all possible assignments of ranks to g samples with $n_1, n_2, \ldots, n_g$ observations, and compute the value of H or H_c for each possible assignment.

If there are no ties and each rank appears exactly once, then there are

$$\frac{n!}{n_1! n_2! \ldots n_g!}$$

possible assignments of the n ranks to the g samples. Each of these assignments will lead to a certain value for the test statistic H. This is illustrated in the next example.

An important insight is that every possible assignment of ranks is equally likely when the null hypothesis is true. Each possible assignment then has a probability of

$$\frac{n_1! n_2! \ldots n_g!}{n!}.$$

Example 13.1.2 *Suppose that the location of three populations is compared, and that each of the samples from these populations contains two observations, so that $g = 3$, $n_1 = n_2 = n_3 = 2$, and $n = 6$. In that case, there are*

$$\frac{6!}{2! 2! 2!} = 90$$

possible assignments of the ranks 1, 2, 3, 4, 5, and 6 to the three samples. Each assignment has a probability of $1/90$ when the null hypothesis is true. The first possible assignment of the six ranks is as follows:

Sample 1	Sample 2	Sample 3
1	3	5
2	4	6

In this scenario, the two smallest observations (with ranks 1 and 2) belong to the first sample, the two largest observations (with ranks 5 and 6) belong to the third sample, and the two middle observations (with ranks 3 and 4) belong to the second sample. In this scenario, the average rank for the first sample, $\bar{r}_1$, is 1.5, while the average ranks for the second and the third samples, $\bar{r}_2$ and $\bar{r}_3$, equal 3.5 and 5.5, respectively. The computed test statistic then is

$$h = \frac{12}{6(6+1)} \left[2\left(1.5 - \frac{6+1}{2}\right)^2 + 2\left(3.5 - \frac{6+1}{2}\right)^2 + 2\left(5.5 - \frac{6+1}{2}\right)^2 \right],$$

$$= \frac{2}{7} \left(2 \times (-2)^2 + 2 \times 0^2 + 2 \times 2^2\right),$$

$$= \frac{32}{7} \approx 4.57.$$

The second possible assignment is as follows:

Sample 1	Sample 2	Sample 3
1	3	4
2	5	6

In this scenario, the two smallest observations (with ranks 1 and 2) again belong to the first sample, the observations with ranks 3 and 5 belong to the second sample, and the observations with ranks 4 and 6 belong to the third sample. In this scenario, the average rank for the first sample, $\bar{r}_1$, is again 1.5, and the average ranks for the second and the third samples, $\bar{r}_2$ and $\bar{r}_3$, equal 4 and 5, respectively. The computed test statistic then is 3.71.

In the last scenario, the 90th, the assignment of ranks to the three samples is as follows:

Sample 1	Sample 2	Sample 3
5	3	1
6	4	2

In this scenario, the two smallest observations (with ranks 1 and 2) belong to the third sample. The two largest observations (with ranks 5 and 6) belong to the first sample, and the two middle observations (with ranks 3 and 4) belong to the second sample. In this scenario, the average rank for the first sample, $\bar{r}_1$, is 5.5, while the average ranks for the second and the third sample, $\bar{r}_2$ and $\bar{r}_3$, are equal to 3.5 and 1.5, respectively. The computed test statistic then is 4.57, as in the first scenario.

If we were to enumerate all 90 scenarios, we would see that there are nine different values for the test statistic H. These nine values are shown in the first column of Table 13.3. The values 0, 2.57, and 4.57 each occur in six of the 90 scenarios. Therefore, they have a probability of $6/90 = 1/15$. The other values, 0.29, 0.86, 1.14, 2, 3.43, and 3.71, each occur in 12 of the 90 scenarios. Therefore, they have a probability of $12/90 = 2/15$. The nine probabilities are

Table 13.3 The probability distribution for the test statistic H in a Kruskal–Wallis test for three samples of two observations without ties.

h	Probability		p-value
0	1/15	1	1.0000
0.29	2/15	14/15	0.9333
0.86	2/15	12/15	0.8000
1.14	2/15	10/15	0.6667
2	2/15	8/15	0.5333
2.57	1/15	6/15	0.4000
3.43	2/15	5/15	0.3333
3.71	2/15	3/15	0.2000
4.57	1/15	1/15	0.0667

shown in the second column of Table 13.3. Together with the nine possible values of H, they form the discrete probability distribution of H. This probability distribution is the basis for the calculation of the exact p-values.

The last two columns in Table 13.3 contain the exceedance probabilities

$$P(H \geq h).$$

These probabilities are the only possible p-values for the Kruskal–Wallis test in this small example with six observations. Because the smallest p-value in the table is 0.0667, the null hypothesis of the Kruskal–Wallis test for this small example can never be rejected when a significance level of 5% is used. A data set with six observations is simply not large enough for powerful statistical decisions.

The researchers J. Patrick Meyer (University of Virginia) and Michael A. Seaman (University of South Carolina) calculated potential p-values for the study of three populations with a maximum of 105 sample observations, and for four populations with a maximum of 40 observations. They also established tables with critical values h_α for the test statistic H and exact p-values. These tables are available on the website *http://faculty.virginia.edu/kruskal-wallis/*. A highly simplified version of the table for three populations and up to eight observations per sample can be found in Appendix L. The critical values in the appendix correspond to significance levels of about 10%, 5%, and 1% (in most cases, it is impossible to find critical values that exactly match the significance levels 10%, 5%, and 1%, because in each scenario, only a limited number of p-values exist).

Example 13.1.3   *The table in Appendix L contains one of the possible values of the test statistic as well as one of the possible p-values for $n_1 = n_2 = n_3 = 2$ that we have listed in Table 13.3. More specifically, it is the value 4.57 and the corresponding p-value of 0.0667 that appear in Appendix L. The table does not contain critical values for $\alpha \approx 0.05$ and $\alpha \approx 0.01$, because larger values than 4.57 are impossible for the test statistic, and p-values smaller than 0.0667 therefore do not exist.*

In order to use the table in Appendix L, it is necessary to determine a significance level α, and compare the computed test statistic h with the critical value h_α corresponding to the chosen significance level. If the computed test statistic h exceeds the critical value in Appendix L, then the null hypothesis can be rejected.

Example 13.1.4   *In Example 13.1.1, the test statistic was computed for a Kruskal–Wallis test for the life span of lamps, in which the sample sizes n_1, n_2, and n_3 were all equal to 5. The computed test statistic h was equal to 6.66. Now, suppose that a significance level of 5% is chosen.*

The table entry for $n_1 = n_2 = n_3 = 5$ in Appendix L does not report a critical value that exactly matches a significance level of 5%. However, there is a critical value corresponding to an exceedance probability of 0.04878, namely 5.78. Now, the computed test statistic, 6.66, is larger than the critical value, 5.78. At a significance level of 4.878%, we can therefore reject the null hypothesis and conclude that there is a significant difference in life span between the three brands of lamps.

The table in Appendix L also contains the critical value 8, corresponding to an exceedance probability of 0.00946 (about 1%). Since the computed test statistic, 6.66, lies between 5.78 and 8, we know that the exact p-value for the Kruskal–Wallis test in this example is between 0.00946 and 0.04878.

The tables of Meyer and Seaman have been constructed for situations without ties. In the presence of ties, the probability distribution of the test statistic H (or H_c) has to be determined for the specific pattern of ties at hand.

JMP does not provide exact p-values for the Kruskal–Wallis test. Instead, it reports approximate p-values based on the χ^2-distribution.

13.1.4   Approximate p-Values

The calculation of approximate p-values for the Kruskal–Wallis test is based on the fact that the rank sums $R_1, R_2, \ldots, R_g$ and the mean values $\overline{R}_1, \overline{R}_2, \ldots, \overline{R}_g$ are approximately normally distributed, when the sample sizes $n_1, n_2, \ldots, n_g$ are sufficiently large. This follows from the central limit theorem.

In that case, both test statistics, H (if there are no ties) and H_c (if there are ties), are approximately χ^2-distributed with $g - 1$ degrees of freedom. The approximate p-value of the Kruskal–Wallis test can therefore be computed as

$$p = P\left(\chi^2_{g-1} > h\right)$$

in the absence of ties. In the presence of ties, the p-value should be computed as

$$p = P\left(\chi^2_{g-1} > h_c\right).$$

If the p-value is smaller than the chosen significance level, the null hypothesis can be rejected.

Alternatively, an approximate critical value can be determined based on the χ^2-distribution. The approximate critical value is equal to $\chi^2_{\alpha;g-1}$. If the computed test statistic h or h_c is larger than the approximate critical value $\chi^2_{\alpha;g-1}$, then the null hypothesis should be rejected.

Example 13.1.5   *In Example 13.1.1, the test statistic was computed for a Kruskal–Wallis test concerning the life span of lamps. Based on three samples of five observations (thus, $g = 3$ and $n_1 = n_2 = n_3 = 5$), a value of 6.66 was obtained for the computed test statistic h. The significance level was set to 5% in Example 13.1.4.*

The required approximate critical value from the χ^2-distribution is $\chi^2_{0.05;3-1} = \chi^2_{0.05;2} = 5.991$. This value can be determined in JMP using the command "Chi Square Quantile(0.95, 2)", and found in Appendix C. In any case, the computed test statistic, 6.66, is larger than the approximate critical value. Therefore, we can reject the null hypothesis.

We can determine the approximate p-value using the command "1-Chi Square Distribution(6.66, 2)". It is equal to 0.0358. This value is smaller than the chosen significance level, and thus leads to a rejection of the null hypothesis.

Wilcoxon/Kruskal-Wallis Tests (Rank Sums)					
			Expected		
Level	Count	Score Sum	Score	Score Mean	(Mean-Mean0)/Std0
Brand 1	5	28.000	40.000	5.6000	−1.408
Brand 2	5	61.000	40.000	12.2000	2.511
Brand 3	5	31.000	40.000	6.2000	−1.041

1-way Test, ChiSquare Approximation		
ChiSquare	DF	Prob>ChiSq
6.6600	2	0.0358*

Small sample sizes. Refer to statistical tables for tests, rather than large-sample approximations.

Figure 13.1 The JMP output for the Kruskal–Wallis test in Example 13.1.5 with an approximate p-value based on the χ^2-distribution.

Therefore, we conclude that there is a significant difference between the three brands in terms of the life span of the lamps. The conclusion based on the approximate p-value matches the conclusion drawn from the exact p-value in Example 13.1.4.

The JMP output with the approximate p-value is shown in Figure 13.1. The relevant elements in this output are the computed values of R_1, R_2, and R_3 in the column named "Score Sum", the computed values of $\overline{R}_1$, $\overline{R}_2$ and $\overline{R}_3$ in the column named "Score Mean", the test statistic of 6.66, the required number of degrees of freedom of 2, and the approximate p-value of 0.0358.

It is noteworthy that JMP itself suggests that it is necessary to consult tables to find an exact p-value. In fact, the last piece of output contains the following sentences: "Small sample sizes. Refer to statistical tables for tests, rather than large-sample approximations".

13.2 The van der Waerden Test

The Kruskal–Wallis test works well in many cases. However, one criticism is that when the populations under study are normally distributed, a one-way analysis of variance has a larger power than the Kruskal–Wallis test. In order to remedy this problem, the van der Waerden test can be used. This test combines the advantage of the Kruskal–Wallis test, namely an appropriate treatment of nonnormally distributed populations, with the advantage of analysis of variance; that is, a large power for normally distributed populations.

If there are no ties, the van der Waerden test uses the same ranks as the Kruskal–Wallis test. It then associates a quantile of the standard normal distribution with each rank. This is done using the following formula:

$$A_{ij} = z_{1-R_{ij}/(n+1)}.$$

The A_{ij} values are called the van der Waerden scores[2]. The average van der Waerden score for sample i is

$$\overline{A}_i = \frac{1}{n_i} \sum_{j=1}^{n_i} A_{ij},$$

---

[2] The van der Waerden scores are also used in the construction of quantile diagrams in Chapter 6.

and the test statistic is given by

$$V = \frac{1}{S_A^2} \sum_{i=1}^{g} n_i \overline{A}_i^2,$$

where S_A^2 is the sample variance of the scores A_{ij}.

As with the Kruskal–Wallis test, one can determine exact and approximate p-values for the van der Waerden test. Ideally, exact p-values are used, but, as with the Kruskal–Wallis test, this requires the construction of the entire probability distribution of the test statistic based on all possible assignments of ranks to the observations in all g samples. Because this is computationally very demanding and the van der Waerden test is less popular than the Kruskal–Wallis test, no detailed tables exist with exact p-values and exact quantiles for the test statistic V. Moreover, it turns out that the approximate p-values based on the χ^2-distribution with $g - 1$ degrees of freedom work very well. These approximate p-values can be computed as

$$p = P(\chi_{g-1}^2 > v).$$

where v represents the computed statistic.

Example 13.2.1   *In Examples 13.1.1, 13.1.4, and 13.1.5, a Kruskal–Wallis test was performed for a data set involving 15 observations of the life span of lamps from three different brands. Table 13.4 shows this data set, along with the 15 ranks for the observations. Based on these ranks, the van der Waerden scores have been determined. For example, the*

Table 13.4   The ranks and van der Waerden scores for the measurements of the life span in Example 13.1.1.

Life span	Rank	$r_{ij}/(n+1)$	$a_{ij} = z_{1-r_{ij}/(n+1)}$	Brand
836	1	0.0625	−1.5341	Brand 1
857	2	0.1250	−1.1503	Brand 3
898	3	0.1875	−0.8871	Brand 1
913	4	0.2500	−0.6745	Brand 3
952	5	0.3125	−0.4888	Brand 1
987	6	0.3750	−0.3186	Brand 3
988	7	0.4375	−0.1573	Brand 2
1003	8	0.5000	0.0000	Brand 1
1045	9	0.5625	0.1573	Brand 3
1079	10	0.6250	0.3186	Brand 3
1084	11	0.6875	0.4888	Brand 1
1126	12	0.7500	0.6745	Brand 2
1324	13	0.8125	0.8871	Brand 2
1334	14	0.8750	1.1503	Brand 2
1538	15	0.9375	1.5341	Brand 2

van der Waerden score for the one but two smallest observation, corresponding to the rank number 3, is

$$z_{1-3/(n+1)} = z_{0.8125} = -0.8871.$$

This value can be derived from Appendix B for the standard normal distribution. It can also be computed using the JMP command "Normal Quantile(0.1875)". Note that the sum of all 15 van der Waerden scores is equal to zero.

The sum of the van der Waerden scores for all observations in the first sample is -2.4213. The sum of the van der Waerden scores of the observations in the second sample is 4.0888, while it is -1.6675 for all observations in the third sample. The mean scores for the three samples, $\bar{a}_1$, $\bar{a}_2$, and $\bar{a}_3$, are -0.4843, 0.8178, and -0.3335, respectively. The sample variance s_A^2 of the 15 van der Waerden scores in Table 13.4 is 0.7549. As a result, the computed test statistic is

$$v = \frac{1}{0.7549} \left[5(-0.4843)^2 + 5(0.8178)^2 + 5(-0.3335)^2 \right],$$
$$= 6.7196.$$

The corresponding approximate p-value is the probability that a χ^2-distributed random variable V with $g-1 = 2$ degrees of freedom takes values larger than 6.7196,

$$P(\chi^2_{g-1} > 6.7196) = 0.0347.$$

Based on this approximate p-value, we can reject the null hypothesis at a significance level of 5% and conclude that there is a significant difference in life span for the three brands of lamps. The approximate critical value is again $\chi^2_{0.05;3-1} = \chi^2_{0.05;2} = 5.991$. As the computed test statistic, $v = 6.7196$, exceeds this approximate critical value, we reject the null hypothesis, as with the Kruskal–Wallis test.

The JMP output with the approximate p-value for the van der Waerden test is shown in Figure 13.2. The relevant elements in this output are the three sums of van der Waerden scores

Van der Waerden Test (Normal Quantiles)

Level	Count	Score Sum	Expected Score	Score Mean	(Mean-Mean0)/Std0
Brand 1	5	−2.421	0.000	−0.48425	−1.526
Brand 2	5	4.089	0.000	0.81776	2.578
Brand 3	5	−1.668	0.000	−0.33351	−1.051

1-way Test, ChiSquare Approximation

ChiSquare	DF	Prob>ChiSq
6.7196	2	0.0347*

Small sample sizes. Refer to statistical tables for tests, rather than large-sample approximations.

Figure 13.2 The JMP output for the van der Waerden test in Example 13.2.1 with an approximate *p*-value based on the χ^2-distribution.

in the column named "Score Sum", the average van der Waerden scores $\bar{a}_1$, $\bar{a}_2$ and $\bar{a}_3$ in the column named "Score Mean", the test statistic 6.7196, the required number of degrees of freedom of 2, and the approximate p-value, 0.0347.

In the output, JMP again indicates that it is necessary to consult tables with exact p-values. The last bit of output contains the following sentences: "Small sample sizes. Refer to statistical tables for tests, rather than large-sample approximations". However, tables with exact p-values for the van der Waerden test are difficult to find.

If ties are present, there are two alternative methods for the determination of the van der Waerden scores. A first method uses average ranks and the corresponding quantiles of the standard normal distribution. When this method is used, the sum of all van der Waerden scores is generally different from zero. When using the second method, which is the one implemented in JMP, the sum of the van der Waerden scores remains zero. The second method to deal with ties does not use average ranks. Despite the presence of ties, it uses the integers from 1 to n as ranks, and it converts them into van der Waerden scores. Finally, for each group of tied observations, the average van der Waerden score is computed. The following example illustrates this approach.

Example 13.2.2   *Table 13.5 shows a slightly modified data set for the example on the life span of lamps. The difference between the new data and the original data is that the one but smallest observation of 857 days was replaced by the value 836. As a result, the modified data set contains two observations with a life span of 836 days. These two observations are the two smallest ones, so that they are assigned the ranks 1 and 2. The corresponding van der Waerden scores are*

$$z_{1-1/(n+1)} = z_{0.9375} = -1.5341$$

Table 13.5   The determination of the van der Waerden scores for the modified life span data in Example 13.2.2.

Life span	Rank	$r_{ij}/(n+1)$	$a_{ij} = z_{1-r_{ij}/(n+1)}$	Adjusted score	Brand
836	1	0.0625	−1.5341	−1.3422	Brand 1
836	2	0.1250	−1.1503	−1.3422	Brand 3
898	3	0.1875	−0.8871	−0.8871	Brand 1
913	4	0.2500	−0.6745	−0.6745	Brand 3
952	5	0.3125	−0.4888	−0.4888	Brand 1
987	6	0.3750	−0.3186	−0.3186	Brand 3
988	7	0.4375	−0.1573	−0.1573	Brand 2
1003	8	0.5000	0.0000	0.0000	Brand 1
1045	9	0.5625	0.1573	0.1573	Brand 3
1079	10	0.6250	0.3186	0.3186	Brand 3
1084	11	0.6875	0.4888	0.4888	Brand 1
1126	12	0.7500	0.6745	0.6745	Brand 2
1324	13	0.8125	0.8871	0.8871	Brand 2
1334	14	0.8750	1.1503	1.1503	Brand 2
1538	15	0.9375	1.5341	1.5341	Brand 2

and

$$z_{1-2/(n+1)} = z_{0.875} = -1.1503.$$

Because the two smallest observations are tied, these two van der Waerden scores are replaced by their average $(-1.5341 + (-1.1503))/2 = -1.3422$. *This is shown in the column named "Adjusted score" in Table 13.5. By using the average van der Waerden scores, the sum of all scores remains zero.*

    The sum of the van der Waerden scores for all observations in the first sample is -2.2293. *The sum of the van der Waerden scores for the observations in the second sample is* 4.0888, *while it is* -1.8594 *for all observations in the third sample. The mean scores for the three samples,* $\bar{a}_1$, $\bar{a}_2$, *and* $\bar{a}_3$, *are* -0.4459, 0.8178, *and* -0.3719, *respectively. The sample variance* s_A^2 *of the 15 (adjusted) van der Waerden scores in Table 13.5 is* 0.7496. *As a result, the computed test statistic is*

$$v = \frac{1}{0.7496} \left[5\,(-0.4459)^2 + 5\,(0.8178)^2 + 5\,(-0.3719)^2 \right],$$
$$= 6.7092.$$

The approximate p-value is

$$P\left(\chi^2_{g-1} > 6.7092 \right) = 0.0349.$$

Based on this approximate p-value, we can reject the null hypothesis at a significance level of 5% and conclude that there is a significant difference in life span for the three brands of lamps. The JMP output with the approximate p-value for the van der Waerden test is shown in Figure 13.3.

Van der Waerden Test (Normal Quantiles)					
Level	**Count**	**Score Sum**	**Expected Score**	**Score Mean**	**(Mean-Mean0)/Std0**
Brand 1	5	−2.229	0.000	−0.44588	−1.410
Brand 2	5	4.089	0.000	0.81776	2.587
Brand 3	5	−1.859	0.000	−0.37188	−1.176

1-way Test, ChiSquare Approximation		
ChiSquare	**DF**	**Prob> ChiSq**
6.7092	2	0.0349*

Small sample sizes. Refer to statistical tables for tests, rather than large-sample approximations.

Figure 13.3 The JMP output for the van der Waerden test in Example 13.2.2 with an approximate *p*-value based on the χ^2-distribution.

13.3　The Median Test

A condition for the validity of both the Kruskal–Wallis test and the van der Waerden test is that the populations under study have probability densities with the same shape. If this condition is not met, it is better to use the median test. The median test remains valid for populations that have probability densities with different shapes.

The median test does not use ranks, but compares each observation with the median of the entire data set. In principle, an observation gets a score of 1 if it is larger than the overall median, and a score of 0 otherwise. We denote the score for the jth observation in the ith sample by M_{ij}.

Example 13.3.1　*The overall median of the* 15 *observations of the life span of lamps in Example 13.1.1 is equal to* 1003. *This median corresponds to the observation with rank* 8 *in Table 13.2. The seven observations that are larger than* 1003 *get a score of* 1, *while the remaining eight observations get a score of* 0. *This is shown in Table 13.6.*

The test statistic is based on the sum of the scores in each sample,

$$M_i = \sum_{j=1}^{n_i} M_{ij},$$

and the sample variance of all n scores, S_M^2. The test statistic is computed as

$$W = \frac{1}{S_M^2} \left(\sum_{i=1}^{8} \frac{M_i^2}{n_i} - \frac{1}{n} \left(\sum_{i=1}^{8} M_i \right)^2 \right).$$

Table 13.6　The assignment of the scores 0 and 1 to the original life span measurements in Example 13.1.1 for the median test. The overall median is the middle observation.

Life span	Rank	Score	Brand
836	1	0	Brand 1
857	2	0	Brand 3
898	3	0	Brand 1
913	4	0	Brand 3
952	5	0	Brand 1
987	6	0	Brand 3
988	7	0	Brand 2
1003	8	0	Brand 1
1045	9	1	Brand 3
1079	10	1	Brand 3
1084	11	1	Brand 1
1126	12	1	Brand 2
1324	13	1	Brand 2
1334	14	1	Brand 2
1538	15	1	Brand 2

Median Test (Number of Points Above Median)					
Level	Count	Score Sum	Expected Score	Score Mean	(Mean-Mean0)/Std0
Brand 1	5	1.000	2.333	0.200000	−1.414
Brand 2	5	4.000	2.333	0.800000	1.768
Brand 3	5	2.000	2.333	0.400000	−0.354

1-way Test, ChiSquare Approximation		
ChiSquare	DF	Prob>ChiSq
3.5000	2	0.1738

Small sample sizes. Refer to statistical tables for tests, rather than large-sample approximations.

Figure 13.4 The JMP output for the median test in Examples 13.3.1 and 13.3.2 with an approximate p-value based on the χ^2-distribution.

This test statistic is approximately χ^2-distributed with $g - 1$ degrees of freedom, so that an approximate p-value is given by the exceedance probability

$$p = P(\chi^2_{g-1} > w),$$

where w is the computed test statistic.

Example 13.3.2   *Starting from the scores in Table 13.6 in Example 13.3.1, we find that $m_1 = 1$, $m_2 = 4$, and $m_3 = 2$. This means that one of the large observations belongs to the first sample, four of the large observations belong to the second sample, and the remaining two large observations belong to the third sample. The sample variance of the 15 scores m_{ij} is equal to $4/15$. The computed test statistic is*

$$w = \frac{1}{4/15} \left(\frac{1^2}{5} + \frac{4^2}{5} + \frac{2^2}{5} - \frac{1}{15}(1 + 4 + 2)^2 \right),$$

$$= \frac{15}{4} \left(\frac{21}{5} - \frac{49}{15} \right),$$

$$= \frac{7}{2} = 3.5.$$

The corresponding approximate p-value can be computed using the JMP command "1-Chi Square Distribution(3.5, 2)", and is equal to 0.1738. The JMP output for the median test is shown in Figure 13.4. It contains the sum of the scores, the computed test statistic of 3.5, and the approximate p-value. Based on the median test, we cannot reject the null hypothesis that there is no difference between the brands in terms of the life span of lamps.

The procedure for assigning scores to individual observations outlined above has to be adjusted when multiple observations are equal to the overall median. More specifically, these observations obtain a score between 0 and 1, depending on their number. This is illustrated in the following example.

Example 13.3.3   *A physics professor grades the exams of 30 students. The students come from five different high schools (therefore, $g = 5$). There are six students from each of the*

Table 13.7 The exam results of 30 students from five different schools.

School 1	School 2	School 3	School 4	School 5
15	16	8	5	12
18	17	7	6	19
17	21	10	13	18
19	16	15	11	12
19	19	14	9	17
20	17	14	10	14

five schools, so that $n_1 = n_2 = \cdots = n_5 = 6$. The data is shown in Table 13.7. The professor wonders whether the results differ significantly between the schools from which the students were recruited.

To perform the median test, we need to rank the 30 observations from small to large, and assign scores to them. A difficulty here is that two observations coincide with the overall median. In principle, the 15 largest observations would get a score of 1, because, generally, 15 observations are larger than the median in a data set with 30 observations. The median of the 30 exam results is equal to 15, and two observations are equal to 15. These two observations both get a score of 0.5. The complete assignment of the scores m_{ij} to the 30 observations is shown in Table 13.8.

After the assignment of the 30 scores, we obtain $m_1 = 5.5$, $m_2 = 6$, $m_3 = 0.5$, $m_4 = 0$, and $m_5 = 3$. The sample variance s_M^2 of the 30 scores m_{ij} is equal to 0.2414. The computed test statistic is

$$w = \frac{1}{0.2414}\left(\frac{(5.5)^2}{6} + \frac{6^2}{6} + \frac{(0.5)^2}{6} + \frac{0^2}{6} + \frac{3^2}{6} - \frac{1}{30}(5.5 + 6 + 0.5 + 0 + 3)^2\right),$$

$$= \frac{1}{0.2414}\left(\frac{75.5}{6} - \frac{225}{30}\right),$$

$$= 21.0595.$$

The corresponding approximate p-value is equal to the probability that a χ^2-distributed random variable with $g - 1 = 4$ degrees of freedom exceeds the value 21.0595, namely 0.0003. Based on this p-value, we can conclude that the exam results differ significantly from school to school.

For the analysis of the exam results, we can also use the Kruskal–Wallis test and the van der Waerden test. The required computations are visualized in Table 13.9. For the Kruskal–Wallis test, we have $r_1 = 141.5$, $r_2 = 132.5$, $r_3 = 55$, $r_4 = 33.5$ and $r_5 = 102.5$. Consequently,

$$h = \left(\frac{12}{30(30 + 1)}\left\{\frac{(141.5)^2}{6} + \frac{(132.5)^2}{6} + \ldots + \frac{(102.5)^2}{6}\right\}\right) - 3(30 + 1),$$

$$= 19.3269.$$

Table 13.8 The scores for the median test for the exam result data in Table 13.7.

Rank	Result	School	Score
1	5	4	0
2	6	4	0
3	7	3	0
4	8	3	0
5	9	4	0
6	10	3	0
7	10	4	0
8	11	4	0
9	12	5	0
10	12	5	0
11	13	4	0
12	14	3	0
13	14	3	0
14	14	5	0
15	15	1	0.5
16	15	3	0.5
17	16	2	1
18	16	2	1
19	17	1	1
20	17	2	1
21	17	2	1
22	17	5	1
23	18	1	1
24	18	5	1
25	19	1	1
26	19	1	1
27	19	2	1
28	19	5	1
29	20	1	1
30	21	2	1

This value needs to be corrected because several sets of tied observations appear in the ranking. In total, there are eight groups of ties, so that $v = 8$. The eight group sizes, $t_1, t_2, \ldots, t_8$, are equal to 2, 2, 3, 2, 2, 4, 2, and 4, respectively, so that the correction factor for the ties is

$$c = 1 - \frac{(2^3 - 2) + (2^3 - 2) + \cdots + (4^3 - 4)}{30^3 - 30} = 0.9935.$$

As a result, the computed value for the corrected test statistic is

$$h_c = \frac{h}{c} = \frac{19.3269}{0.9935} = 19.4524.$$

The approximate p-value for the Kruskal–Wallis test here is 0.0006.

For the van der Waerden test, we have $a_1 = 4.7796$, $a_2 = 4.0811$, $a_3 = -3.6480$, $a_4 = -6.1861$, and $a_5 = 0.9734$. Consequently, the average scores for the five samples are

Table 13.9   The ranks and scores for the Kruskal–Wallis test and the van der Waerden test for the exam result data in Table 13.7.

Result	School	Initial rank	Final rank	Original van der Waerden score	Final van der Waerden score
5	4	1	1	−1.8486	−1.8486
6	4	2	2	−1.5179	−1.5179
7	3	3	3	−1.3002	−1.3002
8	3	4	4	−1.1310	−1.1310
9	4	5	5	−0.9892	−0.9892
10	3	6	6.5	−0.8649	−0.8088
10	4	7	6.5	−0.7527	−0.8088
11	4	8	8	−0.6493	−0.6493
12	5	9	9.5	−0.5524	−0.5065
12	5	10	9.5	−0.4605	−0.5065
13	4	11	11	−0.3723	−0.3723
14	3	12	13	−0.2869	−0.2040
14	3	13	13	−0.2035	−0.2040
14	5	14	13	−0.1216	−0.2040
15	1	15	15.5	−0.0404	0.0000
15	3	16	15.5	0.0404	0.0000
16	2	17	17.5	0.1216	0.1626
16	2	18	17.5	0.2035	0.1626
17	1	19	20.5	0.2869	0.4180
17	2	20	20.5	0.3723	0.4180
17	2	21	20.5	0.4605	0.4180
17	5	22	20.5	0.5524	0.4180
18	1	23	23.5	0.6493	0.7010
18	5	24	23.5	0.7527	0.7010
19	1	25	26.5	0.8649	1.0713
19	1	26	26.5	0.9892	1.0713
19	2	27	26.5	1.1310	1.0713
19	5	28	26.5	1.3002	1.0713
20	1	29	29	1.5179	1.5179
21	2	30	30	1.8486	1.8486

$\bar{a}_1 = 0.7966$, $\bar{a}_2 = 0.6802$, $\bar{a}_3 = -0.6080$, $\bar{a}_4 = -1.0310$, and $\bar{a}_5 = 0.1622$. The variance of the 30 van der Waerden scores is 0.8329. As a result, the computed test statistic is

$$v = \frac{1}{0.8329}(6 \times (0.7966)^2 + 6 \times (0.6802)^2 + \cdots + 6 \times (0.1622)^2),$$
$$= 18.3620.$$

The approximate p-value for the van der Waerden test is 0.0010.

In summary, the three nonparametric tests all lead to the conclusion that there is a significant difference in the exam results between the different schools. It is important to point out here,

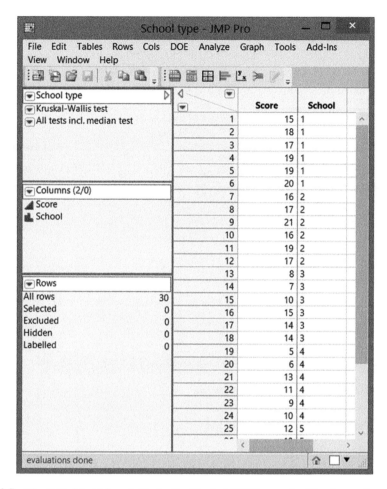

Figure 13.5 The JMP data table suitable for the Kruskal–Wallis, van der Waerden, and median tests.

however, that only the use of the median test is fully justified in this example. This is because the exam grades are bounded by 0 and 25, and the Kruskal–Wallis test as well as the van der Waerden test should only be used when the distributions of the populations under study have the same shapes and are shifted versions of each other. If the distribution of grades for one school were to be a shifted version of the grade distribution for another school, then this would mean that the professor would have used different lower and upper bounds for the various schools. Obviously, this would make no sense.

13.4 JMP

To perform the three nonparametric tests described in this chapter in JMP, we first have to enter the data in an appropriate format. Figure 13.5 shows a suitable JMP table for the data in Table 13.7 (Example 13.3.3). It is crucial that one variable is quantitative (the grade for the

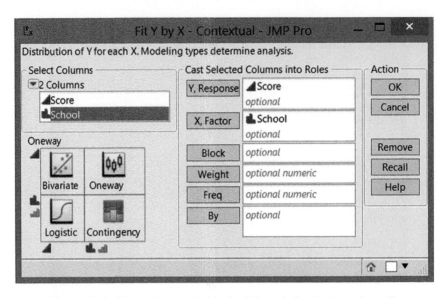

Figure 13.6   The entries required in the dialog window "Fit Y by X".

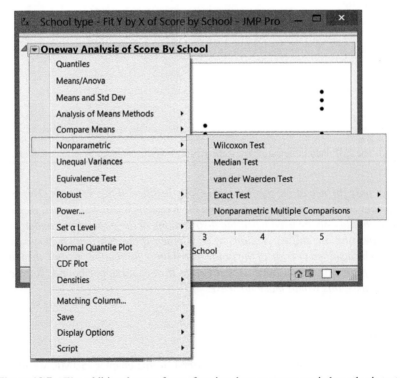

Figure 13.7   The additional menu for performing three nonparametric hypothesis tests.

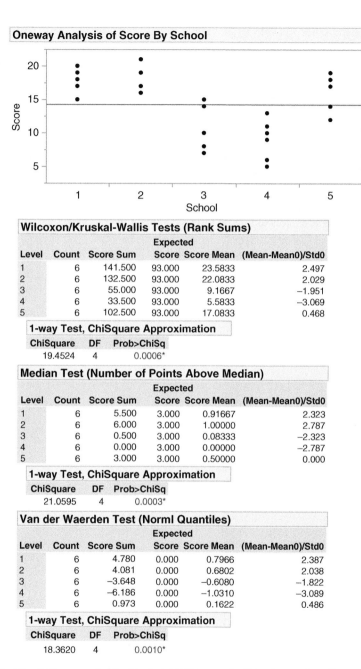

Figure 13.8 The JMP output for the Kruskal–Wallis, van der Waerden, and median tests in Example 13.3.3 with approximate *p*-values based on the χ^2-distribution.

exam) and that the other variable is qualitative (the school). In the "Analyze" menu, we then have to select the option "Fit Y by X". In the resulting dialog window, we need to enter the quantitative variable and the qualitative variable in the fields "Y, Response" and "X, Factor", respectively (see Figure 13.6). This initially only produces a graphical display of the data.

To proceed to the nonparametric tests, we then have to click the hotspot (red triangle) next to the term "Oneway Analysis of Score By School" in the initial output, and select the option "Nonparametric". We can then successively choose "Wilcoxon Test", "Median Test", and "van der Waerden Test" (see Figure 13.7). The first option, "Wilcoxon Test", returns the result of the Kruskal–Wallis test when more than two populations are studied. The option "Exact Test" does not work for more than two populations. JMP can only compute approximate p-values based on the χ^2-distribution.

After selecting the three nonparametric tests, you obtain the output shown in Figure 13.8.

14

Hypothesis Tests for More Than Two Population Variances

The biweekly checkups involve diet consultations, and we monitor any variances in your overall health. This is key for early detection, for calibrating any meds you might be on, for seeing any problems a few miles away, as opposed to after they've run over you.

(from *The Circle*, Dave Eggers, pp. 152–153)

In Chapter 12, we discussed the one-way analysis of variance or ANOVA in detail. The initial aim of an analysis of variance is to test whether the means of g populations differ. The one-way analysis of variance approach is based on the assumption that the variances of the populations under study are all equal. The present chapter deals with tests that allow us to verify whether this assumption is justified. Therefore, the hypotheses tested in this chapter are

$$H_0 : \sigma_1^2 = \sigma_2^2 = \cdots = \sigma_g^2$$

and

$$H_a : \text{at least one population variance } \sigma_i^2 \text{ differs from the others.}$$

We study four different methods for testing the homogeneity of g population variances. The first test, Bartlett's test, is the classic one. The test is intuitive because it uses the sample variances of the g samples obtained from the populations under study. The other three tests are all based on a transformation of the original data and do not use the sample variances. Of these three tests, Levene's test is the best-known one, but O'Brien's test is considered the best. The Brown–Forsythe test is better than Levene's test, but worse than O'Brien's.

As in Chapter 12, we denote the total number of observations by n, the number of populations studied by g, and the number of observations in the ith sample (from population i) by n_i. The sample mean for sample i is denoted by $\overline{X}_i$, while the overall mean is denoted by $\overline{X}$. The sample variance of sample i is denoted by S_i^2.

Statistics with JMP: Hypothesis Tests, ANOVA and Regression, First Edition. Peter Goos and David Meintrup.
© 2016 John Wiley & Sons, Ltd. Published 2016 by John Wiley & Sons, Ltd.
Companion Website: http://www.wiley.com/go/goosandmeintrup/JMP

14.1 Bartlett's Test

Bartlett's test is the best possible one for normally distributed populations, but is particularly poor when the populations under study are not normally distributed. Therefore, the use of the test is not recommended when there is doubt about the normality of the populations studied. The test was published in 1937 and is based on the variances of g samples from the populations studied.

14.1.1 The Test Statistic

The test statistic in Bartlett's test is

$$B = \frac{(n-g)\ln\left(S_p^2\right) - \sum_{i=1}^{g}(n_i-1)\ln\left(S_i^2\right)}{1 + \dfrac{1}{3(g-1)}\left(\displaystyle\sum_{i=1}^{g}\dfrac{1}{n_i-1} - \dfrac{1}{n-g}\right)},$$

where

$$S_p^2 = \frac{1}{n-g}\sum_{i=1}^{g}(n_i-1)S_i^2$$

is the pooled sample variance.

This test statistic takes small values when the sample variances $S_1^2, S_2^2, \ldots, S_g^2$ are all almost equal. When the null hypothesis that the population variances are identical is true, Bartlett's test statistic is χ^2-distributed with $g-1$ degrees of freedom.

14.1.2 The Technical Background

The test statistic B looks very complicated. The denominator of the test statistic is a correction factor, which is always positive, and which ensures that the expected value of the test statistic is closer to the expected value of a χ^2-distributed random variable with $g-1$ degrees of freedom. In other words, the correction factor ensures that the probability density of B is better approximated by the χ^2-distribution.

The numerator of the test statistic B compares the arithmetic mean of all sample variances,

$$\frac{1}{g}\sum_{i=1}^{g}S_i^2,$$

with the geometric mean

$$\left(\prod_{i=1}^{g}S_i^2\right)^{1/g} = \sqrt[g]{\prod_{i=1}^{g}S_i^2}.$$

The arithmetic mean of a set of positive numbers (e.g., g sample variances $S_1^2, S_2^2, \ldots, S_g^2$) is always greater than or equal to the geometric mean. Both means are identical if and only if all numbers in the set are equal. The larger the variation in the numbers, the larger is the difference between the arithmetic mean and the geometric mean[1].

As

$$\frac{1}{g} \sum_{i=1}^{g} S_i^2 \geq \left(\prod_{i=1}^{g} S_i^2 \right)^{1/g},$$

we have that

$$\frac{\frac{1}{g} \sum_{i=1}^{g} S_i^2}{\left(\prod_{i=1}^{g} S_i^2 \right)^{1/g}} \geq 1$$

and therefore

$$\ln \left(\frac{\frac{1}{g} \sum_{i=1}^{g} S_i^2}{\left(\prod_{i=1}^{g} S_i^2 \right)^{1/g}} \right) \geq 0,$$

regardless of the exact values of the sample variances $S_1^2, S_2^2, \dots, S_g^2$.

To see that the numerator of Bartlett's test statistic, B, indeed compares the arithmetic mean with the geometric mean, it is convenient to assume equal sample sizes. Therefore, we assume that $n_1 = n_2 = \cdots = n_g = n/g$. In that case, the pooled sample variance is

$$S_p^2 = \frac{1}{n-g} \sum_{i=1}^{g} \left(\frac{n}{g} - 1 \right) S_i^2 = \frac{1}{n-g} \sum_{i=1}^{g} \left(\frac{n-g}{g} \right) S_i^2 = \frac{1}{g} \sum_{i=1}^{g} S_i^2,$$

and the numerator of the test statistic B is

$$B^* = (n-g) \ln \left(S_p^2 \right) - \sum_{i=1}^{g} \left(\frac{n}{g} - 1 \right) \ln \left(S_i^2 \right),$$

$$= (n-g) \ln \left(S_p^2 \right) - \left(\frac{n}{g} - 1 \right) \sum_{i=1}^{g} \ln \left(S_i^2 \right),$$

---

[1] The arithmetic mean of the numbers 2, 4, 6, and 8 is equal to 5, while the geometric mean is equal to 4.43. The arithmetic mean of the numbers 1, 3, 7, and 9 is also equal to 5, but the geometric mean is only 3.71. Therefore, the difference between the two types of means increases with the variation in the numbers studied.

$$= (n - g) \ln \left(S_p^2 \right) - \left(\frac{n - g}{g} \right) \sum_{i=1}^{g} \ln \left(S_i^2 \right),$$

$$= (n - g) \left[\ln \left(S_p^2 \right) - \frac{1}{g} \sum_{i=1}^{g} \ln \left(S_i^2 \right) \right],$$

$$= (n - g) \left[\ln \left(\frac{1}{g} \sum_{i=1}^{g} S_i^2 \right) - \frac{1}{g} \ln \left(\prod_{i=1}^{g} S_i^2 \right) \right],$$

$$= (n - g) \left[\ln \left(\frac{1}{g} \sum_{i=1}^{g} S_i^2 \right) - \ln \left(\prod_{i=1}^{g} S_i^2 \right)^{1/g} \right],$$

$$= (n - g) \ln \left(\frac{\frac{1}{g} \sum_{i=1}^{g} S_i^2}{\left(\prod_{i=1}^{g} S_i^2 \right)^{1/g}} \right),$$

$$= (n - g) \ln \left(\frac{\text{arithmetic mean of } S_i^2}{\text{geometric mean of } S_i^2} \right).$$

This expression is always positive, so that B is always positive. Strong positive values indicate sample variances that differ substantially. Strong positive values therefore point in the direction of the alternative hypothesis, and will result in small p-values. If the null hypothesis is true, then the computed arithmetic mean and the computed geometric mean of the sample variances will generally be nearly identical. The numerator of the computed test statistic b is then close to $(n - g) \ln(1) = 0$.

14.1.3   The p-Value

The p-value of Bartlett's test indicates how extreme the computed value of the test statistic B is. It is computed as

$$p = P\left(\chi_{g-1}^2 > b \right),$$

where b represents the computed Bartlett test statistic. If the p-value is smaller than the significance level chosen, the null hypothesis is rejected.

Alternatively, we can use the critical value $\chi_{\alpha;g-1}^2$. If the computed test statistic exceeds this critical value, the null hypothesis is rejected.

Example 14.1.1 *Example 13.3.3 deals with a physics professor who grades the exams of 30 students. The students come from five different schools, so that $g = 5$. There are six students from each school, so that $n_1 = n_2 = \cdots = n_5 = 6$. The five sample variances are $s_1^2 = 3.2$,*

$s_2^2 = 3.8667$, $s_3^2 = 11.8667$, $s_4^2 = 9.2$ and $s_5^2 = 9.4667$. *The arithmetic mean of the sample variances is 7.52, while their geometric mean is equal to 6.6277. Consequently, the numerator of the test statistic B takes the value*

$$(n - g)\ln\left(\frac{7.52}{6.6277}\right) = (30 - 5) \times 0.1263 = 3.1579.$$

The correction factor in the denominator of the test statistic B is

$$1 + \frac{1}{3(5 - 1)}\left(5 \cdot \frac{1}{6 - 1} - \frac{1}{30 - 5}\right) = 1.08.$$

The computed value of the test statistic is thus

$$b = \frac{3.1579}{1.08} = 2.9240.$$

The corresponding p-value is

$$P\left(\chi_{5-1}^2 > 2.9240\right) = P\left(\chi_4^2 > 2.9240\right) = 0.5706.$$

The critical value for the test is $\chi_{0.05;5-1}^2 = \chi_{0.05;4}^2 = 9.4877$, if we utilize a significance level of 5%. This critical value is not exceeded by the computed test statistic. Therefore, based on Bartlett's test, the null hypothesis cannot be rejected. So, Bartlett's test does not provide evidence that the variances of the exam results vary from school to school.

14.2 Levene's Test

Levene's test does not use the sample variances, but assigns a score D_{ij} to each original observation X_{ij}. The steps required for the test are as follows:

(1) For each sample i, compute the sample mean $\overline{X}_i$.
(2) Subtract the sample mean $\overline{X}_i$ from each observation X_{ij}.
(3) Determine the absolute value of all the differences:

$$D_{ij} = \left|X_{ij} - \overline{X}_i\right|.$$

(4) Perform a one-way analysis of variance of the scores D_{ij} instead of the original observations X_{ij}.
(5) Reject the null hypothesis and conclude that the variances differ if the p-value of this analysis of variance is smaller than the significance level chosen.

The logic behind Levene's test is that populations with large variances will give rise to large absolute deviations from the sample means. In other words, the scores D_{ij} will be strongly

Table 14.1   The calculation of the scores d_{ij} from the original observations x_{ij} in Example 14.2.1 by subtracting the sample means 18, 17.67, 11.33, 9, and 15.33 and determining the absolute value of the differences.

School 1		School 2		School 3		School 4		School 5	
x_{1j}	d_{1j}	x_{2j}	d_{2j}	x_{3j}	d_{3j}	x_{4j}	d_{4j}	x_{5j}	d_{5j}
15	3.00	16	1.67	8	3.33	5	4.00	12	3.33
18	0.00	17	0.67	7	4.33	6	3.00	19	3.67
17	1.00	21	3.33	10	1.33	13	4.00	18	2.67
19	1.00	16	1.67	15	3.67	11	2.00	12	3.33
19	1.00	19	1.33	14	2.67	9	0.00	17	1.67
20	2.00	17	0.67	14	2.67	10	1.00	14	1.33

positive for populations with large variances. Indeed, the individual observations X_{ij} from such populations will deviate substantially from the corresponding sample means $\overline{X}_i$, so that the scores $D_{ij} = \left| X_{ij} - \overline{X}_i \right|$ will be large positive numbers.

The main weakness of Levene's test is that the probability of type I errors is generally larger than the chosen significance level α. In contrast to Bartlett's test, Levene's test works quite well for populations that are not normally distributed, as long as the distributions of the populations are symmetrical.

Example 14.2.1   *In order to apply Levene's test to the exam result data in Example 13.3.3, we first have to assign a score d_{ij} to each observation x_{ij}, by subtracting the corresponding sample mean $\overline{x}_i$ and taking the absolute value of the resulting difference. The result of this operation is shown in Table 14.1. Then, we have to carry out a one-way analysis of variance of the scores d_{ij}. This produces a between-group sum of squares ssb of 12.24 and a within-group sum of squares ssw of 33.48, resulting in a between-group variance msb of 3.06 and a within-group variance msw of 1.34. The computed test statistic f is therefore $3.06/1.34 = 2.28$. The corresponding p-value is $P(F_{4;25} > 2.28) = 0.0885$. This p-value is obtained using an F-distribution with $g - 1 = 4$ degrees of freedom for the numerator and $n - g = 25$ degrees of freedom for the denominator. The analysis of variance produced by JMP for the scores d_{ij} is shown in Figure 14.1. If a significance level of 5% is used, then the null hypothesis that the variances of the exam results for the five schools are equal cannot be rejected.*

Analysis of Variance

Source	DF	Sum of Squares	Mean Square	F Ratio	Prob > F
School	4	12.237037	3.05926	2.2843	0.0885
Error	25	33.481481	1.33926		
C.Total	29	45.718519			

Figure 14.1   The JMP output of the analysis of variance of the scores d_{ij} in Table 14.1 in the context of Levene's test.

14.3   The Brown–Forsythe Test

The Brown–Forsythe test is similar to Levene's test, but it uses the g sample medians $Me_1, Me_2, \ldots, Me_g$ instead of the g sample means $\overline{X}_1, \overline{X}_2, \ldots, \overline{X}_g$. Consequently, the steps required for the Brown–Forsythe test are as follows:

(1) For each sample i, compute the sample median Me_i.
(2) Subtract the sample median Me_i from each observation X_{ij}.
(3) Determine the absolute value of all the differences:

$$D_{ij}^* = \left| X_{ij} - Me_i \right|.$$

(4) Perform a one-way analysis of variance of the scores D_{ij}^* instead of the original observations X_{ij}.
(5) Reject the null hypothesis and conclude that the variances differ if the p-value of this analysis of variance is smaller than the significance level chosen.

If a score D_{ij}^* turns out to be zero (which is inevitable in samples with an odd number of observations), this zero is replaced by the smallest strictly positive score in sample i.

The logic behind the Brown–Forsythe test is that populations with large variances will give rise to large absolute deviations from the median. In other words, the scores D_{ij}^* will be strongly positive for populations with large variances. The Brown–Forsythe test works well if the data do not come from normally distributed populations, even if the populations have skewed distributions. This is an advantage over Levene's and Bartlett's tests.

Example 14.3.1   *To apply the Brown–Forsythe test to the exam result data in Example 13.3.3, we first assign a score d_{ij}^* to each observation x_{ij} by subtracting the corresponding sample median Me_i and taking the absolute value of the differences. The result of this operation is shown in Table 14.2. Next, we conduct a one-way analysis of variance of the new scores d_{ij}^*. This leads to a between-group sum of squares* ssb *of 14.13 and a within-group sum of squares* ssw *of 45.83, resulting in a between-group variance* msb *of 3.53 and a within-group variance* msw *of 1.83. The computed test statistic f is therefore 3.53/1.83 = 1.93. The corresponding*

Table 14.2   The calculation of the scores d_{ij}^* from the original observations x_{ij} in Example 14.3.1 by subtracting the sample medians 18.5, 17, 12, 9.5, and 15.5 and determining the absolute value of the differences.

School 1		School 2		School 3		School 4		School 5	
x_{1j}	d_{1j}^*	x_{2j}	d_{2j}^*	x_{3j}	d_{3j}^*	x_{4j}	d_{4j}^*	x_{5j}	d_{5j}^*
15	3.5	16	1.0	8	4.0	5	4.5	12	3.5
18	0.5	17	0.0	7	5.0	6	3.5	19	3.5
17	1.5	21	4.0	10	2.0	13	3.5	18	2.5
19	0.5	16	1.0	15	3.0	11	1.5	12	3.5
19	0.5	19	2.0	14	2.0	9	0.5	17	1.5
20	1.5	17	0.0	14	2.0	10	0.5	14	1.5

Analysis of Variance

Source	DF	Sum of Squares	Mean Square	F Ratio	Prob > F
School	4	14.133333	3.53333	1.9273	0.1371
Error	25	45.833333	1.83333		
C.Total	29	59.966667			

Figure 14.2   The JMP output for the analysis of variance of the scores d^*_{ij} in Table 14.2 in the context of the Brown–Forsythe test.

p-value is $P(F_{4;25} > 1.93) = 0.1371$. *The ANOVA table produced by JMP for the new scores* d^*_{ij} *is shown in Figure 14.2. If a significance level of 5% is used, then the null hypothesis that the variances of the exam results from the five schools are equal again cannot be rejected.*

14.4   O'Brien's Test

The Brown–Forsythe test works well for testing the homogeneity of g population variances, but the transformation of the data used is not suitable for more sophisticated analyses (such as, for example, pairwise comparisons of variances). Therefore, O'Brien's transformation is recommended, even though it is much more complicated than the Levene and Brown–Forsythe transformations. A characteristic of O'Brien's method is that the resulting scores are related to the sample variances $S_1^2, S_2^2, \ldots, S_g^2$.

The steps required for O'Brien's test are as follows:

(1) For each sample i, compute the sample mean $\overline{X}_i$.
(2) For each sample i, also compute the sum of squares

$$\mathrm{SS}_i = \sum_{j=1}^{n_i}(X_{ij} - \overline{X}_i)^2 = (n_i - 1)S_i^2.$$

(3) For each observation X_{ij}, calculate the score

$$
\begin{aligned}
S_{ij} &= \frac{n_i(n_i - 1.5)(X_{ij} - \overline{X}_i)^2 - \mathrm{SS}_i/2}{(n_i - 1)(n_i - 2)} \\
&= \frac{n_i(n_i - 1.5)(X_{ij} - \overline{X}_i)^2 - (n_i - 1)S_i^2/2}{(n_i - 1)(n_i - 2)}.
\end{aligned}
\tag{14.1}
$$

(4) Perform a one-way analysis of variance of the scores S_{ij} instead of the original observations X_{ij}.
(5) Reject the null hypothesis and conclude that the variances differ if the p-value of this analysis of variance is smaller than the significance level chosen.

An important feature of the scores in O'Brien's test is that the mean of all scores S_{ij} in a sample i is equal to the sample variance S_i^2:

$$\bar{S}_i = \frac{1}{n_i} \sum_{j=1}^{n_i} S_{ij},$$

$$= \frac{1}{n_i} \sum_{j=1}^{n_i} \frac{n_i(n_i - 1.5)(X_{ij} - \bar{X}_i)^2 - (n_i - 1)S_i^2/2}{(n_i - 1)(n_i - 2)},$$

$$= \frac{1}{n_i} \sum_{j=1}^{n_i} \frac{n_i(n_i - 1.5)(X_{ij} - \bar{X}_i)^2}{(n_i - 1)(n_i - 2)} - \frac{1}{n_i} \sum_{j=1}^{n_i} \frac{(n_i - 1)S_i^2/2}{(n_i - 1)(n_i - 2)},$$

$$= \frac{1}{n_i} \sum_{j=1}^{n_i} \frac{n_i(n_i - 1.5)(X_{ij} - \bar{X}_i)^2}{(n_i - 1)(n_i - 2)} - \frac{1}{n_i} \frac{n_i(n_i - 1)S_i^2/2}{(n_i - 1)(n_i - 2)},$$

$$= \frac{n_i - 1.5}{n_i - 2} \sum_{j=1}^{n_i} \frac{(X_{ij} - \bar{X}_i)^2}{n_i - 1} - \frac{0.5}{n_i - 2}S_i^2,$$

$$= \frac{n_i - 1.5}{n_i - 2}S_i^2 - \frac{0.5}{n_i - 2}S_i^2,$$

$$= \frac{n_i - 1.5 - 0.5}{n_i - 2}S_i^2,$$

$$= S_i^2.$$

This feature means that O'Brien's test implicitly uses the sample variances. An additional advantage of the test is that it minimizes the risk of a type I or type II error.

Unlike Bartlett's test, O'Brien's test also works properly if the populations studied are not normally distributed.

Example 14.4.1 *To apply O'Brien's test to the exam result data in Example 13.3.3, we have to assign a score s_{ij} to each observation x_{ij}. This should be done using Equation (14.1). The result of this operation is shown in Table 14.3. To check whether the computed scores*

Table 14.3 The calculation of the scores s_{ij} from the original observations x_{ij} in Example 14.4.1.

School 1		School 2		School 3		School 4		School 5	
x_{1j}	s_{1j}	x_{2j}	s_{2j}	x_{3j}	s_{3j}	x_{4j}	s_{4j}	x_{5j}	s_{5j}
15	11.75	16	3.27	8	13.52	5	20.45	12	13.82
18	−0.40	17	0.12	7	23.87	6	11.00	19	16.97
17	0.95	21	14.52	10	0.92	13	20.45	18	8.42
19	0.95	16	3.27	15	16.67	11	4.25	12	13.82
19	0.95	19	1.92	14	8.12	9	−1.15	17	2.57
20	5.00	17	0.12	14	8.12	10	0.20	14	1.22

Analysis of Variance

Source	DF	Sum of Squares	Mean Square	F Ratio	Prob > F
School	4	345.1937	86.2984	1.7284	0.1753
Error	25	1248.2100	49.9284		
C.Total	29	1593.4037			

Figure 14.3 The JMP output for the analysis of variance of the scores s_{ij} in Table 14.3 in the context of O'Brien's test.

s_{ij} *are correct, we can calculate the mean score* $\bar{s}_i$ *for each sample and compare it to the sample variance* s_i^2 *of the original observations in the sample. If the mean scores* $\bar{s}_i$ *and the sample variances* s_i^2 *are not identical, then this indicates that a mistake has been made when calculating the scores. As an illustration, for the first sample, the average score is*

$$\bar{s}_1 = \frac{1}{5}(11.75 + (-0.40) + 0.95 + 0.95 + 0.95 + 5.00) = 3.2,$$

which is identical to the sample variance s_1^2.

Next, we perform a one-way analysis of variance of the scores s_{ij}. *This results in a between-group sum of squares* ssb *of 345.19 and a within-group sum of squares* ssw *of 1248.21, and a between-group variance* msb *of 86.30 and a within-group variance* msw *of 49.93. The computed test statistic f is therefore 86.30/49.93 = 1.73. The corresponding p-value is* $P(F_{4;25} > 1.73) = 0.1753$. *The ANOVA table produced by JMP for the scores* s_{ij} *is shown in Figure 14.3. If a significance level of 5% is used, then it is again not possible to reject the null hypothesis that the variances of the exam results for the five schools are equal.*

14.5 JMP

All four tests described in this chapter can be performed in JMP. This can be done using the "Fit Y by X" platform in the "Analyze" menu. The required data table and the corresponding input in the dialog window of the "Fit Y by X" platform are shown in Figures 13.5 and 13.6. Via the hotspot (red triangle) next to the term "Oneway Analysis of Score By School", we need to select the option "Unequal Variances" if we want to perform the tests for equal variances. This is illustrated in Figure 14.4 and yields the output shown in Figure 14.5.

In the JMP output of the option "Unequal Variances" in Figure 14.5, we first see a graphical representation of the sample standard deviations of the different samples, followed by a table that contains the numerical values of the sample standard deviations. The next table in the output summarizes the results of the four tests for the homogeneity of the variances. The O'Brien, Brown–Forsythe, Levene, and Bartlett tests provide computed test statistics of 1.7279, 1.9273, 2.2843, and 0.7310, respectively. In the output, all these values appear in a column named "F Ratio", although Bartlett's test is not based on the F-distribution, but on the χ^2-distribution. The table also indicates the numbers of degrees of freedom required for

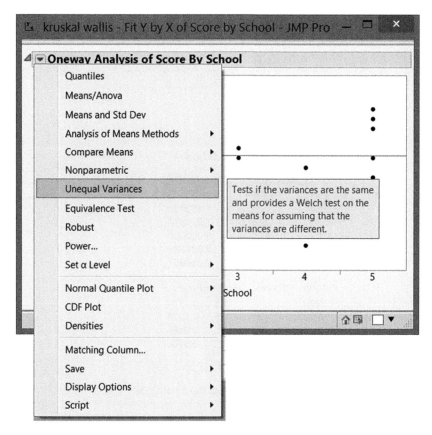

Figure 14.4 The option "Unequal Variances" in the "Fit Y by X" platform.

the F-tests and the χ^2-test. In the output, the column labeled "DFNum" shows the number of degrees of freedom for the numerator, while the column labeled "DFDen" shows the number of degrees of freedom for the denominator. The first number is $g - 1 = 5 - 1 = 4$, while the second is $n - g = 30 - 5 = 25$. These degrees of freedom are the same as those in an analysis of variance, because the O'Brien, Brown–Forsythe, and Levene tests involve an analysis of variance of transformed data based on g samples and n observations in total. For Bartlett's test, only one number of degrees of freedom is reported in the JMP output, because the χ^2-distribution has only one number of degrees of freedom. The p-values in the output equal 0.1754, 0.1371, 0.0885, and 0.5706, and match the p-values that we obtained in the examples in this chapter. None of the four tests allows us to reject the null hypothesis that the variances are equal.

Note that the computed test statistic reported by JMP for Bartlett's test equals 0.7310, while we obtained a computed test statistic of 2.9240 in Example 14.1.1. The reason for this discrepancy is that JMP divides our test statistic of 2.9240 by the number of degrees of freedom, $g - 1 = 4$. The computed test statistic and the p-value for O'Brien's test in Figure 14.5

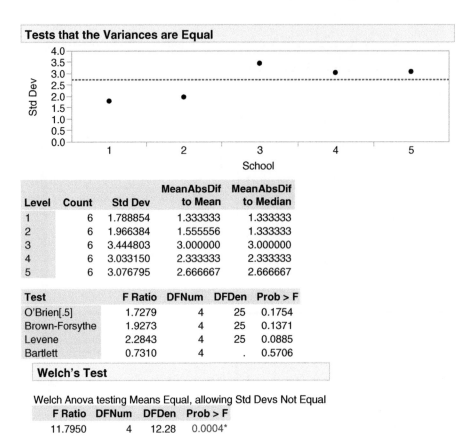

Figure 14.5 The JMP output for the Bartlett, Levene, Brown–Forsythe, and O'Brien tests for equal variances, and the Welch test for equal means.

differ slightly from those in Example 14.4.1. This is simply due to rounding effects in the intermediate calculations.

14.6   The Welch Test

The last part of the output in Figure 14.5 contains the results of the Welch test. This output is only relevant when the null hypothesis that the variances are equal is rejected. In that case, it is not appropriate to use the classical analysis of variance approach from Chapter 12 to test the hypotheses

$$H_0 : \mu_1 = \mu_2 = \cdots = \mu_g$$

and

$$H_a : \text{ at least one population mean } \mu_i \text{ differs from the others.}$$

As a matter of fact, in that case, the assumption of equal variances, on which traditional analysis of variance is based, is not fulfilled, and it is better to use the Welch test instead. The Welch test uses the test statistic

$$F = \frac{\dfrac{1}{g-1} \displaystyle\sum_{i=1}^{g} \dfrac{n_i(\overline{X}_i - A)^2}{S_i^2}}{1 + \dfrac{2(g-2)}{g^2-1} \displaystyle\sum_{i=1}^{g} H_i},$$

where

$$H_i = \sum_{i=1}^{g} \frac{1}{n_i - 1} \left(1 - \frac{n_i/S_i^2}{\sum_{j=1}^{g} \left(n_j/S_j^2 \right)} \right)^2,$$

and

$$A = \frac{1}{\displaystyle\sum_{i=1}^{g} \dfrac{n_i}{S_i^2}} \sum_{i=1}^{g} \frac{n_i \overline{X}_i}{S_i^2}$$

is a weighted average of the sample means $\overline{X}_i$. The weights in the weighted average are based on the (unequal) variances and ensure that A is the best possible estimator of the common population mean if the null hypothesis that $\mu_1 = \mu_2 = \cdots = \mu_g$ is true and the variances are different. Under the null hypothesis, the Welch test statistic is F-distributed with $g - 1$ degrees of freedom for the numerator and

$$v = \frac{g^2 - 1}{3 \sum_{i=1}^{g} h_i}$$

degrees of freedom for the denominator, where h_i is the computed value of H_i. The complicated number of degrees of freedom for the denominator ensures that the numerator and the denominator of the test statistic have the same expected value when the null hypothesis that $\mu_1 = \mu_2 = \cdots = \mu_g$ is true, even when the variances are different. If we denote the computed test statistic by f, we find the p-value for the Welch test as

$$p = P(F_{g-1;v} > f).$$

If this p-value is smaller than the significance level α, then we reject the null hypothesis that $\mu_1 = \mu_2 = \cdots = \mu_g$. We then conclude that at least one of the population means differs from the others.

The logic behind the Welch test is that whenever the null hypothesis is incorrect, the individual sample means $\overline{X}_1, \overline{X}_2, \ldots, \overline{X}_g$ will take values that differ substantially from the value taken by A, the estimator of the overall mean for all populations. This leads to large values for the squared deviations $(\overline{X}_i - A)^2$, and thus to a large value for the test statistic F. A

Table 14.4   The calculations required for determining the Welch test statistic in Example 14.6.1.

n_i	$\bar{x}_i$	s_i^2	n_i/s_i^2	$n_i\bar{x}_i/s_i^2$	h_i	$\bar{x}_i - a$	$n_i(\bar{x}_i - a)^2/s_i^2$
6	18.0000	3.2000	1.8750	33.7500	0.0821	2.1938	9.0236
6	17.6667	3.8667	1.5517	27.4138	0.0987	1.8605	5.3710
6	11.3333	11.8667	0.5056	5.7303	0.1631	−4.4729	10.1160
6	9.0000	9.2000	0.6522	5.8696	0.1531	−6.8062	30.2119
6	15.3333	9.4667	0.6338	9.7183	0.1544	−0.4729	0.1418

large value for the test statistic ultimately results in a small p-value and a rejection of the null hypothesis.

Example 14.6.1   *For the exam result data in Example 13.3.3, the outcome of the Welch test is displayed at the bottom of Figure 14.5. The reconstruction of this outcome requires a substantial amount of work. Table 14.4 shows various intermediate steps. More specifically, it shows how the values n_i/s_i^2, $n_i\bar{x}_i/s_i^2$, h_i, $\bar{x}_i - a$, and $n_i(\bar{x}_i - a)^2/s_i^2$ are determined for each sample i. These calculations lead to the following intermediate results: a = 15.8062,*

$$\sum_{i=1}^{8} h_i = 0.6515,$$

and

$$\sum_{i=1}^{8} \frac{n_i(\bar{x}_i - a)^2}{s_i^2} = 54.8642.$$

Finally, the computed test statistic is

$$f = \frac{\frac{1}{5-1} \cdot 54.8642}{1 + \frac{2(5-2)}{5^2-1} \cdot 0.6515} = 11.7950.$$

As

$$v = \frac{5^2 - 1}{3 \cdot 0.6515} = 12.2802,$$

the p-value is

$$p = P(F_{4;12.2802} > 11.7950) = 0.00036 \approx 0.0004.$$

This p-value can be determined in JMP using the command "1 − F Distribution (11.7950, 4, 12.2802)" and leads to a rejection of the null hypothesis that $\mu_1 = \mu_2 = \cdots = \mu_5$ for the exam result data. We conclude that at least one population mean differs from the others.

Part Five

Additional Useful Tests and Procedures

15

The Design of Experiments and Data Collection

Almost everything in social life is produced by rare but consequential shocks and jumps; all the while almost everything studied about social life focuses on the "normal", particularly with "bell curve" methods that tell you close to nothing. Why? Because the bell curve ignores large deviations, cannot handle them, yet makes us confident that we have tamed uncertainty.

(from *The Black Swan*, Nassim Nicholas Taleb, p. xxiv)

This book deals with estimating population parameters and testing hypotheses. Whenever we estimate a population parameter, we, of course, prefer to obtain a precise estimate. Likewise, whenever we perform a hypothesis test, we like this test to have a large power. The number of observations obviously has a large impact on both the precision of the parameter estimates and the power of the hypothesis tests: if we want to increase the precision of estimates or the power of tests, then we should collect more observations. This generally requires more time and effort, and usually costs more money.

In many cases, a researcher not only has to decide how many observations are desirable, but also how these observations should be collected. In this chapter, we study, for the sake of illustration, a situation in which a researcher wants to quantify the difference between two population means μ_1 and μ_2 by means of a point estimate and an interval estimate of the difference $\mu_1 - \mu_2$. We consider three different scenarios, each of which produces a different result. In the first two scenarios, all observations are equally expensive. This is not the case in the third scenario.

15.1 Equal Costs for All Observations

Example 15.1.1 *A manufacturer of steel wires uses two machines to produce cables for car tires. The thicknesses of the cables produced on the two machines have to be equal, because customers should not notice any difference in quality between cables made on one machine*

Statistics with JMP: Hypothesis Tests, ANOVA and Regression, First Edition. Peter Goos and David Meintrup.
© 2016 John Wiley & Sons, Ltd. Published 2016 by John Wiley & Sons, Ltd.
Companion Website: http://www.wiley.com/go/goosandmeintrup/JMP

and cables made on the other. Therefore, at regular time intervals, the manufacturer's quality control department checks whether the thickness of the cables produced on the two machines is the same. For this purpose, a confidence interval for the difference in mean thickness of the cables produced on the two machines is determined.

As a result of extensive experience with the two machines, the variances and standard deviations of the thicknesses for both machines are known. The manufacturer has also observed over the years that the thicknesses produced are normally distributed.

Of course, the quality control department would like to estimate the difference in mean thickness between the two machines as accurately as possible. To this end, the variance of the estimator should be as small as possible, and the confidence interval should be as narrow as possible. This would result in a clear picture of any existing differences in thickness between the cables from the two machines.

In the example, the aim is to estimate the difference in the population means, $\mu_1 - \mu_2$. Here, μ_1 represents the mean of the first population (i.e., the mean thickness of the cables produced on machine 1), while μ_2 is the mean of the second population (i.e., the mean thickness of the cables produced on machine 2). Because the corresponding variances σ_1^2 and σ_2^2 are known and the populations are normally distributed, the formula required to calculate the confidence interval of $\mu_1 - \mu_2$ is

$$\left[\overline{X}_1 - \overline{X}_2 - z_{\alpha/2} \sqrt{\frac{\sigma_1^2}{n_1} + \frac{\sigma_2^2}{n_2}} \,,\; \overline{X}_1 - \overline{X}_2 + z_{\alpha/2} \sqrt{\frac{\sigma_1^2}{n_1} + \frac{\sigma_2^2}{n_2}} \right],$$

where n_1 and n_2 represent the numbers of observations from the first and the second population, respectively. This expression was derived in Section 8.1.4, and is based on the fact that $\overline{X}_1 - \overline{X}_2$ is an unbiased estimator of $\mu_1 - \mu_2$ with variance

$$\frac{\sigma_1^2}{n_1} + \frac{\sigma_2^2}{n_2}.$$

The width of the confidence interval is

$$2z_{\alpha/2} \sqrt{\frac{\sigma_1^2}{n_1} + \frac{\sigma_2^2}{n_2}}.$$

Ideally, the variance of $\overline{X}_1 - \overline{X}_2$ and the width of the confidence interval are small.

Now, suppose that the quality department's budget allows for the daily measurement of the thickness of 12 steel cables. If we were to follow our intuition, we would spread these 12 observations evenly over the two populations studied. In the example, this would mean that we would select six cables to measure from the first machine as well as six cables from the second machine. In that case, n_1 and n_2 would both be equal to 6.

A question that can be asked in this context is whether an equal distribution of the observations across the two machines leads to the most precise estimator of $\mu_1 - \mu_2$ and the narrowest confidence interval for it. After all, there exist 11 different ways in which to collect 12 observations from two populations (i.e., from the two machines):

- $n_1 = 1$ and $n_2 = 11$: one observation is made for machine 1, while 11 observations are made for machine 2;
- $n_1 = 2$ and $n_2 = 10$: two observations are made for machine 1, while 10 observations are made for machine 2;
- ...
- $n_1 = 6$ and $n_2 = 6$: six observations are performed for machine 1 as well as for machine 2;
- ...
- $n_1 = 11$ and $n_2 = 1$: 11 observations are made for machine 1, while only one observation is made for machine 2.

Which of these 11 options is the best depends on the values of σ_1^2 and σ_2^2. We start by studying the scenario in which σ_1^2 and σ_2^2 are equal.

15.1.1   Equal Variances

Example 15.1.2   *The two machines in Example 15.1.1 are of the same type and have the same age. Both machines have been used equally intensively and have been maintained equally well. In that case, it is justifiable to assume that the two machines are operating with the same precision, and that the variance of the thickness is the same for both machines.*

If σ_1^2 and σ_2^2 are equal, then we can denote the variance of the two machines by means of a common symbol, σ^2, and we can simplify the expressions for the variance of $\overline{X}_1 - \overline{X}_2$ and the width of the confidence interval for $\mu_1 - \mu_2$ to

$$\frac{\sigma^2}{n_1} + \frac{\sigma^2}{n_2} = \sigma^2 \left(\frac{1}{n_1} + \frac{1}{n_2} \right)$$

and

$$2z_{\alpha/2} \sqrt{\frac{\sigma^2}{n_1} + \frac{\sigma^2}{n_2}} = 2z_{\alpha/2}\sigma \sqrt{\frac{1}{n_1} + \frac{1}{n_2}},$$

respectively. Using these expressions, we can now examine which of the 11 options for the collection of the 12 observations is the best. It should be clear that the option with the smallest value for

$$\frac{1}{n_1} + \frac{1}{n_2}$$

will lead to the smallest variance for $\overline{X}_1 - \overline{X}_2$ and to the narrowest confidence interval, for any value of the common variance σ^2 and for any significance level α chosen. In Table 15.1, the

Table 15.1   The evaluation of the different options for the sample sizes n_1 and n_2 when $\sigma_1^2 = \sigma_2^2 = \sigma^2 = 1$.

n_1	n_2	$\text{var}(\overline{X}_1 - \overline{X}_2)$	Efficiency (%)	Width CI (%)
1	11	1.0909	30.6	180.91
2	10	0.6000	55.6	134.16
3	9	0.4444	75.0	115.47
4	8	0.3750	88.9	106.07
5	7	0.3429	97.2	101.42
6	6	0.3333	100.0	100.00
7	5	0.3429	97.2	101.42
8	4	0.3750	88.9	106.07
9	3	0.4444	75.0	115.47
10	2	0.6000	55.6	134.16
11	1	1.0909	30.6	180.91

11 options are listed along with the corresponding variances of $\overline{X}_1 - \overline{X}_2$. When calculating the variances, we have assumed for simplicity that the common variance of the two machines, σ^2, is equal to 1. The actual value of σ^2 does not impact the choice of data collection option.

Table 15.1 shows that distributing the 12 observations evenly over the two populations or machines yields the smallest variance for $\overline{X}_1 - \overline{X}_2$. Indeed, the option with $n_1 = n_2 = 6$ results in the smallest variance, which equals $1/3 \approx 0.3333$ when $\sigma^2 = 1$. The two worst possible options involve one observation from one of the two populations, and 11 observations from the other population. These options lead to a variance of 1.0909, which is more than three times as large as the smallest possible variance.

In order to quantify the relative performance of the 11 different options, we can proceed in two ways. The first method uses relative efficiencies. A relative efficiency is a ratio of two variances. For example, the efficiency of the two worst possible options in Table 15.1 relative to the best one is $0.3333/1.0909 = 0.3055 = 30.6\%$. The fourth column in Table 15.1 shows all of the relative efficiencies. The best option has a relative efficiency of 100%. The two next best options are both 97.2% efficient.

A second way to evaluate the 11 different options is to compare the widths of the resulting confidence intervals. The worst possible options in Table 15.1 yield confidence intervals that are $\sqrt{1.0909/0.3333} \approx 1.81$ times as wide as those produced by the best option. In other words, the interval for the worst possible options is 81% wider than the interval for the best possible method. The extent to which the width of the confidence intervals is inflated is shown in the last column of Table 15.1.

In any case, we can conclude that, when the variances of the two populations under study are equal, it is optimal to evenly distribute the observations over the two populations.

15.1.2   Unequal Variances

It is not always realistic to assume that the variances of the two populations under study are equal. This is illustrated by the following example.

Example 15.1.3 *Suppose that the two machines in Example 15.1.1 are of a different type. The first machine is brand new and works very precisely, while the second machine is several years old and becoming obsolete. Due to the high demand for steel cables, the second machine cannot be discarded, even though it works substantially less precisely. A study of historical data has shown that the variance of the thicknesses produced by machine 2 is nine times as large as the variance of the thicknesses produced by machine 1. As a result, $\sigma_2^2 = 9\sigma_1^2$.*

If $\sigma_2^2 = 9\sigma_1^2$, then the variance of $\overline{X}_1 - \overline{X}_2$ and the width of the confidence interval for $\mu_1 - \mu_2$ are

$$\frac{\sigma_1^2}{n_1} + \frac{9\sigma_1^2}{n_2} = \sigma_1^2 \left(\frac{1}{n_1} + \frac{9}{n_2} \right)$$

and

$$2z_{\alpha/2}\sqrt{\frac{\sigma_1^2}{n_1} + \frac{9\sigma_1^2}{n_2}} = 2z_{\alpha/2}\sigma_1\sqrt{\frac{1}{n_1} + \frac{9}{n_2}},$$

respectively. Using these expressions, we can now reexamine which of the 11 options for the collection of 12 observations is the best one. This time, it is the option with the smallest value of

$$\frac{1}{n_1} + \frac{9}{n_2}$$

that will lead to the smallest variance for $\overline{X}_1 - \overline{X}_2$ and to the narrowest confidence interval, for any value of the variance σ_1^2 and for any significance level α chosen. In Table 15.2, the 11 options are listed along with the corresponding variances of $\overline{X}_1 - \overline{X}_2$. When calculating the

Table 15.2 The evaluation of the different options for the sample sizes n_1 and n_2 when $\sigma_2^2 = 9\sigma_1^2$ and $\sigma_1^2 = 1$.

n_1	n_2	var($\overline{X}_1 - \overline{X}_2$)	Efficiency (%)	Width CI (%)
1	11	1.8182	73.3	116.77
2	10	1.4000	95.2	102.47
3	9	1.3333	100.0	100.00
4	8	1.3750	97.0	101.55
5	7	1.4857	89.7	105.56
6	6	1.6667	80.0	111.80
7	5	1.9429	68.6	120.71
8	4	2.3750	56.1	133.46
9	3	3.1111	42.9	152.75
10	2	4.6000	29.0	185.74
11	1	9.0909	14.7	261.12

variances, we assumed for simplicity that the variance of machine 1, σ_1^2, is equal to 1. The actual value of σ_1^2 does not impact the choice of data collection option.

Table 15.2 shows that a uniform distribution of the observations over the two populations or machines is no longer optimal. The option in which three observations are made for the first population (with the smaller variance) and nine observations are made for the second population (with the larger variance) is optimal and leads to a variance of $4/3 \approx 1.3333$ for the estimator $\overline{X}_1 - \overline{X}_2$. The worst possible option involves 11 observations from the first population and only one observation from the second population. That option leads to a variance of 9.0909 for $\overline{X}_1 - \overline{X}_2$. The efficiency of the worst possible option relative to the best one is $1.3333/9.0909 = 14.7\%$. The relative efficiency of the most intuitive approach, with $n_1 = n_2 = 6$, is $1.3333/1.6667 = 80\%$.

It is a waste of resources to use the most intuitive approach with $n_1 = n_2 = 6$ in the case of unequal variances. The variance of $\overline{X}_1 - \overline{X}_2$ then is equal to 1.6667. However, we can obtain variances smaller than 1.6667 with fewer observations. If, for example, we were to collect 10 observations with $n_1 = 2$ and $n_2 = 8$, then the variance of $\overline{X}_1 - \overline{X}_2$ would be equal to 1.6250. If we were to collect 10 observations with $n_1 = 3$ and $n_2 = 7$, then the variance would be equal to 1.6190. In other words, if we could be satisfied with a variance of 1.6667, it would not be necessary to make 12 observations. It would suffice to have 10 observations. This means that the intuitive approach wastes two observations. when $\sigma_2^2 = 9\sigma_1^2$. In applications where the collection of observations is time-consuming and expensive, this is highly undesirable.

15.2   Unequal Costs for the Observations

In the previous section, we assumed that the cost of an observation from the first population was equal to the cost of an observation from the second population. This is not always the case, as illustrated by the following example.

Example 15.2.1   *A medical study is performed to compare a new, experimental treatment for a specific type of malignant tumor to an existing treatment. Due to the high cost of research and development, the new treatment is twice as expensive as the existing one. The cost of the new treatment, called treatment 1, is €20 000. The cost of the existing treatment, which we call treatment 2, is €10 000. The total budget available for the study is €240 000.*

The study is designed to measure the extent to which the number of months that a patient survives is a result of his treatment, and to estimate the difference in mean survival times between the two treatments. Given the budget of €240 000, there are several possible ways in which to conduct the medical study. A first possible option is to use the first and second treatments for eight patients each. The overall cost of this approach is $8 \times 20\,000 + 8 \times 10\,000 = €240\,000$. With the available budget, this approach is just feasible. Alternatively, one could use the first treatment for 10 patients and the second treatment for four patients. The cost of this option is $10 \times 20\,000 + 4 \times 10\,000 = €240\,000$. In total, there are 11 different ways to fully use the available budget:

- $n_1 = 1$ *and* $n_2 = 22$: *one observation is made with the expensive treatment 1, while 22 observations are made with the less expensive treatment 2;*

- $n_1 = 2$ and $n_2 = 20$: *two observations are made with the expensive treatment 1, while 20 observations are made with the less expensive treatment 2;*
- ...
- $n_1 = 8$ and $n_2 = 8$: *eight observations are made with treatment 1 as well as treatment 2;*
- ...
- $n_1 = 11$ and $n_2 = 2$: *11 observations are made with the expensive treatment 1, while two observations are made with the less expensive treatment 2.*

For each of these methods, the total cost is $n_1 \times 20\,000 + n_2 \times 10\,000 = €240\,000$. The question that arises is which option is the best from a statistical point of view.

If we assume that the variance of the observations with the first treatment is equal to the variance of the observations with the second treatment, and that the variable under study (the number of months that a patient survives) is normally distributed, then we can use the expressions

$$\frac{\sigma^2}{n_1} + \frac{\sigma^2}{n_2} = \sigma^2 \left(\frac{1}{n_1} + \frac{1}{n_2} \right)$$

and

$$2z_{\alpha/2} \sqrt{\frac{\sigma^2}{n_1} + \frac{\sigma^2}{n_2}} = 2z_{\alpha/2}\sigma \sqrt{\frac{1}{n_1} + \frac{1}{n_2}}$$

for the variance of the estimator $\overline{X}_1 - \overline{X}_2$ and the width of the confidence interval for $\mu_1 - \mu_2$, respectively, to make a choice.

Table 15.3 shows the variances for the 11 possible options with a total cost of €240 000, when $\sigma^2 = \sigma_1^2 = \sigma_2^2 = 1$. The option involving seven observations for the first, expensive, and experimental treatment and 10 observations for the second, less expensive, existing treatment ($n_1 = 7$ and $n_2 = 10$) produces the smallest variance, namely 0.2429. However, there are two

Table 15.3 An evaluation of the different options for the sample sizes n_1 and n_2 for the medical study in Example 15.2.1 when $\sigma^2 = \sigma_1^2 = \sigma_2^2 = 1$.

n_1	n_2	var($\overline{X}_1 - \overline{X}_2$)	Efficiency (%)	Width CI (%)
1	22	1.0455	23.2	207.48
2	20	0.5500	44.2	150.49
3	18	0.3889	62.4	126.54
4	16	0.3125	77.7	113.44
5	14	0.2714	89.5	105.72
6	12	0.2500	97.1	101.46
7	10	0.2429	100.0	100.00
8	8	0.2500	97.1	101.46
9	6	0.2778	87.4	106.95
10	4	0.3500	69.4	120.05
11	2	0.5909	41.1	155.99

other options that result in an almost equally low variance, namely the option with $n_1 = 6$ and $n_2 = 12$ and the option with $n_1 = n_2 = 8$. The variance of the estimator of $\overline{X}_1 - \overline{X}_2$ for these methods is 0.2500. Their efficiency relative to the optimal approach is $0.2429/0.2500 = 97.1\%$.

A striking aspect of this example is that it is not optimal to perform as many observations as possible. The option with the largest number of observations is the one with $n_1 = 1$ and $n_2 = 22$. This option thus involves 23 observations in total, six more than the optimal option, which involves 17 observations. With a variance of 1.0455, the option with the largest number of observations is actually the worst of all. Its efficiency relative to the optimal approach is merely $0.2429/1.0455 = 23.2\%$. Using the maximum number of observations results in a confidence interval for $\mu_1 - \mu_2$ that is more than twice as wide as the interval produced by the optimal approach with $n_1 = 7$ and $n_2 = 10$.

The calculations in Table 15.3 show that there are three methods that are virtually equivalent from a statistical point of view. A possible argument in favor of the option with $n_1 = n_2 = 8$ is that it involves one additional observation with the new treatment. Therefore, this option results in a smaller variance for $\overline{X}_1$, and a slightly more precise estimate of μ_1 than the option with $n_1 = 7$ and $n_2 = 10$. In this study, it might indeed be more important to obtain a better estimate of μ_1 than of μ_2, because less knowledge will be available about the new treatment than about the existing one.

This chapter has given a taste of a particular area in statistics, namely the optimal design of experiments. This specialization deals with designing experiments and other studies in such a way that one or more (functions of) parameters can be estimated as precisely as possible. In other words, statistical design of experiments is about planning experiments or studies so that the resulting data contains as much information as possible. The "optimal design of experiments" approach is explained in great detail in the book *Optimal Design of Experiments: A Case Study Approach*, by Peter Goos and Bradley Jones.

16

Testing Equivalence

Not that the lavender produce passed on to Hopsom would necessarily be inferior; to the contrary, every effort would be made, as always, to ensure that all crops were perfect, and every manufacture flawless. It might well be that, nine times out of ten, there would be no difference anyone could tell between (for example) the lavender water bearing Hopsom's label and that which bore Rackham's.
(from *The Crimson Petal and the White*, Michel Faber, pp. 293–294)

In Chapter 3, the principles of hypothesis testing have been explained. Hypothesis tests use a null hypothesis and an alternative hypothesis. The alternative hypothesis is sometimes called the **research hypothesis** because it typically contains a statement that a researcher seeks to prove. For example, researchers often want to prove that "something new" is better than "something existing", that a particular product does not comply with a legal requirement, that a relationship does exist between two variables, and so on. In contrast, the null hypothesis generally involves the statement that there is no effect, no difference, or no relationship. The purpose of hypothesis tests is then to demonstrate that certain differences or deviations are statistically significant. Oftentimes, when performing a hypothesis test, the researcher hopes that the null hypothesis can be rejected.

The hypothesis tests that we have encountered in the previous chapters are generally intended to show that differences do exist. Sometimes, however, the null hypothesis in such a test cannot be rejected. In that case, the conclusion is that there is no significant difference. It is tempting to interpret this outcome as evidence for the absence of a difference. This interpretation is wrong, because hypothesis tests are based on the assumption that the null hypothesis is true. In other words, hypothesis tests are based on the assumption that there is no difference. Only if the evidence against the null hypothesis is sufficiently large is the null hypothesis rejected. That hypothesis tests are based on the null hypothesis can be seen from the fact that the p-values and the critical values are computed from the distribution of the test statistic when the null hypothesis is true.

For this reason, it can be inadvisable to use the phrase that a null hypothesis is accepted, because accepting the null hypothesis could be misinterpreted as claiming that the null hypothesis is true (see also Section 3.6.2 for a discussion on this topic). Instead of saying that you accept a null hypothesis, it is more correct to say that you cannot reject the null hypothesis.

Statistics with JMP: Hypothesis Tests, ANOVA and Regression, First Edition. Peter Goos and David Meintrup.
© 2016 John Wiley & Sons, Ltd. Published 2016 by John Wiley & Sons, Ltd.
Companion Website: http://www.wiley.com/go/goosandmeintrup/JMP

In many cases, there is even a wide range of null hypotheses that cannot be rejected. This is illustrated in Section 16.1 below. In summary, not rejecting the null hypothesis should be interpreted as concluding that the null hypothesis is one of the possible truths.

In some cases, however, the researcher's intention is to demonstrate that no difference (or no practically important difference) exists between two population means, between two population variances, or between one population mean and a specified value. In these cases, the research hypothesis, the statement that the researcher seeks to prove, is that there is no difference. As a result, it is the alternative hypothesis that should contain the statement that there is no difference. The resulting hypothesis test is called an **equivalence test**. The exact procedure for an equivalence test is discussed in detail in Section 16.2 for a scenario involving one population mean. Two scenarios involving two populations are discussed in Section 16.3. The next few examples describe typical scenarios where an equivalence test is desired.

Example 16.0.1 *A customer wants to order cables from Metal Inc., with an electrical resistance of* 50 Ω *(ohms). The customer asks Metal Inc. to demonstrate that the cables they produce actually do have, on average, an electrical resistance of* 50 Ω.

Example 16.0.2 *A pharmaceutical company has deployed an expensive production process for manufacturing a medical drug. After a major research and development effort, the company is now able to manufacture the drug in a cheaper way. The European Medicines Agency and the US Food and Drug Administration, however, require the company to demonstrate that both production processes lead to the same amount of the active ingredient in the drug.*

16.1 Shortcomings of Classical Hypothesis Tests

In classical hypothesis tests, there are two possible decisions: either the null hypothesis is rejected or not. If the null hypothesis is rejected, then this corresponds to an acceptance of the alternative hypothesis. In that case, the conclusion is that there is a significant difference. If the null hypothesis cannot be rejected, it is tempting to interpret this as a proof that there is no difference; that is, as a proof that the null hypothesis is true. The following example shows that this interpretation makes no sense.

Example 16.1.1 *In order to demonstrate that the cables they produce have a resistance of* 50 Ω, *the engineers of Metal Inc. carry out measurements of the electrical resistance of 40 randomly selected cables. The results of these measurements are shown in Table 16.1. The histogram, the box plot, and the descriptive statistics in Figure 16.1 tell us that the 40 measurements are all close to* 50 Ω. *For example, the 95% confidence interval for the mean resistance is* [49.3126, 50.5671]. *The variation in the data is due to two facts: (1) that the quality of the cables is somewhat variable (as a result of an imperfect manufacturing process, and small fluctuations in the composition of the metal used); and (2) that the measurement instrument for the electrical resistance may exhibit an error of about* 0.2 Ω.

The engineers also perform a classical two-tailed hypothesis test for a population mean with a significance level α *of 5% and the following null hypothesis and alternative hypothesis:*

$$H_0 : \mu = 50 \ \Omega,$$

Table 16.1 The measurements of the electrical resistance (in ohms) for the 40 cables in Example 16.1.1.

48.88	46.73	50.15	50.32
48.49	53.57	49.54	49.37
48.91	48.25	48.30	49.57
49.39	51.42	49.81	51.14
46.17	47.73	49.67	53.55
50.16	52.42	53.35	50.25
48.84	52.27	52.79	51.74
51.97	48.80	47.99	49.80
49.74	48.52	49.05	47.74
53.50	51.21	46.40	50.09

$$H_a : \mu \neq 50 \ \Omega.$$

This kind of hypothesis test was discussed in detail in Chapters 3 and 4. The JMP output for the two-tailed test is shown in Figure 16.2. The output indicates that the p-value for the two-tailed test is equal to 0.8472. This p-value is based on the test statistic

$$t = \frac{\bar{x} - 50}{s/\sqrt{n}} = \frac{49.9399 - 50}{1.96118/\sqrt{40}} = -0.1940.$$

The p-value is computed as

$$p = 2\,P(t_{n-1} > |t|) = 2\,P(t_{39} > 0.1940) = 2 \times 0.4236 = 0.8472.$$

Based on this p-value, the engineers cannot reject the null hypothesis. When they present their results to their department head, the engineers tell her that they have shown that the electrical resistance of the cables is, on average, 50 Ω.

However, the department head is not convinced. In the presence of the engineers, she performs some additional analyses. More specifically, she performs two additional classical two-tailed hypothesis tests. First, she tests the hypotheses

$$H_0 : \mu = 49.5 \ \Omega,$$

$$H_a : \mu \neq 49.5 \ \Omega.$$

Next, she tests the hypotheses

$$H_0 : \mu = 50.5 \ \Omega,$$

$$H_a : \mu \neq 50.5 \ \Omega.$$

The corresponding JMP outputs are shown in Figure 16.3. The outputs show that the p-values for the two additional two-tailed hypothesis tests are 0.1640 and 0.0786. In other words, the null hypothesis cannot be rejected in either of the two additional tests. The department head

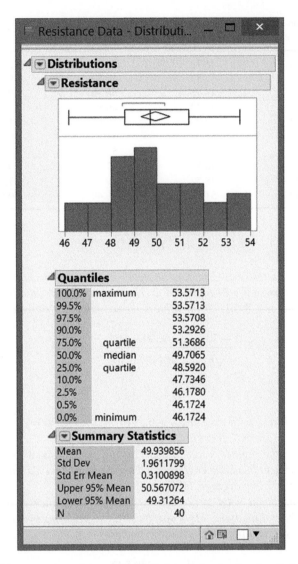

Figure 16.1 The histogram, box plot, and descriptive statistics for the electrical resistance of the 40 cables in Example 16.1.1.

concludes with the comment: "If I follow your reasoning, I have now shown that the electrical resistance is, on average, 49.5 Ω as well as that it is, on average, 50.5 Ω. So something's wrong with your method. Classical hypothesis tests are not designed to prove the null hypothesis. As you can see, there are always plenty of null hypotheses that cannot be rejected. What you need is an equivalence test, which is intended to show that the mean resistance is equal to or equivalent to the specified value, 50 Ω".

This example illustrates that it is wrong to blindly believe the null hypothesis if it cannot be rejected. There are many possible μ_0 values that can be specified in the null hypothesis and

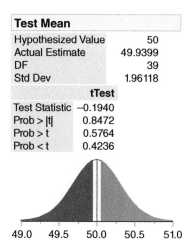

Figure 16.2 The JMP output for the first two-tailed hypothesis test in Example 16.1.1 with null hypothesis $\mu = 50 \, \Omega$.

for which the null hypothesis cannot be rejected. More precisely, the set of all μ_0 values for which the null hypothesis of a two-tailed test cannot be rejected is given by the confidence interval. In fact, if we were to use the lower limit 49.3126 or the upper limit 50.5671 of the 95% confidence interval as the value for μ_0 in the classical hypothesis test, then the p-value of the hypothesis test would be exactly 0.05. For all values of μ_0 between these confidence limits, the p-value will exceed 0.05, and, as a result, the null hypothesis cannot be rejected.

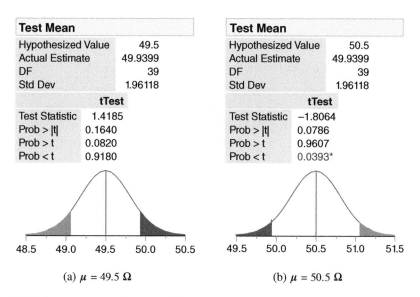

Figure 16.3 The JMP outputs for the additional two-tailed hypothesis tests with the null hypotheses $\mu = 49.5 \, \Omega$ and $\mu = 50.5 \, \Omega$.

The conclusion is that classical hypothesis tests are not suitable for demonstrating equality or equivalence; for example, to show that the mean electrical resistance of the cables produced is indeed equal to 50 Ω.

16.2   The Principles of Equivalence Tests

16.2.1   The Use of Two One-Sided Tests

The previous section made clear that we need an alternative method to demonstrate, for example, the true properties of electrical cables. Ideally, we would like to have a method by means of which we can prove that the resistance is 50 Ω. However, this is impossible. After all, if our measurement instrument were accurate enough, then we would always be able to detect some difference between the actual electrical resistance and the required 50 Ω.

This situation, however, is not as bad as it seems: instead of demonstrating that the resistance is exactly 50 Ω, the really important issue is to prove that there is no practically important deviation from this value. In other words, it is important to demonstrate that the deviation from 50 Ω is so small that it has no practical importance. The aim of the statistical analysis is then to show that the mean electrical resistance is located in a narrow interval around the value of 50 Ω. One way of writing down this interval is

$$[50 - \varepsilon, 50 + \varepsilon],$$

where ε is a small positive number that represents the maximal allowable deviation.

The hypotheses to be tested are then

$$H_0 : \mu \text{ does not belong to the interval } [50 - \varepsilon, 50 + \varepsilon]$$

and

$$H_a : \mu \text{ does belong to the interval } [50 - \varepsilon, 50 + \varepsilon],$$

or, alternatively,

$$H_0 : \mu \leq 50 - \varepsilon \quad \text{or} \quad \mu \geq 50 + \varepsilon$$

and

$$H_a : 50 - \varepsilon < \mu < 50 + \varepsilon.$$

Another alternative way of writing these hypotheses is

$$H_0 : |\mu - 50| \geq \varepsilon$$

and

$$H_a : |\mu - 50| < \varepsilon.$$

The null hypothesis states that the absolute value of the deviation from 50 Ω is greater than ε, while the alternative hypothesis is that the absolute deviation is less than or equal to ε.

There is no simple method to test these hypotheses all at once. Instead, the test is split into two separate one-sided hypothesis tests:

$$H_0 : \mu \geq 50 + \varepsilon, \qquad H_0 : \mu \leq 50 - \varepsilon,$$
$$H_a : \mu < 50 + \varepsilon, \qquad H_a : \mu > 50 - \varepsilon.$$

If the null hypotheses H_0 are rejected in both one-tailed tests, then it has been shown that $\mu < 50 + \varepsilon$ as well as that $\mu > 50 - \varepsilon$. In other words, the rejection of both null hypotheses leads to the conclusion that $50 - \varepsilon < \mu < 50 + \varepsilon$, and thus that the population mean is not substantially different from 50 Ω.

In general, we write the tested hypotheses as

$$H_0 : |\mu - \mu_0| \geq \varepsilon,$$
$$H_a : |\mu - \mu_0| < \varepsilon.$$

In practice, we test these hypotheses by combining a right-tailed and a left-tailed test. The hypotheses to be tested in this procedure are the following:

$$H_0 : \mu \geq \mu_0 + \varepsilon, \qquad H_0 : \mu \leq \mu_0 - \varepsilon,$$
$$H_a : \mu < \mu_0 + \varepsilon, \qquad H_a : \mu > \mu_0 - \varepsilon.$$

If H_0 is rejected in both one-tailed tests, it has been shown that $\mu_0 - \varepsilon < \mu < \mu_0 + \varepsilon$ and thus that the population mean μ is not substantially different from μ_0. We then say that the population mean μ is equivalent to the given value μ_0. In the example on the electrical resistance of cables, we would say that the mean resistance of the manufactured cables is equivalent to 50 Ω.

In the statistical jargon, this approach in which two one-sided tests are combined is known as **TOST**, which is an abbreviation of **two one-sided tests**.

Example 16.2.1 *In order to carry out a correct equivalence test, the engineers at Metal Inc. need to fix a value for ε, the maximum permissible difference between the mean electrical resistance of the cables and the value 50 Ω. Since the error in the measurement of the resistance is equal to 0.2 Ω, the engineers decide that 0.6 Ω (three times the measurement error) is a suitable value for ε. Consequently, the hypotheses tested are*

$$H_0 : \mu \geq 50.6, \qquad H_0 : \mu \leq 49.4,$$
$$H_0 : \mu < 50.6, \qquad H_a : \mu > 49.4.$$

The left-tailed hypothesis test yields a computed test statistic of

$$t = \frac{\bar{x} - 50.6}{s/\sqrt{n}} = \frac{49.9399 - 50.6}{1.96118/\sqrt{40}} = -2.1289.$$

The corresponding p-value is

$$p = P(t_{n-1} < t) = P(t_{39} < -2.1289) = 0.0198.$$

This p-value is smaller than the significance level of 5% used in Example 16.1.1, so that the null hypothesis of the first one-sided test can be rejected. The right-tailed hypothesis test yields a computed test statistic of

$$t = \frac{\bar{x} - 49.4}{s/\sqrt{n}} = \frac{49.9399 - 49.4}{1.96118/\sqrt{40}} = 1.7410.$$

The corresponding p-value is

$$p = P(t_{n-1} > t) = P(t_{39} > 1.7410) = 0.0448.$$

This p-value is also smaller than the significance level, so that the null hypothesis of the second one-sided test can also be rejected.

Based on the results of the two one-sided hypothesis tests, the engineers of Metal Inc. can conclude that the mean electrical resistance is not substantially different from 50 Ω. In other words, the mean electrical resistance is equivalent to 50 Ω.

To perform an equivalence test for one population mean in JMP, you have to carry out the two one-sided tests consecutively and specify the values $\mu_0 - \varepsilon = 49.4$ and $\mu_0 + \varepsilon = 50.6$ in the two null hypotheses. This leads to the outputs shown in Figure 16.4. The next step is to identify the relevant p-values in the outputs. In Figure 16.4, these p-values are marked with a rectangular box.

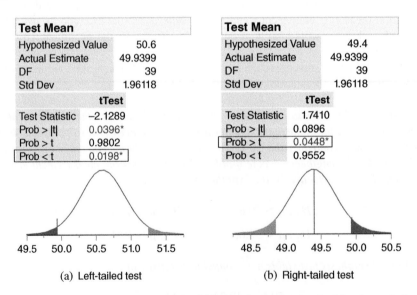

(a) Left-tailed test   (b) Right-tailed test

Figure 16.4 The JMP outputs for the two one-sided hypothesis tests required for the equivalence test in Example 16.2.1.

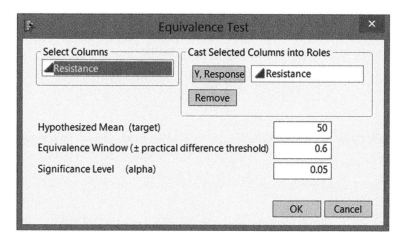

Figure 16.5 The input for the JMP script "Equivalence Test with a Standard" to carry out the equivalence test in Example 16.2.1.

To facilitate the implementation of the equivalence test described here for one population mean, a JMP script was created to automatically perform both one-sided tests, select the correct p-values, and make a decision regarding the equivalence. The script is called "Equivalence Test with a Standard". The input for the script is shown in Figure 16.5, while the additional output is shown in Figure 16.6.

16.2.2   The Use of a Confidence Interval

In Section 16.2.1, we argued that the purpose of an equivalence test is to demonstrate that the population mean μ lies in a narrow interval around the value μ_0. A general expression for this interval is

$$[\mu_0 - \varepsilon, \mu_0 + \varepsilon],$$

where ε is a small positive number and represents the maximal allowable deviation from μ_0.

Equivalence Test for mu = 50 ± 0.6	
Test	**P-value**
mu > 49.4	0.04479*
mu < 50.6	0.01982*
Mean is equivalent to 50	
(Both tests must be significant at the 0.05 level)	

Figure 16.6 The output of the JMP script "Equivalence Test with a Standard" for Example 16.2.1.

In the equivalence test, we first have to perform a left-tailed test to show that μ is smaller than $\mu_0 + \varepsilon$. To be able to draw this conclusion, the p-value of the test has to be smaller than the chosen significance level. Alternatively, the test statistic

$$\frac{\bar{x} - (\mu_0 + \varepsilon)}{s/\sqrt{n}}$$

has to be smaller than the critical value $-t_{\alpha;n-1}$. The latter condition can be rewritten as

$$\bar{x} + t_{\alpha;n-1}\frac{s}{\sqrt{n}} < \mu_0 + \varepsilon. \tag{16.1}$$

The second step of the equivalence test is to perform a right-tailed test to show that μ is larger than $\mu_0 - \varepsilon$. To reach this conclusion, the p-value of the test has to be smaller than the chosen significance level. Alternatively, the test statistic

$$\frac{\bar{x} - (\mu_0 - \varepsilon)}{s/\sqrt{n}}$$

has to be larger than the critical value $t_{\alpha;n-1}$. This condition can be rewritten as

$$\bar{x} - t_{\alpha;n-1}\frac{s}{\sqrt{n}} > \mu_0 - \varepsilon. \tag{16.2}$$

Now, the left-hand sides of Equations (16.1) and (16.2) together form a confidence interval for μ:

$$\left[\bar{x} - t_{\alpha;n-1}\frac{s}{\sqrt{n}} , \ \bar{x} + t_{\alpha;n-1}\frac{s}{\sqrt{n}}\right].$$

If this interval lies entirely between $\mu_0 - \varepsilon$ and $\mu_0 + \varepsilon$, then the conditions in Equations (16.1) and (16.2) are automatically satisfied and we can conclude that μ is equivalent to μ_0 at a significance level of α.

A special feature of the confidence interval is that it has a confidence level of $1 - 2\alpha$ (and thus a "significance level" of 2α). This can be seen from the values $-t_{\alpha;n-1}$ and $t_{\alpha;n-1}$ in the lower and upper limits of the interval. Therefore, to carry out an equivalence test with a significance level α, a confidence interval with a confidence level of $1 - 2\alpha$ (i.e., with a significance level 2α) has to be constructed.

Example 16.2.2 *In Example 16.2.1, an equivalence test was performed with a significance level of 5%. The values $\mu_0 - \varepsilon$ and $\mu_0 + \varepsilon$ in this example are 49.4 Ω and 50.6 Ω, while the sample mean $\bar{x}$ is equal to 49.9399 Ω.*

Since the significance level for the equivalence test is 5%, we determine a 90% confidence interval for the population mean μ. This is because, when the desired significance level for *the equivalence test is* $\alpha = 0.05$, *the required confidence level for the confidence interval is* $1 - 2\alpha = 1 - 2 \times 0.05 = 0.90$. *The 90% confidence interval is equal to*

$$\left[49.9399 - 1.6849 \cdot \frac{1.96118}{\sqrt{40}}, \ 49.9399 + 1.6849 \cdot \frac{1.96118}{\sqrt{40}} \right] = [49.4174, 50.4623].$$

This interval lies entirely between $\mu_0 - \varepsilon = 49.4$ *and* $\mu_0 + \varepsilon = 50.6$. *As a result, we can conclude that* μ *is equivalent to* μ_0.

16.3 An Equivalence Test for Two Population Means

Equivalence tests are often used to show that two population parameters are equivalent. Example 16.0.2 illustrates this in the context of a pharmaceutical company. In this section, we study the procedure for an equivalence test for two population means μ_1 and μ_2. An equivalence test for two population means can be performed using paired samples as well as independent samples. The distinction between paired samples and independent samples was explained in Chapter 7. Confidence intervals for $\mu_1 - \mu_2$ and classical hypothesis tests for paired samples both make use of a difference variable (see Chapters 10 and 11), while this is not the case for independent samples (see Chapters 8 and 9). Equivalence tests involving paired samples also employ a difference variable, while equivalence tests based on independent samples do not.

For both types of data, the hypotheses tested in an equivalence test for two population means are

$$H_0 : |\mu_1 - \mu_2| \geq \varepsilon,$$
$$H_a : |\mu_1 - \mu_2| < \varepsilon.$$

In practice, these hypotheses are again tested by combining a left-tailed and a right-tailed test. The null hypotheses and the alternative hypotheses for these two one-sided tests are the following:

$$H_0 : \mu_1 - \mu_2 \geq \varepsilon, \qquad H_0 : \mu_1 - \mu_2 \leq -\varepsilon,$$
$$H_a : \mu_1 - \mu_2 < \varepsilon, \qquad H_a : \mu_1 - \mu_2 > -\varepsilon.$$

The exact procedure for testing these hypotheses depends on the type of data. The value ε is sometimes referred to as the **minimal relevant difference**.

16.3.1 Independent Samples

In the case of independent samples, we need to distinguish between a scenario with equal variances and a scenario with unequal variances. In this section, we assume that the variances of the populations under study are equal. If this assumption is not justified, we need to slightly modify the procedure to take the two different variances into account (for an explanation on how to deal with unequal variances in a test for two population means, see Section 8.1.3).

The test statistic for the left-tailed alternative hypothesis in the context of the equivalence test is equal to

$$T = \frac{\overline{X}_1 - \overline{X}_2 - \varepsilon}{S_p \sqrt{\dfrac{1}{n_1} + \dfrac{1}{n_2}}}.$$

This test statistic is t-distributed with $n_1 + n_2 - 2$ degrees of freedom if $\mu_1 - \mu_2 = \varepsilon$, for samples from normally distributed populations or large samples from nonnormally distributed populations (see also the generalized hypothesis test for two population means with unknown, but equal variances in Section 8.1.3). In the expression for the test statistic T, S_p is the estimator of the common standard deviation $\sigma = \sigma_1 = \sigma_2$ of the two studied populations. This estimator is the square root of the **pooled estimator**

$$S_p^2 = \frac{(n_1 - 1)S_1^2 + (n_2 - 1)S_2^2}{n_1 + n_2 - 2}$$

of the common population variance $\sigma^2 = \sigma_1^2 = \sigma_2^2$. The p-value for the left-tailed alternative hypothesis is

$$p = P(t_{n_1+n_2-2} < t) = P\left(t_{n_1+n_2-2} < \frac{\overline{x}_1 - \overline{x}_2 - \varepsilon}{s_p \sqrt{\dfrac{1}{n_1} + \dfrac{1}{n_2}}}\right).$$

The test statistic for the right-tailed alternative hypothesis in the context of the equivalence test is

$$T = \frac{\overline{X}_1 - \overline{X}_2 - (-\varepsilon)}{S_p \sqrt{\dfrac{1}{n_1} + \dfrac{1}{n_2}}} = \frac{\overline{X}_1 - \overline{X}_2 + \varepsilon}{S_p \sqrt{\dfrac{1}{n_1} + \dfrac{1}{n_2}}}.$$

This test statistic is t-distributed with $n_1 + n_2 - 2$ degrees of freedom if $\mu_1 - \mu_2 = -\varepsilon$, for samples from normally distributed populations or large samples from nonnormally distributed populations. The p-value for the right-tailed alternative hypothesis is

$$p = P(t_{n_1+n_2-2} > t) = P\left(t_{n_1+n_2-2} > \frac{\overline{x}_1 - \overline{x}_2 - (-\varepsilon)}{s_p \sqrt{\dfrac{1}{n_1} + \dfrac{1}{n_2}}}\right),$$

$$= P\left(t_{n_1+n_2-2} > \frac{\overline{x}_1 - \overline{x}_2 + \varepsilon}{s_p \sqrt{\dfrac{1}{n_1} + \dfrac{1}{n_2}}}\right).$$

If both p-values are smaller than the chosen significance level α, the two alternative hypotheses can be accepted. In that case, the conclusion of the equivalence test is that μ_1 and μ_2 are equivalent.

Example 16.3.1   *A hotel manager wants to reward regular guests by offering them free wine when they dine in the hotel restaurant. The manager wants to investigate whether the guests notice a difference between French wine and Spanish wine. If no significant difference is observed, then the management may opt for the cheaper wine.*

In the study conducted by the hotel manager, 23 customers received French wine, while 19 customers were served Spanish wine. To quantify the level of customer satisfaction, the waiters recorded for how many minutes the 42 customers stayed in the restaurant. The manager now wants to demonstrate that neither of the wines outperforms the other in terms of customer satisfaction. In other words, the manager wants to prove that there is no relevant difference between the two types of wines in terms of the time that hotel guests spend in the restaurant. The manager decides that, to be meaningful, the difference in time spent in the restaurant should be at least 10 minutes. The significance level used is 5%. The descriptive statistics and graphical representations of the data are shown in Figure 16.7.

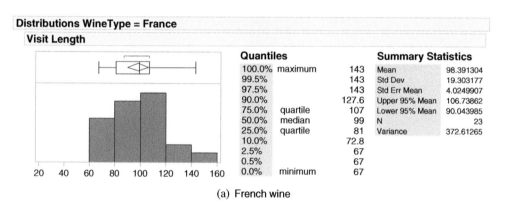

(a) French wine

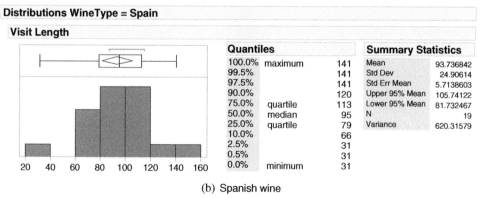

(b) Spanish wine

Figure 16.7   The descriptive statistics and graphical representations of the data in Example 16.3.1.

It is appropriate to perform an equivalence test with $\varepsilon = 10$. The hypotheses tested are then

$$H_0 : |\mu_1 - \mu_2| \geq 10,$$

$$H_a : |\mu_1 - \mu_2| < 10,$$

where μ_1 represents the mean time spent in the restaurant by hotel guests who were served French wine, and μ_2 represents the mean time spent in the restaurant by guests who received Spanish wine. Another way of writing the hypotheses is

$$H_0 : \mu_1 - \mu_2 \geq 10, \qquad\qquad H_0 : \mu_1 - \mu_2 \leq -10,$$

$$H_a : \mu_1 - \mu_2 < 10, \qquad\qquad H_a : \mu_1 - \mu_2 > -10.$$

The mean number of minutes spent in the restaurant is 98.4 for the French wine, while it is 93.7 for the Spanish wine. The variances are, respectively, 372.6 and 620.3 minutes^2. These values can be found in Figure 16.7. The pooled sample variance is therefore

$$s_p^2 = \frac{(23 - 1)372.6 + (19 - 1)620.3}{23 + 19 - 2} = 484.1,$$

so that the pooled estimate for the standard deviation, s_p, is $\sqrt{484.1} = 22$. The computed test statistic for the left-tailed test is then

$$\frac{98.4 - 93.7 - 10}{22\sqrt{\frac{1}{23} + \frac{1}{19}}} = -0.7837,$$

resulting in a p-value of

$$p = P(t_{n_1+n_2-2} < -0.7837) = 0.2189.$$

The computed test statistic for the right-tailed test is

$$\frac{98.4 - 93.7 - (-10)}{22\sqrt{\frac{1}{23} + \frac{1}{19}}} = \frac{98.4 - 93.7 + 10}{22\sqrt{\frac{1}{23} + \frac{1}{19}}} = 2.1485,$$

resulting in a p-value of

$$p = P(t_{n_1+n_2-2} > 2.1485) = 0.0189.$$

As the former p-value is larger than the significance level used, the manager cannot conclude that the wines are equivalent.

Instead of computing two p-values, it is also possible to determine a $(1 - 2\alpha) \times 100\%$ confidence interval for $\mu_1 - \mu_2$. If this interval lies entirely between $-\varepsilon$ and ε, then we can conclude that the two population means are equivalent. The required confidence interval is

$$\left[\bar{x}_1 - \bar{x}_2 - t_{\alpha;n_1+n_2-2}s_p\sqrt{\frac{1}{n_1} + \frac{1}{n_2}} \, , \, \bar{x}_1 - \bar{x}_2 + t_{\alpha;n_1+n_2-2}s_p\sqrt{\frac{1}{n_1} + \frac{1}{n_2}} \, \right].$$

Example 16.3.2   *The confidence interval required to test the equivalence in Example 16.3.1 is*

$$\left[98.4 - 93.7 - 1.68385 \times 22\sqrt{\frac{1}{23} + \frac{1}{19}} \, , \, 98.4 - 93.7 + 1.68385 \times 22\sqrt{\frac{1}{23} + \frac{1}{19}} \, \right],$$

which is equal to

$$[-6.8309, 16.1399].$$

This interval does not lie between -10 and 10. Consequently, we cannot conclude that the wines are equivalent.

Example 16.3.3   *If, in Examples 16.3.1 and 16.3.2, the hotel manager does not want to assume that the variances of the French and Spanish wines are equal, then he should continue to work with s_1^2 and s_2^2 as separate estimates for σ_1^2 and σ_2^2, instead of with the pooled estimate s_p^2. This would result in the p-values 0.0218 and 0.2249. Therefore, the conclusion of the equivalence test would remain unchanged: the Spanish and French wines are not equivalent. The confidence interval required to test the equivalence is $[-7.1683, 16.4772]$ when treating the variances σ_1^2 and σ_2^2 as unequal. This interval also does not lie between -10 and 10.*

The equivalence test under the assumption of equal variances σ_1^2 and σ_2^2 can be performed in JMP via the option "Fit Y by X" in the "Analyze" menu. Figure 16.8 shows how the variables "VisitLength" and "WineType" should be entered in the dialog window produced by selecting the option "Fit Y by X". The initial output is a plot of the data. The extra option "Equivalence Test" can then be chosen via the hotspot (red triangle) next to the term "Oneway Analysis of VisitLength By WineType". This is illustrated in Figure 16.9. In the resulting dialog window, shown in Figure 16.10, the value of ε needs to be entered. The final output, shown in Figure 16.11, contains the two p-values. The option "Equivalence Test" only produces the p-values, and not the corresponding confidence interval. Note that the test statistic values in Figure 16.11 have signs opposite to those that we obtained in Example 16.3.1. This is due to the fact that JMP works with the difference $\mu_2 - \mu_1$ rather than $\mu_1 - \mu_2$. Also, JMP automatically associates France with population 1 and thus with mean μ_1 because, alphabetically, France precedes Spain.

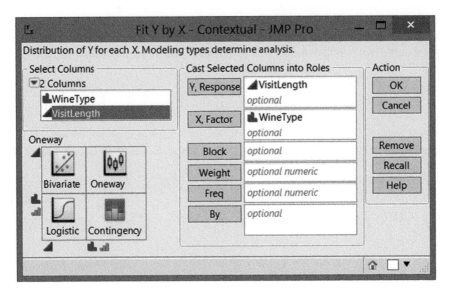

Figure 16.8 The input required for the option "Fit Y by X" in Example 16.3.1.

16.3.2 Paired Samples

In order to perform a hypothesis test for two population means based on paired observations, a difference variable is used. This was discussed in detail in Chapter 10. Table 16.2 visualizes the structure of the data in the case of paired samples as well as the way in which the difference variable D is computed.

Instead of calculating the sample variances for the two samples separately, the sample variance of the difference variable D has to be computed in the case of paired data. We call this sample variance S_D^2.

The test statistic for the left-tailed alternative hypothesis is then

$$T = \frac{\overline{X}_1 - \overline{X}_2 - \varepsilon}{S_D / \sqrt{n}}.$$

Table 16.2 The data structure in the case of paired samples, along with the calculation of the difference variable.

Observation	Sample 1	Sample 2	Difference
1	X_{11}	X_{21}	$D_1 = X_{11} - X_{21}$
2	X_{12}	X_{22}	$D_2 = X_{12} - X_{22}$
3	X_{13}	X_{23}	$D_3 = X_{13} - X_{23}$
$\vdots$	$\vdots$	$\vdots$	$\vdots$
n	X_{1n}	X_{2n}	$D_n = X_{1n} - X_{2n}$

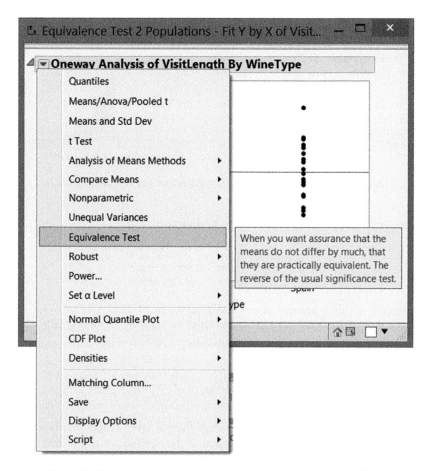

Figure 16.9   The option "Equivalence Test" for two populations.

This test statistic is t-distributed with $n - 1$ degrees of freedom if $\mu_1 - \mu_2 = \varepsilon$, for samples from normally distributed populations or large samples from nonnormally distributed populations. The p-value for the left-tailed alternative hypothesis is

$$p = P(t_{n-1} < t) = P\left(t_{n-1} < \frac{\bar{x}_1 - \bar{x}_2 - \varepsilon}{s_D/\sqrt{n}}\right).$$

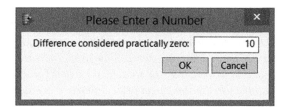

Figure 16.10   Entering the value of ε for the equivalence test in Example 16.3.1.

Practical Equivalence between Spain and France

Specified Practical Difference Threshold	10
Actual Difference in Means	−4.65446
Std Error of Difference	6.820906

Test	t Ratio	p-Value
Upper Threshold	−2.14846	0.0189*
Lower Threshold	0.783699	0.2189
Max over both		0.2189

Figure 16.11   The JMP output for the equivalence test in Example 16.3.1.

The test statistic for the right-tailed alternative hypothesis is

$$T = \frac{\overline{X}_1 - \overline{X}_2 - (-\varepsilon)}{S_D / \sqrt{n}} = \frac{\overline{X}_1 - \overline{X}_2 + \varepsilon}{S_D / \sqrt{n}}.$$

This test statistic is t-distributed with $n - 1$ degrees of freedom if $\mu_1 - \mu_2 = -\varepsilon$, for samples from normally distributed populations or large samples from nonnormally distributed populations. The p-value for the right-tailed alternative hypothesis is

$$p = P(t_{n-1} > t) = P\left(t_{n-1} > \frac{\overline{x}_1 - \overline{x}_2 + \varepsilon}{s_D / \sqrt{n}} \right).$$

If both p-values are smaller than the chosen significance level α, then the two alternative hypotheses can be accepted. The final conclusion of the equivalence test is then that μ_1 and μ_2 are equivalent.

Instead of computing the two p-values, it is also possible to construct a $(1 - 2\alpha) \times 100\%$ confidence interval and to check whether this interval lies between $-\varepsilon$ and ε. If this is the case, the conclusion is that μ_1 and μ_2 are equivalent. The required interval can be computed as

$$\left[\overline{X}_1 - \overline{X}_2 - t_{\alpha;n-1} \frac{S_D}{\sqrt{n}}, \overline{X}_1 - \overline{X}_2 + t_{\alpha;n-1} \frac{S_D}{\sqrt{n}} \right].$$

The next example describes a situation in which an equivalence test for paired samples is required.

Example 16.3.4   *In a small college town, there are two bookstores. The students believe that the prices charged by the two bookstores do not differ substantially. A diligent student randomly selects 50 textbooks, and records the selling price for each of them in the first bookstore and in the second bookstore. For the student, a price difference of €1.5 is relevant. The student can now examine whether the mean price difference is indeed smaller than €1.5, using an equivalence test for paired observations.*

17

The Estimation and Testing of Correlation and Association

The scene in the dining hall was the same as the day before – the mood, the voices, the faces. Only the menu had changed. The balding man in white,...., joined the three of us at our table and talked for a long time about the correlation of brain size to intelligence.

(from *Norwegian Wood*, Haruki Murakami, p. 194)

Many statistical analyses aim to study relationships between two or more variables. A particular branch of statistics is called multivariate statistics, because it deals with the study of several variables at the same time. In this chapter, we restrict ourselves to studying two variables simultaneously. In the book *Statistics with JMP: Graphs, Descriptive Statistics and Probability*, we introduced two statistics to quantify relationships between two quantitative variables. The first of these statistics was the Pearson correlation coefficient, which quantifies the extent to which there is a linear relationship between two variables. The second statistic was Spearman's rank correlation coefficient, which quantifies the extent to which there is a monotonically increasing or decreasing relation (possibly nonlinear) between two variables. In this chapter, we show how to perform hypothesis tests for the Pearson correlation coefficient as well as for Spearman's rank correlation coefficient. We also study the construction of confidence intervals for both correlation coefficients.

There also exist statistics that quantify the relationship between two qualitative variables. These statistics are beyond the scope of this book. However, we do discuss a hypothesis test that investigates whether there is a relation or association between two qualitative variables.

17.1 The Pearson Correlation Coefficient

In *Statistics with JMP: Graphs, Descriptive Statistics and Probability*, the sample correlation coefficient for two variables X and Y was defined as

$$R_{XY} = \frac{S_{XY}}{S_X S_Y} = \frac{\sum_{i=1}^{n}(X_i - \overline{X})(Y_i - \overline{Y})}{\sqrt{\sum_{i=1}^{n}(X_i - \overline{X})^2}\sqrt{\sum_{i=1}^{n}(Y_i - \overline{Y})^2}}.$$

Statistics with JMP: Hypothesis Tests, ANOVA and Regression, First Edition. Peter Goos and David Meintrup.
© 2016 John Wiley & Sons, Ltd. Published 2016 by John Wiley & Sons, Ltd.
Companion Website: http://www.wiley.com/go/goosandmeintrup/JMP

In this expression, S_{XY} represents the sample covariance between the two variables, while S_X and S_Y are the sample standard deviations of X and Y, respectively. The sample correlation coefficient is used as an estimator for a population correlation coefficient, which is indicated by the symbol ρ_{XY}. The correlation coefficient R_{XY} is also known as the Pearson correlation coefficient and takes values between -1 and $+1$.

There are two hypothesis tests for the Pearson correlation coefficient. The first test is intended to determine whether the correlation is different from zero. The second test is intended to find out whether the correlation coefficient is different from a given value other than zero. The execution of the tests requires both variables under study to be normally distributed.

In the remainder of this section, we use a shorthand notation for the sample and population correlation coefficients: for simplicity, we use R and ρ instead of R_{XY} and ρ_{XY}.

17.1.1   A Test for $\rho = 0$

A typical test for a Pearson correlation coefficient investigates whether or not it is different from zero. In that case, the null hypothesis is

$$H_0 : \rho = 0.$$

The possible alternative hypotheses are

$$H_a : \rho > 0,$$
$$H_a : \rho < 0,$$

and

$$H_a : \rho \neq 0.$$

We use these alternative hypotheses for a right-tailed, a left-tailed, and a two-tailed test, respectively.

The required test statistic for the execution of the tests is

$$T = R \sqrt{\frac{n-2}{1-R^2}}.$$

This test statistic is t-distributed with $n - 2$ degrees of freedom if the null hypothesis that $\rho = 0$ is correct and if X and Y both are normally distributed. The test statistic is approximately normally distributed for large samples of nonnormally distributed populations, if the null hypothesis is correct.

The sign of the test statistic is the same as that of the correlation coefficient. If the correlation coefficient is positive, the test statistic is also positive. If the correlation coefficient is negative, the test statistic is negative. The test statistic T is an increasing function of the correlation coefficient R. When the correlation coefficient tends towards 1, the test statistic goes to $+\infty$. When the correlation coefficient tends to -1, the test statistic goes to $-\infty$. If the correlation is zero, then the test statistic will also take the value 0.

If we denote the computed statistic by t, then, making use of the t-distribution, the p-value for the right-tailed test can be calculated as

$$p = P(t_{n-2} > t) = P\left(t_{n-2} > r\sqrt{\frac{n-2}{1-r^2}}\right).$$

This p-value will be small when the computed test statistic is strongly positive; that is, when the correlation coefficient is substantially larger than 0. For the left-tailed test, the p-value is

$$p = P(t_{n-2} < t) = P\left(t_{n-2} < r\sqrt{\frac{n-2}{1-r^2}}\right).$$

This p-value will be small when the computed test statistic is strongly negative; that is, when the correlation coefficient is substantially smaller than 0. For the two-tailed test, finally, the p-value is

$$p = 2\,P(t_{n-2} > |t|) = 2\,P\left(t_{n-2} > \left|r\sqrt{\frac{n-2}{1-r^2}}\right|\right).$$

This p-value will be small when the computed test statistic is strongly positive or strongly negative; that is, when the correlation coefficient is substantially larger than 0 or substantially smaller than 0.

If the p-value obtained is smaller than the chosen significance level α, then the null hypothesis can be rejected.

Example 17.1.1   *Suppose that you have the results of 10 students for exams in English and mathematics, and that you want to quantify the relationship between these two results. You want to determine whether a high (low) score in English is associated with a high (low) score in mathematics. In other words, you want to study whether there is a positive correlation between the results for the two subjects. You opt for a significance level of 5%. The data is shown in Table 17.1. The sample Pearson correlation coefficient for the two sets of results is $r = 0.8038$.*

Based on the sample correlation coefficient, you can now compute the test statistic to investigate whether or not the correlation is larger than zero. Therefore, you want to perform a right-tailed hypothesis test. The computed test statistic is

$$t = r\sqrt{\frac{n-2}{1-r^2}} = 0.8038\sqrt{\frac{10-2}{1-(0.8038)^2}} = 3.8216,$$

because $n = 10$. The p-value is therefore

$$p = P(t_{10-2} > t) = P\left(t_8 > 3.8216\right) = 0.0025.$$

This p-value is smaller than the significance level α of 5%, so you can reject the null hypothesis. As a conclusion, the correlation is significantly larger than zero.

Table 17.1   The results in English and mathematics for 10 students, along with the ranks based on these results.

Student	Result in English	Result in mathematics	Rank in English	Rank in mathematics
A	56	66	9	4
B	75	70	3	2
C	45	40	10	10
D	71	60	4	7
E	61	65	7	5
F	64	56	5	9
G	58	59	8	8
H	80	77	1	1
I	76	67	2	3
J	62	63	6	6

If you were to perform a two-tailed test in order to determine whether the correlation is different from zero, then the p-value would be

$$p = 2\,P(t_{10-2} > |t|) = 2\,P\left(t_8 > |3.8216|\right) = 0.005078.$$

This p-value is also small enough for the null hypothesis to be rejected.

The two-tailed test to determine whether a correlation is different from zero can also be performed in JMP. The data for the two variables under study has to be entered in two separate columns, as shown in Figure 17.1. The Pearson correlation coefficient can then be calculated using the menu "Analyze", where you have to choose "Multivariate Methods", followed

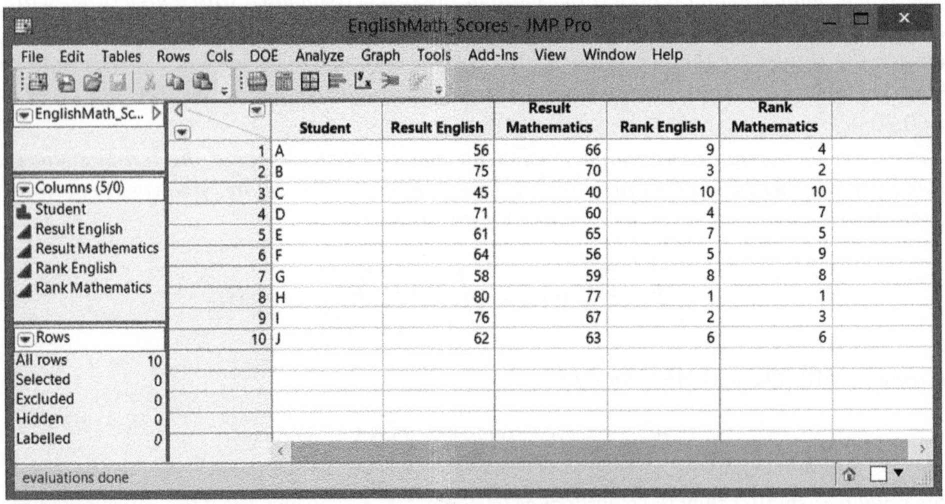

Figure 17.1   The JMP data table with results for English and mathematics in Example 17.1.1.

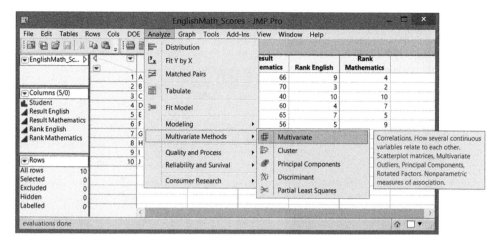

Figure 17.2 The determination of the Pearson correlation coefficient using JMP.

by "`Multivariate`". This is illustrated in Figure 17.2. In the resulting dialog window, you have to enter the names of the variables studied in the field "`Y, Columns`" (see Figure 17.3). JMP then finally shows a correlation matrix with the Pearson correlation coefficient for each pair of variables entered. This output is shown in Figure 17.4.

The last step for performing the two-sided hypothesis test in JMP is to use the hotspot (red triangle) next to the term "`Multivariate`" at the top of the output, and to activate the option "`Pairwise Correlations`". This is illustrated in Figure 17.5. The option "`Pairwise Correlations`" results in the additional output shown in Figure 17.6. This extra output contains the p-value for the two-tailed test in Example 17.1.1. If this p-value is divided by 2, then you obtain the p-value for the right-tailed test.

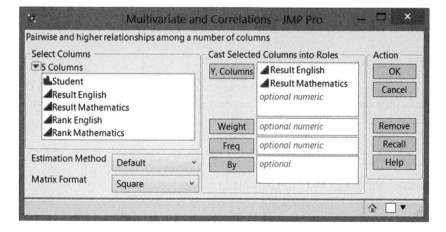

Figure 17.3 The dialog window for the calculation of the Pearson correlation coefficient.

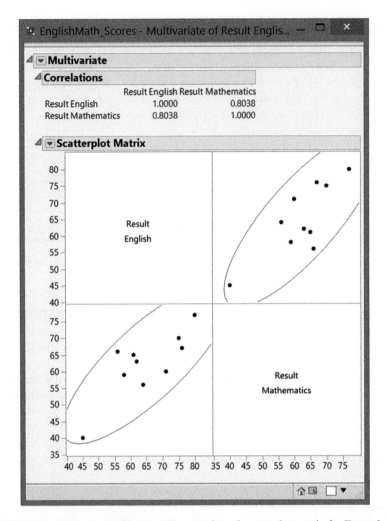

Figure 17.4 The JMP output with correlation matrix and scatterplot matrix for Example 17.1.1.

17.1.2 *A Test for $\rho = \rho_0 \neq 0$*

In some cases, one is not interested in finding out whether the correlation coefficient is different from zero, but whether it is different from a certain nonzero value, ρ_0. The corresponding null hypothesis is

$$H_0 : \rho = \rho_0.$$

The right-tailed, left-tailed, and two-tailed alternative hypotheses are

$$H_a : \rho > \rho_0,$$
$$H_a : \rho < \rho_0,$$

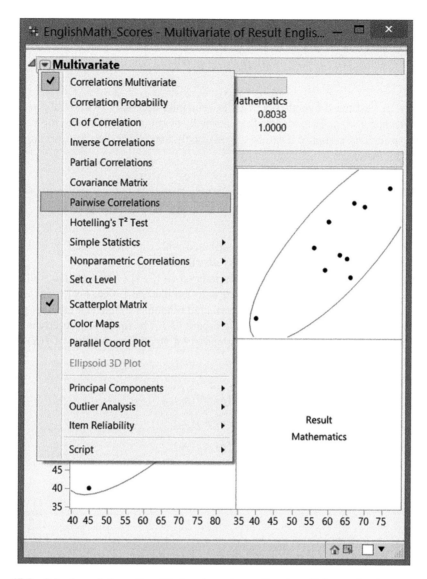

Figure 17.5 Selection of the option "Pairwise Correlations" for performing a two-tailed hypothesis test for the Pearson correlation coefficient and for the determination of a confidence interval for this coefficient. Note also the additional option "CI of Correlation" for determining a confidence interval.

Pairwise Correlations

Variable	by Variable	Correlation	Count	Lower 95%	Upper 95%	Signif Prob	−.8 −.6 −.4 −.2 0 .2 .4 .6 .8
Result Mathematics	Result English	0.8038	10	0.3526	0.9517	0.0051*	

Figure 17.6 The additional output with the result of the two-sided hypothesis test for the Pearson correlation coefficient in Example 17.1.1 and with the confidence interval calculated in Example 17.1.2.

and

$$H_a : \rho \neq \rho_0,$$

respectively. To test these hypotheses, it is, in principle, necessary to know the probability density of R when $\rho = \rho_0$. Even for data from normally distributed populations, this probability density is particularly complex. Therefore, one usually uses the Fisher transformation of the correlation coefficient R, which is given by

$$f(R) = \frac{1}{2} \ln \left(\frac{1+R}{1-R} \right). \tag{17.1}$$

This function is approximately normally distributed with expected value $f(\rho)$ and variance $1/(n-3)$ if the variables studied are normally distributed.

The Fisher transformation ensures that the interval $[-1, +1]$ (which contains all possible values of a correlation coefficient) is mapped onto the interval $]-\infty, +\infty[$ (which contains all of the possible values that a normally distributed random variable can take). A graphical representation of the function $f(R)$ is shown in Figure 17.7. The function is strictly increasing. A small (large) value of R therefore corresponds to a small (large) value of $f(R)$. We can therefore rewrite the hypotheses tested as

$$H_0 : f(\rho) = f(\rho_0),$$
$$H_a : f(\rho) > f(\rho_0),$$
$$H_a : f(\rho) < f(\rho_0),$$

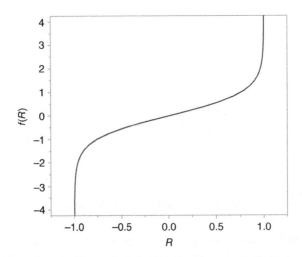

Figure 17.7 The Fisher transformation for a correlation coefficient R.

and

$$H_a : f(\rho) \neq f(\rho_0),$$

and use a z-test based on the standard normal distribution.

The Fisher transformation is necessary because the probability density of R is skewed when $\rho \neq 0$. The transformation results in a new random variable, $f(R)$, which does have a symmetrical distribution and that is approximately normally distributed.

The test statistic for the z-test is

$$Z = \frac{f(R) - f(\rho_0)}{\sqrt{\dfrac{1}{n-3}}}.$$

This test statistic is approximately standard normally distributed if the variables studied are normally distributed and the null hypothesis is true. If the correlation coefficient R takes values greater than ρ_0, then $f(R)$ takes a value greater than $f(\rho_0)$, and the test statistic is positive. If the correlation coefficient R takes values smaller than ρ_0, then $f(R)$ takes a value smaller than $f(\rho_0)$ and the test statistic is negative. If R is identical to ρ_0, the test statistic is zero.

The p-value for the right-tailed test is thus

$$p = P(Z > z) = P\left(Z > \frac{f(r) - f(\rho_0)}{\sqrt{\dfrac{1}{n-3}}}\right),$$

where z represents the computed test statistic. The p-value for the right-tailed test is small when the computed correlation coefficient r is considerably larger than ρ_0, because $f(r)$ is then much larger than $f(\rho_0)$ and z is strongly positive. The p-value for the left-tailed test is

$$p = P(Z < z) = P\left(Z < \frac{f(r) - f(\rho_0)}{\sqrt{\dfrac{1}{n-3}}}\right).$$

This p-value is small when the computed correlation coefficient r is considerably smaller than ρ_0, because $f(r)$ is then much smaller than $f(\rho_0)$ and z is strongly negative. Finally, the p-value for the two-tailed test is

$$p = 2 P(Z > |z|) = 2 P\left(Z > \left|\frac{f(r) - f(\rho_0)}{\sqrt{\dfrac{1}{n-3}}}\right|\right).$$

The p-value for the two-tailed test is small when the computed correlation coefficient r is considerably larger than ρ_0 and when the computed correlation coefficient r is considerably smaller than ρ_0. In both cases, $f(r)$ strongly differs from $f(\rho_0)$, and the absolute value of z is a large positive number.

If the p-value is smaller than the significance level α, then the null hypothesis is rejected in favor of the alternative hypothesis.

JMP does not explicitly offer an option to perform a hypothesis test for the case where ρ_0, the value for ρ specified in the null hypothesis, is different from zero. However, by using a confidence interval, one can implicitly perform such a hypothesis test.

17.1.3 The Confidence Interval

Using the fact that

$$Z = \frac{f(R) - f(\rho_0)}{\sqrt{\dfrac{1}{n-3}}}$$

is approximately standard normally distributed, a confidence interval can be established for the population Pearson correlation coefficient.

It follows from the previous expression that

$$P\left(-z_{\alpha/2} \leq \frac{f(R) - f(\rho)}{\sqrt{\dfrac{1}{n-3}}} \leq z_{\alpha/2} \right) = 1 - \alpha.$$

This expression allows us to derive a confidence interval for $f(\rho)$. The required procedure is similar to the one we used in Section 2.2.2 for determining a confidence interval for a population mean based on the standard normal distribution. This procedure initially yields the following confidence interval for $f(\rho)$:

$$P\left(f(R) - z_{\alpha/2}\sqrt{\frac{1}{n-3}} \leq f(\rho) \leq f(R) + z_{\alpha/2}\sqrt{\frac{1}{n-3}} \right) = 1 - \alpha. \qquad (17.2)$$

To derive a confidence interval for ρ instead of $f(\rho)$, we need the inverse function of the Fisher transformation. Starting from Equation (17.1), we obtain

$$2f(R) = \ln\left(\frac{1+R}{1-R} \right),$$

$$e^{2f(R)} = \frac{1+R}{1-R},$$

$$e^{2f(R)}(1-R) = 1+R,$$

$$e^{2f(R)} - Re^{2f(R)} = 1 + R,$$

$$e^{2f(R)} - 1 = Re^{2f(R)} + R,$$

$$= R(e^{2f(R)} + 1)$$

and

$$R = \frac{e^{2f(R)} - 1}{e^{2f(R)} + 1}.$$

The inverse of the Fisher transformation is thus given by

$$f^{-1}(x) = \frac{e^{2x} - 1}{e^{2x} + 1}.$$

To obtain the desired confidence interval for ρ, we have to apply this inverse function to the three terms of the inequality in Equation (17.2). This yields

$$P\left[f^{-1}\left(f(R) - z_{\alpha/2}\sqrt{\frac{1}{n-3}}\right) \leq \rho \leq f^{-1}\left(f(R) + z_{\alpha/2}\sqrt{\frac{1}{n-3}}\right)\right] = 1 - \alpha$$

and, finally,

$$P\left[\frac{e^{2\left(f(R)-z_{\alpha/2}\sqrt{\frac{1}{n-3}}\right)} - 1}{e^{2\left(f(R)-z_{\alpha/2}\sqrt{\frac{1}{n-3}}\right)} + 1} \leq \rho \leq \frac{e^{2\left(f(R)+z_{\alpha/2}\sqrt{\frac{1}{n-3}}\right)} - 1}{e^{2\left(f(R)+z_{\alpha/2}\sqrt{\frac{1}{n-3}}\right)} + 1}\right] = 1 - \alpha.$$

The $(1 - \alpha) \times 100\%$ confidence interval is thus

$$\left[\frac{e^{2\left(f(R)-z_{\alpha/2}\sqrt{\frac{1}{n-3}}\right)} - 1}{e^{2\left(f(R)-z_{\alpha/2}\sqrt{\frac{1}{n-3}}\right)} + 1}, \frac{e^{2\left(f(R)+z_{\alpha/2}\sqrt{\frac{1}{n-3}}\right)} - 1}{e^{2\left(f(R)+z_{\alpha/2}\sqrt{\frac{1}{n-3}}\right)} + 1}\right]. \tag{17.3}$$

Example 17.1.2   *In Example 17.1.1, we performed a hypothesis test for a Pearson correlation coefficient. The sample correlation coefficient in this example was $r = 0.8038$. We can now establish a 95% confidence interval for the population correlation coefficient ρ. Substituting all of the relevant information in Expression (17.3) initially leads to*

$$\left[\frac{e^{2\left(f(0.8038)-z_{0.025}\sqrt{\frac{1}{10-3}}\right)} - 1}{e^{2\left(f(0.8038)-z_{0.025}\sqrt{\frac{1}{10-3}}\right)} + 1}, \frac{e^{2\left(f(0.8038)+z_{0.025}\sqrt{\frac{1}{10-3}}\right)} - 1}{e^{2\left(f(0.8038)+z_{0.025}\sqrt{\frac{1}{10-3}}\right)} + 1}\right],$$

as $n = 10$ and $\alpha = 0.05$. Since

$$f(0.8038) = \frac{1}{2} \ln \left(\frac{1 + 0.8038}{1 - 0.8038} \right) = 1.1092$$

and $z_{0.025} = 1.96$, we obtain

$$\left[\frac{e^{2\left(1.1092 - 1.96\sqrt{1/7}\right)} - 1}{e^{2\left(1.1092 - 1.96\sqrt{1/7}\right)} + 1}, \frac{e^{2\left(1.1092 + 1.96\sqrt{1/7}\right)} - 1}{e^{2\left(1.1092 + 1.96\sqrt{1/7}\right)} + 1} \right],$$

$$\left[\frac{e^{2\times 0.3684} - 1}{e^{2\times 0.3684} + 1}, \frac{e^{2\times 1.8500} - 1}{e^{2\times 1.8500} + 1} \right],$$

$$\left[\frac{1.0892}{3.0892}, \frac{39.4478}{41.4478} \right],$$

and

$$[0.3526, 0.9517].$$

The interval in Example 17.1.2 can also be computed using JMP. This can be done using the option "`Pairwise Correlations`" (see Figure 17.5), which leads to the output shown in Figure 17.6. In the columns labeled "`Lower 95%`" and "`Upper 95%`", this piece of output reports the lower and upper limits of the computed confidence interval. Another possibility for computing the confidence interval in JMP is to click the option "`CI of Correlation`" in the menu shown in Figure 17.5.

If, for example, we want to perform a two-tailed test to verify whether the population correlation in Example 17.1.2 differs from 0.3, then the confidence interval can be used. Since the 95 % confidence interval does not include the value 0.3, we can conclude that the correlation is significantly different from 0.3, at a significance level of 5%.

17.2 Spearman's Rank Correlation Coefficient

When calculating Spearman's rank correlation coefficient $R_{XY}^{(s)}$ of two variables X and Y (where the superscript (s) refers to Spearman), two sets of ranks are used. One set of ranks comes from ranking all observations according to the first variable under study. The other set comes from ranking all observations according to the second variable under investigation. Finally, the Pearson correlation coefficient is computed for the two sets of ranks. This results in Spearman's rank correlation coefficient. In the case of ties, average ranks have to be used. For more details on the calculation of Spearman's rank correlation coefficient, see *Statistics with JMP: Graphs, Descriptive Statistics and Probability*.

In this section, we omit the subscript XY in the notation $R_{XY}^{(s)}$ and denote Spearman's rank correlation coefficient computed from a sample simply by $R^{(s)}$. We denote the population Spearman's rank correlation coefficient by $\rho^{(s)}$.

Example 17.2.1   *Table 17.1 contains the results of 10 students for an English and a mathematics exam, as well as the rankings of the students according to their results in English and in mathematics. The Pearson correlation coefficient for these two sets of ranks is 0.6606. Spearman's rank correlation coefficient for the results is therefore 0.6606.*

In the literature, different methods for testing Spearman's rank correlation coefficient are described. A distinction can be made between approximate and exact tests. The approximate tests use the t-distribution or the standard normal distribution, while the exact tests are based on the strictly correct probability distribution of the rank correlation coefficient.

17.2.1   The Approximate Test for $\rho^{(s)} = 0$

For large samples, an approximate test can be used based on the t-distribution or on the standard normal distribution. These two methods provide (slightly) different p-values. Instead of choosing one of the methods, it is sometimes recommended to use the average of the two different p-values as the final result for the test (see, e.g., J.H. Zar's 1999 book *Biostatistical Analysis*).

Based on the Standard Normal Distribution

If the variables studied are independent and if the population rank correlation coefficient $\rho^{(s)}$ is equal to 0, then the sample correlation coefficient $R^{(s)}$ is approximately normally distributed, with expected value 0 and variance $1/(n-1)$. Hence,

$$Z = \frac{R^{(s)} - 0}{\sqrt{\dfrac{1}{n-1}}} = \frac{R^{(s)}}{\sqrt{\dfrac{1}{n-1}}} = R^{(s)}\sqrt{n-1}$$

can be used as an approximately standard normally distributed test statistic. The right-tailed test involves the null hypothesis

$$H_0 : \rho^{(s)} = 0$$

and the alternative hypothesis

$$H_a : \rho^{(s)} > 0.$$

The corresponding p-value is

$$p = P(Z > z) = P\left(Z > r^{(s)}\sqrt{n-1}\right).$$

For the left-tailed test with alternative hypothesis

$$H_a : \rho^{(s)} < 0,$$

the *p*-value is

$$p = P(Z < z) = P\left(Z < r^{(s)}\sqrt{n-1}\right).$$

For the two-tailed test with the alternative hypothesis

$$H_a : \rho^{(s)} \neq 0,$$

the *p*-value is

$$p = 2\,P(Z > |z|) = 2\,P\left(Z > \left|r^{(s)}\sqrt{n-1}\right|\right).$$

Based on the *t*-Distribution

An alternative method uses the fact that

$$T = R^{(s)}\sqrt{\frac{n-2}{1-\left(R^{(s)}\right)^2}}$$

is approximately *t*-distributed with $n-2$ degrees of freedom if the null hypothesis that $\rho^{(s)} = 0$ is correct. This test statistic is identical to the one for the Pearson correlation coefficient in Section 17.1.1, except that R is replaced by $R^{(s)}$. The *p*-values can therefore be calculated in the same way as in Section 17.1.1. If we denote the computed test statistic by t, then the *p*-value for the right-tailed test is

$$p = P(t_{n-2} > t) = P\left(t_{n-2} > r^{(s)}\sqrt{\frac{n-2}{1-\left(r^{(s)}\right)^2}}\right).$$

For the left-tailed test, the *p*-value is

$$p = P(t_{n-2} < t) = P\left(t_{n-2} < r^{(s)}\sqrt{\frac{n-2}{1-\left(r^{(s)}\right)^2}}\right).$$

Finally, for the two-tailed test the *p*-value is

$$p = 2\,P(t_{n-2} > |t|) = 2\,P\left(t_{n-2} > \left|r^{(s)}\sqrt{\frac{n-2}{1-\left(r^{(s)}\right)^2}}\right|\right).$$

Example 17.2.2   *In Example 17.2.1, Spearman's rank correlation coefficient was computed to be* $r^{(s)} = 0.6606$ *for the data in Table 17.1. If we want to perform a two-tailed t-test with a* *significance level α of 5% to determine whether this correlation coefficient differs from zero, then the computed t-test statistic is*

$$t = r^{(s)} \sqrt{\frac{n-2}{1-\left(r^{(s)}\right)^2}} = 0.6606 \sqrt{\frac{10-2}{1-(0.6606)^2}} = 2.4888.$$

The corresponding p-value is

$$p = 2\, P(t_{10-2} > 2.4888) = 2\, P(t_8 > 2.4888) = 2 \times 0.018795 = 0.0376.$$

If we use the z-test, then the test statistic is

$$z = r^{(s)} \sqrt{n-1} = 0.6606 \sqrt{10-1} = 1.9818,$$

and the corresponding p-value

$$p = 2\, P(Z > |1.9818|) = 2 \times 0.02375 = 0.0475.$$

These p-values are both smaller than α, which leads to the conclusion that the rank correlation significantly differs from zero. The average p-value is $(0.0376 + 0.0475)/2 = 0.04255$.

JMP offers an option to perform the two-tailed t-test to determine whether the rank correlation differs from zero. The starting point for this test in JMP is a data table such as the one in Figure 17.1. The first few steps required to calculate Spearman's rank correlation coefficient are identical to those for the Pearson correlation coefficient. These steps are shown in Figures 17.2, 17.3, and 17.4. The last step required is to use the hotspot (red triangle) next to the term "Multivariate" at the top of the output in Figure 17.4, and to select "Spearman's ρ" under "Nonparametric Correlations". This is illustrated in Figure 17.8. This leads to the additional output shown in Figure 17.9. This output shows both the computed value 0.6606 for the rank correlation coefficient and the p-value 0.0376 for the two-tailed t-test. The output explicitly states that the approximate p-value must be interpreted with caution, as the sample size n is too small in this example. For such small numbers of observations, it is better to use an exact test.

17.2.2 The Exact Test for $\rho^{(s)} = 0$

For an exact test to determine whether a rank correlation is different from zero, it is necessary to find the exact probability distribution of the sample rank correlation coefficient $R^{(s)}$. This allows the computation of exact p-values and critical values for the rank correlation.

The Probability Distribution of the Sample Rank Correlation Coefficient

The determination of the probability distribution of the sample rank correlation coefficient $R^{(s)}$ starts from the null hypothesis that $\rho^{(s)} = 0$. If this null hypothesis is true, then any ranking of

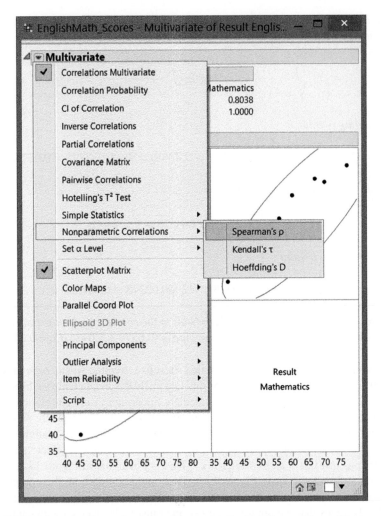

Figure 17.8 The selection of "Spearman's ρ" in the option "Nonparametric Correlations" in JMP for performing a two-tailed hypothesis test for Spearman's rank correlation coefficient.

Nonparametric: Spearman's ρ

Variable	by Variable	Spearman ρ	Prob>|ρ|	–.8 –.6 –.4 –.2 0 .2 .4 .6 .8
Result Mathematics	Result English	0.6606	0.0376*	

Warning: sample size of 10 is too small, P value suspect.

Figure 17.9 The additional output with the results of the two-tailed hypothesis test based on the t-distribution for Spearman's rank correlation coefficient in Example 17.2.2.

the n observations based on one variable, given a particular ranking of the observations based on the other variable, is equally likely.

If we have a total of n observations, there exist $n!$ possible rankings based on the first variable. For each of these rankings, there also exist $n!$ possible rankings based on the second variable. Each of these rankings is equally likely, and therefore has a probability of $1/n!$. This insight allows us to determine the probability distribution of $R^{(s)}$.

We illustrate this by means of an example with only four observations for the variables X and Y. In that case, there are $4! = 24$ possible rankings based on the second variable, for every rankings based on the first variable. The 24 rankings based on the second variable are shown in Table 17.2. In this table, we assume that there are no ties.

In the first scenario, the first observation has the smallest value for X, the second observation has the second smallest value for X, and so on. Therefore, in the first ranking, the first observation has rank 1, the second observation has rank 2, and so on. This explains the vector $(1,2,3,4)$ in the first row and the first column of Table 17.2.

Still in the first scenario, the first observation also has the smallest value for Y, the second observation has the second smallest value for Y, and so on. Therefore, in the second ranking,

Table 17.2 A list of the 24 possible rankings based on the variable Y, for the given ranking $(1,2,3,4)$ based on the variable X. For each scenario, the computed value $r^{(s)}$ of the rank correlation coefficient is given.

Ranking based on X	Ranking based on Y	Probability	$r^{(s)}$
$(1,2,3,4)$	$(1,2,3,4)$	$1/24$	1.0
$(1,2,3,4)$	$(1,2,4,3)$	$1/24$	0.8
$(1,2,3,4)$	$(1,3,2,4)$	$1/24$	0.8
$(1,2,3,4)$	$(1,3,4,2)$	$1/24$	0.4
$(1,2,3,4)$	$(1,4,2,3)$	$1/24$	0.4
$(1,2,3,4)$	$(1,4,3,2)$	$1/24$	0.2
$(1,2,3,4)$	$(2,1,3,4)$	$1/24$	0.8
$(1,2,3,4)$	$(2,1,4,3)$	$1/24$	0.6
$(1,2,3,4)$	$(2,3,1,4)$	$1/24$	0.4
$(1,2,3,4)$	$(2,3,4,1)$	$1/24$	-0.2
$(1,2,3,4)$	$(2,4,1,3)$	$1/24$	0.0
$(1,2,3,4)$	$(2,4,3,1)$	$1/24$	-0.4
$(1,2,3,4)$	$(3,1,2,4)$	$1/24$	0.4
$(1,2,3,4)$	$(3,1,4,2)$	$1/24$	0.0
$(1,2,3,4)$	$(3,2,1,4)$	$1/24$	0.2
$(1,2,3,4)$	$(3,2,4,1)$	$1/24$	-0.4
$(1,2,3,4)$	$(3,4,1,2)$	$1/24$	-0.6
$(1,2,3,4)$	$(3,4,2,1)$	$1/24$	-0.8
$(1,2,3,4)$	$(4,1,2,3)$	$1/24$	-0.2
$(1,2,3,4)$	$(4,1,3,2)$	$1/24$	-0.4
$(1,2,3,4)$	$(4,2,1,3)$	$1/24$	-0.4
$(1,2,3,4)$	$(4,2,3,1)$	$1/24$	-0.8
$(1,2,3,4)$	$(4,3,1,2)$	$1/24$	-0.8
$(1,2,3,4)$	$(4,3,2,1)$	$1/24$	-1.0

the first observation has rank 1, the second observation has rank 2, and so on. This explains the vector $(1, 2, 3, 4)$ in the first row and the second column of Table 17.2. It is clear that, in the first scenario, the two rankings match. As a result, in the first scenario, the rank correlation coefficient is equal to 1.

In the last scenario, the ranking according to the variable X is the exact opposite of the ranking according to the variable Y. Therefore, the rank correlation coefficient equals -1 in this scenario.

Table 17.2 makes clear that there are only 11 possible values for the rank correlation coefficient, if a sample with four observations is studied. Not all of these values are equally likely. The values -1 and 1, for example, appear only once, while the values -0.8 and 0.8 occur three times. The complete probability distribution for the values of the rank correlation coefficient is given in Table 17.3, together with the left- and right-tailed probabilities. The probability distribution for $R^{(s)}$ is perfectly symmetrical around 0.

It is important to emphasize that we would obtain exactly the same probability distribution for the rank correlation coefficient $R^{(s)}$ if we were to use a ranking other than $(1, 2, 3, 4)$ for the first variable X.

p-Values

For right-tailed and left-tailed tests, the left-tailed and right-tailed probabilities in Table 17.3 are the only possible exact p-values when there are no ties. The only exact p-values for a two-tailed test are twice the exceedance probabilities in Table 17.3.

As shown in Table 17.3, the smallest possible one-sided p-value for a sample with four observations is $1/24 \approx 0.04167$, which is just a little smaller than 5%. This p-value corresponds to a rank correlation of -1 or 1. This means that, for $n = 4$ and a significance level of 5%, the null hypothesis can only be rejected if the rank correlation is -1 or 1. If a significance level of 1% is used, then the null hypothesis can never be rejected with a sample of only four observations.

For a two-tailed test, the smallest possible p-value is $2 \times 1/24 \approx 0.0833$. Since this value is larger than 5%, a two-tailed test with $n = 4$ and a significance level of 5% will never result in a rejection of the null hypothesis. In order to be able to reject the null hypothesis in a two-tailed test at a significance level of 5%, larger numbers of observations are needed.

Table 17.3 The probability distribution for the rank correlation coefficient $R^{(s)}$ when $n = 4$ and $\rho^{(s)} = 0$, along with the left-tailed and right-tailed probabilities.

$r^{(s)}$	$P(R^{(s)} = r^{(s)})$	$P(R^{(s)} \leq r^{(s)})$	$P(R^{(s)} \geq r^{(s)})$
-1.0	$1/24$	$1/24$	1
-0.8	$3/24$	$4/24$	$23/24$
-0.6	$1/24$	$5/24$	$20/24$
-0.4	$4/24$	$9/24$	$19/24$
-0.2	$2/24$	$11/24$	$15/24$
0.0	$2/24$	$13/24$	$13/24$
0.2	$2/24$	$15/24$	$11/24$
0.4	$4/24$	$19/24$	$9/24$
0.6	$1/24$	$20/24$	$5/24$
0.8	$3/24$	$23/24$	$4/24$
1.0	$1/24$	1	$1/24$

In general, we compute the p-value for a right-tailed test as

$$p = P(R^{(s)} \geq r^{(s)}).$$

The p-value for a left-tailed test is

$$p = P(R^{(s)} \leq r^{(s)}),$$

and the p-value for a two-tailed test is

$$p = 2\, P\left(R^{(s)} \geq \left|r^{(s)}\right|\right).$$

If such a p-value is smaller than α, then we can reject the null hypothesis, and conclude that the rank correlation is significantly larger than, smaller than, or different from zero, depending on the type of hypothesis test conducted.

Critical Values

JMP does not compute exact p-values. Therefore, we have to rely on tables with critical values for the rank correlation coefficient if we want to perform an exact test. Perhaps the most extensive of such tables is in J.H. Zar's 1999 book *Biostatistical Analysis*. A more concise table of critical values is given in Appendix M.

Example 17.2.3   *In Example 17.2.2, we performed two different tests for the rank correlation coefficient of the data in Table 17.1. These tests led to the approximate p-values 0.0376 and 0.0475 for a rank correlation of 0.6606 and n = 10. The averaged approximate p-value was 0.04255.*

In the table in Appendix M, we can see that the critical values for a two-tailed test for the α-values 0.10, 0.05, 0.02, and 0.01 are equal to 0.564, 0.648, 0.745, and 0.794, respectively. The computed rank correlation coefficient of 0.6606 lies between 0.648 and 0.745. Consequently, the exact p-value lies between 0.02 and 0.05.

Both approximate p-values also lie between 0.02 and 0.05. This shows that the approximate p-values are certainly not completely wrong, not even for a sample of only 10 observations.

17.2.3   The Approximate Test for $\rho^{(s)} = \rho_0^{(s)} \neq 0$

In some situations, one does not want to test whether a population rank correlation coefficient $\rho^{(s)}$ is different from zero but, rather, whether it is different from a certain nonzero value $\rho_0^{(s)}$. The corresponding null hypothesis is then

$$H_0 : \rho^{(s)} = \rho_0^{(s)}.$$

The right-tailed, left-tailed, and two-tailed alternative hypotheses are

$$H_a : \rho^{(s)} > \rho_0^{(s)},$$
$$H_a : \rho^{(s)} < \rho_0^{(s)},$$

and

$$H_a : \rho^{(s)} \neq \rho_0^{(s)},$$

respectively. To perform the tests for these hypotheses, we need the Fisher transformation, just as in Section 17.1.2. The Fisher transformation of the rank correlation coefficient $R^{(s)}$ is

$$f(R^{(s)}) = \frac{1}{2} \ln \left(\frac{1 + R^{(s)}}{1 - R^{(s)}} \right). \tag{17.4}$$

This function is approximately normally distributed, with expected value $f(\rho^{(s)})$ and variance $1.060/(n-3)$ if $n \geq 10$ and $\rho^{(s)} \leq 0.9$, so that a z-test based on the standard normal distribution can be used.

The test statistic in this z-test is

$$Z = \frac{f(R^{(s)}) - f(\rho_0^{(s)})}{\sqrt{\dfrac{1.060}{n-3}}}. \tag{17.5}$$

The p-value is then computed in the same way as in Section 17.1.2.

The only difference between the approach for the Pearson correlation coefficient and the one for Spearman's rank correlation coefficient is that a slightly larger variance is used for the Fisher transformation of the rank correlation coefficient: $1.060/(n-3)$ instead of $1/(n-3)$.

17.2.4   The Confidence Interval

Just as for the Pearson correlation coefficient, one can use the Fisher transformation to establish a confidence interval for Spearman's rank correlation coefficient. As the random variable Z in Equation (17.5) is approximately standard normally distributed, we have

$$P \left(-z_{\alpha/2} \leq \frac{f(R^{(s)}) - f(\rho^{(s)})}{\sqrt{\dfrac{1.060}{n-3}}} \leq z_{\alpha/2} \right) = 1 - \alpha.$$

As in Section 17.1.3, we can now derive the $(1 - \alpha) \times 100\%$ confidence interval for $\rho^{(s)}$:

$$\left[\frac{e^{2\left(f(R^{(s)}) - z_{\alpha/2}\sqrt{\frac{1.060}{n-3}} \right)} - 1}{e^{2\left(f(R^{(s)}) - z_{\alpha/2}\sqrt{\frac{1.060}{n-3}} \right)} + 1}, \; \frac{e^{2\left(f(R^{(s)}) + z_{\alpha/2}\sqrt{\frac{1.060}{n-3}} \right)} - 1}{e^{2\left(f(R^{(s)}) + z_{\alpha/2}\sqrt{\frac{1.060}{n-3}} \right)} + 1} \right]. \tag{17.6}$$

For the use of this interval, it is also required that $n \geq 10$ and $\rho^{(s)} \leq 0.9$.

Example 17.2.4 *In Examples 17.2.2 and 17.2.3, we performed a hypothesis test for Spearman's rank correlation coefficient. The sample rank correlation coefficient in these examples was $r^{(s)} = 0.6606$. We can now compute a 95% confidence interval for the corresponding population rank correlation coefficient $\rho^{(s)}$. Substituting all relevant information in Expression (17.6) initially yields*

$$
\left[\frac{e^{2\left(f(0.6606)-z_{0.025}\sqrt{\frac{1.060}{10-3}}\right)} - 1}{e^{2\left(f(0.6606)-z_{0.025}\sqrt{\frac{1.060}{10-3}}\right)} + 1} \;,\; \frac{e^{2\left(f(0.6606)+z_{0.025}\sqrt{\frac{1.060}{10-3}}\right)} - 1}{e^{2\left(f(0.6606)+z_{0.025}\sqrt{\frac{1.060}{10-3}}\right)} + 1} \right].
$$

As

$$
f(0.6606) = \frac{1}{2} \ln\left(\frac{1 + 0.6606}{1 - 0.6606}\right) = 0.7939
$$

and $z_{0.025} = 1.96$, the confidence interval can be calculated as

$$
\left[\frac{e^{2\left(0.7939-1.96\sqrt{1.060/7}\right)} - 1}{e^{2\left(0.7939-1.96\sqrt{1.060/7}\right)} + 1} \;,\; \frac{e^{2\left(0.7939+1.96\sqrt{1.060/7}\right)} - 1}{e^{2\left(0.7939+1.96\sqrt{1.060/7}\right)} + 1} \right],
$$

$$
\left[\frac{e^{2\times0.0312} - 1}{e^{2\times0.0312} + 1} \;,\; \frac{e^{2\times1.5566} - 1}{e^{2\times1.5566} + 1} \right],
$$

$$
\left[\frac{0.0643}{2.0643} \;,\; \frac{21.4924}{23.4924} \right],
$$

and, finally,

$$
[0.0312, 0.9149].
$$

17.3 A Test for the Independence of Two Qualitative Variables

In some studies, a set of n sample observations is arranged into two types of classes based on two qualitative variables. In this section, we examine whether these two qualitative variables are independent of each other. To this end, we use a test based on the χ^2-distribution. This test can be regarded as a counterpart of the test in Section 17.1 for a correlation of zero for normally distributed data, and of the test in Section 17.2 for a rank correlation of zero for nonnormally distributed data.

17.3.1 The Contingency Table

The classification of n observations into two types of classes can easily be summarized using a **contingency table**. Sometimes, the terms "cross-tabulation" or "$r \times c$ table" are used, where r is the number of rows in the table and c is the number of columns. A contingency table

Table 17.4   The contingency table for the data set of the Spanish red wines in Example 17.3.1.

Price category	Rating				
	E	G/E	G	F/G	Total
Cheap (<€ 6)	2	1	7	21	31
Moderately priced	1	3	5	9	18
Expensive (≥ €10)	0	1	4	5	10
Total	3	5	16	35	59

is generally used for nominal and ordinal data, but it can also be used for quantitative data, provided that categories are defined for the quantitative variables under study.

Example 17.3.1   *Based on data concerning Spanish red wines described in Example 2.1 of* Statistics with JMP: Graphs, Descriptive Statistics and Probability, *a contingency table can be compiled for the variables Rating and Price. Rating is an ordinal variable, but Price is a quantitative variable, so that classes have to be defined. Suppose that three price categories are used: cheap (< € 6), moderately priced, and expensive (≥ € 10). The resulting contingency table for this data is presented in Table 17.4. The variable Rating takes the values E, G/E, G, and F/G, which signify the qualities excellent, good/excellent, good, and fair/good.*

A contingency table can be created in JMP using the "Fit Y by X" platform in the "Analyze" menu. The corresponding dialog window is shown in Figure 17.10. In this dialog

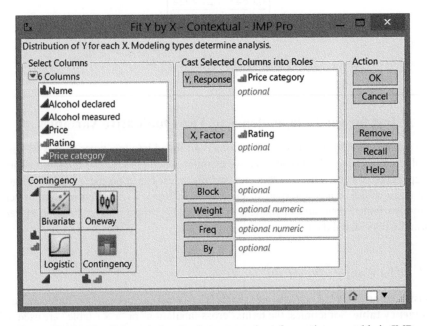

Figure 17.10   The dialog window for the construction of a contingency table in JMP.

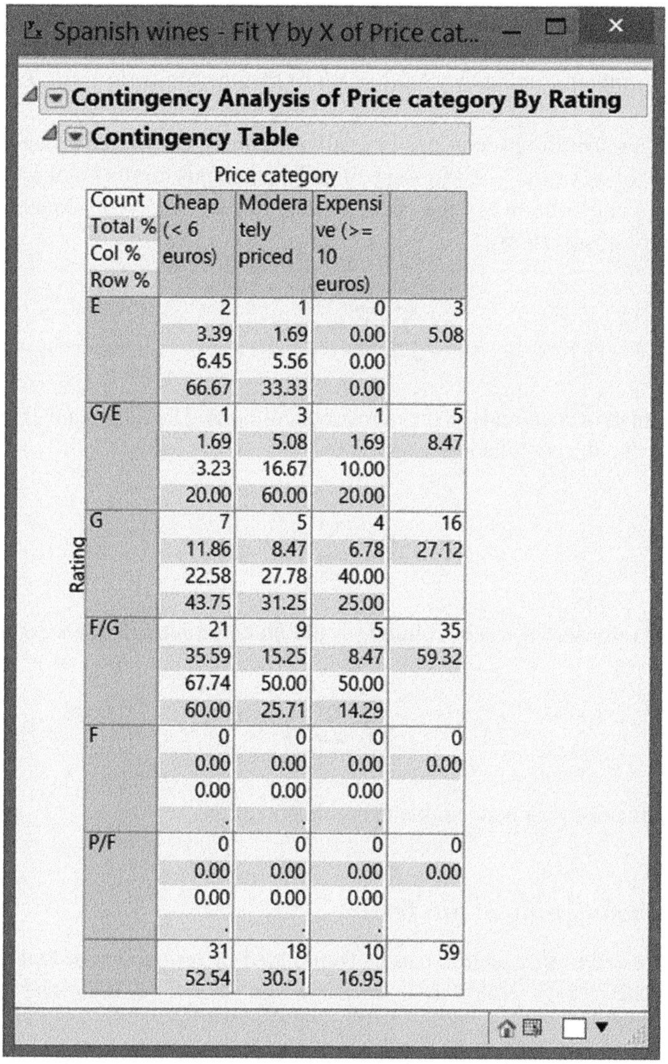

Count Total % Col % Row %	Cheap (< 6 euros)	Moderately priced	Expensive (>= 10 euros)	
E	2	1	0	3
	3.39	1.69	0.00	5.08
	6.45	5.56	0.00	
	66.67	33.33	0.00	
G/E	1	3	1	5
	1.69	5.08	1.69	8.47
	3.23	16.67	10.00	
	20.00	60.00	20.00	
G	7	5	4	16
	11.86	8.47	6.78	27.12
	22.58	27.78	40.00	
	43.75	31.25	25.00	
F/G	21	9	5	35
	35.59	15.25	8.47	59.32
	67.74	50.00	50.00	
	60.00	25.71	14.29	
F	0	0	0	0
	0.00	0.00	0.00	0.00
	0.00	0.00	0.00	
	.	.	.	
P/F	0	0	0	0
	0.00	0.00	0.00	0.00
	0.00	0.00	0.00	
	.	.	.	
	31	18	10	59
	52.54	30.51	16.95	

Figure 17.11 The initial contingency table created by JMP.

window, you need to enter the variable Price category in the field "`Y, Response`*", and the variable Rating in the field "*`X, Factor`*". This produces the output in Figure 17.11. The cells in this table contain four numbers: the absolute frequency for each cell and three relative frequencies. The number* 2 *in the first cell of the table tells us that there are two cheap wines with rating excellent (E). The number* 3.39 *says that* 3.39% *of all* 59 *wines are cheap and were rated as excellent (*$3.39\% = 2/59$*). The number* 6.45 *tells us that* 6.45% *of all* 31 *cheap wines are excellent (*$6.45\% = 2/31$*). Finally, the number* 66.67 *says that* 66.67% *of the three excellent wines are cheap (*$66.67\% = 2/3$*). The last row and the last column of the contingency table contain the column totals and the row totals, respectively, as well as the relative frequency of each price category and of each rating.*

The initial contingency table produced by JMP can be simplified by using the hotspot (red triangle) next to the term "`Contingency Table`*" and unchecking some options. In this way, one can obtain a simpler contingency table, such as the one shown in Table 17.4.*

In general, the number of observations in a cell of a contingency table is denoted by the first letter of the word "observation", O. More specifically, O_{ij} stands for the number of observations in the cell in row i and column j of the table. The sum of all observed frequencies is equal to the total number of observations:

$$\sum_{i=1}^{r}\sum_{j=1}^{c}O_{ij} = n,$$

where r is the number of rows and c is the number of columns. The total number of observations in row i is called the ith row total and is given by

$$O_{i.} = \sum_{j=1}^{c}O_{ij}.$$

The total number of observations in column j is the jth column total and is computed as

$$O_{.j} = \sum_{i=1}^{r}O_{ij}.$$

Table 17.5 is a general contingency table using this notation.

17.3.2   The Functioning of the Test

The hypotheses tested in this section can be formulated in various ways. One formulation in words is as follows:

$$H_0 : \text{The two qualitative variables studied are independent}$$

Table 17.5   A general contingency table with r rows and c columns, indicating row totals and column totals.

	Column				
Row	1	2	...	c	Total
1	O_{11}	O_{12}	...	O_{1c}	$O_{1.}$
2	O_{21}	O_{22}	...	O_{2c}	$O_{2.}$
$\vdots$	$\vdots$	$\vdots$	$\ddots$	$\vdots$	$\vdots$
r	O_{r1}	O_{r2}	...	O_{rc}	$O_{r.}$
Total	$O_{.1}$	$O_{.2}$	...	$O_{.c}$	n

and

$$H_a : \text{The two qualitative variables studied are dependent.}$$

To perform this test, we use insights from probability theory concerning joint discrete probability distributions and marginal probability distributions discussed in *Statistics with JMP: Graphs, Descriptive Statistics and Probability*. Two discrete random variables X and Y are independent if their joint probability distribution can be determined from their marginal distributions. More specifically, for independent discrete random variables, we have

$$p_{XY}(x, y) = p_X(x)p_Y(y)$$

for each combination of values x and y. Applying this definition to the problem that we study here, namely the classification of n observations into r rows and c columns, we can rewrite the joint probability as

$$P(\text{observation is in row } i \text{ and column } j) = P(\text{observation is in row } i)$$
$$\times P(\text{observation is in column } j), \quad (17.7)$$

for each cell in the contingency table.

According to the frequentist definition of probability (see also *Statistics with JMP: Graphs, Descriptive Statistics and Probability*), we can estimate the probability of an event by repeating a random experiment n times and calculating the relative frequency of the event. This means that the joint probability that a random observation is located in the cell in row i and column j of the contingency table is

$$P(\text{observation is in row } i \text{ and column } j) = \frac{\text{number of observations in the cell in row } i \text{ and column } j}{\text{total number of observations}},$$
$$= \frac{O_{ij}}{n}.$$

The marginal probability that a random observation is located in a cell in row i of the contingency table is

$$P(\text{observation is in row } i) = \frac{\text{number of observations in row } i}{\text{total number of observations}},$$
$$= \frac{O_{i.}}{n}.$$

The marginal probability that a random observation is in a cell in column j of the contingency table is

$$P(\text{observation is in column } j) = \frac{\text{number of observations in column } j}{\text{total number of observations}},$$
$$= \frac{O_{.j}}{n}.$$

These new expressions for the joint probability $p_{XY}(x, y)$ and the marginal probabilities $p_X(x)$ and $p_Y(y)$ allow us to rewrite Equation (17.7) as

$$\frac{O_{ij}}{n} = \frac{O_{i.}}{n} \times \frac{O_{.j}}{n}$$

and

$$O_{ij} = \frac{O_{i.} \, O_{.j}}{n}.$$

The left-hand side of this equality is the actual observed number of observations in the cell in row i and column j in the contingency table. The right-hand side is the expected number of observations in this cell if the two variables studied are independent. We call this expected value E_{ij}. In other words, we define E_{ij} as

$$E_{ij} = \frac{O_{i.} \, O_{.j}}{n}.$$

This line of reasoning allows us to reformulate the null hypothesis and the alternative hypothesis as

$$H_0 : O_{ij} = E_{ij} \text{ for each value of } i \ (i = 1, 2 \ldots, r) \text{ and } j \ (j = 1, 2, \ldots, c)$$

and

$$H_a : O_{ij} \neq E_{ij} \text{ for at least one cell of the contingency table.}$$

The test statistic for testing these hypotheses is similar to one from Section 6.1, where we investigated whether a sample came from a population with a given discrete probability distribution. To test the independence of two qualitative variables, the test statistic is

$$\chi = \sum_{i=1}^{r} \sum_{j=1}^{c} \frac{(O_{ij} - E_{ij})^2}{E_{ij}}.$$

This test statistic is approximately χ^2-distributed with $(r - 1)(c - 1)$ degrees of freedom if the null hypothesis is correct. If the null hypothesis is correct, the random variables O_{ij} and E_{ij} will take values that are only marginally different, and all differences $O_{ij} - E_{ij}$ will be close to zero. In that case, all squares $(O_{ij} - E_{ij})^2$ will be small positive values, and the test statistic χ will also be a small positive value.

If the null hypothesis is incorrect, then the two studied random variables are dependent and, as a result,

$$O_{ij} \neq \frac{O_{i.} \, O_{.j}}{n}$$

and $O_{ij} \neq E_{ij}$ for at least one cell in the contingency table. In that case, there will be a substantial difference between the values taken by O_{ij} and by E_{ij} for at least one cell in the contingency table. This leads to a large positive value for the corresponding square $(O_{ij} - E_{ij})^2$ and thus also to a large positive value for the test statistic χ.

Large values for the test statistic χ therefore indicate dependence between the variables studied (and thus point in the direction of the alternative hypothesis), while small values of χ indicate independence (and thus point towards the null hypothesis).

The p-value for the χ^2-test can generally be computed as

$$p = P\left(\chi^2_{(r-1)(c-1)} > x\right),$$

where x is the computed value of the test statistic χ. A small p-value leads to a rejection of the null hypothesis. In order to decide whether or not the null hypothesis should be rejected, one can also make use of a critical value. The critical value needed here is $\chi^2_{\alpha;(r-1)(c-1)}$. If the computed value x of the test statistic χ exceeds this critical value, then it is large enough for the null hypothesis to be rejected and for us to conclude that the two variables under study are dependent.

The p-value can be computed in JMP using the command "$1 -$ ChiSquare Distribution$(x, (r-1)(c-1))$", while the critical value can be obtained using "ChiSquare Quantile$(1 - \alpha, (r-1)(c-1))$".

Example 17.3.2   *We illustrate the χ^2-test for independence based on a data table on the type of car owned by 303 drivers. Figure 17.12 shows a contingency table generated using JMP for the qualitative variables Country and Size, representing the country of origin and the size of the cars. The contingency table contains three rows and three columns because there are three regions (America, Europe, and Japan) and three different sizes. In this example, we therefore have $r = c = 3$. Moreover, $n = 303$.*

We can now wonder whether there is a relation between the size of a car and its origin. To answer this question, it is appropriate to test the following hypotheses:

$$H_0 : \text{The variables Country and Size are independent}$$

and

$$H_a : \text{The variables Country and Size are dependent.}$$

To compute the test statistic, we have to know the value o_{ij} for each cell, the value $o_{i\cdot}$ for each row, and the value $o_{\cdot j}$ for each column. All these values are reported by default in JMP. The value o_{11}, for example, is equal to 36 and represents the number of American cars that are large. The value o_{12} is equal to 4 and corresponds to the number of European cars that are large. All values o_{ij} appear first in each individual cell of the contingency table. The three row totals, $o_{1\cdot}$, $o_{2\cdot}$, and $o_{3\cdot}$, equal 42, 124, and 137, respectively, and appear in the rightmost column of the table. The three column totals, $o_{\cdot 1}$, $o_{\cdot 2}$, and $o_{\cdot 3}$, equal 115, 40, and 148, and appear in the final row of the table.

Contingency Table

size	Count Total % Col % Row %	country American	 European	 Japanese	
	Large	36	4	2	42
		11.88	1.32	0.66	13.86
		31.30	10.00	1.35	
		85.71	9.52	4.76	
	Medium	53	17	54	124
		17.49	5.61	17.82	40.92
		46.09	42.50	36.49	
		42.74	13.71	43.55	
	Small	26	19	92	137
		8.58	6.27	30.36	45.21
		22.61	47.50	62.16	
		18.98	13.87	67.15	
		115	40	148	303
		37.95	13.20	48.84	

Figure 17.12 The initial contingency table created by JMP for Example 17.3.2.

Based on the row and column totals, we can now compute the expected frequencies e_{ij}. This yields

$$
\begin{aligned}
e_{11} &= \frac{o_{1.}\, o_{.1}}{n} = \frac{42 \times 115}{303} = 15.9406, \\
e_{12} &= \frac{o_{1.}\, o_{.2}}{n} = \frac{42 \times 40}{303} = 5.5446, \\
&\vdots \\
e_{33} &= \frac{o_{3.}\, o_{.3}}{n} = \frac{137 \times 148}{303} = 66.9175.
\end{aligned}
$$

With these values, the value x of the test statistic χ can be computed:

$$
\begin{aligned}
x &= \frac{(o_{11} - e_{11})^2}{e_{11}} + \frac{(o_{12} - e_{12})^2}{e_{12}} + \cdots + \frac{(o_{33} - e_{33})^2}{e_{33}}, \\
&= \frac{(36 - 15.9406)^2}{15.9406} + \frac{(4 - 5.5446)^2}{5.5446} + \cdots + \frac{(92 - 66.9175)^2}{66.9175}, \\
&= 25.2425 + 0.4303 + \cdots + 9.4016, \\
&= 66.313.
\end{aligned}
$$

Tests

	N	DF	-LogLike	RSquare (U)
	303	4	36.309616	0.1217

Test	ChiSquare	Prob>ChiSq
Likelihood Ratio	72.619	<.0001*
Pearson	66.313	<.0001*

Figure 17.13   The JMP output with the result of the χ^2-test for independence in Example 17.3.2.

To compute the corresponding p-value, we have to use a χ^2-distributed random variable with $(r-1)(c-1) = (3-1)(3-1) = 4$ degrees of freedom. The p-value is equal to the probability

$$P\left(\chi_4^2 \geq x\right) = P\left(\chi_4^2 \geq 66.313\right) \approx 10^{-13}.$$

The p-value is thus virtually zero, so that we can reject the null hypothesis. We conclude that the two variables studied are dependent. In other words, there is a relation between the size and the origin of the cars.

If you generate a contingency table with JMP, it automatically performs the χ^2-test for independence of the two variables studied. In addition to the contingency table in Figure 17.12, the JMP output even contains the results of two tests. This extra piece of output is shown in Figure 17.13. In this figure, you can see the number of observations (303) and the degrees of freedom (4) for the χ^2-test. Next to the name "Pearson" in the output, you find the computed test statistic of 66.313 from Example 17.3.2. For the p-value, the corresponding entry is < .0001, indicating that the p-value is practically zero. JMP uses the name "Pearson", because, in 1900, Karl Pearson was the first to study the properties of the χ^2-test.

The JMP output also reports an alternative test for the independence of two qualitative variables, namely the likelihood ratio test. For the data in Example 17.3.2, this test is also based on a χ^2-distribution with four degrees of freedom, and also leads to a rejection of the null hypothesis. We do not discuss likelihood ratio tests in this book.

JMP also offers the possibility of displaying all e_{ij} values and the contribution of each cell to the test statistic, $(o_{ij} - e_{ij})^2/e_{ij}$, in the contingency table. To use this option, click the hotspot (red triangle) next to the term "Contingency Table" in the output. In the resulting drop-down menu, check the options "Expected" and "Cell Chi Square", as shown in Figure 17.14. The final contingency table is shown in Figure 17.15. In this final table, you can find the values 15.9406, 5.5446, and 66.9175 for e_{11}, e_{12}, and e_{33}, as well as the values 25.2425, 0.4303, and 9.4016 for $(o_{11} - e_{11})^2/e_{11}$, $(o_{12} - e_{12})^2/e_{12}$, and $(o_{33} - e_{33})^2/e_{33}$. All these values were intermediate results in the computation of the test statistic in Example 17.3.2.

Example 17.3.3   *For the contingency table in Figure 17.11 from Example 17.3.1, JMP provides a large p-value (0.58) for the χ^2-test. This is shown in Figure 17.16, where the extended contingency table contains the different values of e_{ij} and the contribution of each cell to the test statistic, $(o_{ij} - e_{ij})^2/e_{ij}$.*

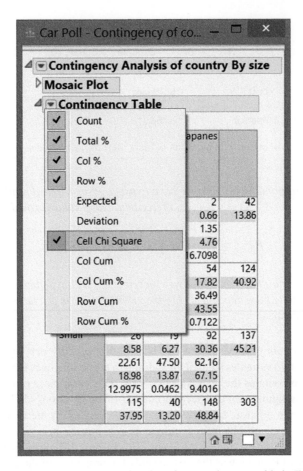

Figure 17.14   The additional options for a contingency table in JMP.

However, the JMP output contains a warning about the reported p-value. This warning says: "20% of cells have expected count less than 5, ChiSquare suspect". This warning refers to the fact that there are too many cells in the contingency table that have e_{ij} values smaller than 5. In that case, the probability distribution of the test statistic is not well approximated by the χ^2-distribution and the p-value is not reliable.

This final example illustrates that the χ^2-test cannot always be used. In order for the test to be reliable, all expected values e_{ij} must be at least 1, and at most 20% of all cells can have an expected value smaller than 5. In principle, a similar recommendation applies to the test in Section 6.1. If too many cells in a contingency table have too small values for e_{ij}, it is recommended to combine two or more rows and/or columns in the table.

For a contingency table with only two rows and two columns, it is better to use the so-called Fisher's exact test than the χ^2-test to test the (in)dependence of the two variables under study. Fisher's exact test is based on the hypergeometric distribution (see *Statistics with JMP: Graphs, Descriptive Statistics and Probability*) and can only be performed using JMP Pro.

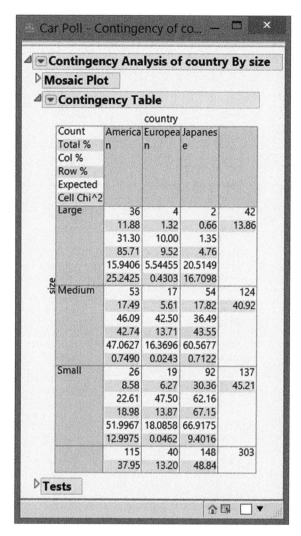

Figure 17.15 A contingency table containing the values e_{ij} and $(o_{ij} - e_{ij})^2/e_{ij}$ for each cell, required for computing the value of the test statistic χ in Example 17.3.2.

17.3.3 The Homogeneity Test

The χ^2-test for independence of two qualitative variables as introduced above, is intended for the study of one sample from one population. This sample contains data on (at least) two qualitative variables. The different steps of the χ^2-test can, however, also be used in a different context, namely when we study two or more populations and we want to determine whether a certain discrete probability distribution is common to all these populations. In that case, we denote the number of populations studied by c, while the number of possible outcomes of the discrete probability distribution under consideration is denoted by r. This scenario also leads to a contingency table with r rows and c columns.

Price category

Rating	Count / Total % / Col % / Row % / Expected / Cell Chi^2	Cheap (< 6 euros)	Moderately priced	Expensive (>= 10 euros)	
E	Count	2	1	0	3
	Total %	3.39	1.69	0.00	5.08
	Col %	6.45	5.56	0.00	
	Row %	66.67	33.33	0.00	
	Expected	1.57627	0.91525	0.50847	
	Cell Chi^2	0.1139	0.0078	0.5085	
G/E	Count	1	3	1	5
	Total %	1.69	5.08	1.69	8.47
	Col %	3.23	16.67	10.00	
	Row %	20.00	60.00	20.00	
	Expected	2.62712	1.52542	0.84746	
	Cell Chi^2	1.0078	1.4254	0.0275	
G	Count	7	5	4	16
	Total %	11.86	8.47	6.78	27.12
	Col %	225.8	27.78	40.00	
	Row %	43.75	31.25	25.00	
	Expected	8.40678	4.88136	2.71186	
	Cell Chi^2	0.2354	0.0029	0.6119	
F/G	Count	21	9	5	35
	Total %	35.59	15.25	8.47	59.32
	Col %	67.74	50.00	50.00	
	Row %	60.00	25.71	14.29	
	Expected	18.3898	10.678	5.9322	
	Cell Chi^2	0.3705	0.2637	0.1465	
F	Count	0	0	0	0
	Total %	0.00	0.00	0.00	0.00
	Col %	0.00	0.00	0.00	
	Row %	.	.	.	
	Expected	0.00781	7e-310	3e-315	
	Cell Chi^2	0.0078	0.0000	0.0000	
P/F	Count	0	0	0	0
	Total %	0.00	0.00	0.00	0.00
	Col %	0.00	0.00	0.00	
	Row %	.	.	.	
	Expected	4e-315	0	0	
	Cell Chi^2	0.0000	0.0000	0.0000	
	Count	31	18	10	59
	Total %	52.54	30.51	16.95	

(a)

Tests

N	DF	-LogLike	RSquare (U)
59	6	2.5799864	0.0437

Test	ChiSquare	Prob>ChiSq
Likelihood Ratio	5.160	0.5235
Pearson	4.722	0.5800

Warning: 20% of cells have expected count less than 5, ChiSquare suspect.
Warning: Average cell count less than 5, LR ChiSquare suspect.

(b)

Figure 17.16   JMP output with the result of the χ^2-test for independence in Examples 17.3.1 and 17.3.3.

Example 17.3.4   *We want to investigate whether boys and girls make similar study choices. For this purpose, we randomly interview 200 female students and 200 male students. Based on the students' answers, we can create a contingency table with rows for the study choices ("Literature and philosophy", "Law", "Economics", "Bioscience Engineering", etc.) and columns for the two genders "Male" and "Female". The following hypotheses can then be tested with a χ^2-test:*

$$H_0 : \text{The distribution of study choices is the same for boys and girls}$$

and

$$H_a : \text{The distribution of study choices is different for boys and girls.}$$

Contingency Table

Gender

Count Total % Col % Row % Expected Cell Chi^2	Boys	Girls	
Boomerang	12	15	27
	12.00	15.00	27.00
	24.00	30.00	
	44.44	55.56	
	13.5	13.5	
	0.1667	0.1667	
Disney	18	10	28
	18.00	10.00	28.00
	36.00	20.00	
	64.29	35.71	
	14	14	
	1.1429	1.1429	
KIKA	13	11	24
	13.00	11.00	24.00
	26.00	22.00	
	54.17	45.83	
	12	12	
	0.0833	0.0833	
Nickelodeon	7	14	21
	7.00	14.00	21.00
	14.00	28.00	
	33.33	66.67	
	10.5	10.5	
	1.1667	1.1667	
	50	50	100
	50.00	50.00	

Channel

(a)

Tests

N	DF	-LogLike	RSquare (U)
100	3	2.5986720	0.0375

Test	ChiSquare	Prob>ChiSq
Likelihood Ratio	5.197	0.1579
Pearson	5.119	0.1633

(b)

Figure 17.17   The JMP output with the result of the χ^2-test for independence in Example 17.3.5.

Example 17.3.5 *One hundred children (50 boys and 50 girls) were asked what their favorite TV channel is. The children could choose between Boomerang, Disney Channel, Nickelodeon, and KIKA. The result of the query is displayed in the contingency table in Figure 17.17a. For this data, $r = 4$ (the number of TV channels) and $c = 2$ (the number of populations), and a χ^2-test can be performed to test the following hypotheses:*

H_0 : *The preference for TV channels is distributed equally among boys and girls*

and

H_a : *The preference for TV channels is not distributed equally among boys and girls.*

The result of the χ^2-test is shown in Figure 17.17b. The p-value of the χ^2-test is equal to 0.1633. As a result, there is no indication that the preferences of boys and girls differ. It is a good exercise to carry out the χ^2-test for this example by yourself. Some of the required intermediate results – in particular, the values of e_{ij} and $(o_{ij} - e_{ij})^2/e_{ij}$ – can be found in Figure 17.17a.

If a χ^2-test is used in a context as presented in the last two examples of this section, it is called a test for homogeneity of probability distributions across several populations.

18

An Introduction to Regression Modeling

Sexual attraction is Gene's area of expertise. "There's a correlation?" I asked.
"People with long earlobes are more likely to choose partners with long earlobes. It's a better predictor than IQ".

(from The Rosie Project, Graeme Simsion, p. 75)

As mentioned at the start of Chapter 17, many statistical analyses study relationships between two or more variables. Often, these types of analyses are referred to as multivariate statistics. In Chapters 18 and 19, we focus on a technique called regression. More specifically, we focus on simple linear regression. First, however, we provide a general introduction to regression analysis. In doing so, we explain the differences between linear and nonlinear regression and between simple regression and multiple regression.

Regression techniques are heavily used in virtually every scientific domain. Economists, for instance, frequently use regression analyses, and the research area dealing with regression techniques for economic data has a special name, "econometrics". Similarly, "chemometrics" is the typical name for multivariate statistical methodology, including regression techniques, used in certain branches of chemistry. In the context of food research, the term "sensometrics" is used, while "psychometrics" is the name used by psychologists. Regression is also one of the most commonly used methods in medicine, engineering, and the political and social sciences.

A key concept in regression analysis is the **model**. In the context of regression, this is a mathematical equation intended to explain the behavior or the variation of a certain variable of interest. The challenge is to find an appropriate equation that fits the available data. We call the process of fitting a model to data **model estimation**.

18.1 From a Theory to a Model

A theory often gives us an understanding of the relationship between different variables. Economic theory, for example, provides us with insights about relationships between certain

Statistics with JMP: Hypothesis Tests, ANOVA and Regression, First Edition. Peter Goos and David Meintrup.
© 2016 John Wiley & Sons, Ltd. Published 2016 by John Wiley & Sons, Ltd.
Companion Website: http://www.wiley.com/go/goosandmeintrup/JMP

economic variables. These relationships are expressed with the aid of a mathematical equation, a model. An example of a model is

$$c = f(x),$$

where c is the level of a household's consumption and x represents the disposable income. Another example is given by

$$q = f(p, p^s, p^c),$$

where q represents the demand for a car, p the price of the car, p^s the price of a substitute (such as, for example, a motorcycle), and p^c the price of a complementary good (such as fuel).

In a similar way, there is a link between, on the one hand, the fuel consumption of a car, v, and on the other, the weight (g), the length (l), the width (b), and the power (p):

$$v = f(g, l, b, p).$$

When extracting starch from corn or rice, the yield (y) of the extraction process depends on the concentration of specific kinds of salts (s) used during the steeping phase. Hence,

$$y = f(s).$$

Each of the four models involves one variable the behavior of which we are trying to explain. In other words, we are trying to find out why that variable goes up or down; that is, we are trying to explain its variation. For example, we are trying to explain why certain cars consume much fuel, whereas others consume little. We name that variable the **response**, the **response variable**, or the **dependent variable**. The tradition in statistics is to denote that variable by y. The name "dependent variable" is utilized because the value of that variable depends on, or is a function of, the value(s) of one or more other variables. The name "response variable" is used because it is as if its value reacts to, or responds to, changes in one or more other variables.

Explaining the variation means that we attempt to figure out what other variables are related to the response. We name these variables the **explanatory variables** or the **independent variables**. Another name that is often used for the explanatory variables is **predictors**, because they can be used to predict the value of the response variable. If there is just one explanatory variable, we generally denote it by x. When there are k explanatory variables, we denote them by $x_1, x_2, \ldots, x_k$.

When we can establish a relationship between the response and one or more explanatory variables, we also want to find out whether the relationship is positive or negative, or perhaps a more complicated nonmonotonic relationship. In starch extraction, for example, it is known that using a very low concentration of salt (hence, a small value for the explanatory variable) leads to a poor yield (a small value for the response variable). The use of a concentration of about 5% leads to a good yield, and going beyond that concentration makes the yield go down again.

In addition to identifying the nature of the relationship between the response and the explanatory variable(s), it is also important to quantify its strength. In other words, we want to quantify how strong a positive or a negative relationship is. For example, we know from economic theory that there is a positive relationship between a household's income and its consumption, but we do not know to what extent the consumption increases as a result of an increased income. However, based on sample data of incomes and consumption levels, we can estimate a model; that is, we can estimate the strength of the relationship.

In general, a regression model involving one explanatory variable x is written as

$$y = f(x).$$

Such a model is a **simple regression model**. A regression model involving k explanatory variables $x_1, x_2, \ldots, x_k$ is written as

$$y = f(x_1, x_2, \ldots, x_k).$$

Such a model is called a **multiple regression model**.

18.2   A Statistical Model

The models from the previous section are deterministic. Consider, for example, the model

$$v = f(g, l, b, p)$$

for the fuel consumption of a car. What this model suggests is that every car with the same weight (g), the same length (l), the same width (b), and the same power (p) will consume exactly the same amount of fuel. Obviously, this is nonsense. One car may, for example, be more aerodynamic than another or use more modern technology, so that it consumes less fuel. As a result, since factors other than the weight, the length, the width, and the power play a role, the model will not explain the variation in the fuel consumption perfectly.

One reaction to this could be to improve the model, and take into account the aerodynamic characteristics or the technology used. It is, however, extremely hard to identify and measure all relevant variables. So, there will virtually always be variables that we overlook or for which we have no data.

We should not view this as problematic, since what we usually want is to get a good picture of a population rather than of every individual. In the context of regression analysis, this means that we are satisfied if we can find a good model for a "typical" or an "average" car with a given weight, length, width, and power.

Another reason why the deterministic model may not work perfectly well is that the response variable is inherently stochastic. In other words, the response or dependent variable is a random variable. For example, it is not hard to think of two similar households that have the same income but quite different consumption patterns. The occupants of one household might spend a lot of money because they happen to like fast, expensive cars and exotic holidays in luxurious hotels, whereas those in another household might have an ordinary car and a preference for local hiking holidays. Thus, the consumption varies at random due to random variation in the tastes or preferences. In a similar way, the yield of a starch extraction process varies at random. One reason for this could be that in one run of the process, the batch of corn utilized might be of high quality because it was grown on fertile soil, while in another run, the corn might be of poorer quality because it happens to have originated from an area with less fertile soil.

Often, there is also a measurement error when recording the value of the response variable. For example, measuring a household's exact annual consumption is cumbersome. Similarly, quantifying the exact yield of a starch extraction process is not trivial. Moreover, measurement equipment only has a certain precision, which leads to rounding errors in the measurement process.

These considerations render deterministic models impractical. Instead, it makes sense to use a statistical model, which treats the response variable as a random variable and which takes into account the fact that certain unobserved variables may have an impact on the response. Statistical models that are often appropriate are

$$Y = f(x) + U,$$

when there is a single explanatory variable, or

$$Y = f(x_1, x_2, \ldots, x_k) + U,$$

when there are multiple explanatory variables.

In these models, an uppercase letter is used for the response, Y, to stress that it is a random variable. The uppercase letter U represents an error term. That error term is a random variable, which captures the impact of all kinds of unobserved variables on the response. In the example on households' consumption patterns, with income as the explanatory variable, the error term U describes the impact of unobserved variables such as the number of members of the household or the type of accommodation. In the starch extraction example, the error term U describes the impact of unobserved variables such as the corn variety, the exact steeping time, and the ambient temperature, all of which impact the yield.

The statistical model consists of a systematic part, $f(x)$ or $f(x_1, x_2, \ldots, x_k)$, and a random part, U. We assume that the systematic part provides a good description of the mean response at any given value of the explanatory variables. This is expressed as

$$E(Y \mid x) = f(x)$$

for simple regression, and as

$$E(Y \mid x_1, x_2, \ldots, x_k) = f(x_1, x_2, \ldots, x_k)$$

for multiple regression. This implies that, on average, the error in the model is zero:

$$E(U) = 0.$$

One of the key goals in regression is to identify the nature and the strength of the relationship of each explanatory variable with the response variable. This requires the estimation of model parameters. These parameters are often denoted by $\beta_0, \beta_1, \beta_2, \ldots, \beta_p$. The number of parameters, p, may or may not be equal to the number of explanatory variables, k. To make explicit that the exact nature of the model depends on the parameters $\beta_0, \beta_1, \beta_2, \ldots, \beta_p$, we can also write the model as

$$Y = f(x_1, x_2, \ldots, x_k; \beta_1, \beta_2, \ldots, \beta_p) + U, \qquad (18.1)$$

The purpose of regression methods is to estimate the values of the unknown parameters based on sample data. Once the model has been estimated (i.e., once its parameters have been estimated), it can be used for interpretation, to gain insight, as a confirmation for a theory or hypothesis developed, for decision-making, and for prediction.

18.3 Causality

Estimating a model using regression methods allows us to establish relationships between different variables. After such relationships have been quantified, it is always tempting to assume causality between the explanatory variable(s) and the response variable. However, a strong empirical relationship between these variables is not sufficient to assume a causal relationship. This is demonstrated by the following example.

Example 18.3.1 *Table 18.1 contains data on the number of inhabitants, the surface area, the birth rate, and the number of pairs of storks in 17 European countries. There is a significant positive correlation between the number of pairs of storks and the birth rate. If we represent*

Table 18.1 Data from 17 European countries.

Country	Surface area (km^2)	Number of stork pairs	Inhabitants (millions)	Births (thousands)
Albania	28 750	100	3.2	83
Belgium	30 520	1	9.9	118
Bulgaria	111 000	5 000	9	117
Denmark	43 100	9	5.1	59
Germany	357 000	3 300	78	901
France	544 000	140	56	774
Greece	132 000	2 500	10	106
Hungary	93 000	5 000	11	124
Italy	301 280	5	57	551
The Netherlands	41 900	4	15	188
Austria	83 860	300	7.6	87
Poland	312 680	30 000	38	610
Portugal	92 390	1 500	10	120
Romania	237 500	5 000	23	367
Spain	504 750	8 000	39	439
Turkey	779 450	25 000	56	1 576
Switzerland	41 290	150	6.7	82

Source: Matthews, 2001. Reproduced with permission of Wiley.

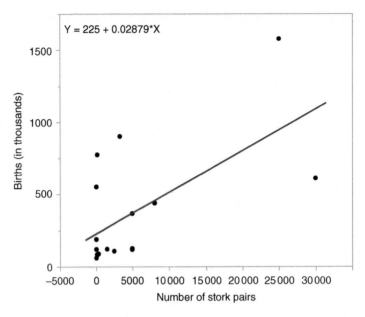

Figure 18.1   A scatter plot of the number of pairs of storks and the birth rate in 17 European countries, along with the estimation of a possible model.

the birth rate by y and the number of pairs of storks by x, then one of the possible models estimated from the data in Table 18.1 is

$$y = 225.03 + 0.0288x.$$

This model, which is graphically shown in Figure 18.1, suggests that the birth rate y increases with the number of stork pairs. Of course, there is no scientific or theoretical support for a causal relationship between the number of storks and the birth rate, and therefore there is no causality.

One may wonder, then, why there is a positive correlation between the two variables. The reason for this is that there is a third variable, the surface area of the 17 countries, which is positively correlated with both the number of births and the number of pairs of storks. The larger the surface area of a country, the more stork pairs it houses. Also, the larger a country's surface area, the greater is the number of births. This explains the positive correlation between the birth rate and the number of stork pairs.

Obviously, often a researcher's goal is not just to show that a relationship exists. Typically, a researcher wants to claim that a causal relationship exists. In other words, the researcher wants to claim that a change in an explanatory variable causes the response variable to change as well. To this end, the researcher has two possibilities:

(1) Theoretical arguments are needed to be able to speak of a causal relationship, in addition to a strong empirical relationship. For example, economic theory predicts that the demand for a good will go down when its price goes up.

(2) Well-conducted, randomized designed experiments, in which (i) the value(s) of the explanatory variable(s) is (are) changed on purpose and the response variable is measured after each change and (ii) everything else is controlled so that the results are not impacted, also allow a researcher to establish a causal relationship.

At this point, we can make a distinction between experimental data and observational data. When collecting observational data, the researcher does not intervene in the system that he is studying. Essentially, the researcher just records what is happening. When collecting experimental data, the researcher is more active in the sense that he deliberately changes the levels of the explanatory variables to measure how this impacts the response variable. Conducting an experiment requires the researcher to have control over the value(s) of the explanatory variable(s).

Example 18.3.2 *An experiment can be designed to measure the impact of the duration of a commercial and the frequency with which it is shown. Several videotapes are then created in which a number of TV programs alternate with commercials. The tapes differ in the duration of the commercials and the frequency with which each commercial is shown. Each tape is then shown to a different group of test persons. The next day, each test person is asked how many products were advertised and whether he remembers the products' brand names. By estimating a regression model, the researcher can then quantify the effect of the duration and the frequency of the commercial, as well as the joint effect of the duration and the frequency.*

Example 18.3.3 *To find out which concentration of salt leads to an acceptable yield or to an optimal yield, a researcher can try different salt concentrations and measure the yield for each of these. Afterwards, the researcher can estimate a regression model to see how the yield depends on the salt concentration. The resulting regression model can then be used to identify the salt concentration that produces the highest yield.*

It is often impossible for the researcher to perform a controlled experiment. In that case, he has to rely on observational data.

Example 18.3.4 *To determine the extent to which salary depends on educational level, number of years of work experience, gender, and sector of employment, data is collected from a large number of randomly selected employees. This data can then be used, for example, to determine whether men and women receive equal pay for equal work, taking into account their qualifications and experience.*

Many people believe that only controlled experiments allow causal relationships to be established, and that observational data can only be used to detect patterns that are useful for prediction. For example, in a paper in the scientific journal *Significance* in 2011, S. Stanley Young and Alan Karr wrote that "Any claim coming from an observational study is most likely to be wrong. Startling, but true. Coffee causes pancreatic cancer. Type A personality causes heart attacks. Trans-fat is a killer. Women who eat breakfast cereal give birth to more boys. All these claims come from observational studies; yet when the studies are carefully examined, the claimed links appear to be incorrect". Young and Karr found that many claims made based on observational studies cannot be confirmed by means of properly designed experiments. In fact, it even happens sometimes that experiments show the opposite of the original claim based on observational data. Young and Karr discuss various reasons why things go wrong. It would lead us too far astray to discuss these reasons here. We merely want to say that, in most cases, claiming causality based on observational data is too optimistic.

Despite the fact that observational data do not allow causal claims to be made, carrying out regression analyses remains interesting. Banks, for instance, successfully use regression models to make predictions concerning the creditworthiness of potential customers. Similarly, insurance companies use regression models to predict whether they should offer a contract to a particular driver, and, if so, what price they have to charge.

18.4   Linear and Nonlinear Regression Models

For the estimation of a regression model, it is necessary to specify the shape of the function $f()$ in the model in Equation (18.1). Various kinds of models exist. The simplest models to estimate belong to the class of linear regression models. In some cases, however, linear regression models do not capture the relationship between a response variable and one or more explanatory variables well. A nonlinear model is then needed instead.

In a **simple linear regression model**, the systematic part $E(Y \mid x)$ is a linear combination of two parameters β_0 and β_1. The simplest linear regression model is

$$Y = \beta_0 + \beta_1 x + U.$$

In that model, we use the original variables x and Y themselves. In that case, the simple linear regression model describes a linear relationship between x and Y.

For some data sets, however, a better model is obtained by working with a transformation of x and/or Y. If we decide to transform the explanatory variable x, the simple linear regression model could become, for example,

$$Y = \beta_0 + \beta_1 \ln(x) + U.$$

If working with x^2 instead of x or $\ln(x)$ is preferred, the simple linear regression model becomes

$$Y = \beta_0 + \beta_1 x^2 + U.$$

Finally, it is also possible to use transformations of both x and Y, as in the model

$$\ln(Y) = \beta_0 + \beta_1 \ln(x) + U.$$

In models involving a transformation of x and/or Y, the simple linear regression model describes a nonlinear relationship between x and Y. That nonlinear relationship can be quadratic or logarithmic, as in the above examples. However, many other nonlinear relationships exist as well.

In a **multiple linear regression model**, the systematic part $E(Y \mid x_1, x_2, \ldots, x_k)$ is a linear combination of the parameters $\beta_0, \beta_1, \beta_2, \ldots$. Examples of such models are

$$Y = \beta_0 + \beta_1 x_1 + \beta_2 x_2 + \cdots + \beta_k x_k + U,$$

$$\frac{1}{Y} = \beta_0 + \beta_1 x_1 + \beta_2 x_2 + \cdots + \beta_k x_k + U,$$

and

$$Y = \beta_0 + \beta_1 x_1 + \beta_2 x_2 + \beta_3 x_1^2 + \beta_4 x_2^2 + \beta_5 x_1 x_2 + U.$$

These example models illustrate that, just like simple linear regression models, multiple linear regression models may or may not involve transformations of the explanatory variables and/or the response variable. Their key feature is that the response, regardless of whether or not it is transformed, is a linear function of the unknown parameters.

Examples of nonlinear models are

$$Y = \beta_0 + \beta_1 x_1^{\beta_2} U,$$

and

$$Y = \beta_0 x_1^{\beta_1} x_2^{\beta_2} U.$$

Nonlinear models are much harder to estimate than linear models, and the theory of nonlinear regression models is much harder as well. In recent years, however, the availability of powerful PCs and good statistical software has made the estimation and use of nonlinear models by nonexperts possible.

It should be pointed out that some nonlinear models can be linearized. For example, taking the natural logarithms of the left- and right-hand sides of the latter equation yields

$$\ln(Y) = \ln\left(\beta_0 x_1^{\beta_1} x_2^{\beta_2} U\right),$$

which can be rewritten as

$$\ln(Y) = \ln(\beta_0) + \ln\left(x_1^{\beta_1}\right) + \ln\left(x_2^{\beta_2}\right) + \ln(U),$$

and, subsequently, as

$$\ln Y = \ln(\beta_0) + \beta_1 \ln(x_1) + \beta_2 \ln(x_2) + \ln(U).$$

The first term on the right-hand side of this expression, $\ln(\beta_0)$, is the logarithm of the unknown parameter β_0. The logarithm of an unknown parameter is itself an unknown parameter, for which we can use a dedicated symbol; for instance, $\tilde{\beta}_0$. In a similar way, we can use the symbol

$\tilde{U}$ instead of $\ln(U)$, since the logarithm of a random variable is itself a random variable. This results in the following expression for the model:

$$\ln Y = \tilde{\beta}_0 + \beta_1 \ln x_1 + \beta_2 \ln x_2 + \tilde{U}.$$

The systematic part of this model, $\tilde{\beta}_0 + \beta_1 \ln x_1 + \beta_2 \ln x_2$, is a linear combination of the parameters $\tilde{\beta}_0$, β_1 and β_2. Therefore, it is a linear regression model. As a result, the logarithmic transformation has turned the nonlinear regression model into a linear one. Therefore, the transformation has turned a model that is hard to estimate (the nonlinear one) into a model that is relatively easy to estimate (a linear one).

In the next chapter of this book, we focus on the simple linear regression model. Multiple linear regression model and nonlinear regression models are outside the scope of this book. These models, along with an in-depth treatment of the simple linear regression model, are discussed in detail in the book *Statistics with JMP: Linear and Generalized Linear Models*.

19

Simple Linear Regression

I didn't know what the odds were. All I could tell him was that they were somewhere between 9.9 and 27.2 per cent based on the batting average and percentage of home runs listed in the profile I had read. I had not had time to memorise the statistics for doubles and triples. Fat Baseball Fan nevertheless seemed impressed […]. He showed me […] how the more sophisticated statistics worked. I had no idea sport could be so intellectually stimulating.

(from *The Rosie Project*, Graeme Simsion, pp. 235–236)

As already mentioned at the start of Chapter 17, many statistical analyses study relationships between two or more variables. Oftentimes, these types of analyses are referred to as multivariate statistics. In this chapter, we focus on simple linear regression. As discussed in Chapter 18, it is sometimes necessary to use transformations of the two variables under study for the simple linear regression model to be appropriate.

19.1 The Simple Linear Regression Model

19.1.1 Examples

A simple linear regression model examines the relationship between a response variable or a dependent variable, on the one hand, and an explanatory variable, independent variable, or predictor, on the other. The two variables are quantitative.

Example 19.1.1 *The experience of a vendor shows that the sales of soft drinks on the Belgian coast increase with the temperature. Estimation of the simple linear regression model*

$$\text{Sales} = \beta_0 + \beta_1 \cdot \text{Temperature} + U$$

Statistics with JMP: Hypothesis Tests, ANOVA and Regression, First Edition. Peter Goos and David Meintrup.
© 2016 John Wiley & Sons, Ltd. Published 2016 by John Wiley & Sons, Ltd.
Companion Website: http://www.wiley.com/go/goosandmeintrup/JMP

gives him an idea of the extent to which the sales increase when the temperature rises by one degree. In addition, the estimated model allows the vendor to make a prediction of the sales for any given temperature in the future.

Example 19.1.2   *The management of a multinational company, concerned about the efficiency of its production processes, is interested in quantifying the dependence of the yield of its starch extraction process on the salt concentration used during the steeping phase. One of the company's bioscience engineers therefore conducts an experiment to estimate the following regression model:*

$$\text{Yield} = \beta_0 + \beta_1 \cdot \text{Concentration} + U.$$

Example 19.1.3   *An economist wants to investigate the relation between a household's income and its consumption. To this end, she estimates a linear regression model, with the natural logarithm of the consumption level as response variable and the natural logarithm of the household's income as explanatory variable:*

$$\ln(\text{Consumption}) = \beta_0 + \beta_1 \cdot \ln(\text{Income}) + U.$$

19.1.2   The Formal Description of the Model

In general, the simple linear regression model can be written as

$$Y = \beta_0 + \beta_1 x + U. \tag{19.1}$$

If we assume that, on average, the model provides a good description of the variation in the response variable, then the expected value of the error term, $E(U)$, is zero. So,

$$E(Y|x) = \beta_0 + \beta_1 x. \tag{19.2}$$

The unknown parameters in this model are the regression coefficients β_0 and β_1. The parameter β_0 is called the **intercept**, while β_1 is called the **slope** or the **effect** of the explanatory variable x on the response Y. Sometimes, β_1 is also called the linear effect or the main effect. The latter names are especially useful in the context of multiple linear regression, where quadratic effects and/or interaction effects may be needed.

The meanings of the two model parameters are as follows:

- The intercept β_0 represents the expected response when the explanatory variable x takes the value 0. In other words, $\beta_0 = E(Y|0)$. For instance, in the starch extraction example, it represents the expected yield, or the average yield, when the salt concentration used in the steeping phase is zero. In the economist's study on the consumption level, the intercept β_0 is the expected value of the logarithm of the consumption when the logarithm of the income is zero.

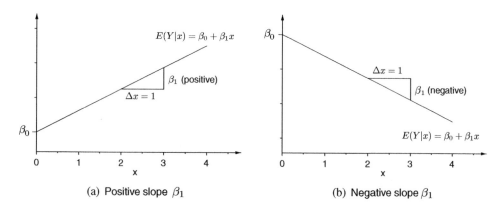

Figure 19.1 A simple linear regression model with intercept β_0 and slope β_1.

- The slope β_1 represents the expected change in the response when the explanatory variable increases by one unit. The expected response for any given x value is given by Equation (19.2). The expected response after an increase of x by one unit is

$$E(Y|x+1) = \beta_0 + \beta_1(x+1) = \beta_0 + \beta_1 x + \beta_1.$$

The change in the expected response due to the increase of x is therefore

$$E(Y|x+1) - E(Y|x) = \beta_0 + \beta_1 x + \beta_1 - (\beta_0 + \beta_1 x) = \beta_1.$$

If β_1 is positive, an increase in the value of the explanatory variable x is accompanied by an increase in the expected Y value. If β_1 is negative, an increase in the value of the explanatory variable x is accompanied by an decrease in the expected Y value.

Graphical representations of the model for a positive and a negative slope are shown in Figures 19.1a and 19.1b, respectively.

19.2   Estimation of the Model

The most commonly used method for estimating the unknown parameters β_0 and β_1 of a simple linear regression model is the least squares method. This method is also called **ordinary least squares estimation** or simply **OLS**. Many other estimation techniques exist. For example, one can use weighted least squares estimation or generalized least squares estimation. Also, nonparametric estimation techniques exist. The nonparametric techniques do not assume any specific model form a priori. Another family of estimation techniques is known as robust estimation techniques. The results of a robust estimation are insensitive to outliers in the data, as opposed to least squares estimation approaches. These advanced techniques fall outside the scope of this book. Some of them will be discussed in the book *Statistics with JMP: Linear and Generalized Linear Models*.

Table 19.1   A small data set for
a simple linear regression.

y_i	1	1	2	2	4
x_i	1	2	3	4	5

19.2.1   Intuition and Important Concepts

In this chapter on simple linear regression, we study the relationship between two quantitative variables, the response and the independent variable. When studying two quantitative variables, it is always useful to create a scatter plot of the data.

Example 19.2.1   *Suppose that we want to estimate the simple linear model in Equation (19.1) for the data in Table 19.1. The data points are represented graphically in Figure 19.2 by means of a scatter plot, with the response variable on the vertical axis and the explanatory variable on the horizontal axis.*

Note that in the table, we denote the ith observation of the explanatory variable by x_i. The ith observation of the response variable is denoted by y_i. We use a lowercase letter for the response here, because we are dealing with observed data in this example. Once the data has been collected, the response is no longer a random variable, so that we can use a lowercase letter instead of an uppercase letter.

The figure shows a clear positive relationship between the variables x and y. Therefore, we should expect a positive estimate for β_1. The challenge, however, is to estimate the value of the slope β_1, as well as the value of the intercept β_0, from the data. In other words, we should find a simple linear regression model that fits the data well. The graphical representation of a simple linear regression model is a straight line. Therefore, we should identify a straight line that summarizes the points in the scatter plot well.

For the data in this example, we can draw several straight lines through the data points in the scatter plot and try to find a good one. One attractive option is shown in Figure 19.3. The

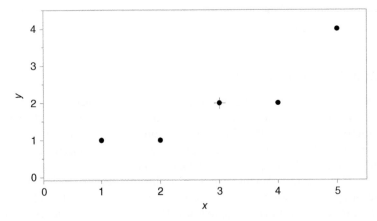

Figure 19.2   A scatter plot of the data in Table 19.1 for the estimation of a simple linear regression model.

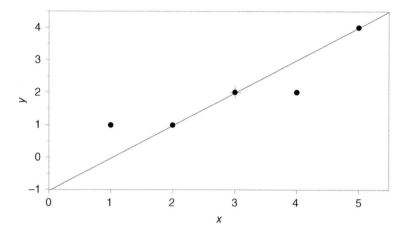

Figure 19.3 One possible regression line for the data in Figure 19.2 and Example 19.2.1.

reason why this regression line is attractive is that it goes through three of the five data points, and it does not seem too far off the remaining two points.

With the knowledge that three points lie on the regression line, we can work out its equation. Doing so results in the following estimated simple linear regression model:

$$\hat{y} = -1 + 1x = -1 + x. \tag{19.3}$$

As a result, the intercept β_0 is estimated to be −1, while the slope β_1 is estimated to be 1.

Note that on the left-hand side of the estimated model, we now use the symbol $\hat{y}$, which is pronounced as "y hat". We use this symbol to stress that now we have estimated a regression model, and we can use it for predicting the response value for any value of x. Therefore, $\hat{y}$ indicates a predicted value of y. For example, using the estimated model, we can predict the response value for an x value of 3.5. To this end, we need to substitute the value 3.5 for x in Equation (19.3). This results in a prediction of

$$\hat{y} = -1 + 1 \times 3.5 = 2.5.$$

Figure 19.3 shows that the regression line provides a reasonably good summary of the relation between the explanatory variable x and the response y. A critical note on the model, however, is that it is perfect for three of the five data points while being quite poor for the remaining two data points, which do not lie on our regression line. In fact, the vertical distances between these two points and the regression line are equal to 1. This is shown graphically in Figure 19.4.

One way to quantify the vertical distances is to calculate the difference between the observed response value and the response value predicted by the estimated model. For the first data point in Table 19.1, this yields

$$u_1 = y_1 - \hat{y}_1 = 1 - (-1 + 1x_1) = 1 - (-1 + 1 \times 1) = 1 - 0 = 1.$$

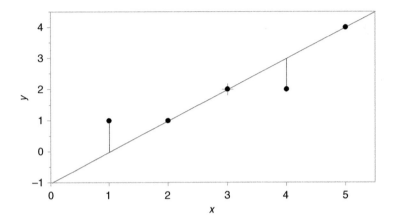

Figure 19.4 A graph showing the vertical distances between the data points and our initial regression line.

For the four other data points, we obtain

$$u_2 = y_2 - \hat{y}_2 = 1 - (-1 + 1x_2) = 1 - (-1 + 1 \times 2) = 1 - 1 = 0,$$

$$u_3 = y_3 - \hat{y}_3 = 2 - (-1 + 1x_3) = 2 - (-1 + 1 \times 3) = 2 - 2 = 0,$$

$$u_4 = y_4 - \hat{y}_4 = 2 - (-1 + 1x_4) = 2 - (-1 + 1 \times 4) = 2 - 3 = -1,$$

and

$$u_5 = y_5 - \hat{y}_5 = 4 - (-1 + 1x_5) = 4 - (-1 + 1 \times 5) = 4 - 4 = 0.$$

The values u_1, u_2, u_3, u_4, and u_5 are called residuals. A residual is strictly positive if the observed response value is larger than the one predicted. Graphically, this means that the corresponding data point lies above the regression line. For example, the positive residual u_1 corresponds to the leftmost data point in Figure 19.4, which lies above the regression line. A residual is exactly zero if the observed response value is equal to the one predicted. For instance, the zero residuals u_2, u_3, and u_5 correspond to the three data points that lie on the regression line in Figure 19.4. Finally, a negative residual corresponds to a data point that lies below the regression line. For example, the negative residual u_4 corresponds to the only data point below the regression line in Figure 19.4. The absolute values of the residuals give the vertical distances between the data points and the regression line.

Given the fact that two of the vertical distances in Figure 19.4 are 1, one may wonder if it is possible to find a regression line that results in smaller vertical distances. Perhaps there is a regression line that goes through none of the data points, but that comes very close to each of the points, so that none of the vertical distances is as large as 1. In the remainder of this section, we will show that a better regression model indeed exists.

The goal of a simple linear regression is to estimate the model in Equation (19.1). This requires estimating β_0 and β_1. In general, we call the estimates for these parameters b_0 and b_1, respectively. The estimated model is then written as

$$\hat{y} = b_0 + b_1 x,$$

where $\hat{y}$ symbolizes the predicted value of the response. In general, the residual for a given observation i is calculated as

$$u_i = y_i - \hat{y}_i = y_i - (b_0 + b_1 x_i).$$

Ideally, all the residuals are close to zero, because this means that the regression line passes very close to each data point. In that case, we say that the regression line fits the data well, or that we have a good model fit. The question now is how we can find the regression line with the best fit.

In technical terms, a regression line has a good fit if all the residuals are close to zero. There are various ways to achieve this goal. For example, if we were to find a regression line minimizing the sum of the absolute values of all residuals,

$$\sum_{i=1}^{n} |u_i|,$$

then all individual residuals would be small too. Similarly, if we were to find a regression line minimizing the sum of all squared residuals,

$$\sum_{i=1}^{n} u_i^2,$$

then again all individual residuals would be small. Minimizing the sum of the absolute residuals results in the **least absolute deviation** estimation of the regression model. Minimizing the sum of the squared residuals yields the **least squares** estimation of the regression model.

The least squares estimation is by far the most popular one. To a large extent, this is because it is a mathematically convenient approach. First, analytical expressions exist for calculating the estimates. Second, the least squares estimates of the unknown model parameters turn out to be linear combinations of the values of the response variable. The calculation of expected values and variances of linear combinations of random variables is quite easy. Moreover, due to the central limit theorem, linear combinations of independent random variables are approximately normally distributed. In addition, linear combinations of normally distributed random variables are also normally distributed. Therefore, the least squares estimation approach is mathematically more attractive than the least absolute deviation estimation approach. As explained in Section 19.3.5, the so-called Gauss–Markov theorem provides another good argument for using the least squares approach.

The main weakness of the least squares approach is that it is sensitive to outliers in the data, unlike the least absolute deviation approach. As it is less sensitive to outliers, the least absolute deviation approach is called a robust estimation method. Whenever using least squares estimation, it is thus important to verify that there are no outliers in the data, as they may severely affect the quality of the estimated model.

19.2.2   The Least Squares Method

The Normal Equations

The ordinary least squares estimation approach seeks the regression line that minimizes the sum of the squared residuals. If we represent the estimated regression line by

$$\hat{y} = b_0 + b_1 x,$$

then the least squares method requires determination of the values for b_0 and b_1 that minimize

$$S(b_0, b_1) = \sum_{i=1}^{n} u_i^2,$$

$$= \sum_{i=1}^{n} (y_i - \hat{y}_i)^2,$$

$$= \sum_{i=1}^{n} \{y_i - (b_0 + b_1 x_i)\}^2,$$

where n represents the number of observations.

Example 19.2.2   *In Example 19.2.1, the sum of the squared residuals amounts to*

$$S(b_0, b_1) = S(-1, 1) = \sum_{i=1}^{n} u_i^2 = (-1)^2 + 0^2 + 0^2 + 1^2 + 0^2 = 2,$$

since $b_0 = -1$ and $b_1 = 1$. The question now is whether this is the smallest possible sum for the squared residuals. Do other values exist for b_0 and b_1 that result in an $S(b_0, b_1)$ value smaller than 2?

To minimize a continuous differentiable function such as $S(b_0, b_1)$, we calculate the function's first derivatives with respect to the unknowns and set them to zero. This yields a set of equations. Solving that set of equations results in the values for the unknowns that potentially minimize the function. This is exactly the route we take here. The sum of squared residuals, $S(b_0, b_1)$, is a function of two unknowns, b_0 and b_1. We will therefore differentiate $S(b_0, b_1)$ with respect to b_0 and b_1, and set the two resulting derivatives to zero.

The derivatives of $S(b_0, b_1)$ with respect to the unknowns b_0 and b_1 are as follows:

$$\frac{\partial S(b_0, b_1)}{\partial b_0} = \sum_{i=1}^{n} 2(y_i - b_0 - b_1 x_i)(-1) = -2 \sum_{i=1}^{n} (y_i - b_0 - b_1 x_i),$$

$$\frac{\partial S(b_0, b_1)}{\partial b_1} = \sum_{i=1}^{n} 2(y_i - b_0 - b_1 x_i)(-x_i) = -2 \sum_{i=1}^{n} x_i(y_i - b_0 - b_1 x_i).$$

Setting these two derivatives to zero yields the following system of two linear equations with two unknowns:

$$-2\sum_{i=1}^{n}(y_i - b_0 - b_1 x_i) = 0,$$

$$-2\sum_{i=1}^{n} x_i(y_i - b_0 - b_1 x_i) = 0.$$

Dividing the left- and right-hand sides of both equations by 2 and rearranging the sums initially yields

$$-\sum_{i=1}^{n} y_i + \sum_{i=1}^{n} b_0 + \sum_{i=1}^{n} b_1 x_i = 0,$$

$$-\sum_{i=1}^{n} x_i y_i + \sum_{i=1}^{n} b_0 x_i + \sum_{i=1}^{n} b_1 x_i^2 = 0.$$

Finally, bringing the first term in each of the two equations to the right-hand side and taking into account that b_0 and b_1 are constants, we obtain the following pair of equations:

$$n b_0 + b_1 \sum_{i=1}^{n} x_i = \sum_{i=1}^{n} y_i, \tag{19.4}$$

$$b_0 \sum_{i=1}^{n} x_i + b_1 \sum_{i=1}^{n} x_i^2 = \sum_{i=1}^{n} x_i y_i. \tag{19.5}$$

These two equations are called the **normal equations**. Solving these two equations for b_0 and b_1 is not really difficult, but a bit cumbersome.

Solving the Normal Equations

To solve the system of normal equations, we can start by solving Equation (19.4) for b_0. This yields the following expression for b_0:

$$b_0 = \frac{\sum_{i=1}^{n} y_i}{n} - b_1 \frac{\sum_{i=1}^{n} x_i}{n} = \bar{y} - b_1 \bar{x}, \tag{19.6}$$

where $\bar{x}$ and $\bar{y}$ represent the mean value of the explanatory variable and the response variable in the sample.

Using this result, we can rewrite Equation (19.5) as

$$\left(\frac{\sum_{i=1}^{n} y_i}{n} - b_1 \frac{\sum_{i=1}^{n} x_i}{n} \right) \sum_{i=1}^{n} x_i + b_1 \sum_{i=1}^{n} x_i^2 = \sum_{i=1}^{n} x_i y_i, \tag{19.7}$$

$$\frac{1}{n} \left(\sum_{i=1}^{n} x_i \right) \left(\sum_{i=1}^{n} y_i \right) - b_1 \frac{1}{n} \left(\sum_{i=1}^{n} x_i \right)^2 + b_1 \sum_{i=1}^{n} x_i^2 = \sum_{i=1}^{n} x_i y_i, \tag{19.8}$$

$$b_1 \sum_{i=1}^{n} x_i^2 - b_1 \frac{1}{n} \left(\sum_{i=1}^{n} x_i \right)^2 = \sum_{i=1}^{n} x_i y_i - \frac{1}{n} \left(\sum_{i=1}^{n} x_i \right) \left(\sum_{i=1}^{n} y_i \right), \tag{19.9}$$

$$b_1 \left\{ \sum_{i=1}^{n} x_i^2 - \frac{1}{n} \left(\sum_{i=1}^{n} x_i \right)^2 \right\} = \sum_{i=1}^{n} x_i y_i - \frac{1}{n} \left(\sum_{i=1}^{n} x_i \right) \left(\sum_{i=1}^{n} y_i \right), \tag{19.10}$$

and, finally, as

$$b_1 = \frac{\sum_{i=1}^{n} x_i y_i - \frac{1}{n} \left(\sum_{i=1}^{n} x_i \right) \left(\sum_{i=1}^{n} y_i \right)}{\sum_{i=1}^{n} x_i^2 - \frac{1}{n} \left(\sum_{i=1}^{n} x_i \right)^2}, \tag{19.11}$$

and

$$b_1 = \frac{n \sum_{i=1}^{n} x_i y_i - \left(\sum_{i=1}^{n} x_i \right) \left(\sum_{i=1}^{n} y_i \right)}{n \sum_{i=1}^{n} x_i^2 - \left(\sum_{i=1}^{n} x_i \right)^2}. \tag{19.12}$$

The final step in this mathematical derivation requires the multiplication of the numerator and the denominator of Equation (19.11) by n.

Using the second derivative of the sum of the squared residuals, $S(b_0, b_1)$, it can be shown that the solution for b_0 and b_1 indeed corresponds to a minimum.

Example 19.2.3 *We now estimate the intercept and the slope for the data in Example 19.2.1 using the least squares method. For this example, all the required calculations can be done by hand. This can best be done using a table. Table 19.2 shows how the sums $\sum_{i=1}^{n} x_i$ and $\sum_{i=1}^{n} y_i$ are computed first, followed by the squares x_i^2 and their sum $\sum_{i=1}^{n} x_i^2$, and the individual cross-products $x_i y_i$ and their sum $\sum_{i=1}^{n} x_i y_i$. Obviously, since there are five data points in the example, n equals 5.*

The table shows that $\sum_{i=1}^{5} x_i = 15$, $\sum_{i=1}^{5} y_i = 10$, $\sum_{i=1}^{5} x_i^2 = 55$, and $\sum_{i=1}^{5} x_i y_i = 37$. Substituting these results in Equation (19.12) yields

$$b_1 = \frac{5 \times 37 - 15 \times 10}{5 \times 55 - (15)^2} = \frac{185 - 150}{275 - 225} = \frac{35}{50} = \frac{7}{10} = 0.7.$$

Table 19.2 The calculations required to find the least squares estimates b_0 and b_1 of the intercept and the slope in Example 19.2.3.

i	x_i	y_i	x_i^2	x_iy_i
1	1	1	1	1
2	2	1	4	2
3	3	2	9	6
4	4	2	16	8
5	5	4	25	20
Sum	15	10	55	37

As a result, we obtain a positive least squares estimate b_1 for the slope β_1. This makes sense given the positive relationship between the explanatory variable and the response variable in the scatter plot in Figure 19.2. Using Equation (19.6), we can also obtain the least squares estimate b_0 for the intercept β_0:

$$b_0 = \frac{10}{5} - b_1\frac{15}{5} = 2 - 0.7 \times 3 = -0.1.$$

In summary, the estimated least squares regression line for the example is

$$\hat{y} = b_0 + b_1x = -0.1 + 0.7x.$$

A graphical representation of that regression line is shown in Figure 19.5.

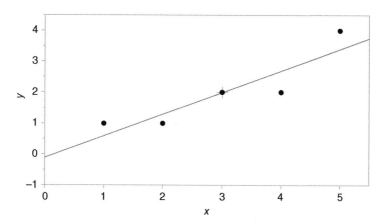

Figure 19.5 The least squares regression line for the data in Figure 19.2 and in Examples 19.2.1 and 19.2.3.

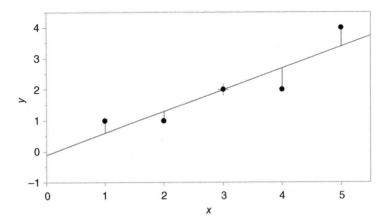

Figure 19.6 A graph showing the vertical distances between the data points and the least squares regression line in Example 19.2.3.

The five residuals for the least squares regression line are

$$u_1 = y_1 - \hat{y}_1 = 1 - (-0.1 + 0.7 \times 1) = 1 - 0.6 = 0.4,$$

$$u_2 = y_2 - \hat{y}_2 = 1 - (-0.1 + 0.7 \times 2) = 1 - 1.3 = -0.3,$$

$$u_3 = y_3 - \hat{y}_3 = 2 - (-0.1 + 0.7 \times 3) = 2 - 2 = 0,$$

$$u_4 = y_4 - \hat{y}_4 = 2 - (-0.1 + 0.7 \times 4) = 2 - 2.7 = -0.7,$$

and

$$u_5 = y_5 - \hat{y}_5 = 4 - (-0.1 + 0.7 \times 5) = 4 - 3.4 = 0.6.$$

Only one of the residuals is zero, which implies that only one of the five data points lies on the regression line. This is confirmed by Figure 19.6. The figure also shows that two data points (the first and the fifth) appear above the regression line, and that the two remaining points (the second and the fourth) lie below the line. The former two points correspond to the positive residuals u_1 and u_5, while the latter correspond to the negative residuals u_2 and u_4.

The sum of the squared residuals corresponding to the least squares regression line is

$$S(b_0, b_1) = S(-0.1, 0.7) = \sum_{i=1}^{n} u_i^2,$$

$$= (0.4)^2 + (-0.3)^2 + 0^2 + (-0.7)^2 + (0.6)^2 = 1.1.$$

This value is clearly smaller than that for $S(-1, 1)$ in Example 19.2.2. Therefore, when looking at the sum of squared residuals, the least squares regression line outperforms the regression line in Figure 19.4.

Alternative Expressions for the Estimate of the Slope

There are several alternative expressions for the least squares estimate of the slope β_1; for example,

$$b_1 = \frac{\sum_{i=1}^{n} (x_i - \bar{x})(y_i - \bar{y})}{\sum_{i=1}^{n} (x_i - \bar{x})^2}. \tag{19.13}$$

If we divide the numerator and the denominator in that alternative expression by $n - 1$, we obtain

$$b_1 = \frac{\dfrac{\sum_{i=1}^{n} (x_i - \bar{x})(y_i - \bar{y})}{n - 1}}{\dfrac{\sum_{i=1}^{n} (x_i - \bar{x})^2}{n - 1}}. \tag{19.14}$$

This formula shows that the least squares estimate of β_1 is equal to the sample covariance of the explanatory variable x and the response variable y,

$$s_{XY} = \frac{\sum_{i=1}^{n} (x_i - \bar{x})(y_i - \bar{y})}{n - 1},$$

divided by the sample variance of the explanatory variable x,

$$s_X^2 = \frac{\sum_{i=1}^{n} (x_i - \bar{x})^2}{n - 1}.$$

This relationship between simple linear regression and the sample covariance should not come as a surprise. In fact, the sample covariance quantifies the extent to which two variables x and y are linearly related, and simple linear regression attempts to express the relationship between x and y using a line. In any case, a simple expression for the least squares estimate of β_1 is

$$b_1 = \frac{s_{XY}}{s_X^2}. \tag{19.15}$$

Because the sample variance s_X^2 is always positive, b_1 and s_{XY} will always have the same sign. In other words, if the sample covariance between x and y is positive, the estimate for the

Table 19.3 The calculations required to find the sample covariance s_{XY} and the sample variance s_X^2 in Example 19.2.4.

i	x_i	y_i	$x_i - \bar{x}$	$(x_i - \bar{x})^2$	$y_i - \bar{y}$	$(x_i - \bar{x})(y_i - \bar{y})$
1	1	1	-2	4	-1	2
2	2	1	-1	1	-1	1
3	3	2	0	0	0	0
4	4	2	1	1	0	0
5	5	4	2	4	2	4
Sum	15	10	0	10	0	7

slope, b_1, will also be positive. Similarly, if the sample covariance between the two variables is negative, the estimate for the slope will also be negative.

Example 19.2.4   *The sample covariance s_{XY} between the explanatory variable x and the response variable y in Example 19.2.1 amounts to 1.75, while the sample variance of x is $s_X^2 = 2.5$. Therefore, the least squares estimate of the slope of the regression line is*

$$b_1 = \frac{s_{XY}}{s_X^2} = \frac{1.75}{2.5} = 0.7.$$

The calculations required to find the sample covariance between x and y and the sample variance of x are shown in Table 19.3. Note that both the sample covariance s_{XY} and the slope estimate b_1 are positive.

An interesting aspect of Equation (19.15) is that it shows that we need at least two different values for the explanatory variable x to be able to estimate the slope β_1. Indeed, if we had n observations, each time with the same value for x, then the sample variance s_X^2 would be zero, and the least squares estimate would be undefined because it would then involve a division by zero.

Before moving on to the next useful expression for b_1, we show that Equation (19.13) always produces the same result for b_1 as Equation (19.12). This is not obvious. We start by rewriting the numerator of Equation (19.13):

$$\sum_{i=1}^{n}(x_i - \bar{x})(y_i - \bar{y}) = \sum_{i=1}^{n} x_i y_i - \sum_{i=1}^{n} \bar{x} y_i - \sum_{i=1}^{n} x_i \bar{y} + \sum_{i=1}^{n} \bar{x}\,\bar{y},$$

$$= \sum_{i=1}^{n} x_i y_i - \bar{x} \sum_{i=1}^{n} y_i - \bar{y} \sum_{i=1}^{n} x_i + n\bar{x}\,\bar{y},$$

$$= \sum_{i=1}^{n} x_i y_i - \left(\frac{\sum_{i=1}^{n} x_i}{n}\right) \sum_{i=1}^{n} y_i$$

$$-\left(\frac{\sum_{i=1}^{n} y_i}{n}\right) \sum_{i=1}^{n} x_i + n\left(\frac{\sum_{i=1}^{n} x_i}{n}\right)\left(\frac{\sum_{i=1}^{n} y_i}{n}\right),$$

$$= \sum_{i=1}^{n} x_i y_i - \frac{1}{n}\left(\sum_{i=1}^{n} x_i\right)\left(\sum_{i=1}^{n} y_i\right).$$

The denominator of Equation (19.13) can be rewritten as

$$\sum_{i=1}^{n}(x_i - \bar{x})^2 = \sum_{i=1}^{n} x_i^2 + \sum_{i=1}^{n}(-2x_i\bar{x}) + \sum_{i=1}^{n} \bar{x}^2,$$

$$= \sum_{i=1}^{n} x_i^2 - 2\bar{x}\sum_{i=1}^{n} x_i + n\bar{x}^2,$$

$$= \sum_{i=1}^{n} x_i^2 - 2\left(\frac{\sum_{i=1}^{n} x_i}{n}\right)\sum_{i=1}^{n} x_i + n\left(\frac{\sum_{i=1}^{n} x_i}{n}\right)^2,$$

$$= \sum_{i=1}^{n} x_i^2 - 2\frac{1}{n}\left(\sum_{i=1}^{n} x_i\right)^2 + \frac{1}{n}\left(\sum_{i=1}^{n} x_i\right)^2,$$

$$= \sum_{i=1}^{n} x_i^2 - \frac{1}{n}\left(\sum_{i=1}^{n} x_i\right)^2.$$

Dividing the new expression for the numerator by the new expression for the denominator yields

$$b_1 = \frac{\sum_{i=1}^{n} x_i y_i - \frac{1}{n}\left(\sum_{i=1}^{n} x_i\right)\left(\sum_{i=1}^{n} y_i\right)}{\sum_{i=1}^{n} x_i^2 - \frac{1}{n}\left(\sum_{i=1}^{n} x_i\right)^2}. \tag{19.16}$$

Multiplying the numerator and denominator of this fraction by n reproduces Equation (19.12). This proves that Equations (19.12) and (19.13) are equivalent.

For the derivation of the properties of the ordinary least squares estimates, it is important to rewrite the expression for b_1 in Equation (19.13) in yet another way:

$$b_1 = \frac{\sum_{i=1}^{n}(x_i - \bar{x})(y_i - \bar{y})}{\sum_{i=1}^{n}(x_i - \bar{x})^2},$$

$$= \frac{\sum_{i=1}^{n} (x_i - \bar{x})y_i - \sum_{i=1}^{n} (x_i - \bar{x})\bar{y}}{\sum_{i=1}^{n} (x_i - \bar{x})^2},$$

$$= \frac{\sum_{i=1}^{n} (x_i - \bar{x})y_i}{\sum_{i=1}^{n} (x_i - \bar{x})^2} - \frac{\sum_{i=1}^{n} (x_i - \bar{x})\bar{y}}{\sum_{i=1}^{n} (x_i - \bar{x})^2},$$

$$= \frac{\sum_{i=1}^{n} (x_i - \bar{x})y_i}{\sum_{i=1}^{n} (x_i - \bar{x})^2} - \frac{\bar{y}\sum_{i=1}^{n} (x_i - \bar{x})}{\sum_{i=1}^{n} (x_i - \bar{x})^2}.$$

Since $\sum_{i=1}^{n} (x_i - \bar{x})$ always equals zero, we can drop the second term in this expression. As a consequence,

$$b_1 = \frac{\sum_{i=1}^{n} (x_i - \bar{x})y_i}{\sum_{i=1}^{n} (x_i - \bar{x})^2} = \sum_{i=1}^{n} \left\{ \frac{(x_i - \bar{x})}{\sum_{i=1}^{n} (x_i - \bar{x})^2} \right\} y_i. \tag{19.17}$$

The importance of this expression lies in the fact that it clearly shows that the least squares estimate b_1 is a linear combination of all n response values. The coefficient of each y_i value in the linear combination is

$$\frac{(x_i - \bar{x})}{\sum_{i=1}^{n} (x_i - \bar{x})^2}.$$

The sum of squared deviations of the values of the explanatory variable from the mean is often denoted by SS_{xx}:

$$SS_{xx} = \sum_{i=1}^{n} (x_i - \bar{x})^2.$$

Using this new notation, we obtain a final expression for b_1:

$$b_1 = \sum_{i=1}^{n} \left\{ \frac{(x_i - \bar{x})}{SS_{xx}} \right\} y_i. \tag{19.18}$$

An Alternative Expression for the Estimate of the Intercept

First, we show that the least squares estimate for the intercept is also a linear combination of the response values. To this end, we rewrite Equation (19.6) as follows:

$$\begin{aligned}
b_0 &= \bar{y} - b_1 \bar{x}, \\
&= \frac{1}{n} \sum_{i=1}^{n} y_i - b_1 \bar{x}, \\
&= \frac{1}{n} \sum_{i=1}^{n} y_i - \left[\sum_{i=1}^{n} \left\{ \frac{(x_i - \bar{x})}{SS_{xx}} \right\} y_i \right] \bar{x}, \\
&= \sum_{i=1}^{n} \frac{1}{n} y_i - \sum_{i=1}^{n} \left\{ \frac{(x_i - \bar{x})\bar{x}}{SS_{xx}} \right\} y_i, \\
&= \sum_{i=1}^{n} \left\{ \frac{1}{n} - \frac{(x_i - \bar{x})\bar{x}}{SS_{xx}} \right\} y_i.
\end{aligned} \tag{19.19}$$

Rewriting Equation (19.6) in yet another way reveals an interesting property of the least squares regression line. Rearranging the terms in Equation (19.6) yields

$$\bar{y} = b_0 + b_1 \bar{x}.$$

This expression is of the same form as the equation for the regression line,

$$\hat{y} = b_0 + b_1 x.$$

This implies that the least squares regression line crosses the point $(\bar{x}, \bar{y})$, which is the mean data point. Since the regression line is supposed to provide a good summary of the bivariate data, it makes sense that it passes right through the average data point.

Some Properties of the Least Squares Residuals

When the least squares estimation approach is used for β_0 and β_1, the sum of the residuals u_i is always zero and the residuals are orthogonal to the independent variable. To show that the residuals sum to zero, we need the following derivation:

$$\sum_{i=1}^{n} u_i = \sum_{i=1}^{n} (y_i - \hat{y}_i),$$

$$= \sum_{i=1}^{n} \left\{ y_i - (b_0 + b_1 x_i) \right\},$$

$$= \sum_{i=1}^{n} (y_i - b_0 - b_1 x_i),$$

$$= \sum_{i=1}^{n} y_i - \sum_{i=1}^{n} b_0 - \sum_{i=1}^{n} b_1 x_i,$$

$$= n\bar{y} - nb_0 - b_1 \sum_{i=1}^{n} x_i,$$

$$= n\bar{y} - nb_0 - nb_1 \bar{x},$$

$$= n(\bar{y} - b_0 - b_1 \bar{x}),$$

which is always zero because, as shown above,

$$\bar{y} = b_0 + b_1 \bar{x}.$$

Note that the derivation used to show that the residuals sum to zero makes use of the facts that

$$n\bar{y} = \sum_{i=1}^{n} y_i$$

and

$$n\bar{x} = \sum_{i=1}^{n} x_i.$$

This follows from the definition of the mean. Obviously, since the sum of the residuals is zero, the average of all residuals, $\bar{u}$, is also zero.

An immediate consequence of the fact that the residuals sum to zero is that the sum of the observed response values is identical to the sum of the predicted response values:

$$\sum_{i=1}^{n} y_i = \sum_{i=1}^{n} \hat{y}_i.$$

To show that the residuals and the explanatory variable are orthogonal, we need to prove that the sum of all cross-products $u_i x_i$ is zero. This can be done in the following way:

$$\sum_{i=1}^{n} u_i x_i = \sum_{i=1}^{n} (y_i - b_0 - b_1 x_i) x_i,$$

$$= \sum_{i=1}^{n} x_i y_i - b_0 \sum_{i=1}^{n} x_i - b_1 \sum_{i=1}^{n} x_i^2,$$

$$= \sum_{i=1}^{n} x_i y_i - (\bar{y} - b_1 \bar{x}) \sum_{i=1}^{n} x_i - b_1 \sum_{i=1}^{n} x_i^2,$$

$$= \sum_{i=1}^{n} x_i y_i - \bar{y} \sum_{i=1}^{n} x_i + b_1 \bar{x} \sum_{i=1}^{n} x_i - b_1 \sum_{i=1}^{n} x_i^2,$$

$$= \sum_{i=1}^{n} x_i y_i - \frac{1}{n} \left(\sum_{i=1}^{n} x_i \right) \left(\sum_{i=1}^{n} y_i \right) + b_1 \frac{1}{n} \left(\sum_{i=1}^{n} x_i \right)^2 - b_1 \sum_{i=1}^{n} x_i^2,$$

$$= \sum_{i=1}^{n} x_i y_i - \frac{1}{n} \left(\sum_{i=1}^{n} x_i \right) \left(\sum_{i=1}^{n} y_i \right) - b_1 \left\{ \sum_{i=1}^{n} x_i^2 - \frac{1}{n} \left(\sum_{i=1}^{n} x_i \right)^2 \right\}.$$

Now, due to Equation (19.16), we have that

$$\sum_{i=1}^{n} x_i y_i - \frac{1}{n} \left(\sum_{i=1}^{n} x_i \right) \left(\sum_{i=1}^{n} y_i \right) = b_1 \left\{ \sum_{i=1}^{n} x_i^2 - \frac{1}{n} \left(\sum_{i=1}^{n} x_i \right)^2 \right\},$$

so that

$$\sum_{i=1}^{n} u_i x_i = 0.$$

The interpretation of this result is that the information contained within the residuals u_i is completely different from the information contained within the explanatory variable. In other words, the information contained within the residuals u_i reflects the remaining variation in the response values y_i after all variation that can be explained by the x_i values has been filtered out.

Based on the previous two results, it is not very difficult to show that

$$\sum_{i=1}^{n} u_i \hat{y}_i = 0$$

as well.

Table 19.4   The calculations required to verify the properties of the residuals in Example 19.2.5.

i	x_i	y_i	$\hat{y}_i$	u_i	$u_i x_i$
1	1	1	0.6	0.4	0.4
2	2	1	1.3	−0.3	−0.6
3	3	2	2.0	0.0	0.0
4	4	2	2.7	−0.7	−2.8
5	5	4	3.4	0.6	3.0
Sum	15	10	10	0.0	0.0

Example 19.2.5   *We can verify the newly derived properties of the residuals by calculating all residuals u_i and the cross-products $u_i x_i$ for the least squares regression line and the data in Example 19.2.3. The steps required to do so are shown in Table 19.4. Note that the predicted responses $\hat{y}_i$ are calculated as $-0.1 + 0.7 x_i$.*

The columns for u_i and $u_i x_i$ in the table indeed sum to zero. The sums of the observed responses y_i and the predicted responses $\hat{y}_i$ are equal. Both of them turn out to be 10.

It can also be shown that the explanatory variable and the residuals are uncorrelated. This follows from the fact that the covariance s_{XU} between the two variables is zero, as we show next:

$$s_{XU} = \frac{1}{n-1} \sum_{i=1}^{n} (u_i - \bar{u})(x_i - \bar{x}),$$

$$= \frac{1}{n-1} \sum_{i=1}^{n} (u_i - 0)(x_i - \bar{x}),$$

$$= \frac{1}{n-1} \left\{ \sum_{i=1}^{n} u_i x_i - \sum_{i=1}^{n} u_i \bar{x} \right\},$$

$$= \frac{1}{n-1} \left\{ 0 - \bar{x} \sum_{i=1}^{n} u_i \right\},$$

$$= \frac{1}{n-1} (-\bar{x} \times 0),$$

$$= 0.$$

In a similar fashion, it can be shown that the correlation between the residuals and the predicted responses is zero:

$$s_{U\hat{Y}} = \frac{1}{n-1} \sum_{i=1}^{n} (u_i - \bar{u})(\hat{y}_i - \bar{\hat{y}}),$$

$$= \frac{1}{n-1} \sum_{i=1}^{n} (u_i - 0)(\hat{y}_i - \bar{y}),$$

$$= \frac{1}{n-1} \left\{ \sum_{i=1}^{n} u_i \hat{y}_i - \sum_{i=1}^{n} u_i \bar{y} \right\},$$

$$= \frac{1}{n-1} \left\{ \sum_{i=1}^{n} u_i (b_0 + b_1 x_i) - \bar{y} \sum_{i=1}^{n} u_i \right\},$$

$$= \frac{1}{n-1} \left\{ \sum_{i=1}^{n} u_i b_0 + \sum_{i=1}^{n} u_i b_1 x_i - \bar{y} \times 0 \right\},$$

$$= \frac{1}{n-1} \left\{ b_0 \sum_{i=1}^{n} u_i + b_1 \sum_{i=1}^{n} u_i x_i \right\},$$

$$= \frac{1}{n-1} \left\{ b_0 \times 0 + b_1 \times 0 \right\},$$

$$= 0.$$

The Starch Extraction Example

Example 19.2.6 *When extracting starch from corn or rice, the yield (y) of the extraction process depends on the concentration of specific kinds of salts (x) used during the steeping phase. Starch is an important ingredient in food products and beverages.*

To maximize the yield of its starch extraction process, a multinational company has performed an experiment involving 54 tests. For these tests, corn was used from five different locations. In this way, the company is trying to ensure that the conclusions from the experiment will be valid for a wide variety of corn batches.

For each test, the yield, expressed in %, was determined. The company was interested in quantifying the dependence of the yield on the salt concentration using linear regression techniques. The data produced by the experiment is shown in Table 19.5. The table shows that

Table 19.5 The data for the starch extraction experiment in Example 19.2.6.

Concentration (x, %)	Yield (y, %)
2	40.5, 39.4, 42.3, 40.8, 39.4, 38.6, 49.3, 48.5, 38.1
3	43.2, 41.7, 41.3, 40.1, 43.5, 44.4, 51.0, 52.8, 44.7
4	50.4, 43.6, 46.6, 45.7, 42.8, 48.4, 55.8, 62.2, 50.6
	47.4, 45.0, 48.5, 47.6, 48.9, 45.8, 59.4, 55.5, 52.1
5	49.1, 53.1, 51.3, 46.9, 47.6, 50.8, 56.4, 60.5, 51.3
6	50.0, 48.3, 50.1, 49.1, 48.3, 50.2, 60.0, 60.1, 52.6

Source: Goos and Vandebroek, 2005. Reproduced with permission of Taylor and Francis.

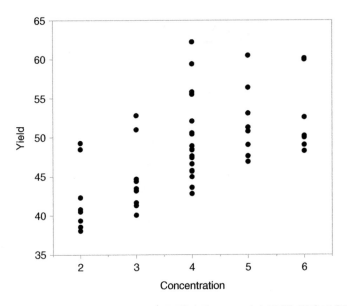

Figure 19.7   A scatter plot of the data on the starch extraction process in Table 19.5 and Example 19.2.6.

five different concentrations were tested in the experiment, ranging from 2% to 6%. The middle concentration, 4%, was used more frequently in the experiment.

A scatter plot of the 54 data points is shown in Figure 19.7. The plot shows that there is a positive relationship between the salt concentration and the yield of the extraction process. However, for any given salt concentration, the range of possible yields is quite large. In other words, there is a substantial spread in the data.

Using the "Graph Builder" in the "Graph" menu of JMP, the least squares regression line can easily be added to the scatter plot, along with the equation for the estimated model. The result is shown in Figure 19.8. The estimated model is

$$\hat{y} = 37.35 + 2.754x,$$

where $\hat{y}$ represents the predicted yield and x represents the salt concentration. The least squares estimate of the slope, $b_1 = 2.754$, means that the expected yield goes up by 2.754% each time the salt concentration is increased by 1%. The meaning of the intercept estimate, $b_0 = 37.35$, is that the expected yield is 37.35% if no salt is used; that is, if the salt concentration is 0%.

To produce Figure 19.8 using the "Graph Builder" in JMP, you have to create a scatter plot first. To this end, drag the variable "Concentration" to the "X" zone on the horizontal axis and the variable "Yield" to the "Y" zone on the vertical axis. Next, click the third button at the top of the "Graph Builder". That button shows a straight line through a scatter plot (see Figure 19.9). At the bottom left, you can then tick the option named "Equation", so that the equation for the least squares model appears in the top left corner of the plot.

There are other ways to obtain the least squares estimates of the intercept and slope of the simple linear regression model for the starch extraction data. One way is to use the

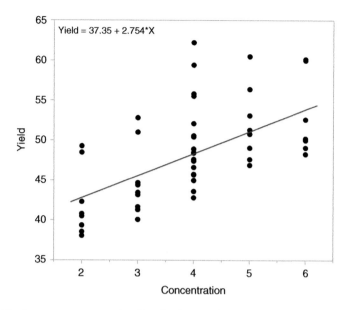

Figure 19.8   The least squares regression line for the starch extraction data in Table 19.5 and Example 19.2.6.

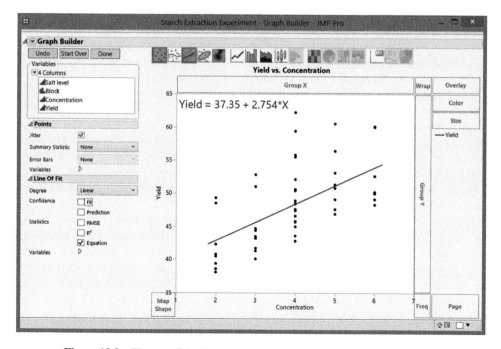

Figure 19.9   The use of the "Graph Builder" in JMP to create Figure 19.8.

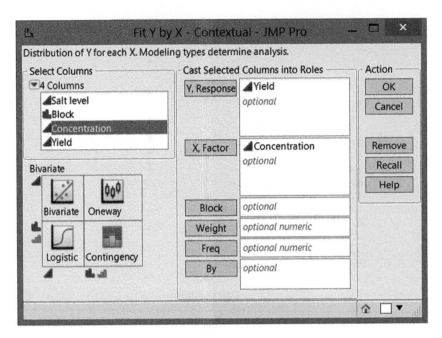

Figure 19.10   The input required to perform a simple linear regression analysis of the data in Table 19.5 and Example 19.2.6 using the "Fit Y by X" platform in the "Analyze" menu.

"Fit Y by X" *platform in the* "Analyze" *menu. The use of that platform produces the dialog window shown in Figure 19.10, in which we have to enter the variable "Yield" in the field called* "Y, Response" *and the variable "Concentration" in the field called* "X, Factor". *Initially, this only results in a scatter plot. However, next to the term* "Bivariate Fit of Yield By Concentration" *in the initial output, you can select the hotspot (red triangle) menu. This menu is shown in Figure 19.11. The option we need in that menu is named* "Fit Line". *This option produces substantial extra output, most of which we cannot interpret yet. The output, shown in Figure 19.12, shows the estimated model right under the graphical display of the data and the regression line. At the bottom of the output, the estimates b_0 and b_1, 37.3452 and 2.7544, are shown once more, next to the words "Intercept" and "Concentration".*

The output in Figure 19.12 shows another hotspot (red triangle) next to the term "Linear Fit", just below the picture of the regression line. Clicking that hotspot brings up another menu, as shown in Figure 19.13. The options that are of interest to us now are called "Save Predicteds" *and* "Save Residuals". *Selecting these options creates two extra columns in the JMP data table, containing the predicted response $\hat{y}_i$ and the residual u_i for each of the 54 data points from the starch extraction experiment. Figure 19.14 shows the modified data table with the columns for the predicted responses and the residuals.*

One interesting observation is that only five different predicted response values appear in the data table in Figure 19.14. This is due to the fact that the starch extraction experiment only involved five different salt concentrations. The predicted responses are 42.85, 45.61, 48.36,

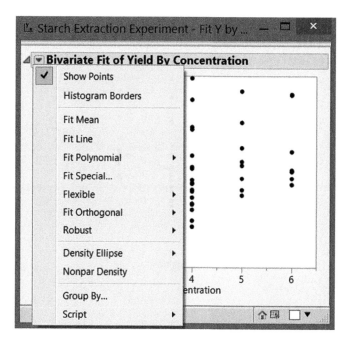

Figure 19.11   The additional options after using the "Fit Y by X" platform in Example 19.2.6.

51.12, and 53.87, and they correspond to salt concentrations of 2%, 3%, 4%, 5%, and 6%. *Right-clicking on the column header "Predicted Yield" and selecting "Formula" reveals how JMP calculates the predicted responses. The formula that JMP uses is shown in Figure 19.15 and matches the equation for the least squares regression line.*

The new data table we have created allows us to calculate the correlations between the values of the explanatory variable x_i, the observed values of the response variable y_i, the predicted values of the response variable $\hat{y}_i$, and the residuals u_i. This can be done using the "Multivariate" platform in the "Analyze" menu. The correlation matrix of the four variables is depicted in Figure 19.16.

The correlation r_{XY} between the explanatory variable and the response variable is identical to the correlation $r_{Y\hat{Y}}$ between the observed response variable and the predicted response variable, 0.6036. The correlation between the explanatory variable and the residuals is zero, as is the correlation between the predicted response and the residuals. The predicted response and the explanatory variable are perfectly positively correlated because the predicted responses are linear transformations $b_0 + b_1 x_i$ of the explanatory variables' values x_i and b_1 is positive.

It is a useful exercise to show that, for any simple linear regression, $r_{XY} = r_{Y\hat{Y}}$.

19.3   The Properties of Least Squares Estimators

As long as we have not yet collected the data, we cannot calculate the least squares estimates. In that case, we consider the individual response values to be random variables, and we

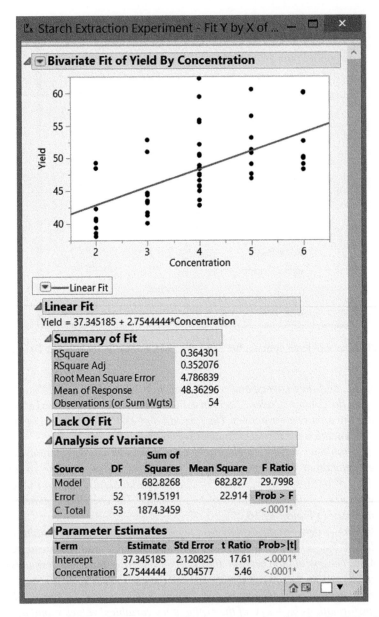

Figure 19.12   The additional output generated by the "Fit Line" option for Example 19.2.6.

denote them using uppercase letters: $Y_1, Y_2, \ldots, Y_n$. If we substitute this in Equations (19.6), (19.12), (19.13), (19.18), and (19.19), this results in the so-called least squares estimators of the intercept β_0 and β_1. We denote the least squares estimator for the intercept by $\hat{\beta}_0$ and the least squares estimator for the slope by $\hat{\beta}_1$. The least squares estimators $\hat{\beta}_0$ and $\hat{\beta}_1$ are functions of the random variables $Y_1, Y_2, \ldots, Y_n$. Therefore, they are random variables themselves. The

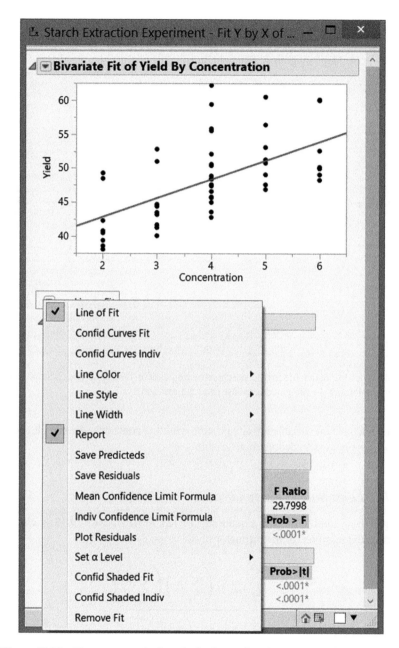

Figure 19.13 The menu required to obtain the predicted responses and the residuals.

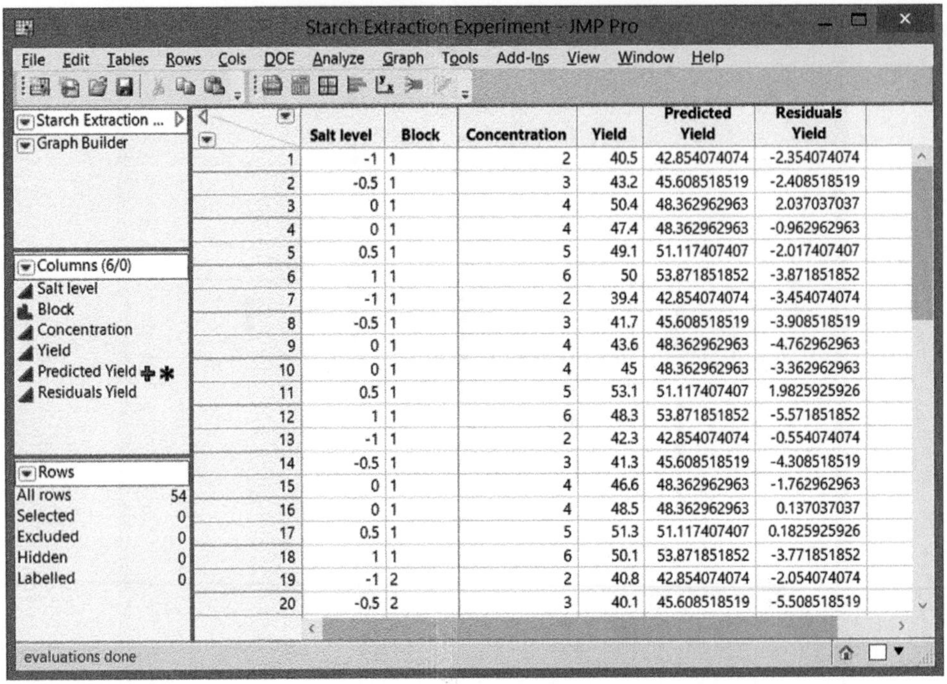

	Salt level	Block	Concentration	Yield	Predicted Yield	Residuals Yield
1	-1	1	2	40.5	42.854074074	-2.354074074
2	-0.5	1	3	43.2	45.608518519	-2.408518519
3	0	1	4	50.4	48.362962963	2.037037037
4	0	1	4	47.4	48.362962963	-0.962962963
5	0.5	1	5	49.1	51.117407407	-2.017407407
6	1	1	6	50	53.871851852	-3.871851852
7	-1	1	2	39.4	42.854074074	-3.454074074
8	-0.5	1	3	41.7	45.608518519	-3.908518519
9	0	1	4	43.6	48.362962963	-4.762962963
10	0	1	4	45	48.362962963	-3.362962963
11	0.5	1	5	53.1	51.117407407	1.9825925926
12	1	1	6	48.3	53.871851852	-5.571851852
13	-1	1	2	42.3	42.854074074	-0.554074074
14	-0.5	1	3	41.3	45.608518519	-4.308518519
15	0	1	4	46.6	48.362962963	-1.762962963
16	0	1	4	48.5	48.362962963	0.137037037
17	0.5	1	5	51.3	51.117407407	0.1825925926
18	1	1	6	50.1	53.871851852	-3.771851852
19	-1	2	2	40.8	42.854074074	-2.054074074
20	-0.5	2	3	40.1	45.608518519	-5.508518519

Figure 19.14 The JMP data table for the starch extraction data in Table 19.5 and Example 19.2.6, with extra columns containing the predicted responses and the residuals.

values of the independent variable are always considered as constants. As a result, we continue to use a lowercase letter x to denote that variable.

19.3.1 The Least Squares Estimators

Substituting Y_i for y_i and $\overline{Y}$ for $\overline{y}$ in Equations (19.12), (19.13), and (19.18) yields the three most important expressions for the estimator of the slope β_1:

$$\hat{\beta}_1 = \frac{n\sum_{i=1}^{n} x_i Y_i - \left(\sum_{i=1}^{n} x_i\right)\left(\sum_{i=1}^{n} Y_i\right)}{n\sum_{i=1}^{n} x_i^2 - \left(\sum_{i=1}^{n} x_i\right)^2}, \tag{19.20}$$

$$\hat{\beta}_1 = \frac{\sum_{i=1}^{n} (x_i - \overline{x})(Y_i - \overline{Y})}{\sum_{i=1}^{n} (x_i - \overline{x})^2}, \tag{19.21}$$

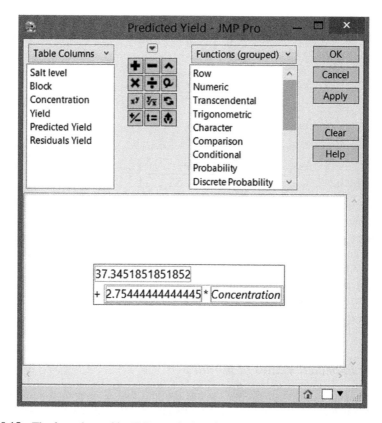

Figure 19.15 The formula used by JMP to calculate the predicted response values in Figure 19.14.

and

$$\hat{\beta}_1 = \sum_{i=1}^{n} \left(\frac{x_i - \bar{x}}{SS_{xx}} \right) Y_i. \tag{19.22}$$

Correlations

	Yield	Concentration	Predicted Yield	Residuals Yield
Yield	1.0000	0.6036	0.6036	0.7973
Concentration	0.6036	1.0000	1.0000	−0.0000
Predicted Yield	0.6036	1.0000	1.0000	−0.0000
Residuals Yield	0.7973	−0.0000	−0.0000	1.0000

Figure 19.16 The correlation matrix of the observed values of the response variable y_i, the values of the explanatory variable x_i, the predicted values of the response variable $\hat{y}_i$, and the residuals u_i in Example 19.2.6.

There are two important expressions for the least squares estimator of the intercept. They are obtained by substituting $\overline{Y}$ for $\overline{y}$ and $\hat{\beta}_1$ for b_1 in Equation (19.6) and by substituting Y_i for y_i in Equation (19.19):

$$\hat{\beta}_0 = \overline{Y} - \hat{\beta}_1 \overline{x}, \tag{19.23}$$

and

$$\hat{\beta}_0 = \sum_{i=1}^{n} \left\{ \frac{1}{n} - \frac{(x_i - \overline{x})\overline{x}}{SS_{xx}} \right\} Y_i. \tag{19.24}$$

In this section, we will make extensive use of Equations (19.22) and (19.24) because they express the least squares estimators of the intercept and the slope as linear combinations of the response values Y_i. The least squares estimators $\hat{\beta}_0$ and $\hat{\beta}_1$ are called **linear estimators** because they are linear functions of the values Y_i.

The fact that the least squares estimators $\hat{\beta}_0$ and $\hat{\beta}_1$ are linear combinations of random variables $Y_1, Y_2, \ldots, Y_n$ and, therefore, are random variables themselves, implies that they have a certain expected value, a certain variance, and a certain distribution.

19.3.2   The Expected Values of $\hat{\beta}_0$ and $\hat{\beta}_1$

To derive the expected value of $\hat{\beta}_1$, we exploit the fact that an expected value of a linear combination of random variables is the linear combination of the expected values of the random variables (see the book *Statistics with JMP: Graphs, Descriptive Statistics and Probability*). When applying this result to Equation (19.22), we obtain

$$E\left(\hat{\beta}_1\right) = E\left[\frac{1}{SS_{xx}} \sum_{i=1}^{n} (x_i - \overline{x}) Y_i\right],$$

$$= \frac{1}{SS_{xx}} \sum_{i=1}^{n} (x_i - \overline{x}) E(Y_i). \tag{19.25}$$

To be able to continue, we need to make a first important assumption about our model:

Assumption 1   *We assume that there is a linear relation $Y_i = \beta_0 + \beta_1 x_i + U_i$ between the explanatory variable x_i and the response Y_i with unknown constants β_0 and β_1, and that $E(U_i) = 0$. Equivalently, we assume that $E(Y_i) = \beta_0 + \beta_1 x_i$.*

One way to rephrase this assumption is as follows: our model is correctly specified. This means, for example, that we do not overlook any other explanatory variable(s), and that neither the response variable nor the explanatory variable has to be transformed (e.g., using a logarithmic transformation) for the linear model to be true.

Under the assumption that the model is correctly specified, the expected value of $\hat{\beta}_1$ becomes

$$E\left(\hat{\beta}_1\right) = \frac{1}{SS_{xx}} \sum_{i=1}^{n} (x_i - \bar{x})(\beta_0 + \beta_1 x_i),$$

$$= \frac{1}{SS_{xx}} \left\{ \sum_{i=1}^{n} (x_i - \bar{x})\beta_0 + \sum_{i=1}^{n} (x_i - \bar{x})\beta_1 x_i \right\},$$

$$= \frac{1}{SS_{xx}} \left\{ \beta_0 \sum_{i=1}^{n} (x_i - \bar{x}) + \beta_1 \sum_{i=1}^{n} (x_i - \bar{x})x_i \right\}, \qquad (19.26)$$

$$= \frac{1}{SS_{xx}} \left\{ \beta_0 \times 0 + \beta_1 \sum_{i=1}^{n} (x_i - \bar{x})x_i \right\},$$

$$= \frac{\beta_1}{SS_{xx}} \sum_{i=1}^{n} (x_i - \bar{x})x_i.$$

Because

$$\sum_{i=1}^{n} (x_i - \bar{x})^2 = \sum_{i=1}^{n} (x_i - \bar{x})(x_i - \bar{x}),$$

$$= \sum_{i=1}^{n} (x_i - \bar{x})x_i - \sum_{i=1}^{n} (x_i - \bar{x})\bar{x},$$

$$= \sum_{i=1}^{n} (x_i - \bar{x})x_i - \bar{x} \sum_{i=1}^{n} (x_i - \bar{x}),$$

$$= \sum_{i=1}^{n} (x_i - \bar{x})x_i - \bar{x} \times 0,$$

$$= \sum_{i=1}^{n} (x_i - \bar{x})x_i,$$

this expression can be rewritten as

$$E\left(\hat{\beta}_1\right) = \frac{\beta_1}{SS_{xx}} \sum_{i=1}^{n} (x_i - \bar{x})^2,$$

$$= \frac{\beta_1}{SS_{xx}} SS_{xx}, \qquad (19.27)$$

$$= \beta_1.$$

This result means that the least squares estimator $\hat{\beta}_1$ is an unbiased estimator of the slope β_1.

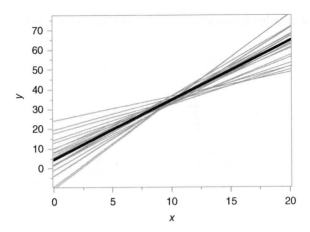

Figure 19.17   Twenty different regression lines obtained by 20 different researchers studying the relation between x and y shown by means of the black line.

In a similar way, it is possible to show that $E(\hat{\beta}_0) = \beta_0$. Therefore, the least squares estimator $\hat{\beta}_0$ is an unbiased estimator of the intercept β_0.

The unbiasedness of the estimators means that, if we were to take infinitely many different samples for the explanatory variable and the response variable, and if we were to calculate the least squares regression line separately for each sample, then the averages of all intercept estimates and all slope estimates would be equal to β_0 and β_1, respectively.

19.3.3   A Demonstration of Unbiasedness by Means of a Simulation Study

Figure 19.17 graphically illustrates the meaning of the unbiasedness of the least squares estimates. The black line in the figure represents the true relationship between the response variable and the explanatory variable. The black line has an intercept of 5 and a slope of 3. Therefore, the "unknown" parameters β_0 and β_1 equal 5 and 3, respectively, in this illustration.

The 20 gray lines in the figure represent the regression lines obtained by 20 fictitious researchers, each of whom (i) took a sample of 10 observations of the explanatory variable and the response variable and (ii) fitted the least squares regression line to the data in that sample. In doing so, each individual researcher obtained a different sample, a different estimate for the intercept, a different estimate for the slope, and, hence, a different regression line.

Figure 19.17 shows that some researchers have been lucky: their estimated regression line is very close to the black line representing the true relationship between the two variables under study. Other researchers have been less lucky. Some researchers have obtained an estimated regression line that is substantially steeper than the black line, while others have obtained an estimated regression line that is much flatter than the black line. Note the use of the word "lucky" here. All the researchers in this demonstration collected data in the same way: they randomly picked 20 objects from a population and measured the values of the x and y variables.

In this way, some researcher happened to get "good" data, which resulted in a regression line close to the black line. Other researchers obtained "poor" data, incorrectly suggesting that the slope was very large or very small.

In Figure 19.17, the largest estimate for the slope was 4.43, while the smallest estimate was 1.23. The largest estimate for the intercept was 24.39, while the smallest estimate was −9.97. The mean estimate for the slope, however, was 2.85, and the mean estimate for the intercept was 6.12. Therefore, the mean estimates are reasonably close to the values of β_0 and β_1, 5 and 3.

The unbiasedness of the least squares estimates now means that if we carry out this exercise a large number of times (e.g., if we use 1000 or 10 000 fictitious researchers), then the mean estimates will match the β_0 and β_1 values almost exactly.

In JMP, a script named "Demonstrate Regression" is available to perform the kind of simulation that we have discussed here.

19.3.4   The Variances of $\hat{\beta}_0$ and $\hat{\beta}_1$

To calculate the variance of the least squares estimators $\hat{\beta}_0$ and $\hat{\beta}_1$, we again exploit the fact that both estimators are linear functions of the responses Y_i. In general, the variance of a linear function of random variables depends on the variances of the individual random variables and on their covariances (see *Statistics with JMP: Graphs, Descriptive Statistics and Probability*). In simple linear regression, however, it is usually assumed that all observations are independent. In that case, the covariance of any pair of responses is zero.

In technical terms, the assumption of independence is written as follows:

Assumption 2   *We assume that the random errors U_i are independent, and, hence, that $cov(U_i, U_j) = 0$ whenever $i \neq j$.*

Another way of phrasing this assumption is to say that the sample observations are made independently, so that the responses Y_i, $i = 1, 2, \ldots, n$, are independent. In that case, $cov(Y_i, Y_j) = 0$ whenever $i \neq j$.

In general, it is not easy to ensure that all observations in a sample are independent. For example, if an experiment takes more than one day or more than one batch of raw material, then the response values obtained on the same day or using the same batch are generally correlated, and therefore statistically dependent. In economics, it is common to collect data at different time points. Typically, responses obtained at two successive time points are correlated, whereas responses observed at two substantially different points in time are uncorrelated. It is important to recognize these kinds of scenarios and to move to more advanced regression techniques whenever the responses indeed turn out to be correlated. We discuss these techniques in *Statistics with JMP: Linear and Generalized Linear Models*.

We also make an assumption about the variances of the individual error terms and the individual responses. More specifically, we assume that all individual variances are identical. This assumption is called the **homoscedasticity** assumption.

Assumption 3   *We assume that* var $(U_i) = \sigma^2(U_i) = \sigma^2$: *the variance of the error term is unknown, but it is the same for each observation.*

Note that if var $(U_i) = \sigma^2$, then we also have that

$$
\begin{aligned}
\text{var}(Y_i) &= \text{var}(\beta_0 + \beta_1 x_i + U_i), \\
&= \text{var}(\beta_0) + \text{var}(\beta_1 x_i) + \text{var}(U_i), \\
&= 0 + 0 + \sigma^2, \\
&= \sigma^2.
\end{aligned}
$$

The variances var (β_0) and var $(\beta_1 x_i)$ in this derivation are zero because β_0 and β_1 are unknown but constant parameters (so they do not vary), and the values x_i of the explanatory variable are also considered constant. So, in summary, we assume that all individual responses have variance σ^2.

Using Assumptions 2 and 3, we can derive the variance of the least squares estimator $\hat{\beta}_1$ of the slope:

$$
\begin{aligned}
\text{var}(\hat{\beta}_1) = \sigma^2_{\hat{\beta}_1} &= \text{var}\left\{ \sum_{i=1}^{n} \left(\frac{x_i - \bar{x}}{\text{SS}_{xx}} \right) Y_i \right\}, \\
&= \sum_{i=1}^{n} \left(\frac{x_i - \bar{x}}{\text{SS}_{xx}} \right)^2 \text{var}(Y_i), \\
&= \frac{1}{(\text{SS}_{xx})^2} \sum_{i=1}^{n} (x_i - \bar{x})^2 \sigma^2, \\
&= \frac{1}{(\text{SS}_{xx})^2} (\text{SS}_{xx}) \sigma^2, \\
&= \frac{\sigma^2}{\text{SS}_{xx}}.
\end{aligned}
\tag{19.28}
$$

We can also derive the variance of the least squares estimator $\hat{\beta}_0$ of the intercept:

$$
\begin{aligned}
\text{var}(\hat{\beta}_0) = \sigma^2_{\hat{\beta}_0} &= \text{var}\left\{ \sum_{i=1}^{n} \left[\frac{1}{n} - \frac{\bar{x}(x_i - \bar{x})}{\text{SS}_{xx}} \right] Y_i \right\}, \\
&= \sum_{i=1}^{n} \left[\frac{1}{n} - \frac{\bar{x}(x_i - \bar{x})}{\text{SS}_{xx}} \right]^2 \text{var}(Y_i), \\
&= \sum_{i=1}^{n} \left[\frac{1}{n} - \frac{\bar{x}(x_i - \bar{x})}{\text{SS}_{xx}} \right]^2 \sigma^2, \\
&= \sigma^2 \sum_{i=1}^{n} \left[\frac{1}{n^2} - 2\frac{1}{n}\frac{\bar{x}(x_i - \bar{x})}{\text{SS}_{xx}} + \frac{\bar{x}^2(x_i - \bar{x})^2}{\text{SS}^2_{xx}} \right],
\end{aligned}
$$

$$= \sigma^2 \left[n\frac{1}{n^2} - 2\frac{1}{n}\frac{\bar{x}}{\mathrm{SS}_{xx}} \sum_{i=1}^{n}(x_i - \bar{x}) + \frac{\bar{x}^2}{\mathrm{SS}_{xx}^2} \sum_{i=1}^{n}(x_i - \bar{x})^2 \right],$$

$$= \sigma^2 \left[\frac{1}{n} - 0 + \frac{\bar{x}^2 \mathrm{SS}_{xx}}{\mathrm{SS}_{xx}^2} \right],$$

$$= \sigma^2 \left[\frac{1}{n} + \frac{\bar{x}^2}{\mathrm{SS}_{xx}} \right],$$

$$= \sigma^2 \left[\frac{\mathrm{SS}_{xx}}{n\mathrm{SS}_{xx}} + \frac{n\bar{x}^2}{n\mathrm{SS}_{xx}} \right],$$

$$= \sigma^2 \left[\frac{\sum_{i=1}^{n}(x_i - \bar{x})^2 + n\bar{x}^2}{n\mathrm{SS}_{xx}} \right],$$

$$= \sigma^2 \left[\frac{\sum_{i=1}^{n}x_i^2 - 2\bar{x}\sum_{i=1}^{n}x_i + n\bar{x}^2 + n\bar{x}^2}{n\mathrm{SS}_{xx}} \right],$$

$$= \sigma^2 \left[\frac{\sum_{i=1}^{n}x_i^2 - 2n\bar{x}^2 + 2n\bar{x}^2}{n\mathrm{SS}_{xx}} \right],$$

$$= \sigma^2 \left[\frac{\sum_{i=1}^{n}x_i^2}{n\mathrm{SS}_{xx}} \right]. \tag{19.29}$$

It can also be shown that the covariance between $\hat{\beta}_0$ and $\hat{\beta}_1$ is

$$\mathrm{cov}(\hat{\beta}_0, \hat{\beta}_1) = \sigma_{\hat{\beta}_0 \hat{\beta}_1} = -\sigma^2 \frac{\bar{x}}{\mathrm{SS}_{xx}}. \tag{19.30}$$

The proof of this result is fairly complicated, so we do not give it here.

The covariance between $\hat{\beta}_0$ and $\hat{\beta}_1$ is negative if the average value of the explanatory variable in the data set, $\bar{x}$, is positive. This can be observed in Figure 19.17. In that figure, the regression lines that have a large estimate for the intercept are generally quite flat (and so have a small estimate for the slope). The regression lines that have a small estimate for the intercept are generally quite steep (and so have a large estimate for the slope).

When the average value of the explanatory variable in the data set is negative, the covariance between $\hat{\beta}_0$ and $\hat{\beta}_1$ is positive.

19.3.5   The Gauss–Markov Theorem

In Chapter 1, we explained that, ideally, estimators are unbiased and efficient or precise. We already know that if the model is correctly specified, the least squares estimators $\hat{\beta}_0$ and $\hat{\beta}_1$ are unbiased estimators of the intercept β_0 and the slope β_1. Therefore, it is natural to wonder whether the estimators are also efficient. The Gauss–Markov theorem provides an answer to that question: the theorem tells us that the least squares estimators are the most precise estimators that can be written as a linear combination of the response values Y_i. This result is usually phrased as "the least squares estimators $\hat{\beta}_0$ and $\hat{\beta}_1$ are the **best linear unbiased estimators**", often abbreviated as **BLUE**.

Theorem 19.3.1   *(Gauss–Markov theorem) The least squares estimator $\hat{\beta}_1$ $[\hat{\beta}_0]$ has the smallest variance of all unbiased estimators for β_1 $[\beta_0]$ that are linear combinations of the responses Y_i.*

Proof.   We provide a proof for the least squares estimator $\hat{\beta}_1$ of the slope β_1. In general, any linear estimator for β_1 can be written as

$$\hat{\beta}_1^* = c_0 + \sum_{i=1}^{n} c_i Y_i.$$

The expected value of such an estimator is

$$E\left(\hat{\beta}_1^*\right) = E\left(c_0 + \sum_{i=1}^{n} c_i Y_i\right),$$

$$= c_0 + \sum_{i=1}^{n} c_i E(Y_i),$$

$$= c_0 + \sum_{i=1}^{n} c_i(\beta_0 + \beta_1 x_i)$$

$$= c_0 + \beta_0 \sum_{i=1}^{n} c_i + \beta_1 \sum_{i=1}^{n} c_i x_i.$$

For $\hat{\beta}_1^*$ to be an unbiased estimator of β_1, its expected value should be equal to β_1. Therefore, the following equation should hold:

$$c_0 + \beta_0 \sum_{i=1}^{n} c_i + \beta_1 \sum_{i=1}^{n} c_i x_i = \beta_1.$$

So, for $\hat{\beta}_1^*$ to be an unbiased estimator of β_1, the following conditions should be met:

$$c_0 = 0, \tag{19.31}$$

$$\sum_{i=1}^{n} c_i = 0, \tag{19.32}$$

and

$$\sum_{i=1}^{n} c_i x_i = 1. \tag{19.33}$$

Assuming that $\hat{\beta}_1^*$ is unbiased, and, hence, $c_0 = 0$, we can derive its variance as

$$\text{var}\left(\hat{\beta}_1^*\right) = \text{var}\left(\sum_{i=1}^{n} c_i Y_i\right),$$

$$= \sum_{i=1}^{n} c_i^2 \text{var}\left(Y_i\right),$$

$$= \sum_{i=1}^{n} c_i^2 \sigma^2,$$

$$= \sigma^2 \sum_{i=1}^{n} c_i^2.$$

The best linear unbiased estimator has the smallest possible value for this variance. So, if we want to find the best linear unbiased estimator, we need to find the values for c_i that minimize var $(\hat{\beta}_1^*)$. While minimizing, however, we need to take into account that the estimator should be unbiased, which means that Equations (19.32) and (19.33) should hold. Therefore, we should perform a constrained minimization, using Lagrange multipliers[1].

To minimize

$$\sum_{i=1}^{n} c_i^2$$

with respect to the c_i values subject to the constraints

$$\sum_{i=1}^{n} c_i = 0$$

and

$$\sum_{i=1}^{n} c_i x_i = 1,$$

we need to minimize the function

$$\sum_{i=1}^{n} c_i^2 - \mu_1\left(\sum_{i=1}^{n} c_i - 0\right) - \mu_2\left(\sum_{i=1}^{n} c_i x_i - 1\right).$$

---

[1] The method of Lagrange multipliers is a standard technique for optimizing functions subject to one or more equality constraints. For each constraint considered, there is one multiplier. The method is named after the Italian mathematician Joseph-Louis Lagrange.

In this function, μ_1 and μ_2 represent the Lagrange multipliers. We now need to set the first derivatives of this function with respect to the n different c_i values, μ_1 and μ_2 to zero. The first derivatives with respect to $c_1, c_2, \ldots, c_n$ are

$$2c_1 - \mu_1 - \mu_2 x_1 = 0, \tag{19.34}$$

$$2c_2 - \mu_1 - \mu_2 x_2 = 0, \tag{19.35}$$

$$\vdots$$

$$2c_n - \mu_1 - \mu_2 x_n = 0. \tag{19.36}$$

Setting the derivatives with respect to μ_1 and μ_2 to zero yields

$$\sum_{i=1}^{n} c_i = 0 \tag{19.37}$$

and

$$\sum_{i=1}^{n} c_i x_i = 1. \tag{19.38}$$

Summing Equations (19.34)–(19.36) results in

$$2 \sum_{i=1}^{n} c_i - n\mu_1 - \mu_2 \sum_{i=1}^{n} x_i = 0.$$

Due to Equation (19.37), this can be rewritten as

$$-n\mu_1 - \mu_2 \sum_{i=1}^{n} x_i = 0,$$

and

$$\mu_1 = -\mu_2 \bar{x}.$$

Substituting this into Equations (19.34)–(19.36) yields

$$c_1 = \frac{1}{2} \mu_2 (x_1 - \bar{x}), \tag{19.39}$$

$$c_2 = \frac{1}{2} \mu_2 (x_2 - \bar{x}), \tag{19.40}$$

$$\vdots$$

$$c_n = \frac{1}{2} \mu_2 (x_n - \bar{x}). \tag{19.41}$$

Substituting Equations (19.39)–(19.41) in Equation (19.38) yields

$$\sum_{i=1}^{n} \frac{1}{2} \mu_2 (x_i - \bar{x}) x_i = 1,$$

which can be rewritten as

$$\sum_{i=1}^{n} \frac{1}{2} \mu_2 (x_i - \bar{x})^2 = 1,$$

$$\frac{1}{2} \mu_2 \sum_{i=1}^{n} (x_i - \bar{x})^2 = 1,$$

$$\frac{1}{2} \mu_2 (SS_{xx}) = 1,$$

and

$$\mu_2 = \frac{2}{SS_{xx}}. \tag{19.42}$$

Combining Equation (19.42) with Equations (19.39)–(19.41) yields

$$c_1 = \frac{x_1 - \bar{x}}{SS_{xx}}, \tag{19.43}$$

$$c_2 = \frac{x_2 - \bar{x}}{SS_{xx}}, \tag{19.44}$$

$$\vdots$$

$$c_n = \frac{x_n - \bar{x}}{SS_{xx}}. \tag{19.45}$$

In conclusion, when minimizing the variance of the linear estimator $\hat{\beta}_1^*$, the coefficient of any response value Y_i is of the form

$$c_i = \frac{x_i - \bar{x}}{SS_{xx}},$$

which is exactly the coefficient of Y_i when we use the least squares estimator for β_1. This completes the proof that the least squares estimator is the most precise estimator that can be written as a linear combination of the response values. ∎

19.4 The Estimation of σ^2

Equations (19.28) and (19.29) provide an expression for the variances of the least squares estimators $\hat{\beta}_0$ and $\hat{\beta}_1$ that we derived in Section 19.3.4. Both variances are proportional to σ^2, the common variance of the error terms $U_1, U_2, \ldots, U_n$ and the response values $Y_1, Y_2, \ldots, Y_n$. Just like β_0 and β_1, this common variance σ^2 is an unknown parameter. If we want to calculate the variance of the least squares estimators, we have to estimate σ^2 from the sample data as well.

It can be shown that

$$\hat{\sigma}^2 = \frac{\sum_{i=1}^{n}(Y_i - \hat{Y}_i)^2}{n-2}$$

is an unbiased estimator of σ^2, if the regression model is correctly specified. This estimator is similar in structure to the estimator for the population variance in Chapter 1. The numerator of the estimator is the sum of the squared residuals, which is often called the **residual sum of squares** or **sum of squared errors** and denoted by SSE. Therefore, the estimator of the variance σ^2 in regression is often written as

$$\hat{\sigma}^2 = \frac{\text{SSE}}{n-2} = \text{MSE},$$

where MSE is short for **mean squared error**.

The division by $n-2$ in this estimator can be explained by the fact that we start from n observations (n units of information) and, in order to be able to calculate the predicted responses $\hat{Y}_1, \hat{Y}_2, \ldots, \hat{Y}_n$, we need to estimate two parameters, β_0 and β_1 (which costs us two units of information). In the technical jargon, $n-2$ is the number of degrees of freedom associated with the sum of squared errors. That number is often called the **residual degrees of freedom**.

After we have collected the responses and estimated the regression line, we can calculate the estimate s^2 of σ^2:

$$s^2 = \frac{\sum_{i=1}^{n}(y_i - \hat{y}_i)^2}{n-2} = \frac{\text{sse}}{n-2} = \text{mse}.$$

Note that, here, we use lowercase letters for the sum of squared errors and the mean squared error because their values can be calculated as soon as we have collected the data.

Now that we are able to estimate σ^2, we can also estimate the variances of the least squares estimators $\hat{\beta}_0$ and $\hat{\beta}_1$ of the intercept and the slope. The actual variance of $\hat{\beta}_0$ was shown to be

$$\text{var}(\hat{\beta}_0) = \sigma^2_{\hat{\beta}_0} = \sigma^2 \left[\frac{\sum_{i=1}^{n} x_i^2}{n\text{SS}_{xx}} \right],$$

while the variance of $\hat{\beta}_1$ was shown to be

$$\text{var}(\hat{\beta}_1) = \sigma^2_{\hat{\beta}_1} = \frac{\sigma^2}{\text{SS}_{xx}}.$$

Unbiased estimators for these variances are

$$\hat{\sigma}^2_{\hat{\beta}_0} = \hat{\sigma}^2 \left[\frac{\sum_{i=1}^n x_i^2}{n\text{SS}_{xx}} \right] = \text{MSE} \left[\frac{\sum_{i=1}^n x_i^2}{n\text{SS}_{xx}} \right]$$

and

$$\hat{\sigma}^2_{\hat{\beta}_1} = \frac{\hat{\sigma}^2}{\text{SS}_{xx}} = \frac{\text{MSE}}{\text{SS}_{xx}}.$$

As a result, an appropriate estimate of the variance of $\hat{\beta}_0$ is

$$s^2_{\hat{\beta}_0} = s^2 \left[\frac{\sum_{i=1}^n x_i^2}{n\text{SS}_{xx}} \right] = \text{mse} \left[\frac{\sum_{i=1}^n x_i^2}{n\text{SS}_{xx}} \right],$$

while a suitable estimate of the variance of $\hat{\beta}_1$ is

$$s^2_{\hat{\beta}_1} = \frac{s^2}{\text{SS}_{xx}} = \frac{\text{mse}}{\text{SS}_{xx}}.$$

The square roots of these estimated variances, $s_{\hat{\beta}_0}$ and $s_{\hat{\beta}_1}$, are the estimated standard deviations or standard errors of the least squares estimators $\hat{\beta}_0$ and $\hat{\beta}_1$. They are reported in every output produced by software packages when carrying out a simple linear regression.

The estimator $\hat{\sigma}$ of the standard deviation of the error terms and the responses, σ, is a biased estimator. More specifically,

$$E(\hat{\sigma}) < \sigma,$$

even though

$$E(\hat{\sigma}^2) = \sigma^2.$$

Consequently,

$$E(\hat{\sigma}_{\hat{\beta}_0}) < \sigma_{\hat{\beta}_0}$$

and

$$E(\hat{\sigma}_{\hat{\beta}_1}) < \sigma_{\hat{\beta}_1}.$$

As a result, the estimate s for σ will generally be too small, and the estimated standard errors $s_{\hat{\beta}_0}$ and $s_{\hat{\beta}_1}$ of the least squares estimators for the intercept and the slope will generally also be smaller than the unknown true standard errors $\sigma_{\hat{\beta}_0}$ and $\sigma_{\hat{\beta}_1}$.

Example 19.4.1   *In Example 19.2.3, where the small data set from Table 19.1 was used, we obtained the value 1.1 for the sse, the sum of the squared errors or the sum of the squared residuals. This implies that the estimate s^2 for σ^2 is*

$$s^2 = \text{mse} = \frac{\text{sse}}{n-2} = \frac{1.1}{5-2} = \frac{11}{30} = 0.3667.$$

In Table 19.3, we can see that

$$\text{SS}_{xx} = \sum_{i=1}^{n}(x_i - \bar{x})^2 = 10,$$

so that the estimated variance of $\hat{\beta}_1$ is

$$s_{\hat{\beta}_1}^2 = \frac{s^2}{\text{SS}_{xx}} = \frac{11}{30} \cdot \frac{1}{10} = \frac{11}{300} = 0.03667,$$

and the estimated standard deviation or standard error of $\hat{\beta}_1$ is

$$s_{\hat{\beta}_1} = \sqrt{\frac{11}{300}} = 0.1915.$$

The estimated variance of $\hat{\beta}_0$ is

$$s_{\hat{\beta}_0}^2 = s^2 \left[\frac{\sum_{i=1}^{n} x_i^2}{n\text{SS}_{xx}}\right] = \frac{11}{30}\left[\frac{55}{5\times 10}\right] = \frac{121}{300} = 0.4033,$$

and the estimated standard deviation or standard error of $\hat{\beta}_0$ is

$$s_{\hat{\beta}_0} = \sqrt{\frac{121}{300}} = \frac{11}{10\sqrt{3}} = 0.6351.$$

The estimated standard errors are shown in the JMP output in Figure 19.18, in the column labeled "Std Error" in the last table of the output. The mse value of 0.3667 is also shown in the output, namely in the column labeled "Mean Square" in the table named "Analysis of Variance". The estimate for σ is shown as the "Root Mean Square Error" in the top part of the output. That estimate is

$$s = \sqrt{\text{mse}} = \sqrt{\frac{11}{30}} = 0.6055.$$

Linear Fit

y = −0.1 + 0.7*x

Summary of Fit

RSquare	0.816667
RSquare Adj	0.755556
Root Mean Square Error	0.60553
Mean of Response	2
Observations (or Sum Wgts)	5

Analysis of Variance

Source	DF	Sum of Squares	Mean Square	F Ratio
Model	1	4.9000000	4.90000	13.3636
Error	3	1.1000000	0.36667	**Prob > F**
C. Total	4	6.0000000		0.0354*

Parameter Estimates

Term	Estimate	Std Error	t Ratio	Prob>|t|
Intercept	−0.1	0.635085	−0.16	0.8849
x	0.7	0.191485	3.66	0.0354*

Figure 19.18 The JMP output for the simple linear regression analysis of the data in Table 19.1.

19.5 Statistical Inference for β_0 and β_1

Insightful statistical analyses go beyond the mere estimation of unknown quantities, and involve the construction of confidence intervals and the testing of hypotheses. These additional steps are commonly called statistical inference. They require knowledge of the distributions of the estimators. In the context of simple linear regression, this means that the distributions of $\hat{\beta}_0$ and $\hat{\beta}_1$ need to be known.

19.5.1 The Normal Distribution of the Least Squares Estimators

To construct confidence intervals and perform hypothesis tests for β_0 and β_1, we need to know the probability densities of the least squares estimators $\hat{\beta}_0$ and $\hat{\beta}_1$. By making a fourth assumption concerning our regression model, we are able to derive the probability densities of the two estimators:

Assumption 4 *We assume that the error terms U_i are normally distributed.*

This assumption is not only mathematically convenient, but it also makes sense if we think about the error terms. The error terms U_i describe the fact that the responses Y_i deviate to some extent from their expected value, $\beta_0 + \beta_1 x_i$. These deviations occur because there exist many factors that influence the response to some extent, but that are not incorporated in the model. Therefore, the error term can be thought of as the sum of the influences of all kinds of variables that impact the response and that differ from the explanatory variable. According to the central limit theorem, the sum or the linear combination of a large number of (independent) random variables is approximately normally distributed. Therefore, the assumption that the error terms are normally distributed is justifiable.

Under the assumption of normality for the error terms, the response variables Y_i are also normally distributed. This is because they are sums of an (unknown) constant and a normally distributed variable,

$$Y_i = \beta_0 + \beta_1 x_i + U_i.$$

Each response is therefore a linear transformation of a normally distributed random variable, and such a transformation is also normally distributed (see *Statistics with JMP: Graphs, Descriptive Statistics and Probability*).

Summarizing the four assumptions that we have made, we have that all the error terms are independent normally distributed random variables with zero mean and variance σ^2:

$$U_i \sim N(0, \sigma^2).$$

Equivalently, all the responses are independent normally distributed random variables with mean $\beta_0 + \beta_1 x_i$ and variance σ^2:

$$Y_i \sim N(\beta_0 + \beta_1 x_i, \sigma^2).$$

In previous sections, we have emphasized on multiple occasions that the least squares estimators $\hat{\beta}_0$ and $\hat{\beta}_1$ are linear combinations of the Y_i values. Since linear combinations of normally distributed random variables are also normally distributed (see *Statistics with JMP: Graphs, Descriptive Statistics and Probability*), the least squares estimators are also normally distributed:

$$\hat{\beta}_0 \sim N\left(\beta_0, \sigma^2 \left[\frac{\sum_{i=1}^{n} x_i^2}{n\text{SS}_{xx}}\right]\right)$$

and

$$\hat{\beta}_1 \sim N\left(\beta_1, \frac{\sigma^2}{\text{SS}_{xx}}\right).$$

19.5.2   A χ^2-Distribution for SSE

Another consequence of the fact that the error terms U_i and the responses Y_i are assumed to be normally distributed is that there is a relation between the sum of squared errors, SSE, and the χ^2-distribution. More specifically, the ratio SSE/σ^2 is χ^2- distributed with $n - 2$ degrees of freedom:

$$\frac{\text{SSE}}{\sigma^2} = \frac{(n-2)\text{MSE}}{\sigma^2} \sim \chi^2_{n-2}. \tag{19.46}$$

To understand this result, which is very similar to that for S^2 in Chapter 1, we have to remember that the sum of squared errors, SSE, is the sum of the squared residuals. As long as we do not have data, we cannot compute the residuals, and we consider them as random variables. We denote these random variables by $\hat{U}_i$. They are defined as

$$\hat{U}_i = Y_i - \hat{Y}_i = Y_i - \hat{\beta}_0 - \hat{\beta}_1 x_i.$$

Since $\hat{U}_i$ is a linear combination of normally distributed variables, it is also normally distributed. Its expected value is zero and its variance is σ^2. Therefore, $\hat{U}_i/\sigma$ is standard normally distributed, and

$$\frac{\text{SSE}}{\sigma^2} = \frac{1}{\sigma^2} \sum_{i=1}^{n} \hat{U}_i^2 = \sum_{i=1}^{n} \left(\frac{\hat{U}_i}{\sigma} \right)^2$$

is the sum of n squared standard normally distributed random variables. If all these standard normally distributed random variables were independent, then SSE/σ^2 would be χ^2-distributed with n degrees of freedom. However, the $\hat{U}_i$ variables all depend on the least squares estimators for β_0 and β_1. Because of this dependence on two estimators, two degrees of freedom are lost. As a result, SSE/σ^2 is χ^2-distributed with $n - 2$ degrees of freedom.

19.5.3   The t-Distribution

The normal distributions for $\hat{\beta}_0$ and $\hat{\beta}_1$ imply that

$$\frac{\hat{\beta}_0 - \beta_0}{\sigma_{\hat{\beta}_0}}$$

and

$$\frac{\hat{\beta}_1 - \beta_1}{\sigma_{\hat{\beta}_1}}$$

are standard normally distributed random variables. The denominators of these expressions, the standard errors $\sigma_{\hat{\beta}_0}$ and $\sigma_{\hat{\beta}_1}$ of the least squares estimators, need to be estimated. Their estimators are

$$\hat{\sigma}_{\hat{\beta}_1} = \frac{\hat{\sigma}}{\sqrt{\text{SS}_{xx}}} = \sqrt{\frac{\text{MSE}}{\text{SS}_{xx}}}$$

and

$$\hat{\sigma}_{\hat{\beta}_0} = \hat{\sigma} \sqrt{\frac{\sum_{i=1}^{n} x_i^2}{n \text{SS}_{xx}}} = \sqrt{\text{MSE} \left(\frac{\sum_{i=1}^{n} x_i^2}{n \text{SS}_{xx}} \right)}.$$

Substituting these estimators in the above fractions yields

$$\frac{\hat{\beta}_0 - \beta_0}{\hat{\sigma}_{\hat{\beta}_0}}$$

and

$$\frac{\hat{\beta}_1 - \beta_1}{\hat{\sigma}_{\hat{\beta}_1}}.$$

These fractions are no longer standard normally distributed. Instead, they are t-distributed with $n - 2$ degrees of freedom. To show this, for example, for the latter expression, we need to rewrite it as follows:

$$\frac{\hat{\beta}_1 - \beta_1}{\hat{\sigma}_{\hat{\beta}_1}} = \frac{\frac{\hat{\beta}_1 - \beta_1}{\sigma_{\hat{\beta}_1}}}{\frac{\hat{\sigma}_{\hat{\beta}_1}}{\sigma_{\hat{\beta}_1}}} = \frac{\frac{\hat{\beta}_1 - \beta_1}{\sigma_{\hat{\beta}_1}}}{\hat{\sigma}/\sqrt{SS_{xx}}} = \frac{\frac{\hat{\beta}_1 - \beta_1}{\sigma_{\hat{\beta}_1}}}{\sqrt{MSE}} = \frac{\frac{\hat{\beta}_1 - \beta_1}{\sigma_{\hat{\beta}_1}}}{\sqrt{\frac{SSE}{(n-2)\sigma^2}}}.$$

This is the ratio of a standard normally distributed random variable and the square root of a χ^2-distributed random variable divided by its degrees of freedom. It can be demonstrated that the two random variables are independent, and, therefore, we can conclude that

$$\frac{\hat{\beta}_1 - \beta_1}{\hat{\sigma}_{\hat{\beta}_1}} \sim t_{n-2}. \tag{19.47}$$

In the same way, it can be shown that

$$\frac{\hat{\beta}_0 - \beta_0}{\hat{\sigma}_{\hat{\beta}_0}} \sim t_{n-2}. \tag{19.48}$$

19.5.4 Confidence Intervals

Equations (19.47) and (19.48) allow us to construct confidence intervals for the slope β_1 and for the intercept β_0 in simple linear regression. We demonstrate the construction of the interval for the slope. The construction for the confidence interval of the intercept is analogous.

From Equation (19.47), it follows that

$$P\left(-t_{\alpha/2;n-2} \leq \frac{\hat{\beta}_1 - \beta_1}{\hat{\sigma}_{\hat{\beta}_1}} \leq t_{\alpha/2;n-2}\right) = 1 - \alpha,$$

where $1 - \alpha$ denotes the confidence level of the confidence interval. To obtain a confidence interval, we need the following intermediate steps:

$$P\left(-t_{\alpha/2;n-2}\,\hat{\sigma}_{\hat{\beta}_1} \le \hat{\beta}_1 - \beta_1 \le t_{\alpha/2;n-2}\,\hat{\sigma}_{\hat{\beta}_1}\right) = 1 - \alpha,$$

$$P\left(-\hat{\beta}_1 - t_{\alpha/2;n-2}\,\hat{\sigma}_{\hat{\beta}_1} \le -\beta_1 \le -\hat{\beta}_1 + t_{\alpha/2;n-2}\,\hat{\sigma}_{\hat{\beta}_1}\right) = 1 - \alpha,$$

and

$$P\left(\hat{\beta}_1 - t_{\alpha/2;n-2}\,\hat{\sigma}_{\hat{\beta}_1} \le \beta_1 \le \hat{\beta}_1 + t_{\alpha/2;n-2}\,\hat{\sigma}_{\hat{\beta}_1}\right) = 1 - \alpha.$$

Therefore, the confidence interval for β_1 is given by

$$\left[\hat{\beta}_1 - t_{\alpha/2;n-2}\,\hat{\sigma}_{\hat{\beta}_1}, \hat{\beta}_1 + t_{\alpha/2;n-2}\,\hat{\sigma}_{\hat{\beta}_1}\right]. \tag{19.49}$$

The confidence interval for β_0 is

$$\left[\hat{\beta}_0 - t_{\alpha/2;n-2}\,\hat{\sigma}_{\hat{\beta}_0}, \hat{\beta}_0 + t_{\alpha/2;n-2}\,\hat{\sigma}_{\hat{\beta}_0}\right]. \tag{19.50}$$

Example 19.5.1 *In Example 19.4.1, we calculated the estimated standard deviations, or standard errors, of the least squares estimators for β_0 and β_1 for the data in Table 19.1. These standard errors, along with the estimates $b_0 = -0.1$ and $b_1 = 0.7$ of the intercept and the slope, are essential inputs when calculating 95% confidence intervals for β_0 and β_1 using Equations (19.49) and (19.50). Substituting b_0 and b_1 for $\hat{\beta}_0$ and $\hat{\beta}_1$, respectively, and $s_{\hat{\beta}_0}$ and $s_{\hat{\beta}_1}$ for $\hat{\sigma}_{\hat{\beta}_0}$ and $\hat{\sigma}_{\hat{\beta}_1}$, we obtain*

$$\left[b_0 - t_{\alpha/2;n-2}\,s_{\hat{\beta}_0}, b_0 + t_{\alpha/2;n-2}\,s_{\hat{\beta}_0}\right]$$

$$= \left[-0.1 - t_{\alpha/2;n-2}\,\frac{11}{10\sqrt{3}}, -0.1 + t_{\alpha/2;n-2}\,\frac{11}{10\sqrt{3}}\right],$$

$$= \left[-0.1 - t_{0.025;5-2}\,\frac{11}{10\sqrt{3}}, -0.1 + t_{0.025;5-2}\,\frac{11}{10\sqrt{3}}\right],$$

$$= \left[-0.1 - 3.1824\,\frac{11}{10\sqrt{3}}, -0.1 + 3.1824\,\frac{11}{10\sqrt{3}}\right],$$

$$= [-2.1211, 1.9211],$$

as the 95% confidence interval for β_0, and

$$\left[b_1 - t_{\alpha/2;n-2}\,s_{\hat{\beta}_1}, b_1 + t_{\alpha/2;n-2}\,s_{\hat{\beta}_1}\right]$$

$$= \left[0.7 - 3.1824\,\sqrt{\frac{11}{300}}, 0.7 + 3.1824\,\sqrt{\frac{11}{300}}\right],$$

$$= [0.0906, 1.3094],$$

as the 95% confidence interval for β_1.

Note that the required quantiles or percentiles from the t-distribution, $-t_{\alpha/2;n-2}$ and $t_{\alpha/2;n-2}$, can be computed in JMP using the formulas "t Quantile(0.025, 3)" and "t Quantile(0.975, 3)", or using the "Distribution Calculator" script.

The interval for β_0 contains the value 0. Therefore, we can conclude that the intercept is not significantly different from 0 at a significance level of 5% (and a confidence level of 95%). The interval for β_1 does not contain the value 0. Therefore, we can conclude that the slope is significantly different from 0 at a significance level of 5% (and a confidence level of 95%). We are able to draw these conclusions because of the relationship between two-sided hypothesis tests and confidence intervals (see Section 3.2.3).

The 95% confidence intervals for β_0 and β_1 are not shown in the JMP output by default. They can be obtained by right-clicking on the table named "Parameter Estimates" at the bottom of the output in Figure 19.18, and selecting "Columns", followed by the options "Lower 95%" and "Upper 95%". The required selections are shown in Figure 19.19. The new "Parameter Estimates" table, now including the confidence intervals, is shown in Figure 19.20.

19.5.5 Hypothesis Tests

In general, the null hypothesis for a test on the slope of a regression line is

$$H_0 : \beta_1 = \beta_1^*.$$

The hypothesis test for the slope can be right-tailed, left-tailed, and two-tailed. The corresponding alternative hypotheses are

$$H_a : \beta_1 > \beta_1^*,$$
$$H_a : \beta_1 < \beta_1^*,$$

and

$$H_a : \beta_1 \neq \beta_1^*,$$

respectively. The three tests can be performed in the same way as the *t*-tests for a population mean in Section 4.1. The test statistic is the ratio

$$T = \frac{\hat{\beta}_1 - \beta_1^*}{\hat{\sigma}_{\hat{\beta}_1}},$$

which is *t*-distributed with $n - 2$ degrees of freedom if the null hypothesis is correct.

The *p*-value for the right-tailed test is computed as

$$p = P\left(t_{n-2} > \frac{b_1 - \beta_1^*}{s_{\hat{\beta}_1}}\right) = P(t_{n-2} > t),$$

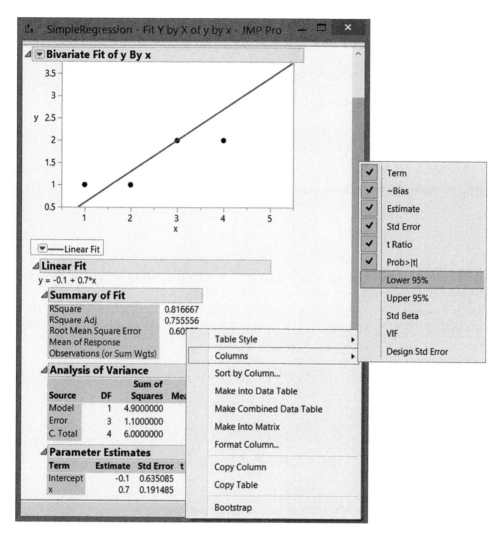

Figure 19.19 The JMP menu for obtaining the confidence intervals for β_0 and β_1 when performing a simple linear regression.

Parameter Estimates

Term	Estimate	Std Error	t Ratio	Prob>|t|	Lower 95%	Upper 95%
Intercept	−0.1	0.635085	−0.16	0.8849	−2.121125	1.9211249
x	0.7	0.191485	3.66	0.0354*	0.0906079	1.3093921

Figure 19.20 The new JMP output for the simple linear regression analysis of the data in Table 19.1, including the 95% confidence intervals for β_0 and β_1.

where

$$t = \frac{b_1 - \beta_1^*}{s_{\hat{\beta}_1}}$$

is the computed value of the test statistic. For the left-tailed test, the p-value is

$$p = P\left(t_{n-2} < \frac{b_1 - \beta_1^*}{s_{\hat{\beta}_1}}\right) = P\left(t_{n-2} < t\right).$$

Finally, for the two-tailed test, the p-value is calculated as

$$p = 2\,P\left(t_{n-2} > \left|\frac{b_1 - \beta_1^*}{s_{\hat{\beta}_1}}\right|\right) = 2\,P\left(t_{n-2} > |t|\right).$$

If the p-value is larger than the significance level α, then the null hypothesis is accepted. Otherwise, it is rejected in favor of the alternative hypothesis.

Often, researchers are specifically interested in testing whether the slope of their regression line is positive, negative, or different from zero. In that case, $\beta_1^* = 0$, and the hypotheses tested are

$$H_0 : \beta_1 = 0,$$

on the one hand, and

$$H_a : \beta_1 > 0,$$
$$H_a : \beta_1 < 0,$$

or

$$H_a : \beta_1 \neq 0,$$

on the other. The corresponding test statistic is

$$T = \frac{\hat{\beta}_1}{\hat{\sigma}_{\hat{\beta}_1}},$$

and the p-values for the right-tailed, left-tailed, and two-tailed tests are calculated as

$$p = P\left(t_{n-2} > \frac{b_1}{s_{\hat{\beta}_1}}\right) = P(t_{n-2} > t),$$

$$p = P\left(t_{n-2} < \frac{b_1}{s_{\hat{\beta}_1}}\right) = P(t_{n-2} < t),$$

and

$$p = 2\,P\left(t_{n-2} > \left|\frac{b_1}{s_{\hat{\beta}_1}}\right|\right) = 2\,P(t_{n-2} > |t|),$$

respectively. The quantity

$$t = \frac{b_1}{s_{\hat{\beta}_1}}$$

in these expressions is the computed value of the test statistic. It is usually called the **t-ratio** in the output of statistical software packages.

By default, most software packages report the p-value of a two-tailed test for $\beta_1 = 0$. JMP is no exception to that. The "Fit Y by X" platform in JMP does not allow the user to carry out a two-tailed test for a nonzero value of β_1^*. The "Fit Model" platform in the "Analyze" menu, however, does offer that option. We discuss the use of the "Fit Model" platform, which can handle simple linear regression, multiple linear regression, logistic regression, and several other types of regression, in detail in *Statistics with JMP: Linear and Generalized Linear Models*.

Example 19.5.2   *A two-tailed test of the hypothesis $H_0 : \beta_1 = 0$ for the data in Table 19.1 and Example 19.2.3 results in the following p-value:*

$$p = 2\,P\left(t_{n-2} > \left|\frac{b_1}{s_{\hat{\beta}_1}}\right|\right) = 2\,P\left(t_{5-2} > \left|\frac{0.7}{0.1915}\right|\right),$$

$$= 2\,P(t_3 > |3.6556|) = 0.0354.$$

We can therefore reject the null hypothesis and conclude that the slope in Example 19.2.3 is significantly different from zero.

The p-value of 0.0354 is shown in the last table of Figure 19.18. It is colored in red and indicated with an asterisk because it is smaller than the commonly used significance level of 5%. The t-ratio is also shown in that table.

Example 19.5.3   *A two-tailed test of the hypothesis $H_0 : \beta_1 = 0$ for the starch extraction data in Example 19.2.6 results in the following p-value:*

$$p = 2\,P\left(t_{n-2} > \left|\frac{b_1}{s_{\hat{\beta}_1}}\right|\right) = 2\,P\left(t_{54-2} > \left|\frac{2.7544}{0.5046}\right|\right),$$

$$= 2\,P\left(t_{52} > |5.4589|\right) = 0.00000068.$$

We can therefore reject the null hypothesis at a significance level of 0.05 and conclude that the slope of the regression line in the starch extraction experiment is significantly different from zero. The p-value is smaller than 0.0001, which is why it is indicated by $< .0001$ in the JMP output in Figure 19.12.

Custom Test

Parameter	
Intercept	0
Concentration	1
=	3

Value	−0.245555556
Std Error	0.5045771582
t Ratio	−0.486656107
Prob>|t|	0.6285470096
SS	5.4267777778

Sum of Squares	5.4267777778
Numerator DF	1
F Ratio	0.2368341666
Prob > F	0.6285470096

Figure 19.21   The JMP output for testing whether $\beta_1 = 3$ in Example 19.5.3, obtained using the "Fit Model" platform in the "Analyze" menu.

A two-tailed test of the hypothesis $H_0 : \beta_1 = 3$ for the starch extraction data results in the following p-value:

$$p = 2\,P\left(t_{n-2} > \left|\frac{b_1 - 3}{s_{\hat{\beta}_1}}\right|\right) = 2\,P\left(t_{54-2} > \left|\frac{2.7544 - 3}{0.5046}\right|\right),$$

$$= 2\,P\left(t_{52} > \left|\frac{-0.2456}{0.5046}\right|\right) = 2\,P\left(t_{52} > |-0.4867|\right) = 0.6285.$$

Now, we cannot reject the null hypothesis. Therefore, we conclude that the slope in Example 19.2.6 is not significantly different from 3. The output produced by the "Fit Model" platform in the "Analyze" menu for this test is shown in Figure 19.21.

In the same way as for the slope β_1, we can perform hypothesis tests for the intercept β_0 in the context of simple linear regression. Usually, however, we are not that interested in the intercept. The main goal of simple linear regression is to find out whether the explanatory variable really has explanatory value; that is, whether the impact of the explanatory variable on the response variable is significant. To this end, we need to perform a hypothesis test for the slope. In fact, when the slope is zero (or not significantly different from zero), this means that the response variable has the same value, no matter what the value of the explanatory variable is. The explanatory variable then has no explanatory or predictive value whatsoever for the response values.

19.6   The Quality of the Simple Linear Regression Model

In this section, we explain how the quality of a simple linear regression model can be assessed. The first method involves the coefficient of determination, denoted by R^2. The second method involves a formal hypothesis test based on the F-distribution.

19.6.1   The Coefficient of Determination

The goal of a simple linear regression model is to explain the variation in the response variable. The variation of the n observed responses around their mean is captured by the deviations

$$y_i - \bar{y}.$$

A deviation $y_i - \bar{y}$ can be rewritten as

$$y_i - \bar{y} - \hat{y}_i + \hat{y}_i = (\hat{y}_i - \bar{y}) + (y_i - \hat{y}_i).$$

Therefore, we have the following partitioning of the ith response's deviation from the mean:

$$y_i - \bar{y} = (\hat{y}_i - \bar{y}) + (y_i - \hat{y}_i).$$

Squaring both sides of this equation results in

$$(y_i - \bar{y})^2 = (\hat{y}_i - \bar{y})^2 + (y_i - \hat{y}_i)^2 + 2(\hat{y}_i - \bar{y})(y_i - \hat{y}_i).$$

Summing the squared deviations of all the observations yields

$$\sum_{i=1}^{n}(y_i - \bar{y})^2 = \sum_{i=1}^{n}(\hat{y}_i - \bar{y})^2 + \sum_{i=1}^{n}(y_i - \hat{y}_i)^2 + 2\sum_{i=1}^{n}(\hat{y}_i - \bar{y})(y_i - \hat{y}_i).$$

This can be simplified to

$$\sum_{i=1}^{n}(y_i - \bar{y})^2 = \sum_{i=1}^{n}(\hat{y}_i - \bar{y})^2 + \sum_{i=1}^{n}(y_i - \hat{y}_i)^2, \tag{19.51}$$

because

$$\sum_{i=1}^{n}(\hat{y}_i - \bar{y})(y_i - \hat{y}_i) = \sum_{i=1}^{n}\hat{y}_i(y_i - \hat{y}_i) - \sum_{i=1}^{n}\bar{y}(y_i - \hat{y}_i),$$

$$= \sum_{i=1}^{n}\hat{y}_i(y_i - \hat{y}_i) - \bar{y}\sum_{i=1}^{n}(y_i - \hat{y}_i),$$

$$= \sum_{i=1}^{n}\hat{y}_i u_i - \bar{y}\sum_{i=1}^{n}u_i,$$

$$= 0.$$

The final step in this derivation follows from the fact that the sum of the least squared residuals u_i is zero and from the fact that the sum of all cross-products $\hat{y}_i u_i$ is also zero. These properties of the residuals were discussed on pages 483–485.

The left-hand side of Equation (19.51) is called the **total sum of squares**. It measures the total variation in the response values. It is usually abbreviated as ssto. The second term on the right-hand side of the equation, the sum of the squared residuals $u_i = y_i - \hat{y}_i$, is the sum of squared errors sse that we have already introduced on page 506. The sse measures the variation in the response that is not explained by the model. Therefore, another name for it is the **unexplained variation**.

The first term on the right-hand side of Equation (19.51) is called the **regression sum of squares** or the **explained variation**, because it measures the variation in the response variable that is explained by the regression model. The abbreviation ssr is usually used for that sum of squares.

In conclusion, the total variation in the response values can be split into two parts. The first part is the variation that we can explain, while the second part is the variation that we cannot explain. Using the three abbreviations ssto, ssr, and sse, we can express the partitioning of the total sum of squares as

$$\text{ssto} = \text{ssr} + \text{sse}.$$

The **coefficient of determination** r^2 of the estimated or fitted model is defined as

$$r^2 = \frac{\text{ssr}}{\text{ssto}} = 1 - \frac{\text{sse}}{\text{ssto}}, \qquad (19.52)$$

when the data is available and the sums of squares ssto, ssr, and sse can be calculated. It is written as

$$R^2 = \frac{\text{SSR}}{\text{SSTO}} = 1 - \frac{\text{SSE}}{\text{SSTO}}, \qquad (19.53)$$

when the data has not yet been collected and the sums of squares are still random variables. The coefficient of determination is thus equal to the fraction of the variation in the response variable that is explained by the regression model. The coefficient always lies between 0 and 1.

The coefficient of determination is equal to zero if ssr $= 0$ or ssto $=$ sse. This is only possible if $\hat{y}_i = \bar{y}$ for each observation. Since

$$\text{ssr} = \sum_{i=1}^{n} (\hat{y}_i - \bar{y})^2,$$

$$= \sum_{i=1}^{n} (b_0 + b_1 x_i - b_0 - b_1 \bar{x})^2,$$

$$= b_1^2 \sum_{i=1}^{n} (x_i - \bar{x})^2,$$

$$= b_1^2 \text{SS}_{xx},$$

a zero value of ssr implies that the estimate for the slope of the regression model, b_1, equals 0. This is logical because the explanatory variable then does not explain changes in the response variable at all. When we do not have any data yet, we rewrite this expression as

$$SSR = \hat{\beta}_1^2 SS_{xx}.$$

The coefficient of determination is equal to 1 if sse $= 0$ or ssto $=$ ssr. The sum of squared errors can only be zero when $\hat{y}_i = y_i$ for each observation. In that case, each predicted response value matches the corresponding observed value, and all residuals $u_i = y_i - \hat{y}_i$ are zero. In that case, all data points lie on the regression line. The estimated model then describes the data perfectly well.

There are some interesting identities between the coefficient of determination and some specific correlation coefficients. First, for simple linear regression, the coefficient of determination is equal to the square of the correlation between the explanatory variable and the response variable:

$$r^2 = \frac{ssr}{ssto},$$

$$= \frac{b_1^2 SS_{xx}}{\sum_{i=1}^{n}(y_i - \bar{y})^2},$$

$$= b_1^2 \frac{\sum_{i=1}^{n}(x_i - \bar{x})^2}{\sum_{i=1}^{n}(y_i - \bar{y})^2},$$

$$= \left\{ \frac{\sum_{i=1}^{n}(x_i - \bar{x})(y_i - \bar{y})}{\sum_{i=1}^{n}(x_i - \bar{x})^2} \right\}^2 \frac{\sum_{i=1}^{n}(x_i - \bar{x})^2}{\sum_{i=1}^{n}(y_i - \bar{y})^2},$$

$$= \frac{\left\{ \sum_{i=1}^{n}(x_i - \bar{x})(y_i - \bar{y}) \right\}^2}{\sum_{i=1}^{n}(x_i - \bar{x})^2 \sum_{i=1}^{n}(y_i - \bar{y})^2},$$

$$= \left\{ \frac{\sum_{i=1}^{n}(x_i - \bar{x})(y_i - \bar{y})}{\sqrt{\sum_{i=1}^{n}(x_i - \bar{x})^2}\sqrt{\sum_{i=1}^{n}(y_i - \bar{y})^2}} \right\}^2,$$

$$= r_{XY}^2,$$

which shows that, indeed, the coefficient of determination r^2 is equal to r^2_{XY}, the square of the correlation between x and y. This finding explains the notation R^2 or r^2 used for the coefficient of determination.

On pages 486 and 487, we have already proven that the correlation $r_{U\hat{Y}}$ between the residuals and the predicted responses is zero. However, this is not the case for the correlation r_{UY} between the residuals and the original response values. It is a useful exercise to show that

$$\text{cov}(U, Y) = s_{UY} = s_U^2.$$

As a result,

$$r_{UY} = \frac{s_{UY}}{s_U s_Y} = \frac{s_U^2}{s_U s_Y} = \frac{s_U}{s_Y} = \frac{\sqrt{\dfrac{1}{n-1}\sum\limits_{i=1}^{n}(u_i - \bar{u})^2}}{\sqrt{\dfrac{1}{n-1}\sum\limits_{i=1}^{n}(y_i - \bar{y})^2}} = \sqrt{\frac{\text{sse}}{\text{ssto}}} = \sqrt{1 - r^2},$$

and

$$r^2_{UY} = 1 - r^2$$

This value is in some sense the opposite of r^2. Therefore, the squared correlation r^2_{UY} can be interpreted as a statistic that measures how poor the fit is. Note that the correlation r_{UY} is always positive.

Finally, recall that, for a simple linear regression model, $r_{XY} = r_{Y\hat{Y}}$. As a result,

$$r^2 = r^2_{Y\hat{Y}}.$$

This latter equality is not only valid for simple linear regression, but also for multiple linear regression. This will be shown in *Statistics with JMP: Linear and Generalized Linear Models*.

Example 19.6.1   *For the data in Table 19.1, the total sum of squares is*

$$\text{ssto} = \sum_{i=1}^{n}(y_i - \bar{y})^2 = (1 - 2)^2 + (1 - 2)^2 + \cdots + (4 - 2)^2 = 6,$$

while the sum of the squared residuals is

$$\text{sse} = 1.1.$$

Therefore, the regression sum of squares is

$$\text{ssr} = \text{ssto} - \text{sse} = 4.9.$$

As a result, the coefficient of determination is

$$r^2 = \frac{4.9}{6} = 0.8167.$$

The simple linear regression model thus explains 81.67% *of the total variation in the response variable. The value for the coefficient of determination is the first summary statistic displayed in the JMP output in Figure 19.18.*

The value 0.8167 *is the square of the sample correlations* r_{XY} *and* $r_{Y\hat{Y}}$, *which are both equal to* 0.9037.

Note also that

$$\text{ssr} = b_1^2 \sum_{i=1}^{n}(x_i - \bar{x})^2 = b_1^2 SS_{xx} = (0.7)^2 \times 10 = 0.49 \times 10 = 4.9.$$

Example 19.6.2 *The coefficient of determination for the simple linear regression model for the starch extraction data in Table 19.5 and Example 19.2.6 is* $r^2 = 0.3643$. *This is because* ssto = 1874.35, sse = 1191.52, *and* ssr = 682.83. *The simple linear regression model thus explains* 36.43% *of the total variation in the response variable. The value for the coefficient of determination as well as the* ssto, sse, *and* ssr *values are shown in the JMP output in Figure 19.12.*

The value 0.3643 *is the square of the sample correlations* r_{XY} *and* $r_{Y\hat{Y}}$, *which are both equal to* 0.6036.

19.6.2 Testing the Significance of the Model

A second way to evaluate the quality of a simple linear regression model is to perform a formal hypothesis test to figure out whether the model explains a significant amount of the total variation in the response variable. In words, the hypotheses tested are

$$H_0 : \text{The model has no explanatory value}$$

and

$$H_a : \text{The model does have explanatory value.}$$

In statistical terms, the two hypotheses can be expressed as

$$H_0 : \beta_1 = 0$$

and

$$H_a : \beta_1 \neq 0.$$

For this test, the explained variation ssr is compared with the unexplained variation sse. If the explained variation is large in comparison to the unexplained variation, this suggests that the simple linear regression model indeed explains part of the variation in the response. To judge how large the ssr value should be compared to the sse value for the null hypothesis to be rejected, we need to find a reference distribution.

We already know that SSE/σ^2 has a χ^2-distribution with $n-2$ degrees of freedom. This result is valid regardless of whether or not the null hypothesis is true.

It turns out that SSR is also linked to a χ^2-distribution. To see this, remember that

$$SSR = \hat{\beta}_1^2 SS_{xx}$$

and that $\hat{\beta}_1$ is normally distributed with mean β_1 and variance σ^2/SS_{xx}. As a result,

$$\frac{\hat{\beta}_1 - \beta_1}{\sqrt{\dfrac{\sigma^2}{SS_{xx}}}} = \frac{(\hat{\beta}_1 - \beta_1)\sqrt{SS_{xx}}}{\sigma}$$

has a standard normal distribution. If the null hypothesis that $\beta_1 = 0$ is true, then

$$\frac{\hat{\beta}_1 \sqrt{SS_{xx}}}{\sigma}$$

is standard normally distributed, and its square

$$\frac{\hat{\beta}_1^2 \, SS_{xx}}{\sigma^2} = \frac{SSR}{\sigma^2}$$

has a χ^2-distribution with one degree of freedom. This result is only valid when the null hypothesis is true.

It can be shown that SSR and SSE are independent random variables. As a result, the ratio

$$F = \frac{\dfrac{SSR}{\sigma^2}\Big/ 1}{\dfrac{SSE}{\sigma^2}\Big/ (n-2)} = \frac{SSR}{SSE/(n-2)}$$

is F-distributed with one numerator degree of freedom and $n-2$ denominator degrees of freedom if the null hypothesis is true. In fact, the numerator of the ratio is a χ^2-distributed random variable with one degree of freedom divided by 1 and its denominator is a χ^2-distributed random variable with $n-2$ degrees of freedom divided by $n-2$. The quantity

$$F = \frac{SSR}{SSE/(n-2)}$$

serves as test statistic. It is often written as

$$F = \frac{MSR}{MSE},$$

because $MSR = SSR/1 = SSR$ and $MSE = SSE/(n-2)$: dividing a sum of squares by its number of degrees of freedom always results in mean squares[2].

To conduct the significance test, we can therefore use the F-distribution with one numerator degree of freedom and $n-2$ denominator degrees of freedom to find a critical value and to determine the p-value. The critical value for a test with significance level α is $F_{\alpha;1;n-2}$. The p-value of the hypothesis test is calculated as

$$p = P\left(F_{1;n-2} > \frac{ssr}{sse/(n-2)}\right) = P\left(F_{1;n-2} > \frac{msr}{mse}\right) = P\left(F_{1;n-2} > f\right),$$

where ssr, sse, msr, mse, and f are the computed versions of the random variables SSR, SSE, MSR, MSE, and F, respectively.

The above results are usually displayed in an analysis of variance or ANOVA table, of the kind we introduced in Section 12.2.4. In the context of simple linear regression, the ANOVA table takes the following form:

Source of variation	Sum of squares	Degrees of freedom	Mean sum of squares	F-test statistic	p-value
Model	SSR	1	$MSR = SSR$	$F = MSR/MSE$	p
Error	SSE	$n-2$	MSE		
Total	SSTO	$n-1$			

The F-test performed in this ANOVA table yields exactly the same result as the two-tailed t-test for $\beta_1 = 0$ derived on pages 516 and 517 for the value of the slope parameter β_1. To show the equivalence of the F-test and the two-tailed t-test, we demonstrate that the p-values for both tests are the same.

The p-value of the two-tailed t-test for $\beta_1 = 0$ can be rewritten as

$$p = 2\,P\left(t_{n-2} > \left|\frac{b_1}{s_{\hat{\beta}_1}}\right|\right),$$

$$= P\left(t_{n-2}^2 > \frac{b_1^2}{s_{\hat{\beta}_1}^2}\right),$$

$$= P\left(F_{1,n-2} > \frac{b_1^2}{s_{\hat{\beta}_1}^2}\right),$$

---

[2] In simple linear regression, the MSR and SSR values are always equal. This is no longer true for more complicated models involving more than one explanatory variable. This is discussed in *Statistics with JMP: Linear and Generalized Linear Models*.

since the square of a t-distributed random variable with $n - 2$ degrees of freedom is F-distributed with one numerator degree of freedom and $n - 2$ denominator degrees of freedom. Moreover,

$$\frac{b_1^2}{s_{\hat{\beta}_1}^2} = \frac{b_1^2}{s^2/\text{SS}_{xx}} = \frac{b_1^2 \, \text{SS}_{xx}}{s^2} = \frac{\text{ssr}}{\text{mse}} = \frac{\text{msr}}{\text{mse}},$$

so that

$$p = 2 \, P\left(t_{n-2} > \left|\frac{b_1}{s_{\hat{\beta}_1}}\right|\right) = P\left(F_{1, n-2} > \frac{\text{msr}}{\text{mse}}\right).$$

Hence, the p-values of the two hypothesis tests are identical in the case of a simple linear regression model.

Example 19.6.3   *For the data in Table 19.1, we have already computed that* sse $= 1.1$ *and that* ssr $= 4.9$. *The computed value for the F test statistic is therefore*

$$f = \frac{\text{ssr}}{\text{sse}/(n-2)} = \frac{4.9}{0.3667} = 13.3636,$$

and the p-value of the F-test is

$$p = P\left(F_{1,3} > 13.3636\right) = 0.035.$$

Hence, the ANOVA table has the following look:

Source of variation	Sum of squares	Degrees of freedom	Mean sum of squares	F-test statistic	p-value
Model	4.9	1	4.9	13.3636	0.035
Error	1.1	3	0.3667		
Total	6	4			

   This ANOVA table can also be found in the JMP output in Figure 19.18. The p-value can be obtained using the JMP command "1 - F Distribution(13.3636, 1, 3)". The value 13.3636 for the test statistic of the F-test is the square of the t-ratio of 3.6556 in Example 19.5.2. The p-value in this example equals the p-value for the two-tailed t-test in Example 19.5.2.

Example 19.6.4   *For the data from the starch extraction experiment in Table 19.5 and Example 19.2.6, the ANOVA table looks as follows:*

Source of variation	Sum of squares	Degrees of freedom	Mean sum of squares	F-test statistic	p-value
Model	1	682.83	682.83	29.7998	$< .0001$
Error	52	1191.52	22.91		
Total	53	1874.35			

This ANOVA table can also be found in the JMP output in Figure 19.12. The value 29.7998 for the test statistic of the F-test is the square of the t-ratio of 5.4589 in Example 19.5.3.

19.7   Predictions

The estimated simple linear regression model can be used to make predictions of the response variable for any given value of the explanatory variable. Two types of predictions are possible for a value x_h of the explanatory variable:

(1) an estimation of the expected value $E(Y \mid x_h) = \beta_0 + \beta_1 x_h$, and
(2) a prediction of the value of the random variable Y_h itself.

In both cases,

$$\hat{y}_h = b_0 + b_1 x_h$$

is used as a point prediction. If, however, an interval estimate or a prediction interval is desired, the result depends on the type of prediction.

At a first glance, the difference between these two kinds of prediction might be unclear. To explain the difference, consider the starch extraction example. We might be interested in predicting the yield of the starch extraction process at a salt concentration of, say, $x_h = 5.5\%$. Two possible scenarios are as follows:

(1) If we are interested in using the salt concentration of 5.5% on many future occasions, we might want to predict what the expected or average yield is on all of these future occasions. In that case, we need a confidence interval for $E(Y \mid x_h)$.
(2) If we are interested in using the salt concentration of 5.5% on a particular occasion, we might want to predict what the yield will be on that occasion. In that case, we need the prediction interval for an individual realization of the random variable Y_h, rather than a confidence interval for its expected value.

Another way in which we can formulate the difference between the two kinds of predictions is as follows. The interval for $E\left(Y \mid x_h\right)$ tells us where the estimated regression line might pass when the explanatory variable takes the value x_h. The interval for $\hat{Y}_h$ tells us what the possible spread is of an individual observation around the estimated regression line.

19.7.1 The Estimation of an Expected Response Value

The random variable $\hat{Y}_h = \hat{\beta}_0 + \hat{\beta}_1 x_h$ is a linear function of the two normally distributed random variables $\hat{\beta}_0$ and $\hat{\beta}_1$ and is therefore also normally distributed. The expected value is

$$E(\hat{Y}_h) = E(\hat{\beta}_0 + \hat{\beta}_1 x_h) = \beta_0 + \beta_1 x_h = E(Y \mid x_h).$$

The point prediction $\hat{Y}_h$ is therefore an unbiased estimator of $E(Y \mid x_h)$. The variance of $\hat{Y}_h$ is

$$
\begin{aligned}
\operatorname{var}\left(\hat{Y}_h\right) = \sigma_{\hat{Y}_h}^2 &= \operatorname{var}\left(\hat{\beta}_0 + \hat{\beta}_1 x_h\right), \\
&= \operatorname{var}\left(\hat{\beta}_0\right) + x_h^2 \operatorname{var}\left(\hat{\beta}_1\right) + 2 x_h \operatorname{cov}(\hat{\beta}_0, \hat{\beta}_1), \\
&= \sigma^2 \left(\frac{\sum_{i=1}^n x_i^2}{n \, \mathrm{SS}_{xx}} + \frac{x_h^2}{\mathrm{SS}_{xx}} - \frac{2 x_h \bar{x}}{\mathrm{SS}_{xx}} \right), \\
&= \sigma^2 \left(\frac{\sum_{i=1}^n x_i^2}{n \, \mathrm{SS}_{xx}} + \frac{n x_h^2}{n \, \mathrm{SS}_{xx}} - \frac{2 n x_h \bar{x}}{n \, \mathrm{SS}_{xx}} \right), \\
&= \sigma^2 \left(\frac{\sum_{i=1}^n x_i^2 + n x_h^2 - 2 n x_h \bar{x}}{n \, \mathrm{SS}_{xx}} \right), \\
&= \sigma^2 \left(\frac{\sum_{i=1}^n x_i^2 + n x_h^2 - 2 n x_h \bar{x} + n \bar{x}^2 - n \bar{x}^2}{n \, \mathrm{SS}_{xx}} \right), \\
&= \sigma^2 \left(\frac{\sum_{i=1}^n x_i^2 - n \bar{x}^2 + n \left(x_h^2 - 2 x_h \bar{x} + \bar{x}^2 \right)}{n \, \mathrm{SS}_{xx}} \right), \\
&= \sigma^2 \left(\frac{\sum_{i=1}^n (x_i - \bar{x})^2 + n(x_h - \bar{x})^2}{n \, \mathrm{SS}_{xx}} \right), \\
&= \sigma^2 \left(\frac{1}{n} + \frac{(x_h - \bar{x})^2}{\mathrm{SS}_{xx}} \right).
\end{aligned}
$$

The first step in this derivation makes use of the formula for the variance of a linear function of two random variables, in this case $\hat{\beta}_0$ and $\hat{\beta}_1$ (see *Statistics with JMP: Graphs, Descriptive Statistics and Probability*).

The true variance $\sigma^2_{\hat{Y}_h}$ is unknown because σ^2, the variance of the error terms in the regression model, is not known. Replacing σ^2 by its estimator $\hat{\sigma}^2$ produces the following estimator of $\sigma^2_{\hat{Y}_h}$:

$$\hat{\sigma}^2_{\hat{Y}_h} = \hat{\sigma}^2 \left(\frac{1}{n} + \frac{(x_h - \bar{x})^2}{SS_{xx}} \right).$$

An estimate for $\sigma^2_{\hat{Y}_h}$ is obtained by substituting s^2 for $\hat{\sigma}^2$:

$$s^2_{\hat{Y}_h} = s^2 \left(\frac{1}{n} + \frac{(x_h - \bar{x})^2}{SS_{xx}} \right).$$

To build a confidence interval for $E(Y \mid x_h)$, we make use of the fact that

$$\frac{\hat{Y}_h - E(Y \mid x_h)}{\hat{\sigma}_{\hat{Y}_h}} = \frac{\dfrac{\hat{Y}_h - E(Y|x_h)}{\sigma_{\hat{Y}_h}}}{\dfrac{\hat{\sigma}_{\hat{Y}_h}}{\sigma_{\hat{Y}_h}}},$$

$$= \frac{\dfrac{\hat{Y}_h - E(Y|x_h)}{\sigma_{\hat{Y}_h}}}{\sqrt{\dfrac{\hat{\sigma}^2_{\hat{Y}_h}}{\sigma^2_{\hat{Y}_h}}}},$$

$$= \frac{\dfrac{\hat{Y}_h - E(Y|x_h)}{\sigma_{\hat{Y}_h}}}{\sqrt{\dfrac{\hat{\sigma}^2}{\sigma^2}}},$$

$$= \frac{\dfrac{\hat{Y}_h - E(Y|x_h)}{\sigma_{\hat{Y}_h}}}{\sqrt{\dfrac{SSE}{(n-2)\sigma^2}}}$$

is the ratio of a standard normally distributed random variable and the square root of a χ^2-distributed random variable divided by its degrees of freedom. Since the numerator and denominator of the ratio are independent, it has a t-distribution with $n - 2$ degrees of freedom.

As a result,

$$P\left(-t_{\alpha/2;n-2} \leq \frac{\hat{Y}_h - E(Y \mid x_h)}{\hat{\sigma}_{\hat{Y}_h}} \leq t_{\alpha/2;n-2}\right) = 1 - \alpha,$$

so that

$$P\left(\hat{Y}_h - t_{\alpha/2;n-2}\,\hat{\sigma}_{\hat{Y}_h} \leq E(Y \mid x_h) \leq \hat{Y}_h + t_{\alpha/2;n-2}\,\hat{\sigma}_{\hat{Y}_h}\right) = 1 - \alpha,$$

and a $100(1 - \alpha)\%$ confidence interval for $E(Y \mid x_h)$ is given by

$$\left[\hat{Y}_h - t_{\alpha/2;n-2}\,\hat{\sigma}_{\hat{Y}_h}, \hat{Y}_h + t_{\alpha/2;n-2}\,\hat{\sigma}_{\hat{Y}_h}\right].$$

Example 19.7.1   *For the data in Table 19.1, we can calculate 95% confidence intervals of $E(Y \mid x_h)$ for different values of the explanatory variable. For example, we can calculate the interval for the case where $x_h = 2$. The point estimate for $E(Y \mid 2)$ is*

$$\hat{y}_h = b_0 + b_1 x_h = -0.1 + 0.7 \times 2 = 1.3.$$

The estimated variance of that estimate is

$$s_{\hat{Y}_h}^2 = s^2\left(\frac{1}{n} + \frac{(x_h - \bar{x})^2}{SS_{xx}}\right) = 0.3667\left(\frac{1}{5} + \frac{(2 - 3)^2}{10}\right) = 0.11,$$

so that the estimated standard deviation is

$$s_{\hat{Y}_h} = \sqrt{0.11} = 0.3317.$$

The 95% confidence interval for $E(Y \mid 2)$ is calculated as

$$\left[1.3 - t_{0.025;5-2} \times 0.3317, 1.3 + t_{0.025;5-2} \times 0.3317\right].$$

Since $t_{0.025;5-2} = 3.1824$, we obtain the interval

$$[0.2445, 2.3555].$$

The width of this interval equals 2.1110. In a similar fashion, we can calculate a confidence interval for $E(Y \mid 4.2)$; that is, for a value of 4.2 for the explanatory variable that is not included in the set of five observations in Table 19.1. That interval is symmetric around the point estimate

$$\hat{y}_h = b_0 + b_1 x_h = -0.1 + 0.7 \times 4.2 = 2.84,$$

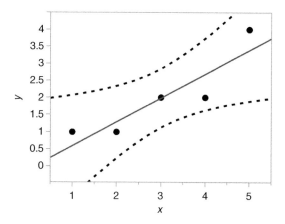

Figure 19.22 A graphical representation of the confidence intervals of $E(Y \mid x_h)$ for the data in Table 19.1 (see Example 19.7.1).

and is equal to

$$[1.7097, 3.9703].$$

The width of that interval is 2.2605.

It is possible to calculate confidence intervals for $E(Y \mid x_h)$ for any value of x_h, and to display all these intervals graphically. This results in a graph such as the one in Figure 19.22. The figure shows that the intervals become wider if we move away from the central value, $\bar{x} = 3$, of the explanatory variable.

Figure 19.22 can be constructed in JMP by selecting the option "Confid Curves Fit" in the hotspot (red triangle) menu next to the term "Linear Fit" shown in Figure 19.13.

In general, the width of the confidence interval of $E(Y \mid x_h)$ is equal to

$$2\, t_{\alpha/2;n-2}\, \hat{\sigma}_{\hat{Y}_h} = 2\, t_{\alpha/2;n-2}\, \hat{\sigma}^2 \left(\frac{1}{n} + \frac{(x_h - \bar{x})^2}{SS_{xx}} \right)$$

This expression shows that there are five factors that influence the width of the confidence interval:

- The confidence level $1 - \alpha$: the larger the desired confidence level, the smaller is the value for α, the larger is the value $t_{\alpha/2;n-2}$, and the wider is the interval.
- The (estimated) variance of the error term: the larger the (estimated) variance, the wider is the interval.
- The distance of x_h to the sample mean $\bar{x}$: the larger the difference between x_h and $\bar{x}$, the wider is the interval.
- The sample size or number of observations n: the more observations, the narrower is the interval.

- The variation in the observed values for the explanatory variable, as expressed by SS_{xx}: the larger the variation in x values, the narrower is the confidence interval.

All of these results make sense:

- If we want to be very confident that the interval estimate of $E(Y \mid x_h)$ will incorporate the true value of $E(Y \mid x_h)$, we should make the interval very wide.
- If there is a lot of variance in the observed responses, then there is a lot of variability in the measurements of the response variable, which prevents us from drawing firm conclusions and making precise statements about estimated parameters, or functions of estimated parameters.
- If we want to make a prediction for a value of the explanatory variable that deviates substantially from the majority of the data, as measured by $\bar{x}$, then that prediction will not be very precise. This results in a wide confidence interval for $E(Y \mid x_h)$.
- If we have a lot of data, we expect to be able to make precise statements.
- If there is a lot of variation in the values of the explanatory variables, this means that our data covers a broad range of scenarios, allowing us to make precise statements. When we are collecting experimental data, this suggests that we should ensure that SS_{xx} is large when planning the experiment.

19.7.2   The Prediction of an Individual Observation of the Response Variable

The prediction of the value of an individual response variable Y_h corresponding to a value of x_h for the explanatory variable involves more uncertainty than estimating the value $E(Y \mid x_h)$. To predict the value of the random variable Y_h, we first observe that

$$E(Y_h) = E(\hat{Y}_h),$$

so that

$$E(Y_h - \hat{Y}_h) = 0.$$

Since a new observation Y_h is independent of the initial data sample from which the regression model has been estimated, and based upon which the prediction $\hat{Y}_h$ is being made, we have that

$$\text{var}\,(Y_h - \hat{Y}_h) = \sigma^2_{Y_h - \hat{Y}_h} = \text{var}\,(Y_h) + \text{var}\,(\hat{Y}_h),$$

$$= \sigma^2 + \sigma^2 \left(\frac{1}{n} + \frac{(x_h - \bar{x})^2}{SS_{xx}} \right),$$

$$= \sigma^2 \left(1 + \frac{1}{n} + \frac{(x_h - \bar{x})^2}{SS_{xx}} \right).$$

This variance has to be estimated from the data. We denote its estimator by $\hat{\sigma}^2_{Y_h - \hat{Y}_h}$. It is given by

$$\hat{\sigma}^2_{Y_h - \hat{Y}_h} = \hat{\sigma}^2 \left(1 + \frac{1}{n} + \frac{(x_h - \bar{x})^2}{SS_{xx}} \right).$$

Now, the ratio

$$\frac{Y_h - \hat{Y}_h - E(Y_h - \hat{Y}_h)}{\hat{\sigma}_{Y_h - \hat{Y}_h}} = \frac{Y_h - \hat{Y}_h}{\hat{\sigma}_{Y_h - \hat{Y}_h}}$$

is t-distributed with $n - 2$ degrees of freedom. As a result,

$$P \left(-t_{\alpha/2;n-2} \leq \frac{Y_h - \hat{Y}_h}{\hat{\sigma}_{Y_h - \hat{Y}_h}} \leq t_{\alpha/2;n-2} \right) = 1 - \alpha,$$

so that

$$P \left(\hat{Y}_h - t_{\alpha/2;n-2} \, \hat{\sigma}_{Y_h - \hat{Y}_h} \leq Y_h \leq \hat{Y}_h + t_{\alpha/2;n-2} \, \hat{\sigma}_{Y_h - \hat{Y}_h} \right) = 1 - \alpha,$$

and a $100(1 - \alpha)\%$ prediction interval for Y_h is given by

$$\left[\hat{Y}_h - t_{\alpha/2;n-2} \, \hat{\sigma}_{Y_h - \hat{Y}_h}, \hat{Y}_h + t_{\alpha/2;n-2} \, \hat{\sigma}_{Y_h - \hat{Y}_h} \right].$$

Example 19.7.2 *For the data in Table 19.1, we can also calculate a 95% prediction interval for individual response values Y_h. For example, we can calculate the interval for the case where $x_h = 2$. The point prediction for Y_h when $x_h = 2$ is*

$$\hat{y}_h = b_0 + b_1 x_h = -0.1 + 0.7 \times 2 = 1.3.$$

The estimate of the variance $\sigma^2_{Y_h - \hat{Y}_h}$ required to construct a prediction interval for Y_h is

$$s^2_{Y_h - \hat{Y}_h} = s^2 \left(1 + \frac{1}{n} + \frac{(x_h - \bar{x})^2}{SS_{xx}} \right) = 0.3667 \left(1 + \frac{1}{5} + \frac{(2 - 3)^2}{10} \right) = 0.4767,$$

so that the estimated standard deviation is

$$s_{Y_h - \hat{Y}_h} = \sqrt{0.4767} = 0.6904.$$

The 95% prediction interval for Y_h is then calculated as

$$\left[1.3 - t_{0.025;5-2} \times 0.6904, 1.3 + t_{0.025;5-2} \times 0.6904 \right].$$

Since $t_{0.025;5-2} = 3.1824$, *we obtain the interval*

$$[-0.8972, 3.4972].$$

The width of this interval equals 4.3944. In a similar fashion, we can calculate a prediction interval for Y_h *for a value of 4.2 for the explanatory variable, which is a value that is not included in the set of five observations in Table 19.1. That interval is symmetric around the point estimate*

$$\hat{y}_h = b_0 + b_1 x_h = -0.1 + 0.7 \times 4.2 = 2.84,$$

and is equal to

$$[0.6059, 5.0741].$$

The width of that interval is 4.4681. Therefore, the interval for $x_h = 4.2$ *is wider than that for* $x_h = 2$. *This is due to the fact that the value 4.2 differs more from* $\bar{x}$ *than the value 2.*

It is possible to calculate prediction intervals for Y_h *for any value of* x_h *and to display all these intervals graphically. This results in a graph such as the one in Figure 19.23. That figure shows that the intervals become wider if we move away from* $\bar{x} = 3$. *Comparing Figures 19.22 and 19.23, we can see that the prediction intervals for* Y_h *are wider than the confidence intervals for* $E(Y \mid x_h)$. *This reflects the fact that more uncertainty is involved when making a prediction for an individual response value than when determining an expected response value.*

Figure 19.23 can be constructed in JMP by selecting the option "Confid Curves Indiv" *in the hotspot (red triangle) menu next to the term* "Linear Fit" *shown in Figure 19.13. The options* "Confid Curves Fit" *and* "Confid Curves Indiv" *can be selected simultaneously, resulting in a graph showing two types of intervals.*

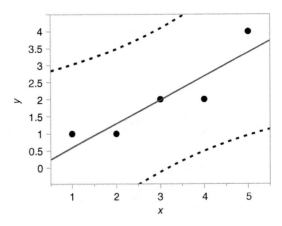

Figure 19.23 A graphical representation of the prediction intervals for Y_h for the data in Table 19.1 (see Example 19.7.2).

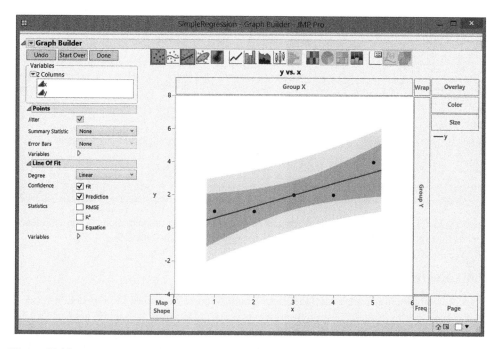

Figure 19.24   A graphical representation of the prediction intervals for Y_h and the confidence intervals for $E(Y \mid x_h)$ for the data in Table 19.1 (see Examples 19.7.1 and 19.7.2), obtained using the "Graph Builder" in the "Graph" menu.

It is also possible to create a similar picture using the "Graph Builder" in the "Graph" menu. To this end, the options "Confidence of Fit" and/or "Confidence of Prediction" need to be selected in the bottom left-hand corner of the "Graph Builder". This is shown in Figure 19.24.

Finally, it is interesting to point out that JMP offers the possibility of calculating the confidence intervals and the prediction intervals from Examples 19.7.1 and 19.7.2, and adding the results to the data table. To this end, we need to select the options "Mean Confidence Limit Formula" and "Indiv Confidence Limit Formula" in the hotspot (red triangle) menu next to the term "Linear Fit" shown in Figure 19.13. We can also add the predicted responses themselves to the data table (for which we need the option "Save Predicteds"). The additional columns are all shown in Figure 19.25.

Note that an extra row was added in the data table in Figure 19.25: to obtain a prediction for a value of 4.2 for the explanatory variable, that value was inserted in the sixth row of the table in the column labeled "x". After selecting the options "Save Predicteds", "Mean Confidence Limit Formula" and "Indiv Confidence Limit Formula", this leads to an automatic calculation of the point prediction and the two types of intervals. Other values of the explanatory variable for which a prediction is desired can be entered in the next few rows in the same way.

		x	y	Predicted y	Lower 95% Mean y	Upper 95% Mean y	Lower 95% Indiv y	Upper 95% Indiv y
●	1	1	1	0.6	-0.892699631	2.0926996305	-1.837568289	3.0375682893
●	2	2	1	1.3	0.244501969	2.355498031	-0.897194364	3.4971943637
●	3	3	2	2	1.1381894665	2.8618105335	-0.110996062	4.110996062
●	4	4	2	2.7	1.644501969	3.755498031	0.5028056363	4.8971943637
●	5	5	4	3.4	1.9073003695	4.8926996305	0.9624317107	5.8375682893
	6	4.2	●	2.84	1.7097460814	3.9702539186	0.6059317605	5.0740682395

Figure 19.25   The additional columns in the data table, showing the point predictions $\hat{y}_h$, the confidence intervals for $E(Y \mid x_h)$, and the prediction intervals for Y_h, for the data in Table 19.1.

19.8   Regression Diagnostics

In the previous sections, we have made four assumptions concerning the regression model, and we have built on these assumptions when deriving appropriate significance tests and constructing confidence and prediction intervals. Checking the validity of these assumptions should be part of every regression analysis. If one of the assumptions is invalid, this can affect the quality of the conclusions drawn from the analysis. In that event, measures are required to remedy this situation; for example, by transforming the response variable or by adding terms to the model. The whole process of checking the assumptions is termed **regression diagnostics**. Typically, this is done after the model has been estimated, because the main tool for the regression diagnostics is the analysis of the residuals.

There exist quite sophisticated statistical and graphical methods for the analysis of residuals. The more complex the statistical model, the greater is the need for elaborate techniques to check the model assumptions. In this chapter on simple linear regression, we restrict our attention to some basic graphical tools. We will cover the topic of regression diagnostics, including formal hypothesis tests, in more detail in *Statistics with JMP: Linear and Generalized Linear Models*.

For convenience, we will restate the four assumptions made in the previous sections here. The first three guaranteed that the least squares estimators are the best possible linear unbiased estimators (see the Gauss–Markov theorem 19.3.1), while the fourth assumption enabled us to construct confidence intervals and hypothesis tests.

Assumption 1   *(Linear relation): We assume that there is a linear relation $Y_i = \beta_0 + \beta_1 x_i + U_i$ between the explanatory variable x_i and the response Y_i with unknown constants β_0 and β_1, and that $E(U_i) = 0$. Equivalently, we assume that $E(Y_i) = \beta_0 + \beta_1 x_i$.*

Assumption 2   *(Independent errors): We assume that the random errors U_i are independent and, hence, that $\mathrm{cov}(U_i, U_j) = 0$ whenever $i \neq j$.*

Assumption 3   *(Homoscedasticity): We assume that $\mathrm{var}(U_i) = \sigma^2(U_i) = \sigma^2$: the variance of the error term is unknown, but it is the same for each observation.*

Assumption 4 *(Normal distribution): We assume that the error terms U_i are normally distributed.*

In the remainder of this section, we discuss some examples in which these assumptions are violated, and how one can respond. We start this section by introducing some helpful plots of residuals.

19.8.1 Analysis of Residuals

Recall that the residual u_i for observation i is the difference between the observed response value y_i and the model's prediction $\hat{y}_i$:

$$u_i = y_i - \hat{y}_i = y_i - (b_0 + b_1 x).$$

A close look at the residuals should be part of every statistical modeling process. In the context of a simple linear regression, the following three graphical outputs are very useful for regression diagnostics:

(1) **Residual-by-Predicted plot**: this display plots the residuals u_i against the predicted value y_i. Ideally, the pattern of the points in this plot looks completely random. Curved patterns or funnel shapes indicate violations of the regression assumptions.
(2) **Residual-by-Row plot**: this display plots the residuals u_i against the row numbers of the observations. Imagine that you perform measurements, and that over time you get better and better at using the measurement device. Then, the first few observations (in the first few rows) will show larger positive and negative residuals, while the last few observations (in the last few rows) will have smaller positive and negative residuals. This phenomenon is known as the **learning effect**. The opposite is possible too. You might become more tired or less able to concentrate over time, resulting in increased residuals for the later observations. The residual-by-row plot may also indicate some drift in the data. For instance, the laboratory in which the data is collected may be heating up, causing the responses to go up or down systematically. This phenomenon is usually referred to as a **time trend**.
(3) **Quantile diagram**: this tool to check the normality of the residuals has already been introduced in Section 6.2.1. Of course, you can add a hypothesis test such as the Shapiro–Wilk test to formally test normality.

Each of these diagnostic plots may also show one or more outlying residuals. They correspond to outliers in the data and require further investigation. These data points may come from mistakes made during the data collection process and should be corrected. If this is not possible, it is preferable to exclude the data points from the analysis.

Example 19.8.1 *When students are asked to report their body weight, they may not be completely honest. Twenty students were asked to report their body weight, after which their real weight was measured. The result of a simple linear regression, using the claimed weight as an explanatory variable for the measured weight, is shown in Figure 19.26. For comparison,*

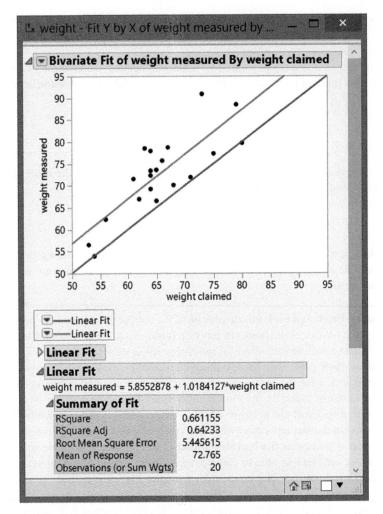

Figure 19.26 The simple linear regression model for Example 19.8.1. The lower line is the diagonal, corresponding to perfect agreement between the claimed and measured weights.

the diagonal line is included in the figure too, which corresponds to complete agreement between the claimed and measured weights. Not too surprisingly, the regression line of the measured weight lies above the diagonal, indicating that the real weight tends to be greater than the one claimed. Note, though, that some individual observations lie on or very close to the diagonal line.

*To obtain the graphs for the regression diagnostics, click on the hotspot (red triangle) next to the term "*Linear Fit*" for the upper regression line in the output in Figure 19.26 and, as shown in Figure 19.27, select the option "*Plot Residuals*". The additional output involves five diagnostic plots and is given in Figure 19.28.*

The first, third, and fourth plots show the residuals against the predicted response variable $\hat{y}$, the row number, and the explanatory variable x, respectively. The idea behind all three plots

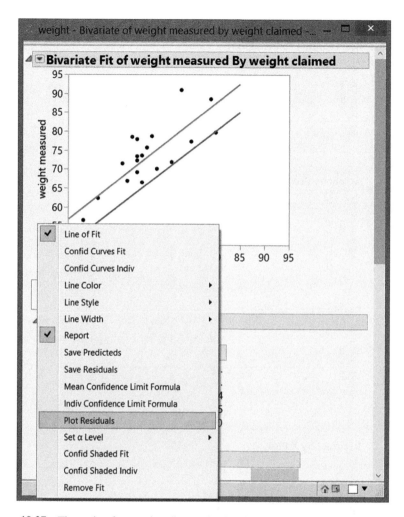

Figure 19.27 The option for creating diagnostic plots in the "Linear Fit" hotspot menu.

is similar: the residuals should not exhibit any systematic relationship with ŷ, the row number, and x. In this example, the points seem to be scattered completely at random. There is no systematic increase or decrease, and no other special structure in the residuals.

The second diagnostic plot produced by the option "Plot Residuals" is called the "Actual by Predicted Plot" and compares the actual value of the response variable (on the vertical axis) and the predicted response value (on the horizontal axis). It is a graph to assess the goodness of fit of the simple linear regression model. If all points are close to the diagonal line in the plot and no special patterns are visible, this indicates a good fit of the model to the data.

Finally, the last diagnostic plot in Figure 19.28 is the normal quantile plot of the residuals. This plot can be reproduced by saving the residuals to the data table (this requires the selection

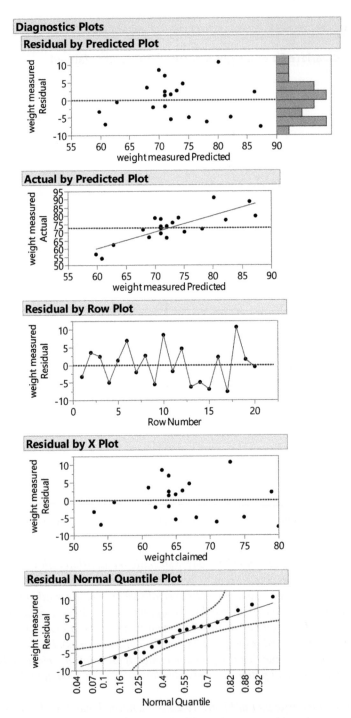

Figure 19.28 The additional output with regression diagnostics produced by the option "Plot Residuals". None of the plots show signs of systematic behavior of the residuals.

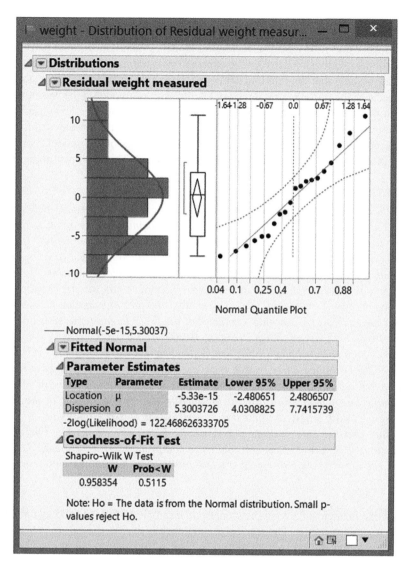

Figure 19.29 The tools to check the normality of the residuals in Example 19.8.1 by means of the "Distribution" platform in the "Analyze" menu. There is no indication for a nonnormal distribution of the residuals.

of the option "Save Residuals" in the hotspot menu in Figure 19.27) and analyzing their distribution. The output shown in Figure 19.29 contains a histogram, a quantile plot, and a fitted normal distribution, as well as the result of the Shapiro–Wilk test for normality. None of these tools shows any sign of a nonnormal distribution of the residuals.

As a conclusion, in this example, the diagnostic tools do not indicate any problems with the assumptions concerning the simple linear regression model.

19.8.2   Nonlinear Relationships

Often, researchers start fitting a linear regression model to their data without considering the exact nature of the relationship between the explanatory variable and the response variable. In doing so, they assume that the relationship between the two variables is linear. Researchers may do so because the linear regression model is the simplest model, or because they are not familiar with other models. In any case, the simple linear regression model is usually a good starting point for an analysis. A careful study of the residuals of the simple linear regression model is required and will generally reveal any shortcomings of the model. For example, the residual plots may indicate that Assumption 1 is violated. This is shown in the following example.

Example 19.8.2   *Whenever the fuel consumption of a car is stated in an advertising brochure, this refers to a very precisely defined test cycle. The car manufacturers optimize the engines with respect to this test cycle so that they can publish low consumption figures. If you drive around with your newly bought car, you will typically be disappointed that you consume much more fuel than the amount mentioned in the brochures and official documents. To obtain an idea about how the fuel consumption depends on its speed, a car was repeatedly driven at a certain constant speed (between 25 and 170 km/h), and the consumption for a distance of 100 km was measured. The resulting scatter plot, the fitted linear regression line, and the Residual-by-Predicted plot are shown in Figure 19.30.*

The Residual-by-Predicted plot clearly shows a nonrandom structure of the residuals. The residuals lie on a U-shaped curve, which indicates that a higher-order effect is required in the model or that a logarithmic transformation of the response would be useful. The lack of fit of the simple linear model can also be detected in the regression plot at the top of the output in Figure 19.30, where the first few as well as the last few points all lie above the regression line and the intermediate points all lie below the regression line. Note that the R^2 value of 0.8374 does not suggest that there is a problem with the model. Discovering the problem requires checking the residuals and plotting the data along with the regression line.

A remedy that is often suitable in the scenario in Example 19.8.2 is to add a quadratic term to the linear regression model. The resulting model then is

$$Y = \beta_0 + \beta_1 x + \beta_2 x^2 + U.$$

This model is quadratic in the explanatory variable x, but it is still a linear regression model because it is a linear function of β_0, β_1, and β_2. Of course, we can add higher-order terms to the model as well:

$$Y = \beta_0 + \beta_1 x + \beta_2 x^2 + \beta_3 x^3 + \cdots + \beta_k x^k + U.$$

The fitting of these types of models is referred to as **polynomial regression**. In some sense, polynomial regression involving one explanatory variable falls in between simple linear regression and multiple linear regression. We still have only one independent variable x (as in simple

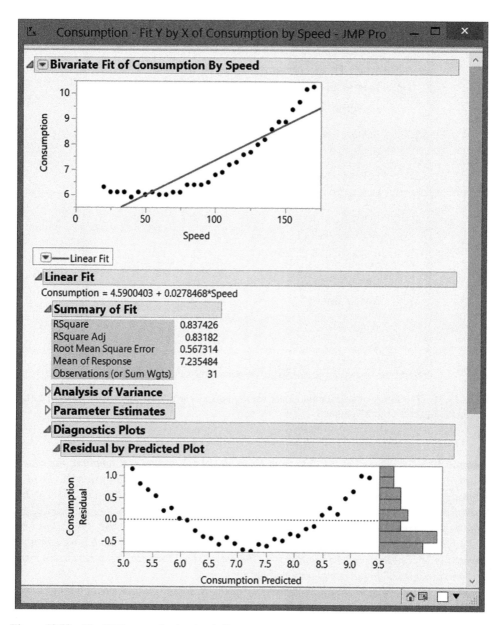

Figure 19.30 The JMP output for the simple linear regression model for Example 19.8.2. The Residual-by-Predicted plot clearly shows a nonrandom structure of the residuals.

linear regression), but the model involves multiple effects, $\beta_1, \beta_2, \ldots, \beta_k$ (as in multiple linear regression). The mathematics for polynomial models is closer to that required for multiple linear regression, which we will cover in detail in *Statistics with JMP: Linear and Generalized Linear Models*. Here, we demonstrate that the fitting of a quadratic model has the potential to remedy diagnostic issues with the residuals.

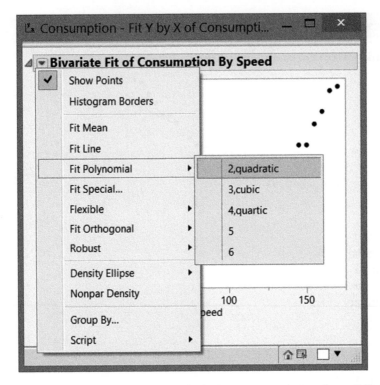

Figure 19.31 The fitting of a quadratic model via the hotspot menu next to the term "`Bivariate Fit`" in the "`Fit Y by X`" output.

Example 19.8.3 *We revisit the data from Example 19.8.2 on fuel consumption, for which the simple linear regression model produced a U-shaped pattern in the residuals. To fit a quadratic model after performing a simple linear regression analysis using the "`Fit Y by X`" platform of JMP, we need to choose "`Fit Polynomial`" followed by "`2, quadratic`" in the hotspot (red triangle) menu next to the term "`Bivariate Fit of Consumption By Speed`". This is shown in Figure 19.31. As for the linear regression model, we can now activate the "`Plot Residuals`" option from the hotspot (red triangle) menu next to the term "`Polynomial Fit`" in the newly obtained output.*

This results in the output shown in Figure 19.32. The Residual-by-Predicted plot now looks perfectly fine: there is no longer any systematic pattern in the residuals. Also note that the R^2 value has increased to 0.9963, and that the straight line in the regression plot in the output has been replaced by a U-shaped curve, which fits the data very well. In conclusion, the quadratic model seems to be a good choice for the data at hand.

Whenever fitting a model to your data, you should always ask whether your model makes sense, independently of the data. This is particularly important for higher-order polynomial models, which may fit the data well, but which may not at all be realistic. In our example on

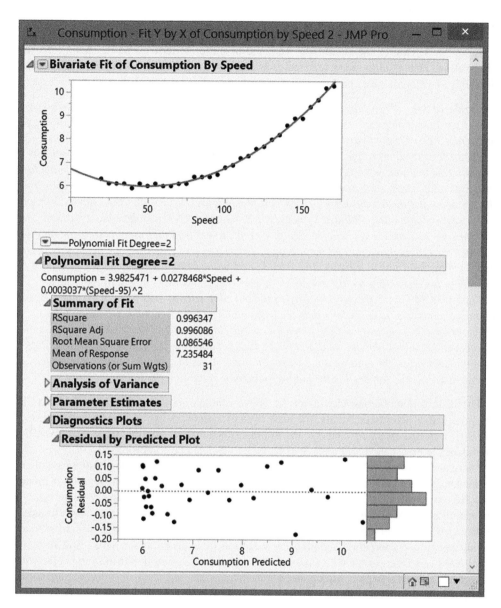

Figure 19.32 The JMP output for the fitted quadratic regression model in Example 19.8.3, including the Residual-by-Predicted plot.

the dependency of fuel consumption on driving speed, a quadratic polynomial model makes perfect sense. The engine in the example has been optimized to have the lowest consumption at about 50 km/h, and driving slower or faster will lead to a greater fuel consumption per kilometer. Therefore, the model should reach a minimum at about 50 km/h.

19.8.3   Heteroscedasticity

Assumption 3 requires the (unknown) variance of the error term to be the same for each observation. This assumption is referred to as the homoscedasticity assumption. For data in the fields of economics and biology, this assumption is violated quite frequently.

Example 19.8.4   *Fruit flies – more specifically, the species* Drosophila melanogaster *– are widely used for biological research. They are easy to care for, breed quickly, lay many eggs, and about 75% of known human disease genes have a match in the genome of fruit flies.*

Their life expectancy ranges from a few weeks to a few months. It can therefore be of interest to predict the life span of fruit flies. A reasonable assumption is that the life span of an individual fruit fly is related to its robustness, expressed by the diameter of its thorax. One study[3] measured the life span and thorax diameter of 125 fruit flies. Figure 19.33 shows the estimated simple linear regression model, obtained by using the thorax diameter as independent variable and the life span as dependent variable. The plot seems to confirm that there is a linear relation between the thorax diameter and the life span. The model has a significant explanatory value, as shown by the F-test in the ANOVA table. The R^2 value of 0.51 also shows the relevance of the thorax for explaining the life span.

*Figure 19.34 shows the Residual-by-Predicted plot corresponding to the simple linear regression analysis. It is obvious that the assumption of homoscedasticity, which requires a constant variance of the residuals over the whole data range, is violated in this example. Instead, the residual plot shows a behaviour that is known as the **funnel effect**. The larger the predicted value, the greater is the variation in the residuals. In other words, the larger the thorax diameter, the less well is the model capable of predicting the life span.*

The lack of homoscedasticity, as in the above example, is called **heteroscedasticity**. In the presence of heteroscedasticity, the least squares estimators are still linear and unbiased, but their standard errors $\hat{\sigma}_{\hat{\beta}_0}$ and $\hat{\sigma}_{\hat{\beta}_1}$ are no longer the smallest possible ones. Moreover, the confidence intervals and hypothesis tests based on them can be compromised.

There are many possible causes for heteroscedasticity. For example, there might be outliers in the data, an important explanatory variable may have been omitted, or the regression model may have been incorrectly specified. The heteroscedasticity might also simply be intrinsic to the nature of the response variable under investigation. For example, asking people to judge distances will result in more accurate values for small distances than for large distances. Also, asking people to report their monthly consumption in a questionnaire will result in more precise answers from people with a low income and a low consumption than from people with a large income and a large consumption.

Sophisticated methods exist to get rid of heteroscedasticity. In the case of a funnel as in Example 19.8.4, which is typical when a life-span variable is involved, it is always worth trying to analyze the logarithmically transformed response variable, $\ln(Y)$, instead of the original response variable Y. In other words, one should estimate the following linear regression model:

$$\ln(Y) = \beta_0 + \beta_1 x.$$

As the next example shows, JMP offers a convenient way to perform this transformation.

---

[3] Source: Partridge L., Farquhar M. (1981) Sexual activity reduces lifespan of male fruitflies. Nature, **294** (5841) 580–582. Reproduced with permission of Macmillan Publishers Ltd.

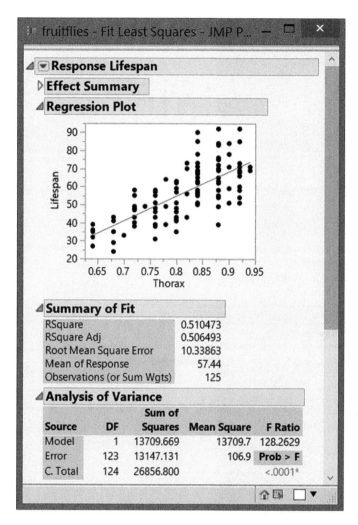

Figure 19.33 The JMP output for the simple linear regression model for the life span of fruit flies in Example 19.8.4. Source: Data from Patridge and Farquhar, 1981. Reproduced with permission of Nature Publishing Group.

Example 19.8.5 *We revisit the data set from Example 19.8.4 but, this time, we model the logarithm of the life span instead of the life span itself, and investigate how this affects the residuals. Here, we use the* "Fit Model" *option in the* "Analyze" *menu, rather than the* "Fit Y by X" *option. In the* "Fit Model" *dialog window, we first need to drag the variable "Thorax" to the* "Construct Model Effects" *field, and the variable "Lifespan" to the* "Y" *field. Next, we need to select the variable "Lifespan" in the* "Y" *field, and choose the option "Log" from the hotspot (red triangle) menu next to the word* "Transform" *at the bottom of the dialog window. This step is shown in Figure 19.35. The resulting new response variable in the* "Y" *field is "Log(Lifespan)". The final step required is to click "Run" in the top right corner of the dialog window. This produces the output shown in*

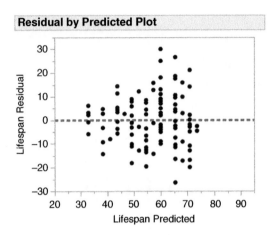

Figure 19.34 The heteroscedasticity (the so-called funnel effect) of the residuals for the simple linear regression in Example 19.8.4. Source: Data from Patridge and Farquhar, 1981. Reproduced with permission of Nature Publishing Group.

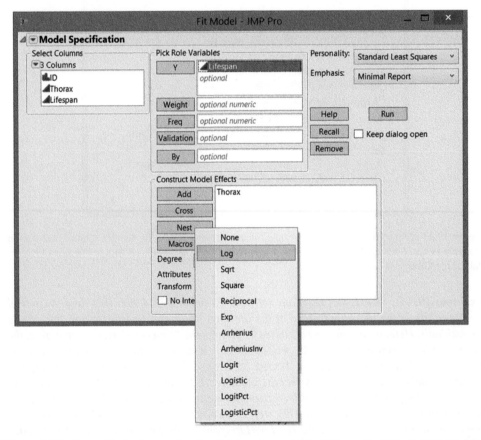

Figure 19.35 Logarithmically transforming the response variable "Lifespan" within the "Fit Model" dialog window.

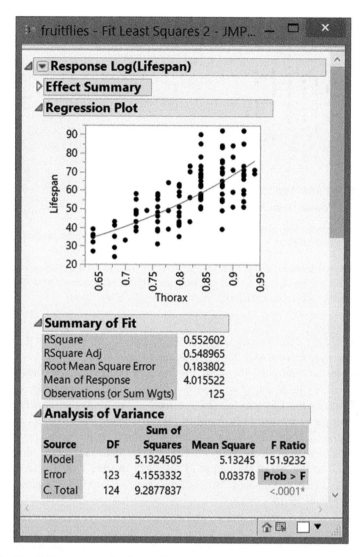

Figure 19.36 The JMP output for the simple linear regression model for the logarithm of the life span of fruit flies in Example 19.8.5.

Figure 19.36. The new model also has a significant explanatory value. The R^2 value of the new model is 0.55. More importantly, the residuals no longer show signs of heteroscedasticity. This can be seen in Figure 19.37a. Therefore, in this example, the logarithmic transformation has successfully removed the heteroscedasticity.

There exist several other ways to perform the simple linear regression analysis of the logarithm of the life span in JMP. One approach would be to create a new column in the data table, containing the logarithms of the life spans, and to enter this new column in the "Y" field in the "Fit Model" dialog window. One advantage of performing the transformation of

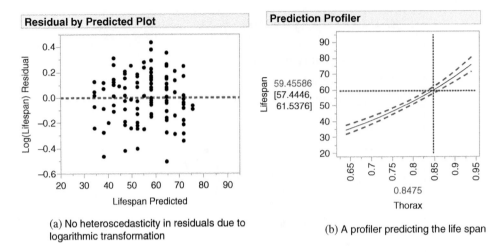

(a) No heteroscedasticity in residuals due to logarithmic transformation

(b) A profiler predicting the life span

Figure 19.37 The residuals and the prediction profiler of the simple linear regression model in Example 19.8.5. Note that the prediction profiler displays the original response variable "Lifespan" on its vertical axis, while the residuals are based on the logarithms of the life spans.

the response within the dialog window of the "Fit Model" option is that JMP remembers that, initially, you were interested in the life span, and not in its logarithm. As a consequence, the regression plot in Figure 19.36 shows the life span itself (i.e., the original response) on its vertical axis. In addition, if you use the "Profiler" to predict a response value, then JMP will predict the life span directly, instead of the logarithm of the life span. The prediction profiler for this example is shown in Figure 19.37b. The "Profiler", which is a dynamic, interactive tool to explore how the response depends on the explanatory variable(s) in regression, can be activated via the hotspot (red triangle) menu next to the term "Response Log(Lifespan)" at the top of the output of the "Fit Model" option. In that menu, the options "Factor Profiling" and "Profiler" have to be chosen successively. This is demonstrated in Figure 19.38.

Another way in which a simple linear regression model for the logarithm of the life span can be estimated is by using the "Fit Y by X" option, which we have almost exclusively used in the current chapter. For the fruit fly data, the "Fit Y by X" option initially produces a scatter plot of the two variables under investigation. Via the hotspot (red triangle) menu next to the term "Bivariate Fit" in the output with the scatter plot, we can perform the desired regression analysis with the logarithm of the life span by activating the option "Fit Special". This option is visible in Figure 19.31, and produces the window shown in Figure 19.39. In that window, the natural logarithm has to be selected for the response variable. This results in an output almost identical to that shown in Figure 19.36.

A comparison of the regression plot in Figure 19.36 with that in Figure 19.33 does not reveal large differences between the two estimated models. As a result, when it comes to making point predictions, there will be almost no difference between the two models. The benefit of logarithmically transforming the response variable mainly lies in the validity of the statistical inference; that is, the hypothesis tests, the confidence intervals, and the prediction intervals.

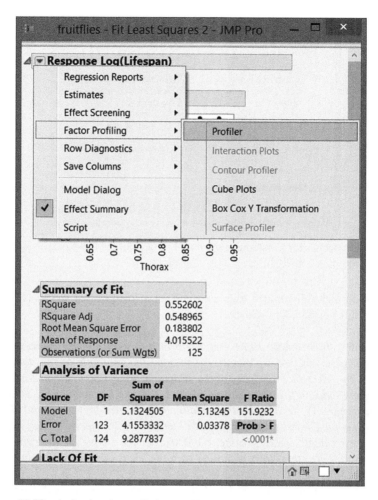

Figure 19.38 Activating the prediction profiler from the "Fit Model" output in JMP.

19.8.4 Nonnormally Distributed Error Terms

Assumption 4, the normality of the error terms, is not required for the Gauss–Markov theorem 19.3.1 to be valid. This means that the least squares estimators are the best linear unbiased estimators even when the error terms are not normally distributed.

In addition, for large sample sizes, we can rely on the central limit theorem to justify our use of the confidence intervals and the hypothesis tests derived under the assumption of normal errors. We should, however, be cautious in using these intervals and tests when we encounter a clear deviation from normality in a small data sample. The problem, however, is that, with a small data set, it is hard to firmly establish the nonnormality of the error terms.

To summarize, nonnormal error terms are only a statistical concern for small data sets. However, as we have already mentioned in Section 19.5, in many situations, we expect the error term to be normal due to the central limit theorem. Therefore, in cases where the error

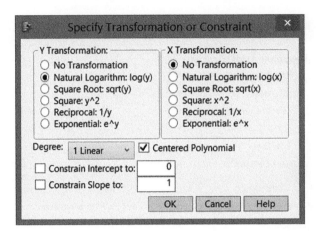

Figure 19.39 The options for transforming the response variable and the explanatory variable when selecting "Fit Special" in the "Fit Y by X" platform.

terms are not normally distributed, this can indicate a misspecified model (e.g., due to a relevant but omitted explanatory variable).

Example 19.8.6   *In Example 19.2.6, we performed a simple linear regression for data on starch extraction, and we explained how to save the residuals to the data table (see Figure 19.14). To assess the normality of the residuals, we generate the histogram and the quantile diagram, and perform the Shapiro–Wilk test (see also Section 6.2.1). The results are shown in Figure 19.40. The histogram already indicates skewness, and the quantile diagram confirms that. Finally, the Shapiro–Wilk test strongly rejects the null hypothesis of normality of the residuals with a p-value of 0.0002. For the simple linear regression model with salt concentration as the only explanatory variable, the residuals are clearly not normally distributed.*

However, we mentioned in Example 19.2.6 that corn was used from five different locations. This information is reflected by the nominal "Block" variable in the data table. This nominal variable is visible in the second column in Figure 19.14, but it has been ignored so far in our regression analysis. Adding this nominal variable to the model leads to a regression model with two explanatory variables, one of which is quantitative (the salt concentration) and one of which is qualitative or categorical (the "Block" variable indicating the origin of the corn). The resulting model is a multiple linear regression model, a type of model covered comprehensively in Statistics with JMP: Linear and Generalized Linear Models. *The distribution of the residuals corresponding to this multiple linear regression model, obtained by adding the "Block" variable to the model, is shown in Figure 19.41. The newly obtained residuals no longer exhibit obvious signs of a nonnormal distribution.*

19.8.5 *Statistically Dependent Responses*

Assumption 3 states that the error terms (and therefore also the responses) are statistically independent. There are several common scenarios in which this assumption is frequently

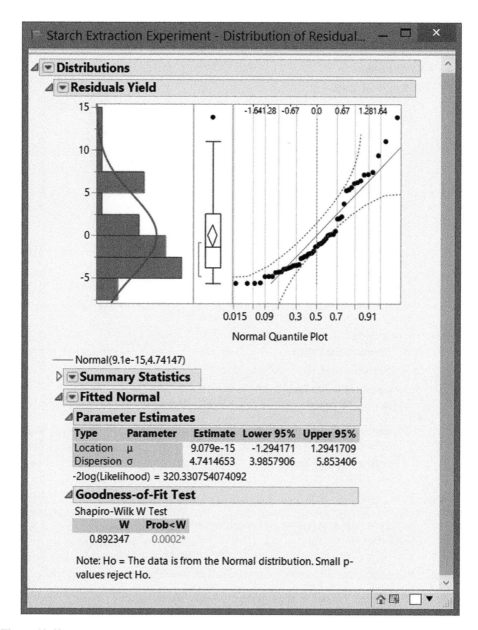

Figure 19.40 The nonnormal distribution of the residuals of the simple linear regression model for the starch extraction data in Example 19.8.6.

invalid. For example, the data may have been collected in groups. Another possibility is that the observations have been made over time. Yet another scenario, which we do not discuss here, is when certain observations were made close to each other in space, while other observations were made at places that are located far away from each other.

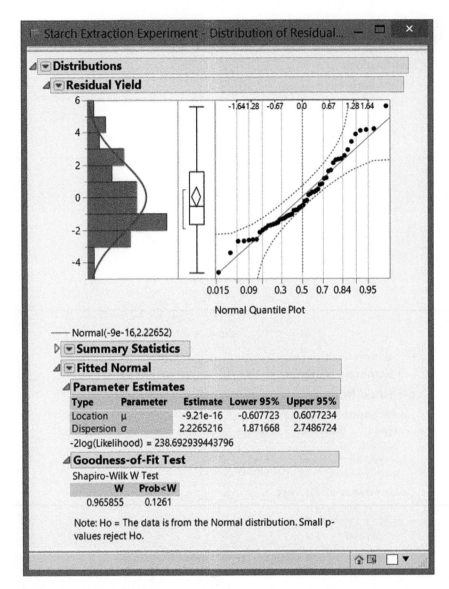

Figure 19.41 The normal distribution of the residuals for the starch extraction data after adding the nominal "Block" variable to the model.

Grouped Data

Sometimes, performing an experiment requires several days or several batches of material. Typically, responses obtained on the same day or using the same batch are correlated. Any proper statistical analysis should take into account the grouped nature of the data in such scenarios. This implies that the statistical model used should reflect the fact that certain observations were made on the same day or using the same batch of material.

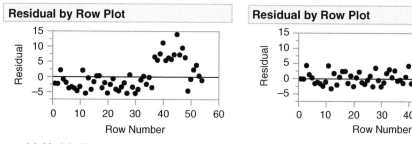

Figure 19.42   Residual-by-Row plots for the linear regression models for the starch extraction data.

Example 19.8.7   *In Example 19.2.6, we performed a simple linear regression for data on starch extraction. The Residual-by-Row plot corresponding to this analysis is shown in Figure 19.42a. The plot shows that the first 36 residuals are almost all negative. The next 12 residuals are all positive, and the final six residuals form a cluster around zero. It is clear that the residuals show a systematic pattern here. The first 36 observations all correspond to the first two batches of corn. The next 12 observations correspond to batches 3 and 4, and the last few observations originate from batch 5. It is clear that all residuals from a given batch behave similarly. In statistical terms, we say that the residuals are correlated, or dependent.*

The dependence of the residuals can be removed in this example by incorporating the "Block" variable in the regression model as an additional explanatory variable, as was already done in Example 19.8.6. The Residual-by-Row plot corresponding to this multiple linear regression analysis is shown in Figure 19.42b. The positive and negative residuals no longer appear in clusters. In conclusion, by taking the grouping of the data into account, the dependence between the residuals is removed.

Grouped data are obtained quite frequently in many disciplines. In agricultural experimentation, it is common to carry out multiple tests on each of a number of fields. Every field then gives rise to multiple correlated observations. Likewise, in medical and veterinary studies, it is common to make repeated measurements for each human or animal in the study. Every human or animal then produces several correlated observations.

In many cases, there are multiple ways in which the data are grouped. For example, in economics and the social sciences, it is common to interview multiple members of one household, select multiple households from one city, and pick multiple cities from each region. The resulting data are grouped in three ways, namely by household, by city, and by region. Typically, observations from the same household are strongly correlated. Observations from the same city but different households are also correlated, but not as strongly as observations from the same household. Finally, observations from the same region but different cities are only weakly correlated.

There are various ways to take into account the grouping of data in statistical analyses. We will discuss several possibilities in *Statistics with JMP: Linear and Generalized Linear Models*.

	Time	Consumption	Income	Price Index	Temperature
1	March 1951	0.386	78	0.27	41
2	April 1951	0.374	79	0.282	56
3	May 1951	0.393	81	0.277	63
4	June 1951	0.425	80	0.28	68
5	July 1951	0.406	76	0.272	69
6	August 1951	0.344	78	0.262	65
7	September 1951	0.327	82	0.275	61
8	October 1951	0.288	79	0.267	47
9	November 1951	0.269	76	0.265	32
10	December 1951	0.256	79	0.277	24
11	January 1952	0.286	82	0.282	28
12	February 1952	0.298	85	0.27	26
13	March 1952	0.329	86	0.272	32
14	April 1952	0.318	83	0.287	40
15	May 1952	0.381	84	0.277	55
16	June 1952	0.381	82	0.287	63
17	July 1952	0.47	80	0.28	72
18	August 1952	0.443	78	0.277	72
19	September 1952	0.386	84	0.277	67

Columns (5/0): Time, Consumption, Income, Price Index, Temperature

Rows: All rows 30, Selected 0, Excluded 0, Hidden 0, Labelled 0

Figure 19.43   The JMP data table showing the ice cream data in Example 19.8.8. Source: Data from Hildreth and Lu, 1960.

Serial Correlation

Observational data is often collected over time. At fixed (typically equally spaced) time points, an observation is made for the response variable and for the explanatory variable under investigation. Successive observations then typically turn out to be correlated.

Example 19.8.8   *Consider the data set shown in Figure 19.43, concerning the monthly ice cream consumption in 30 successive months in the period 1951–1953[4]. Economic theory states that there is a negative relationship between the consumption and the price level. The estimation of a simple linear regression model with consumption as the response variable and price index as the explanatory variable results in the Residual-by-Row plot shown in Figure 19.44a.*

It is clear that the residuals exhibit a nonrandom pattern. In fact, most positive residuals are followed by another positive residual. Likewise, negative residuals are generally followed by other negative residuals. In statistical terms, there appears to be a positive correlation between successive residuals. Therefore, the successive error terms are dependent, meaning that Assumption 3 is violated. The residuals in this example have a cyclical pattern, as can be expected for a seasonal product such as ice cream. In winter, the ice cream consumption

---

[4] Source: Data from Hildreth, C. and Lu, J. (1960) Demand relations with autocorrelated disturbances, Technical Bulletin No 2765, Michigan State University.

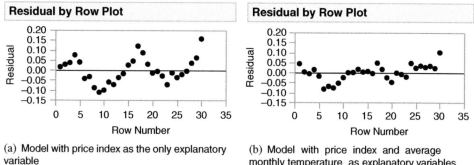

Figure 19.44 Residual-by-Row plots for the linear regression models in Example 19.8.8. Source: Data from Hildreth and Lu, 1960.

is rather low, regardless of the price. In summer, the ice cream consumption tends to be high, regardless of the price.

In Example 19.8.7, the addition of one explanatory variable was enough to remove the correlation in the residuals. For the ice cream data, we can also attempt to remove the dependence in the error terms by adding an explanatory variable. Here, we add the average temperature (expressed in Fahrenheit) to the regression model as an explanatory variable. Doing so transforms our simple linear regression model into a multiple linear regression model. Adding the temperature makes sense because the temperature follows the same kind of cyclical pattern as the residuals in Figure 19.44a.

Figure 19.44b shows the new Residual-by-Row plot corresponding to the multiple linear regression model with the average monthly temperature as an additional explanatory variable. In the new plot, there are still a few clusters of negative residuals, positive residuals, and residuals concentrated around zero. This shows that, in this case, not all of the correlation between the successive residuals was removed by adding an explanatory variable. More sophisticated regression techniques are needed to deal with the correlation in this example.

Appendix A

The Binomial Distribution

This appendix contains exceedance probabilities for the binomial distribution. For example, if $n = 4$ and $\pi = 0.20$, then

$$P(X \geq 2) = 0.1808.$$

n	x	$\pi = 0.05$	0.10	0.15	0.20	0.25	0.30	0.35	0.40	0.45	0.50
2	0	1.0000	1.0000	1.0000	1.0000	1.0000	1.0000	1.0000	1.0000	1.0000	1.0000
	1	0.0975	0.1900	0.2775	0.3600	0.4375	0.5100	0.5775	0.6400	0.6975	0.7500
	2	0.0025	0.0100	0.0225	0.0400	0.0625	0.0900	0.1225	0.1600	0.2025	0.2500

n	x	$\pi = 0.05$	0.10	0.15	0.20	0.25	0.30	0.35	0.40	0.45	0.50
3	0	1.0000	1.0000	1.0000	1.0000	1.0000	1.0000	1.0000	1.0000	1.0000	1.0000
	1	0.1426	0.2710	0.3859	0.4880	0.5781	0.6570	0.7254	0.7840	0.8336	0.8750
	2	0.0073	0.0280	0.0608	0.1040	0.1562	0.2160	0.2818	0.3520	0.4253	0.5000
	3	0.0001	0.0010	0.0034	0.0080	0.0156	0.0270	0.0429	0.0640	0.0911	0.1250

n	x	$\pi = 0.05$	0.10	0.15	0.20	0.25	0.30	0.35	0.40	0.45	0.50
4	0	1.0000	1.0000	1.0000	1.0000	1.0000	1.0000	1.0000	1.0000	1.0000	1.0000
	1	0.1855	0.3439	0.4780	0.5904	0.6836	0.7559	0.8215	0.8704	0.9085	0.9375
	2	0.0140	0.0523	0.1095	0.1808	0.2617	0.3483	0.4370	0.5248	0.6090	0.6875
	3	0.0005	0.0037	0.0120	0.0272	0.0508	0.0837	0.1265	0.1792	0.2415	0.3125
	4		0.0001	0.0005	0.0016	0.0039	0.0081	0.0150	0.0256	0.0410	0.0625

n	x	$\pi = 0.05$	0.10	0.15	0.20	0.25	0.30	0.35	0.40	0.45	0.50
5	0	1.0000	1.0000	1.0000	1.0000	1.0000	1.0000	1.0000	1.0000	1.0000	1.0000
	1	0.2262	0.4095	0.5563	0.6723	0.7627	0.8319	0.8840	0.9222	0.9497	0.9688
	2	0.0226	0.0815	0.1648	0.2627	0.3672	0.4718	0.5716	0.6630	0.7438	0.8125
	3	0.0012	0.0086	0.0266	0.0579	0.1035	0.1631	0.2352	0.3174	0.4069	0.5000
	4		0.0005	0.0022	0.0067	0.0156	0.0308	0.0540	0.0870	0.1312	0.1875
	5			0.0001	0.0003	0.0010	0.0024	0.0053	0.0102	0.0185	0.0313

Statistics with JMP: Hypothesis Tests, ANOVA and Regression, First Edition. Peter Goos and David Meintrup.
© 2016 John Wiley & Sons, Ltd. Published 2016 by John Wiley & Sons, Ltd.
Companion Website: http://www.wiley.com/go/goosandmeintrup/JMP

n	x	π = 0.05	0.10	0.15	0.20	0.25	0.30	0.35	0.40	0.45	0.50
6	0	1.0000	1.0000	1.0000	1.0000	1.0000	1.0000	1.0000	1.0000	1.0000	1.0000
	1	0.2649	0.4686	0.6229	0.7379	0.8220	0.8824	0.9246	0.9533	0.9723	0.9844
	2	0.0328	0.1143	0.2235	0.3446	0.4661	0.5798	0.6809	0.7667	0.8364	0.8906
	3	0.0022	0.0159	0.0473	0.0989	0.1694	0.2557	0.3529	0.4557	0.5585	0.6563
	4	0.0001	0.0013	0.0059	0.0170	0.0376	0.0705	0.1174	0.1792	0.2553	0.3438
	5		0.0001	0.0004	0.0016	0.0046	0.0109	0.0223	0.0410	0.0692	0.1094
	6				0.0001	0.0002	0.0007	0.0018	0.0041	0.0083	0.0156

n	x	π = 0.05	0.10	0.15	0.20	0.25	0.30	0.35	0.40	0.45	0.50
7	0	1.0000	1.0000	1.0000	1.0000	1.0000	1.0000	1.0000	1.0000	1.0000	1.0000
	1	0.3017	0.5217	0.6794	0.7903	0.8665	0.9176	0.9510	0.9720	0.9848	0.9922
	2	0.0444	0.1497	0.2834	0.4233	0.5551	0.6706	0.7662	0.8414	0.8976	0.9375
	3	0.0038	0.0257	0.0738	0.1480	0.2436	0.3529	0.4677	0.5801	0.6836	0.7734
	4	0.0002	0.0027	0.0121	0.0333	0.0706	0.1260	0.1998	0.2898	0.3917	0.5000
	5		0.0002	0.0012	0.0047	0.0129	0.0288	0.0556	0.0963	0.1529	0.2266
	6			0.0001	0.0004	0.0013	0.0038	0.0090	0.0188	0.0357	0.0625
	7					0.0001	0.0002	0.0006	0.0016	0.0037	0.0078

n	x	π = 0.05	0.10	0.15	0.20	0.25	0.30	0.35	0.40	0.45	0.50
8	0	1.0000	1.0000	1.0000	1.0000	1.0000	1.0000	1.0000	1.0000	1.0000	1.0000
	1	0.3366	0.5695	0.7275	0.8322	0.8999	0.9424	0.9681	0.9832	0.9916	0.9961
	2	0.0572	0.1869	0.3428	0.4967	0.6329	0.7447	0.8309	0.8936	0.9368	0.9648
	3	0.0058	0.0381	0.1052	0.2031	0.3215	0.4482	0.5722	0.6846	0.7799	0.8555
	4	0.0004	0.0050	0.0214	0.0563	0.1183	0.1941	0.2936	0.4059	0.5230	0.6367
	5		0.0004	0.0029	0.0104	0.0273	0.0580	0.1061	0.1737	0.2604	0.3633
	6			0.0002	0.0012	0.0042	0.0113	0.0253	0.0498	0.0885	0.1445
	7				0.0001	0.0004	0.0013	0.0036	0.0085	0.0181	0.352
	8						0.0001	0.0002	0.0007	0.0017	0.0039

n	x	π = 0.05	0.10	0.15	0.20	0.25	0.30	0.35	0.40	0.45	0.50
9	0	1.0000	1.0000	1.0000	1.0000	1.0000	1.0000	1.0000	1.0000	1.0000	1.0000
	1	0.3689	0.6126	0.7684	0.8658	0.9249	0.9596	0.9793	0.9899	0.9954	0.9980
	2	0.0712	0.2252	0.4005	0.5638	0.6977	0.8040	0.8789	0.9295	0.9615	0.9805
	3	0.0084	0.0530	0.1409	0.2618	0.3993	0.5372	0.6627	0.7682	0.8505	0.9102
	4	0.0006	0.0083	0.0339	0.0856	0.1657	0.2703	0.3911	0.5174	0.6386	0.7461
	5		0.0009	0.0056	0.0196	0.0489	0.0988	0.1717	0.2666	0.3786	0.5000
	6		0.0001	0.0006	0.0031	0.0100	0.0253	0.0536	0.0994	0.1658	0.2539
	7				0.0003	0.0013	0.0043	0.0112	0.0250	0.0498	0.0898
	8					0.0001	0.0004	0.0014	0.0038	0.0091	0.0195
	9							0.0001	0.0003	0.0008	0.0020

n	x	π = 0.05	0.10	0.15	0.20	0.25	0.30	0.35	0.40	0.45	0.50
10	0	1.0000	1.0000	1.0000	1.0000	1.0000	1.0000	1.0000	1.0000	1.0000	1.0000
	1	0.4013	0.6513	0.8031	0.8926	0.9437	0.9718	0.9865	0.9940	0.9975	0.9990
	2	0.0861	0.2639	0.4557	0.6242	0.7560	0.8507	0.9140	0.9536	0.9767	0.9893
	3	0.0115	0.0702	0.1798	0.3222	0.4744	0.6172	0.7384	0.8327	0.9004	0.9453
	4	0.0010	0.0128	0.0500	0.1209	0.2241	0.3504	0.4862	0.6177	0.7340	0.8281
	5	0.0001	0.0016	0.0099	0.0328	0.0781	0.1503	0.2485	0.3669	0.4956	0.6230
	6		0.0001	0.0014	0.0064	0.0197	0.0473	0.0949	0.1662	0.2616	0.3770
	7			0.0001	0.0009	0.0035	0.0106	0.0260	0.0548	0.1020	0.1719
	8				0.0001	0.0004	0.0016	0.0048	0.0123	0.0274	0.0547
	9						0.0001	0.0005	0.0017	0.0045	0.0107
	10								0.0001	0.0003	0.0010

n	x	π = 0.05	0.10	0.15	0.20	0.25	0.30	0.35	0.40	0.45	0.50
12	0	1.0000	1.0000	1.0000	1.0000	1.0000	1.0000	1.0000	1.0000	1.0000	1.0000
	1	0.4596	0.7176	0.8578	0.9313	0.9683	0.9862	0.9943	0.9978	0.9992	0.9998
	2	0.1184	0.3410	0.5565	0.7251	0.8416	0.9150	0.9576	0.9807	0.9917	0.9968
	3	0.0196	0.1109	0.2642	0.4417	0.6093	0.7472	0.8487	0.9166	0.9579	0.9807
	4	0.0022	0.0256	0.0922	0.2054	0.3512	0.5075	0.6533	0.7747	0.8655	0.9270
	5	0.0002	0.0043	0.0239	0.0726	0.1576	0.2763	0.4167	0.5618	0.6956	0.8062
	6		0.0005	0.0046	0.0197	0.0544	0.1178	0.2127	0.3348	0.4731	0.6128
	7		0.0001	0.0007	0.0039	0.0143	0.0386	0.0846	0.1582	0.2607	0.3872
	8			0.0001	0.0006	0.0028	0.0095	0.0255	0.0573	0.1117	0.1938
	9				0.0001	0.0004	0.0017	0.0056	0.0153	0.0356	0.0730
	10						0.0002	0.0008	0.0028	0.0079	0.0193
	11							0.0001	0.0003	0.0011	0.0032
	12									0.0001	0.0002

n	x	π = 0.05	0.10	0.15	0.20	0.25	0.30	0.35	0.40	0.45	0.50
14	0	1.0000	1.0000	1.0000	1.0000	1.0000	1.0000	1.0000	1.0000	1.0000	1.0000
	1	0.5123	0.7712	0.8972	0.9560	0.9822	0.9932	0.9976	0.9992	0.9998	0.9999
	2	0.1530	0.4154	0.6433	0.8021	0.8990	0.9525	0.9795	0.9919	0.9971	0.9991
	3	0.0301	0.1584	0.3521	0.5519	0.7189	0.8392	0.9161	0.9602	0.9830	0.9935
	4	0.0042	0.0441	0.1465	0.3018	0.4787	0.6448	0.7795	0.8757	0.9368	0.9714
	5	0.0004	0.0092	0.0467	0.1298	0.2585	0.4158	0.5773	0.7207	0.8328	0.9102
	6		0.0015	0.0115	0.0439	0.1117	0.2195	0.3595	0.5141	0.6627	0.7880
	7		0.0002	0.0022	0.0116	0.0383	0.0933	0.1836	0.3075	0.4539	0.6047
	8			0.0003	0.0024	0.0103	0.0315	0.0753	0.1501	0.2586	0.3953
	9				0.0004	0.0022	0.0083	0.0243	0.0583	0.1189	0.2120
	10					0.0003	0.0017	0.0060	0.0175	0.0426	0.0898
	11						0.0002	0.0011	0.0039	0.0114	0.0287
	12							0.0001	0.0006	0.0022	0.0065
	13								0.0001	0.0003	0.0009
	14										0.0001

n	x	$\pi = 0.05$	0.10	0.15	0.20	0.25	0.30	0.35	0.40	0.45	0.50
16	0	1.0000	1.0000	1.0000	1.0000	1.0000	1.0000	1.0000	1.0000	1.0000	1.0000
	1	0.5599	0.8147	0.9257	0.9719	0.9900	0.9967	0.9990	0.9997	0.9999	1.0000
	2	0.1892	0.4853	0.7161	0.8593	0.9365	0.9739	0.9902	0.9967	0.9990	0.9997
	3	0.0429	0.2108	0.4386	0.6482	0.8029	0.9006	0.9549	0.9817	0.9934	0.9979
	4	0.0070	0.0684	0.2101	0.4019	0.5950	0.7541	0.8661	0.9349	0.9719	0.9894
	5	0.0009	0.0170	0.0791	0.2018	0.3698	0.5501	0.7108	0.8334	0.9147	0.9616
	6	0.0001	0.0033	0.0245	0.0817	0.1897	0.3402	0.5100	0.6712	0.8024	0.8949
	7		0.0005	0.0056	0.0267	0.0796	0.1753	0.3119	0.4738	0.6340	0.7728
	8		0.0001	0.0011	0.0070	0.0271	0.0744	0.1594	0.2839	0.4371	0.5982
	9			0.0002	0.0015	0.0075	0.0257	0.2671	0.1423	0.2559	0.4018
	10				0.0002	0.0016	0.0071	0.0229	0.0583	0.1241	0.2272
	11					0.0003	0.0016	0.0062	0.0191	0.0486	0.1051
	12						0.0003	0.0013	0.0049	0.0149	0.0384
	13							0.0002	0.0009	0.0035	0.0106
	14								0.0001	0.0006	0.0021
	15									0.0001	0.0003

n	x	$\pi = 0.05$	0.10	0.15	0.20	0.25	0.30	0.35	0.40	0.45	0.50
18	0	1.0000	1.0000	1.0000	1.0000	1.0000	1.0000	1.0000	1.0000	1.0000	1.0000
	1	0.6028	0.8499	0.9464	0.9820	0.9944	0.9984	0.9996	0.9999	1.0000	1.0000
	2	0.2265	0.5497	0.7759	0.9009	0.9605	0.9858	0.9954	0.9987	0.9997	0.9999
	3	0.0581	0.2662	0.5203	0.7287	0.8647	0.9400	0.9764	0.9918	0.9975	0.9993
	4	0.0109	0.0982	0.2798	0.4990	0.6943	0.8354	0.8917	0.9672	0.9880	0.9962
	5	0.0015	0.0282	0.1206	0.2836	0.4813	0.6673	0.8114	0.9058	0.9589	0.9846
	6	0.0002	0.0064	0.0419	0.1329	0.2825	0.4656	0.6450	0.7912	0.8923	0.9519
	7		0.0012	0.0118	0.0513	0.1390	0.2783	0.4509	0.6257	0.7742	0.8811
	8		0.0002	0.0027	0.0163	0.0569	0.1407	0.2717	0.4366	0.6085	0.7597
	9			0.0005	0.0043	0.0193	0.0596	0.1391	0.2632	0.4222	0.5927
	10			0.0001	0.0009	0.0054	0.0210	0.0597	0.1347	0.2527	0.4073
	11				0.0002	0.0012	0.0061	0.0212	0.0576	0.1280	0.2403
	12					0.0002	0.0014	0.0062	0.0203	0.0537	0.1189
	13						0.0003	0.0014	0.0058	0.0183	0.0481
	14							0.0003	0.0013	0.0049	0.0154
	15								0.0002	0.0010	0.0038
	16									0.0001	0.0007
	17										0.0001

n	x	$\pi = 0.05$	0.10	0.15	0.20	0.25	0.30	0.35	0.40	0.45	0.50
20	0	1.0000	1.0000	1.0000	1.0000	1.0000	1.0000	1.0000	1.0000	1.0000	1.0000
	1	0.6415	0.8784	0.9612	0.9885	0.9968	0.9992	0.9998	1.0000	1.0000	1.0000
	2	0.2642	0.6083	0.8244	0.9308	0.9757	0.9924	0.9979	0.9995	0.9999	1.0000
	3	0.0755	0.3231	0.5951	0.7939	0.9087	0.9645	0.9879	0.9964	0.9991	0.9998
	4	0.0159	0.1330	0.3523	0.5886	0.7748	0.8729	0.9556	0.9840	0.9951	0.9987
	5	0.0026	0.0432	0.1702	0.3704	0.5852	0.7625	0.8818	0.9490	0.9811	0.9941
	6	0.0003	0.0113	0.0673	0.1958	0.3838	0.5836	0.7546	0.8744	0.9447	0.9793
	7		0.0024	0.0219	0.0867	0.2142	0.3920	0.5834	0.7500	0.8701	0.9423
	8		0.0004	0.0059	0.0321	0.1018	0.2277	0.3990	0.5841	0.7480	0.8684
	9		0.0001	0.0013	0.0100	0.0409	0.1133	0.2376	0.4044	0.5857	0.7483
	10			0.0002	0.0026	0.0139	0.0480	0.1218	0.2447	0.4086	0.5881
	11				0.0006	0.0039	0.0171	0.0532	0.1275	0.2493	0.4119
	12				0.0001	0.0009	0.0051	0.0196	0.0565	0.1308	0.2517
	13					0.0002	0.0013	0.0060	0.0210	0.0580	0.1316
	14						0.0003	0.0015	0.0065	0.0214	0.0577
	15							0.0003	0.0016	0.0064	0.0207
	16								0.0003	0.0015	0.0059
	17									0.0003	0.0013
	18										0.0002

Appendix B

The Standard Normal Distribution

This appendix contains exceedance probabilities of the standard normal distribution. For example, $P(Z \geq 1.96) = 0.02500$.

z	0.00	0.01	0.02	0.03	0.04	0.05	0.06	0.07	0.08	0.09
0.0	0.50000	0.49601	0.49202	0.48803	0.48405	0.48006	0.47608	0.47210	0.46812	0.46414
0.1	0.46017	0.45620	0.45224	0.44828	0.44433	0.44038	0.43644	0.43251	0.42858	0.42465
0.2	0.42074	0.41683	0.41294	0.40905	0.40517	0.40129	0.39743	0.39358	0.38974	0.38591
0.3	0.38209	0.37828	0.37448	0.37070	0.36693	0.36317	0.35942	0.35569	0.35197	0.34827
0.4	0.34458	0.34090	0.33724	0.33360	0.32997	0.32636	0.32276	0.31918	0.31561	0.31207
0.5	0.30854	0.30503	0.30153	0.29806	0.29460	0.29116	0.28774	0.28434	0.28096	0.27760
0.6	0.27425	0.27093	0.26763	0.26435	0.26109	0.25785	0.25463	0.25143	0.24825	0.24510
0.7	0.24196	0.23885	0.23576	0.23270	0.22965	0.22663	0.22363	0.22065	0.21770	0.21476
0.8	0.21186	0.20897	0.20611	0.20327	0.20045	0.19766	0.19489	0.19215	0.18943	0.18673
0.9	0.18406	0.18141	0.17879	0.17619	0.17361	0.17106	0.16853	0.16602	0.16354	0.16109

z	0.00	0.01	0.02	0.03	0.04	0.05	0.06	0.07	0.08	0.09
1.0	0.15866	0.15625	0.15386	0.15151	0.14917	0.14686	0.14457	0.14231	0.14007	0.13786
1.1	0.13567	0.13350	0.13136	0.12924	0.12714	0.12507	0.12302	0.12100	0.11900	0.11702
1.2	0.11507	0.11314	0.11123	0.10935	0.10749	0.10565	0.10383	0.10204	0.10027	0.09853
1.3	0.09680	0.09510	0.09342	0.09176	0.09012	0.08851	0.08692	0.08534	0.08379	0.08226
1.4	0.08076	0.07927	0.07780	0.07636	0.07493	0.07353	0.07214	0.07078	0.06944	0.06811
1.5	0.06681	0.06552	0.06426	0.06301	0.06178	0.06057	0.05938	0.05821	0.05705	0.05592
1.6	0.05480	0.05370	0.05262	0.05155	0.05050	0.04947	0.04846	0.04746	0.04648	0.04551
1.7	0.04457	0.04363	0.04272	0.04182	0.04093	0.04006	0.03920	0.03836	0.03754	0.03673
1.8	0.03593	0.03515	0.03438	0.03362	0.03288	0.03216	0.03144	0.03074	0.03005	0.02938
1.9	0.02872	0.02807	0.02743	0.02680	0.02619	0.02559	0.02500	0.02442	0.02385	0.02330

Statistics with JMP: Hypothesis Tests, ANOVA and Regression, First Edition. Peter Goos and David Meintrup.
© 2016 John Wiley & Sons, Ltd. Published 2016 by John Wiley & Sons, Ltd.
Companion Website: http://www.wiley.com/go/goosandmeintrup/JMP

z	0.00	0.01	0.02	0.03	0.04	0.05	0.06	0.07	0.08	0.09
2.0	0.02275	0.02222	0.02169	0.02118	0.02068	0.02018	0.01970	0.01923	0.01876	0.01831
2.1	0.01786	0.01743	0.01700	0.01659	0.01618	0.01578	0.01539	0.01500	0.01463	0.01426
2.2	0.01390	0.01355	0.01321	0.01287	0.01254	0.01222	0.01190	0.01160	0.01130	0.01101
2.3	0.01072	0.01044	0.01017	0.00990	0.00964	0.00939	0.00914	0.00889	0.00866	0.00842
2.4	0.00820	0.00798	0.00776	0.00755	0.00734	0.00714	0.00695	0.00676	0.00657	0.00639
2.5	0.00621	0.00604	0.00587	0.00570	0.00554	0.00539	0.00523	0.00509	0.00494	0.00480
2.6	0.00466	0.00453	0.00440	0.00427	0.00415	0.00403	0.00391	0.00379	0.00368	0.00357
2.7	0.00347	0.00336	0.00326	0.00317	0.00307	0.00298	0.00289	0.00280	0.00272	0.00263
2.8	0.00256	0.00248	0.00240	0.00233	0.00226	0.00219	0.00212	0.00205	0.00199	0.00193
2.9	0.00187	0.00181	0.00175	0.00169	0.00164	0.00159	0.00154	0.00149	0.00144	0.00139

z	0.00	0.01	0.02	0.03	0.04	0.05	0.06	0.07	0.08	0.09
3.0	0.00135	0.00131	0.00126	0.00122	0.00118	0.00114	0.00111	0.00107	0.00104	0.00100
3.1	0.00097	0.00094	0.00090	0.00087	0.00085	0.00082	0.00079	0.00076	0.00074	0.00071
3.2	0.00069	0.00066	0.00064	0.00062	0.00060	0.00058	0.00056	0.00054	0.00052	0.00050
3.3	0.00048	0.00047	0.00045	0.00043	0.00042	0.00040	0.00039	0.00038	0.00036	0.00035
3.4	0.00034	0.00032	0.00031	0.00030	0.00029	0.00028	0.00027	0.00026	0.00025	0.00024
3.5	0.00023	0.00022	0.00022	0.00021	0.00020	0.00019	0.00019	0.00018	0.00017	0.00017
3.6	0.00016	0.00015	0.00015	0.00014	0.00014	0.00013	0.00013	0.00012	0.00012	0.00011
3.7	0.00011	0.00010	0.00010	0.00010	0.00009	0.00009	0.00009	0.00008	0.00008	0.00008
3.8	0.00007	0.00007	0.00007	0.00006	0.00006	0.00006	0.00006	0.00005	0.00005	0.00005
3.9	0.00005	0.00005	0.00004	0.00004	0.00004	0.00004	0.00004	0.00004	0.00004	0.00003

z	0.00	0.01	0.02	0.03	0.04	0.05	0.06	0.07	0.08	0.09
4.0	0.00003	0.00003	0.00003	0.00003	0.00003	0.00002	0.00002	0.00002	0.00002	0.00002

Appendix C

The χ^2-Distribution

This appendix contains the $\chi^2_{\alpha;n}$-values corresponding to the exceedance probability

$$P\left(\chi^2_n \geq \chi^2_{\alpha;n}\right) = \alpha$$

for different numbers of degrees of freedom n.

For example, for $n = 10$ degrees of freedom and an exceedance probability of $\alpha = 0.950$, the corresponding $\chi^2_{0.95;10}$-value is equal to 3.940. Or, if the exceedance probability is $\alpha = 0.05$ and the number of degrees of freedom is 30, then the corresponding $\chi^2_{0.05;30}$-value is equal to 43.773. As an example, the notation 0.0^43927 represents the value 0.00003927.

n	$\alpha = 0.995$	0.990	0.975	0.950	0.050	0.025	0.010	0.005	0.001
1	0.0^43927	0.0^31571	0.0^3982	0.0^2393	3.841	5.024	6.635	7.879	10.828
2	0.01003	0.02010	0.05064	0.1026	5.991	7.378	9.210	10.597	13.816
3	0.07172	0.1148	0.2158	0.3518	7.815	9.348	11.345	12.838	16.266
4	0.2070	0.2971	0.4844	0.7107	9.488	11.143	13.277	14.860	18.467
5	0.4117	0.5543	0.8312	1.145	11.070	12.833	15.086	16.750	20.515
6	0.6757	0.8721	1.237	1.635	12.592	14.449	16.812	18.548	22.458
7	0.9893	1.239	1.690	2.167	14.067	16.013	18.475	20.278	24.322
8	1.344	1.646	2.180	2.733	15.507	17.535	20.090	21.955	26.125
9	1.735	2.088	2.700	3.325	16.919	19.023	21.666	23.589	27.877
n	$\alpha = 0.995$	0.990	0.975	0.950	0.050	0.025	0.010	0.005	0.001
10	2.156	2.558	3.247	3.940	18.307	24.483	23.209	25.188	29.588
11	2.603	3.053	3.816	4.575	19.675	21.920	24.725	26.757	31.264
12	3.074	3.571	4.404	5.226	21.026	23.337	26.217	28.300	32.909
13	3.565	4.107	5.009	5.892	22.362	24.736	27.688	29.819	34.528
14	4.075	4.660	5.629	6.571	23.685	26.119	29.141	31.319	36.123
15	4.601	5.229	6.262	7.261	24.996	27.488	30.578	32.801	37.697
16	5.142	5.812	6.908	7.962	26.296	28.845	32.000	34.267	39.252
17	5.697	6.408	7.564	8.672	27.587	30.191	33.409	35.718	40.790
18	6.265	7.015	8.231	9.390	28.869	31.526	34.805	37.156	42.312
19	6.844	7.633	8.907	10.117	30.143	32.852	36.192	38.582	43.820

Statistics with JMP: Hypothesis Tests, ANOVA and Regression, First Edition. Peter Goos and David Meintrup.
© 2016 John Wiley & Sons, Ltd. Published 2016 by John Wiley & Sons, Ltd.
Companion Website: http://www.wiley.com/go/goosandmeintrup/JMP

n	$\alpha = 0.995$	0.990	0.975	0.950	0.050	0.025	0.010	0.005	0.001
20	7.434	8.260	9.591	10.851	31.410	34.170	37.566	39.997	45.315
21	8.034	8.897	10.283	11.591	32.670	35.479	38.932	41.401	46.797
22	8.643	9.542	10.982	12.338	33.924	36.781	40.289	42.796	48.268
23	9.260	10.196	11.688	13.090	35.172	38.076	41.638	44.181	49.728
24	9.886	10.856	12.401	13.848	36.415	39.364	42.080	45.558	51.179
25	10.520	11.524	13.120	14.611	37.652	40.646	44.314	46.928	52.620
26	11.160	12.198	13.844	15.379	38.885	41.923	45.642	48.290	54.052
27	11.808	12.879	14.573	16.151	40.113	43.194	46.963	49.645	55.476
28	12.461	13.565	15.308	16.928	41.337	44.461	48.278	50.993	56.892
29	13.121	14.256	16.047	17.708	42.557	45.722	49.588	52.336	58.302
n	$\alpha = 0.995$	0.990	0.975	0.950	0.050	0.025	0.010	0.005	0.001
30	13.787	14.954	16.791	18.493	43.773	46.979	50.892	53.672	59.703
40	20.707	22.164	24.433	26.509	55.758	59.342	63.691	66.766	73.402
50	27.991	29.707	32.357	34.764	67.505	71.420	76.154	79.490	86.661
60	35.535	37.485	40.482	43.188	79.082	83.298	88.379	91.952	99.607
70	43.275	45.442	48.758	51.739	90.531	95.023	100.425	104.215	112.317
80	51.172	53.540	57.153	60.391	101.879	106.629	122.329	116.321	124.839
90	59.196	61.754	65.647	69.126	113.145	118.136	124.116	128.299	137.208
100	67.328	70.065	74.222	77.929	124.342	129.561	135.807	140.169	149.449

Appendix D

Student's t-Distribution

This appendix contains the $t_{\alpha;n}$-values corresponding to the exceedance probability

$$P(t_n \geq t_{\alpha;n}) = \alpha$$

of a t-distributed random variable t_n with n degrees of freedom, for different values of n. For example, $t_{0.025;9} = 2.2622$.

n	0.4	0.25	0.15	0.1	0.05	α 0.025	0.01	0.005	0.001	0.0005
1	0.3249	1.0000	1.9626	3.0777	6.3138	12.7062	31.8205	63.6567	318.3087	636.61
2	0.2887	0.8165	1.3862	1.8856	2.9200	4.3027	6.9646	9.9248	22.3271	31.599
3	0.2767	0.7649	1.2498	1.6377	2.3534	3.1824	4.5407	5.8409	10.2145	12.924
4	0.2707	0.7407	1.1896	1.5332	2.1318	2.7764	3.7469	4.6041	7.1732	8.6103
5	0.2672	0.7267	1.1558	1.4759	2.0150	2.5706	3.3649	4.0321	5.8934	6.8688
6	0.2648	0.7176	1.1342	1.4398	1.9432	2.4469	3.1427	3.7074	5.2076	5.9588
7	0.2632	0.7111	1.1192	1.4149	1.8946	2.3646	2.9980	3.4995	4.7853	5.4079
8	0.2619	0.7064	1.1081	1.3968	1.8595	2.3060	2.8965	3.3554	4.5008	5.0413
9	0.2610	0.7027	1.0997	1.3830	1.8331	2.2622	2.8214	3.2498	4.2968	4.7809
10	0.2602	0.6998	1.0931	1.3722	1.8125	2.2281	2.7638	3.1693	4.1437	4.5869
11	0.2596	0.6974	1.0877	1.3634	1.7959	2.2010	2.7181	3.1058	4.0247	4.4370
12	0.2590	0.6955	1.0832	1.3562	1.7823	2.1788	2.6810	3.0545	3.9296	4.3178
13	0.2586	0.6938	1.0795	1.3502	1.7709	2.1604	2.6503	3.0123	3.8520	4.2208
14	0.2582	0.6924	1.0763	1.3450	1.7613	2.1448	2.6245	2.9768	3.7874	4.1405
15	0.2579	0.6912	1.0735	1.3406	1.7531	2.1314	2.6025	2.9467	3.7328	4.0728
16	0.2576	0.6901	1.0711	1.3368	1.7459	2.1199	2.5835	2.9208	3.6862	4.0150
17	0.2573	0.6892	1.0690	1.3334	1.7396	2.1098	2.5669	2.8982	3.6458	3.9651
18	0.2571	0.6884	1.0672	1.3304	1.7341	2.1009	2.5524	2.8784	3.6105	3.9216
19	0.2569	0.6876	1.0655	1.3277	1.7291	2.0930	2.5395	2.8609	3.5794	3.8834

Statistics with JMP: Hypothesis Tests, ANOVA and Regression, First Edition. Peter Goos and David Meintrup.
© 2016 John Wiley & Sons, Ltd. Published 2016 by John Wiley & Sons, Ltd.
Companion Website: http://www.wiley.com/go/goosandmeintrup/JMP

					α					
n	0.4	0.25	0.15	0.1	0.05	0.025	0.01	0.005	0.001	0.0005
20	0.2567	0.6870	1.0640	1.3253	1.7247	2.0860	2.5280	2.8453	3.5518	3.8495
21	0.2566	0.6864	1.0627	1.3232	1.7207	2.0796	2.5179	2.8314	3.5272	3.8193
22	0.2564	0.6858	1.0614	1.3212	1.7171	2.0739	2.5083	2.8188	3.5050	3.7921
23	0.2563	0.6853	1.0603	1.3195	1.7139	2.0687	2.4999	2.8073	3.4850	3.7676
24	0.2562	0.6848	1.0593	1.3178	1.7109	2.0639	2.4922	2.7969	3.4668	3.7454
25	0.2561	0.6844	1.0584	1.3163	1.7081	2.0595	2.4851	2.7874	3.4502	3.7251
26	0.2560	0.6840	1.0575	1.3150	1.7056	2.0555	2.4786	2.7787	3.4350	3.7066
27	0.2559	0.6837	1.0567	1.3137	1.7033	2.0518	2.4727	2.7707	3.4210	3.6896
28	0.2558	0.6834	1.0560	1.3125	1.7011	2.0484	2.4671	2.7633	3.4082	3.6739
29	0.2557	0.6830	1.0553	1.3114	1.6991	2.0452	2.4620	2.7564	3.3962	3.6594
30	0.2556	0.6828	1.0547	1.3104	1.6973	2.0423	2.4573	2.7500	3.3852	3.6460
40	0.2550	0.6807	1.0500	1.3031	1.6839	2.0211	2.4233	2.7045	3.3069	3.5510
45	0.2549	0.6800	1.0485	1.3006	1.6794	2.0141	2.4121	2.6896	3.2815	3.5203
50	0.2547	0.6794	1.0473	1.2987	1.6759	2.0086	2.4033	2.6778	3.2614	3.4960
∞	0.2533	0.6745	1.0364	1.2816	1.6449	1.9600	2.3263	2.5758	3.0902	3.2905

Appendix E

The Wilcoxon Signed-Rank Test

This appendix contains exact p-values for the Wilcoxon signed-rank test when there are no ties. The tables contain probabilities of the type $P(T_+ \leq t_+)$ for performing a left-tailed test, and of the type $P(T_+ \geq t_+)$ for performing a right-tailed test. The tables for $n \leq 15$ contain all possible values of t_+ (and the corresponding p-values), while the tables for $15 < n \leq 20$ only contain some possible values of t_+ (and the corresponding p-values).

Table E.1 Probabilities of type $P(T_+ \leq t_+)$.

														Sample size n					
3		4		5		6		7		8		9	
t_+	P	t_+	P	t_+	P	t_+	P	t_+	P	t_+	P	t_+	P
0	0.125	0	0.062	0	0.031	0	0.016	0	0.008	0	0.004	0	0.002
1	0.250	1	0.125	1	0.062	1	0.031	1	0.016	1	0.008	1	0.004
2	0.375	2	0.188	2	0.094	2	0.047	2	0.023	2	0.012	2	0.006
3	0.625	3	0.312	3	0.156	3	0.078	3	0.039	3	0.020	3	0.010
		4	0.438	4	0.219	4	0.109	4	0.055	4	0.027	4	0.014
		5	0.562	5	0.312	5	0.156	5	0.078	5	0.039	5	0.020
				6	0.406	6	0.219	6	0.109	6	0.055	6	0.027
				7	0.500	7	0.281	7	0.148	7	0.074	7	0.037
						8	0.344	8	0.188	8	0.098	8	0.049
						9	0.422	9	0.234	9	0.125	9	0.064
						10	0.500	10	0.289	10	0.156	10	0.082
								11	0.344	11	0.191	11	0.102
								12	0.406	12	0.230	12	0.125
								13	0.469	13	0.273	13	0.150
								14	0.531	14	0.320	14	0.180
										15	0.371	15	0.213
										16	0.422	16	0.248
										17	0.473	17	0.285
										18	0.527	18	0.326
												19	0.367
												20	0.410
												21	0.455
												22	0.500

Statistics with JMP: Hypothesis Tests, ANOVA and Regression, First Edition. Peter Goos and David Meintrup.
© 2016 John Wiley & Sons, Ltd. Published 2016 by John Wiley & Sons, Ltd.
Companion Website: http://www.wiley.com/go/goosandmeintrup/JMP

Table E.2 Probabilities of type $P(T_+ \leq t_+)$.

							Sample size n						
10		11		12		13		14		15			
t_+	P	t_+	P	t_+	P	t_+	P	t_+	P	t_+	P	t_+	P
0	0.001	1	0.001	2	0.001	3	0.001	5	0.001	7	0.001	55	0.402
1	0.002	2	0.001	3	0.001	4	0.001	6	0.001	8	0.001	56	0.423
2	0.003	3	0.002	4	0.002	5	0.001	7	0.001	9	0.001	57	0.445
3	0.005	4	0.003	5	0.002	6	0.002	8	0.002	10	0.001	58	0.467
4	0.007	5	0.005	6	0.003	7	0.002	9	0.002	11	0.002	59	0.489
5	0.010	6	0.007	7	0.005	8	0.003	10	0.003	12	0.002	60	0.511
6	0.014	7	0.009	8	0.006	9	0.004	11	0.003	13	0.003		
7	0.019	8	0.012	9	0.008	10	0.005	12	0.004	14	0.003		
8	0.024	9	0.016	10	0.010	11	0.007	13	0.005	15	0.004		
9	0.032	10	0.021	11	0.013	12	0.009	14	0.007	16	0.005		
10	0.042	11	0.027	12	0.017	13	0.011	15	0.008	17	0.006		
11	0.053	12	0.034	13	0.021	14	0.013	16	0.010	18	0.008		
12	0.065	13	0.042	14	0.026	15	0.016	17	0.012	19	0.009		
13	0.080	14	0.051	15	0.032	16	0.020	18	0.015	20	0.011		
14	0.097	15	0.062	16	0.039	17	0.024	19	0.018	21	0.013		
15	0.116	16	0.074	17	0.046	18	0.029	20	0.021	22	0.015		
16	0.138	17	0.087	18	0.055	19	0.034	21	0.025	23	0.018		
17	0.161	18	0.103	19	0.065	20	0.040	22	0.029	24	0.021		
18	0.188	19	0.120	20	0.076	21	0.047	23	0.034	25	0.024		
19	0.216	20	0.139	21	0.088	22	0.055	24	0.039	26	0.028		
20	0.246	21	0.160	22	0.102	23	0.064	25	0.045	27	0.032		
21	0.278	22	0.183	23	0.117	24	0.073	26	0.052	28	0.036		
22	0.312	23	0.207	24	0.133	25	0.084	27	0.059	29	0.042		
23	0.348	24	0.232	25	0.151	26	0.095	28	0.063	30	0.047		
24	0.385	25	0.260	26	0.170	27	0.108	29	0.077	31	0.053		
25	0.423	26	0.289	27	0.190	28	0.122	30	0.086	32	0.060		
26	0.461	27	0.319	28	0.212	29	0.137	31	0.097	33	0.068		
27	0.500	28	0.350	29	0.235	30	0.153	32	0.103	34	0.076		
		29	0.382	30	0.259	31	0.170	33	0.121	35	0.084		
		30	0.416	31	0.285	32	0.188	34	0.134	36	0.094		
		31	0.449	32	0.311	33	0.207	35	0.143	37	0.104		
		32	0.483	33	0.339	34	0.227	36	0.163	38	0.115		
		33	0.517	34	0.367	35	0.249	37	0.179	39	0.126		
				35	0.396	36	0.271	38	0.196	40	0.138		
				36	0.425	37	0.294	39	0.213	41	0.151		
				37	0.455	38	0.318	40	0.232	42	0.165		
				38	0.485	39	0.342	41	0.251	43	0.180		
				39	0.515	40	0.368	42	0.271	44	0.195		
						41	0.393	43	0.292	45	0.211		
						42	0.420	44	0.313	46	0.227		
						43	0.446	45	0.335	47	0.244		
						44	0.473	46	0.357	48	0.262		
						45	0.500	47	0.380	49	0.281		
								48	0.404	50	0.300		
								49	0.423	51	0.319		
								50	0.452	52	0.339		
								51	0.476	53	0.360		
								52	0.500	54	0.381		

Table E.3 Probabilities of type $P(T_+ \leq t_+)$.

Sample size n									
16		17		18		19		20	
t_+	P	t_+	P	t_+	P	t_+	P	t_+	P
9	0.001	12	0.001	15	0.001	22	0.001	26	0.001
10	0.001	13	0.001	16	0.001	32	0.005	37	0.005
11	0.001	14	0.001	17	0.001	33	0.005	38	0.005
12	0.001	15	0.001	18	0.001	37	0.009	43	0.010
13	0.001	16	0.001	19	0.001	38	0.010	44	0.011
14	0.002	17	0.002	20	0.001	46	0.025	52	0.024
15	0.002	18	0.002	21	0.002	47	0.027	53	0.027
16	0.003	19	0.002	22	0.002	53	0.048	60	0.049
17	0.003	20	0.003	23	0.002	54	0.052	61	0.053
18	0.004	21	0.003	24	0.003	58	0.072	65	0.072
19	0.005	22	0.004	25	0.003	59	0.078	66	0.077
20	0.006	23	0.005	26	0.004	62	0.098	69	0.095
21	0.007	24	0.006	27	0.005	63	0.105	70	0.101
22	0.008	25	0.006	28	0.005				
23	0.009	26	0.008	29	0.006				
24	0.011	27	0.009	30	0.007				
25	0.013	28	0.010	31	0.008				
26	0.015	29	0.012	32	0.009				
27	0.017	30	0.013	33	0.010				
28	0.019	31	0.015	34	0.012				
29	0.022	32	0.017	35	0.013				
30	0.025	33	0.020	36	0.015				
31	0.029	34	0.022	37	0.017				
32	0.033	35	0.025	38	0.019				
33	0.037	36	0.028	39	0.022				
34	0.042	37	0.032	40	0.024				
35	0.047	38	0.036	41	0.027				
36	0.052	39	0.040	42	0.030				
37	0.058	40	0.044	43	0.033				
38	0.065	41	0.049	44	0.037				
39	0.072	42	0.054	45	0.041				
40	0.080	43	0.060	46	0.045				
41	0.088	44	0.066	47	0.049				
42	0.096	45	0.073	48	0.054				
43	0.106	46	0.080	49	0.059				
		47	0.087	50	0.065				
		48	0.095	51	0.071				
		49	0.103	52	0.077				
				53	0.084				
				54	0.091				
				55	0.098				
				56	0.106				

Table E.4 Probabilities of type $P(T_+ \geq t_+)$.

							Sample size n										
	3		4		5		6		7		8		9	
t_+	P	t_+	P	t_+	P	t_+	P	t_+	P	t_+	P	t_+	P
3	0.625	5	0.562	8	0.500	11	0.500	14	0.531	18	0.527	23	0.500
4	0.375	6	0.438	9	0.406	12	0.422	15	0.469	19	0.473	24	0.455
5	0.250	7	0.312	10	0.312	13	0.344	16	0.406	20	0.422	25	0.410
6	0.125	8	0.188	11	0.219	14	0.281	17	0.344	21	0.371	26	0.367
		9	0.125	12	0.156	15	0.219	18	0.289	22	0.320	27	0.326
		10	0.062	13	0.094	16	0.156	19	0.234	23	0.273	28	0.285
				14	0.062	17	0.109	20	0.188	24	0.230	29	0.248
				15	0.031	18	0.078	21	0.148	25	0.191	30	0.213
						19	0.047	22	0.109	26	0.156	31	0.180
						20	0.031	23	0.078	27	0.125	32	0.150
						21	0.016	24	0.055	28	0.098	33	0.125
								25	0.039	29	0.074	34	0.102
								26	0.023	30	0.055	35	0.082
								27	0.016	31	0.039	36	0.064
								28	0.008	32	0.027	37	0.049
										33	0.020	38	0.037
										34	0.012	39	0.027
										35	0.008	40	0.020
										36	0.004	41	0.014
												42	0.010
												43	0.006
												44	0.004
												45	0.002

Table E.5 Probabilities of type $P(T_+ \geq t_+)$.

										Sample size n							
	10		11		12		13		14			15					
t_+	P	t_+	P	t_+	P	t_+	P	t_+	P	t_+	P	t_+	P				
28	0.500	33	0.517	39	0.515	46	0.500	53	0.500	60	0.511	108	0.002				
29	0.461	34	0.483	40	0.485	47	0.473	54	0.476	61	0.489	109	0.002				
30	0.423	35	0.449	41	0.455	48	0.446	55	0.452	62	0.467	110	0.001				
31	0.385	36	0.416	42	0.425	49	0.420	56	0.428	63	0.445	111	0.001				
32	0.348	37	0.382	43	0.396	50	0.393	57	0.404	64	0.423	112	0.001				
33	0.312	38	0.350	44	0.367	51	0.363	58	0.380	65	0.402	113	0.001				
34	0.278	39	0.319	45	0.339	52	0.342	59	0.357	66	0.381						
35	0.246	40	0.289	46	0.311	53	0.313	60	0.335	67	0.360						
36	0.216	41	0.260	47	0.285	54	0.294	61	0.313	68	0.339						
37	0.188	42	0.232	48	0.259	55	0.271	62	0.292	69	0.319						
38	0.161	43	0.207	49	0.235	56	0.249	63	0.271	70	0.300						
39	0.138	44	0.183	50	0.212	57	0.227	64	0.251	71	0.281						
40	0.116	45	0.160	51	0.190	58	0.207	65	0.232	72	0.262						
41	0.097	46	0.139	52	0.170	59	0.188	66	0.213	73	0.244						
42	0.080	47	0.120	53	0.151	60	0.170	67	0.196	74	0.227						
43	0.065	48	0.103	54	0.133	61	0.153	68	0.179	75	0.211						
44	0.053	49	0.087	55	0.117	62	0.137	69	0.163	76	0.195						
45	0.042	50	0.074	56	0.102	63	0.122	70	0.148	77	0.180						
46	0.032	51	0.062	57	0.088	64	0.108	71	0.134	78	0.165						
47	0.024	52	0.051	58	0.076	65	0.095	72	0.121	79	0.151						
48	0.019	53	0.042	59	0.065	66	0.084	73	0.108	80	0.138						
49	0.014	54	0.034	60	0.055	67	0.073	74	0.097	81	0.126						
50	0.010	55	0.027	61	0.046	68	0.064	75	0.086	82	0.115						
51	0.007	56	0.021	62	0.039	69	0.055	76	0.077	83	0.104						
52	0.005	57	0.016	63	0.032	70	0.047	77	0.068	84	0.094						
53	0.003	58	0.012	64	0.026	71	0.040	78	0.059	85	0.084						
54	0.002	59	0.009	65	0.021	72	0.034	79	0.052	86	0.076						
55	0.001	60	0.007	66	0.017	73	0.029	80	0.045	87	0.068						
		61	0.005	67	0.013	74	0.024	81	0.039	88	0.060						
		62	0.003	68	0.010	75	0.020	82	0.034	89	0.053						
		63	0.002	69	0.008	76	0.016	83	0.029	90	0.047						
		64	0.001	70	0.006	77	0.013	84	0.025	91	0.042						
		65	0.001	71	0.005	78	0.011	85	0.021	92	0.036						
				72	0.003	79	0.009	86	0.018	93	0.032						
				73	0.002	80	0.007	87	0.015	94	0.028						
				74	0.002	81	0.005	88	0.012	95	0.024						
				75	0.001	82	0.004	89	0.010	96	0.021						
				76	0.001	83	0.003	90	0.008	97	0.018						
						84	0.002	91	0.007	98	0.015						
						85	0.002	92	0.005	99	0.013						
						86	0.001	93	0.004	100	0.011						
						87	0.001	94	0.003	101	0.009						
						88	0.001	95	0.003	102	0.008						
								96	0.002	103	0.006						
								97	0.002	104	0.005						
								98	0.001	105	0.004						
								99	0.001	106	0.003						
								100	0.001	107	0.003						

Table E.6   Probabilities of type $P(T_+ \geq t_+)$.

				Sample size n					
	16		17		18		19		20
t_+	P	t_+	P	t_+	P	t_+	P	t_+	P
93	0.106	104	0.103	115	0.106	127	0.105	140	0.101
94	0.096	105	0.095	116	0.098	128	0.098	141	0.095
95	0.088	106	0.087	117	0.091	131	0.078	144	0.077
96	0.080	107	0.080	118	0.084	132	0.072	145	0.072
97	0.072	108	0.073	119	0.077	136	0.052	149	0.053
98	0.065	109	0.066	120	0.071	137	0.048	150	0.049
99	0.058	110	0.060	121	0.065	143	0.027	157	0.027
100	0.052	111	0.054	122	0.059	144	0.025	158	0.024
101	0.047	112	0.049	123	0.054	152	0.010	166	0.011
102	0.042	113	0.044	124	0.049	153	0.009	167	0.010
103	0.037	114	0.040	125	0.045	157	0.005	172	0.005
104	0.033	115	0.036	126	0.041	158	0.005	173	0.005
105	0.029	116	0.032	127	0.037	168	0.001	184	0.001
106	0.025	117	0.028	128	0.033				
107	0.022	118	0.025	129	0.030				
108	0.019	119	0.022	130	0.027				
109	0.017	120	0.020	131	0.024				
110	0.015	121	0.017	132	0.022				
111	0.013	122	0.015	133	0.019				
112	0.011	123	0.013	134	0.017				
113	0.009	124	0.012	135	0.015				
114	0.008	125	0.010	136	0.013				
115	0.007	126	0.009	137	0.012				
116	0.006	127	0.008	138	0.010				
117	0.005	128	0.006	139	0.009				
118	0.004	129	0.006	140	0.008				
119	0.003	130	0.005	141	0.007				
120	0.003	131	0.004	142	0.006				
121	0.002	132	0.003	143	0.005				
122	0.002	133	0.003	144	0.005				
123	0.001	134	0.002	145	0.004				
124	0.001	135	0.002	146	0.003				
125	0.001	136	0.002	147	0.003				
126	0.001	137	0.001	148	0.002				
127	0.001	138	0.001	149	0.002				
		139	0.001	150	0.002				
		140	0.001	151	0.001				
		141	0.001	152	0.001				
				153	0.001				
				154	0.001				
				155	0.001				
				156	0.001				

Appendix F

The Shapiro–Wilk Test

This appendix contains the critical values $c_{\alpha;n}$ for a Shapiro–Wilk test, for different values of n and the significance levels $0.005, 0.01, 0.02, 0.025, 0.05, 0.10, 0.15, 0.2, 0.025, 0.3, 0.4$, and 0.5. Source: Parrish, 1992. Reproduced with permission of Taylor and Francis.

						Significance level, α						
n	0.005	0.010	0.020	0.025	0.050	0.100	0.150	0.200	0.250	0.300	0.400	0.500
3	0.752	0.755	0.759	0.761	0.772	0.794	0.815	0.835	0.854	0.872	0.905	0.933
4	0.674	0.692	0.716	0.726	0.761	0.800	0.826	0.846	0.860	0.872	0.893	0.914
5	0.671	0.700	0.732	0.743	0.777	0.813	0.835	0.853	0.867	0.879	0.899	0.915
6	0.691	0.720	0.750	0.760	0.793	0.828	0.849	0.864	0.877	0.887	0.905	0.919
7	0.709	0.737	0.768	0.778	0.809	0.840	0.860	0.874	0.885	0.894	0.910	0.923
8	0.726	0.754	0.784	0.793	0.821	0.851	0.869	0.882	0.892	0.901	0.915	0.927
9	0.745	0.772	0.799	0.808	0.835	0.863	0.879	0.890	0.899	0.907	0.920	0.932
10	0.760	0.786	0.812	0.821	0.845	0.871	0.886	0.897	0.906	0.913	0.925	0.935
11	0.773	0.798	0.823	0.830	0.854	0.879	0.893	0.903	0.911	0.918	0.929	0.939
12	0.786	0.809	0.832	0.840	0.862	0.885	0.898	0.908	0.916	0.922	0.933	0.942
13	0.797	0.819	0.840	0.847	0.869	0.891	0.904	0.913	0.920	0.926	0.936	0.944
14	0.805	0.827	0.848	0.854	0.875	0.896	0.908	0.917	0.924	0.930	0.939	0.947
15	0.813	0.836	0.856	0.863	0.882	0.902	0.913	0.921	0.928	0.933	0.942	0.949
16	0.823	0.843	0.863	0.869	0.887	0.905	0.916	0.924	0.930	0.935	0.944	0.951
17	0.831	0.850	0.869	0.875	0.892	0.910	0.920	0.928	0.934	0.938	0.946	0.953
18	0.840	0.857	0.874	0.880	0.897	0.913	0.924	0.931	0.936	0.941	0.948	0.955
19	0.844	0.861	0.879	0.884	0.900	0.917	0.926	0.933	0.939	0.943	0.950	0.956
20	0.850	0.867	0.883	0.888	0.905	0.920	0.929	0.936	0.941	0.945	0.952	0.958

Statistics with JMP: Hypothesis Tests, ANOVA and Regression, First Edition. Peter Goos and David Meintrup.
© 2016 John Wiley & Sons, Ltd. Published 2016 by John Wiley & Sons, Ltd.
Companion Website: http://www.wiley.com/go/goosandmeintrup/JMP

					Significance level, α							
n	0.005	0.010	0.020	0.025	0.050	0.100	0.150	0.200	0.250	0.300	0.400	0.500
21	0.857	0.873	0.888	0.893	0.908	0.923	0.932	0.938	0.943	0.947	0.954	0.959
22	0.861	0.876	0.891	0.896	0.911	0.926	0.934	0.940	0.945	0.949	0.955	0.961
23	0.865	0.880	0.894	0.900	0.914	0.928	0.936	0.942	0.947	0.950	0.957	0.962
24	0.870	0.885	0.899	0.903	0.917	0.930	0.938	0.944	0.948	0.952	0.958	0.963
25	0.875	0.888	0.902	0.906	0.919	0.932	0.940	0.945	0.950	0.953	0.959	0.964
26	0.877	0.892	0.905	0.909	0.922	0.934	0.942	0.947	0.951	0.955	0.960	0.965
27	0.881	0.895	0.908	0.912	0.924	0.936	0.944	0.949	0.953	0.956	0.961	0.966
28	0.885	0.898	0.910	0.914	0.926	0.938	0.945	0.950	0.954	0.957	0.962	0.967
29	0.888	0.900	0.912	0.916	0.928	0.939	0.946	0.951	0.955	0.958	0.963	0.968
30	0.892	0.904	0.916	0.919	0.930	0.942	0.948	0.953	0.956	0.959	0.964	0.969
31	0.894	0.906	0.917	0.921	0.932	0.943	0.949	0.954	0.957	0.960	0.965	0.969
32	0.897	0.908	0.919	0.923	0.934	0.944	0.951	0.955	0.958	0.961	0.966	0.970
33	0.899	0.910	0.921	0.925	0.935	0.945	0.952	0.956	0.959	0.962	0.967	0.971
34	0.901	0.912	0.923	0.926	0.937	0.947	0.953	0.957	0.960	0.963	0.968	0.971
35	0.904	0.915	0.925	0.928	0.938	0.948	0.954	0.958	0.961	0.964	0.968	0.972
36	0.908	0.917	0.927	0.930	0.940	0.949	0.955	0.959	0.962	0.965	0.969	0.973
37	0.909	0.918	0.928	0.932	0.941	0.950	0.956	0.960	0.963	0.965	0.970	0.973
38	0.911	0.920	0.930	0.933	0.942	0.951	0.957	0.960	0.964	0.966	0.970	0.974
39	0.912	0.922	0.931	0.934	0.943	0.952	0.958	0.961	0.964	0.967	0.971	0.974
40	0.913	0.924	0.932	0.935	0.944	0.953	0.958	0.962	0.965	0.967	0.971	0.975
41	0.916	0.925	0.935	0.938	0.946	0.954	0.959	0.963	0.966	0.968	0.972	0.975
42	0.917	0.926	0.935	0.938	0.947	0.955	0.960	0.964	0.967	0.969	0.973	0.976
43	0.919	0.928	0.936	0.939	0.948	0.956	0.961	0.964	0.967	0.969	0.973	0.976
44	0.920	0.929	0.938	0.940	0.949	0.957	0.962	0.965	0.968	0.970	0.973	0.976
45	0.922	0.930	0.939	0.942	0.950	0.958	0.962	0.966	0.968	0.970	0.974	0.977
46	0.924	0.932	0.940	0.943	0.951	0.958	0.963	0.966	0.969	0.971	0.974	0.977
47	0.925	0.933	0.941	0.944	0.951	0.959	0.963	0.967	0.969	0.971	0.975	0.978
48	0.927	0.935	0.942	0.945	0.953	0.960	0.964	0.967	0.970	0.972	0.975	0.978
49	0.928	0.936	0.943	0.946	0.953	0.961	0.965	0.968	0.970	0.972	0.976	0.978
50	0.930	0.937	0.944	0.947	0.954	0.961	0.965	0.968	0.971	0.973	0.976	0.979

Appendix G

Fisher's *F*-Distribution

This appendix contains the critical values $F_{\alpha;v_1;v_2}$, with the number of degrees of freedom v_1 for the numerator, and v_2 for the denominator. The critical value corresponds to the exceedance probability

$$P(F_{v_1;v_2} \geq F_{\alpha;v_1;v_2}) = \alpha.$$

For example, $F_{0.05;9;3} = 8.812$.

v_2	α	v_1 1	2	3	4	5	6	7	8	9
1	0.10	39.86	49.50	53.59	55.83	57.24	58.20	58.91	59.44	59.86
	0.05	161.4	199.5	215.7	224.6	230.2	234.0	236.8	238.9	240.5
	0.01	4052	4999	5403	5625	5764	5859	5928	5981	6022
2	0.10	8.526	9.000	9.162	9.243	9.293	9.326	9.349	9.367	9.381
	0.05	18.51	19.00	19.16	19.25	19.30	19.33	19.35	19.37	19.38
	0.01	98.50	99.00	99.17	99.25	99.30	99.33	99.36	99.37	99.39
3	0.10	5.538	5.462	5.391	5.343	5.309	5.285	5.266	5.252	5.240
	0.05	10.13	9.552	9.277	9.117	9.013	8.941	8.887	8.845	8.812
	0.01	34.12	30.82	29.46	28.71	28.24	27.91	27.67	27.49	27.35
4	0.10	4.545	4.325	4.191	4.107	4.051	4.010	3.979	3.955	3.936
	0.05	7.709	6.944	6.591	6.388	6.256	6.163	6.094	6.041	5.999
	0.01	21.20	18.00	16.69	15.98	15.52	15.21	14.98	14.80	14.66
5	0.10	4.060	3.780	3.619	3.520	3.453	3.405	3.368	3.339	3.316
	0.05	6.608	5.786	5.409	5.192	5.050	4.950	4.876	4.818	4.772
	0.01	16.26	13.27	12.06	11.39	10.97	10.67	10.46	10.29	10.16
6	0.10	3.776	3.463	3.289	3.181	3.108	3.055	3.014	2.983	2.958
	0.05	5.987	5.143	4.757	4.534	4.387	4.284	4.207	4.147	4.099
	0.01	13.75	10.92	9.780	9.148	8.746	8.466	8.260	8.102	7.976
7	0.10	3.589	3.257	5.074	2.961	2.883	2.827	2.785	2.752	2.725
	0.05	5.591	4.737	4.347	4.120	3.972	3.866	3.787	3.726	3.677
	0.01	12.25	9.547	8.451	7.847	7.460	7.291	6.993	6.840	6.719

Statistics with JMP: Hypothesis Tests, ANOVA and Regression, First Edition. Peter Goos and David Meintrup.
© 2016 John Wiley & Sons, Ltd. Published 2016 by John Wiley & Sons, Ltd.
Companion Website: http://www.wiley.com/go/goosandmeintrup/JMP

v_2	α	1	2	3	v_1 4	5	6	7	8	9
8	0.10	3.458	3.113	2.924	2.806	2.726	2.668	2.624	2.589	2.561
	0.05	5.318	4.459	4.066	3.838	3.687	3.581	3.500	3.438	3.388
	0.01	11.26	8.649	7.591	7.006	6.632	6.371	6.178	6.029	5.911
9	0.10	3.360	3.006	2.813	2.693	2.611	2.551	2.505	2.469	2.440
	0.05	5.117	4.256	3.863	3.633	3.482	3.374	3.293	3.230	3.179
	0.01	10.56	8.022	6.992	6.422	6.057	5.802	5.613	5.467	5.351
10	0.10	3.285	2.924	2.728	2.605	2.522	2.461	2.414	2.377	2.347
	0.05	4.965	4.103	3.708	3.478	3.326	3.217	3.135	3.072	3.020
	0.01	10.04	7.559	6.552	5.994	5.636	5.386	5.200	5.057	4.942
11	0.10	3.225	2.860	2.660	2.536	2.451	2.389	2.342	2.304	2.274
	0.05	4.844	3.982	3.587	3.357	3.204	3.095	3.012	2.948	2.896
	0.01	9.646	7.206	6.217	5.668	5.316	5.069	4.886	4.774	4.632
12	0.10	3.177	2.807	2.606	2.480	2.384	2.331	2.283	2.245	2.214
	0.05	4.747	3.885	3.490	3.259	3.106	2.996	2.913	2.849	2.796
	0.01	9.330	6.927	5.953	5.412	5.064	4.821	4.640	4.499	4.388
13	0.10	3.136	2.763	2.560	2.434	2.347	2.283	2.234	2.195	2.164
	0.05	4.667	3.906	3.411	3.179	3.025	2.915	2.832	2.767	2.714
	0.01	9.074	6.701	5.739	5.205	4.862	4.620	4.441	4.302	4.191
14	0.10	3.102	2.726	2.522	3.395	2.307	2.243	2.193	2.154	2.122
	0.05	4.600	3.739	3.344	3.112	2.958	2.848	2.764	2.699	2.646
	0.01	8.862	6.515	5.564	5.035	4.695	4.456	4.278	4.140	4.030
15	0.10	3.073	2.695	2.490	2.361	2.273	2.208	2.158	2.119	2.086
	0.05	4.543	3.682	3.287	3.056	2.901	2.790	2.707	2.641	2.588
	0.01	8.683	6.359	5.417	4.893	4.556	4.318	4.142	4.004	3.895
16	0.10	3.048	2.668	2.462	2.333	2.244	2.178	2.128	2.088	2.055
	0.05	4.494	3.634	3.239	3.007	2.852	2.741	2.657	2.591	2.538
	0.01	8.531	6.226	5.292	4.773	4.437	4.202	4.026	3.890	3.780
17	0.10	3.026	2.645	2.437	2.308	2.218	2.152	2.102	2.061	2.028
	0.05	4.451	3.592	3.197	2.965	2.810	2.699	2.614	2.548	2.494
	0.01	8.400	6.112	5.185	4.669	4.336	4.102	3.927	3.791	3.682
18	0.10	3.007	2.624	2.416	2.286	2.196	2.130	2.079	2.038	2.005
	0.05	4.414	3.555	3.160	2.928	2.773	2.661	2.577	2.510	2.456
	0.01	8.285	6.013	5.092	4.579	4.248	4.015	3.841	3.705	3.597
19	0.10	2.990	2.606	2.397	2.266	2.176	2.109	2.058	2.017	1.984
	0.05	4.381	3.522	3.127	2.895	2.740	2.628	2.544	2.477	2.423
	0.01	8.185	5.926	5.010	4.500	4.171	3.939	3.765	3.631	3.523
20	0.10	2.975	2.589	2.380	2.249	2.158	2.091	2.040	1.999	1.965
	0.05	4.351	3.493	3.098	2.866	2.711	2.599	2.514	2.447	2.393
	0.01	8.096	5.849	4.938	4.431	4.103	3.871	3.699	3.564	3.457
21	0.10	2.961	2.575	2.365	2.233	2.142	2.075	2.023	1.982	1.948
	0.05	4.325	3.467	3.072	2.840	2.685	2.573	2.488	2.420	2.366
	0.01	8.017	5.780	4.874	4.369	4.042	3.812	3.640	3.506	3.398

v_2	α	1	2	3	v_1 4	5	6	7	8	9
22	0.10	2.949	2.561	2.351	2.219	2.128	2.060	2.008	1.967	1.933
	0.05	4.301	3.443	3.049	2.817	2.661	2.549	2.464	2.397	2.342
	0.01	7.945	5.719	4.817	4.313	3.988	3.758	3.587	3.453	3.346
23	0.10	2.937	2.549	2.339	2.207	2.115	2.047	1.995	1.935	1.919
	0.05	4.279	3.422	3.028	2.796	2.640	2.528	2.442	2.375	2.320
	0.01	7.881	5.664	4.765	4.264	3.939	3.710	3.539	3.406	3.299
24	0.10	2.927	2.538	2.527	2.195	2.103	2.035	1.983	1.941	1.906
	0.05	4.260	3.403	3.009	2.776	2.621	2.508	2.423	2.355	2.300
	0.01	7.823	5.614	4.718	4.218	3.895	3.667	3.496	3.363	3.256
25	0.10	2.918	2.528	2.317	2.184	2.092	2.024	1.971	1.929	1.895
	0.05	4.242	3.385	2.991	2.759	2.603	2.490	2.405	2.337	2.282
	0.01	7.770	5.568	4.675	4.177	3.855	3.627	3.457	3.324	3.217
26	0.10	2.909	2.519	2.307	2.174	2.082	2.014	1.961	1.919	1.884
	0.05	4.224	3.369	2.975	2.743	2.587	2.474	2.388	2.321	2.265
	0.01	7.721	5.526	4.637	4.140	3.818	3.591	3.421	3.288	3.182
27	0.10	2.901	2.511	2.299	2.165	2.073	2.005	1.952	1.909	1.874
	0.05	4.210	3.354	2.960	2.728	2.572	2.459	2.373	2.305	2.250
	0.01	7.677	5.488	4.601	4.106	3.785	3.558	3.388	3.256	3.149
28	0.10	2.894	2.503	2.291	2.157	2.064	1.996	1.943	1.900	1.865
	0.05	4.196	3.340	2.947	2.714	2.558	2.445	2.359	2.291	2.236
	1.01	7.636	5.453	4.568	4.074	3.754	3.528	3.358	3.226	3.120
29	0.10	2.887	2.495	2.283	2.149	2.057	1.988	1.935	1.892	1.857
	0.05	4.183	3.328	2.934	2.701	2.545	2.432	2.346	2.278	2.223
	0.01	7.598	5.420	4.538	4.045	2.725	3.499	3.330	3.198	3.092
30	0.10	2.881	2.489	2.276	2.142	2.049	1.980	1.927	1.884	1.849
	0.05	4.171	3.316	2.922	2.690	2.534	2.421	2.334	2.266	2.211
	0.01	7.562	5.390	4.510	4.018	3.699	3.473	3.303	3.173	3.067
35	0.10	2.855	2.461	2.247	2.113	2.019	1.950	1.896	1.852	1.817
	0.05	4.121	3.267	2.874	2.641	2.485	2.372	2.285	2.217	2.161
	0.01	7.419	5.268	4.396	3.908	3.592	3.368	3.200	3.069	2.963
40	0.10	2.835	2.440	2.226	2.091	1.997	1.927	1.873	1.829	1.793
	0.05	4.085	3.232	2.839	2.606	2.449	2.336	2.249	2.180	2.124
	0.01	7.314	5.170	4.313	3.828	3.514	3.291	3.124	2.993	2.888
45	0.10	2.820	2.425	2.210	2.074	1.980	1.909	1.855	1.811	1.774
	0.05	4.057	3.204	2.812	2.579	2.422	2.308	2.221	2.152	2.096
	0.01	7.234	5.110	4.249	3.767	3.454	3.232	3.066	2.935	2.830
50	0.10	2.809	2.412	2.197	2.061	1.966	1.895	1.840	1.796	1.760
	0.05	4.034	3.183	2.790	2.557	2.400	2.286	2.199	2.130	2.073
	0.01	7.171	5.057	4.199	3.720	3.408	3.186	3.020	2.890	2.785
60	0.10	2.791	2.393	2.177	2.041	1.946	1.875	1.819	1.775	1.738
	0.05	4.001	3.150	2.758	2.525	2.368	2.254	2.167	2.097	2.040
	0.01	7.077	4.977	4.126	3.649	3.339	3.119	2.953	2.823	2.718

v_2	α	1	2	3	v_1 4	5	6	7	8	9
70	0.10	2.779	2.380	2.164	2.027	1.931	1.860	1.804	1.760	1.723
	0.05	3.978	3.128	2.736	2.503	2.346	2.231	2.143	2.074	2.017
	0.01	7.011	4.922	4.074	3.600	3.291	3.071	2.906	2.777	2.672
80	0.10	2.769	2.370	2.154	2.016	1.921	1.849	1.793	1.748	1.711
	0.05	3.960	3.111	2.719	2.486	2.329	2.214	2.126	2.056	1.999
	0.01	6.963	4.881	4.036	3.563	3.255	3.036	2.871	2.742	2.637
90	0.10	2.762	2.363	2.146	2.008	1.912	1.841	1.785	1.739	1.702
	0.05	3.947	3.098	2.706	2.473	2.316	2.201	2.113	2.043	1.986
	0.01	6.925	4.849	4.007	3.535	3.228	3.009	2.845	2.715	2.611
100	0.10	2.756	2.356	2.139	2.002	1.906	1.834	1.778	1.732	1.695
	0.05	3.936	3.087	2.696	2.463	2.305	2.191	2.103	2.032	1.975
	0.01	6.895	4.824	4.984	3.513	3.206	2.988	2.823	2.694	2.590
120	0.10	2.748	2.347	2.130	1.992	1.896	1.824	1.767	1.722	1.684
	0.05	3.920	3.072	2.680	2.447	2.290	2.175	2.087	2.016	1.956
	0.01	6.851	4.787	3.949	3.480	3.174	2.956	2.792	2.663	2.559
150	0.10	2.739	2.338	2.121	1.983	1.886	1.814	1.757	1.712	1.674
	0.05	3.904	3.056	2.665	2.432	2.274	2.160	2.071	2.001	1.943
	0.01	6.807	4.749	3.915	3.447	3.142	2.924	2.761	2.632	2.528
200	0.10	2.731	2.329	2.111	1.973	1.876	1.804	1.747	1.701	1.663
	0.05	3.888	3.047	2.650	2.417	2.259	2.144	2.056	1.985	1.927
	0.01	6.763	4.713	3.881	3.414	3.110	2.893	2.730	2.601	2.497
500	0.10	2.716	2.313	2.095	1.956	1.859	1.786	1.729	1.683	1.644
	0.05	3.860	3.014	2.623	2.390	2.232	2.117	2.028	1.957	1.899
	0.01	6.686	4.648	3.821	3.357	3.054	2.838	2.675	2.547	2.443
1000	0.10	2.711	2.308	2.089	1.950	1.853	1.780	1.723	1.676	1.638
	0.05	3.851	3.005	2.614	2.381	2.223	2.108	2.019	1.948	1.889
	0.01	6.660	4.625	3.801	3.338	3.036	2.820	2.657	2.529	2.425

v_2	α	10	12	15	v_1 20	25	30	50	100	∞
1	0.10	60.19	60.71	61.22	61.74	62.05	62.26	62.69	63.01	63.33
	0.05	241.9	243.9	245.9	248.0	249.3	250.1	251.8	253.0	254.3
	0.01	6056	6106	6157	6209	6240	6261	6303	6334	6366
2	0.10	9.392	9.408	9.425	9.441	9.451	9.458	9.471	9.481	9.491
	0.05	19.40	19.41	19.43	19.45	19.46	19.46	19.48	19.49	19.50
	0.01	99.40	99.42	99.43	99.45	99.46	99.47	99.48	99.49	99.50
3	0.10	5.230	5.216	5.200	5.184	5.175	5.168	5.155	5.144	5.134
	0.05	8.786	8.745	8.703	8.660	8.634	8.617	8.581	8.554	8.526
	0.01	27.23	27.05	26.87	26.69	26.58	26.50	26.35	26.34	26.13
4	0.10	3.920	3.896	3.870	3.844	3.828	3.817	3.795	3.778	3.761
	0.05	5.964	5.912	5.858	5.803	5.769	5.746	5.699	5.664	5.628
	0.01	14.55	14.37	14.20	14.02	13.91	13.84	13.69	13.58	13.46

v_2	α	10	12	15	v_1 20	25	30	50	100	∞
5	0.10	3.297	3.268	3.238	3.207	3.185	3.174	3.147	3.126	3.105
	0.05	4.735	4.678	4.619	4.558	4.521	4.496	4.444	4.405	4.365
	0.01	10.05	9.888	9.722	9.553	9.449	9.379	9.238	9.130	9.020
6	0.10	2.937	2.905	2.871	2.836	2.815	2.800	2.770	2.746	2.722
	0.05	4.060	4.000	3.938	3.874	3.835	3.808	3.754	3.712	3.669
	0.01	7.874	7.718	7.559	7.396	7.296	7.229	7.091	6.987	6.880
7	0.10	2.703	2.668	2.632	2.595	2.571	2.555	2.523	2.497	2.471
	0.05	3.637	3.575	3.511	3.445	3.404	3.376	3.319	3.275	3.230
	0.01	6.20	6.469	6.314	6.155	6.058	5.992	5.858	5.755	5.650
8	0.10	2.538	2.502	2.464	2.425	2.400	2.383	2.348	2.321	2.293
	0.05	3.347	3.284	3.218	3.150	3.108	3.079	3.020	2.975	2.928
	0.01	5.814	5.667	5.515	5.359	5.263	5.198	5.065	4.963	4.859
9	0.10	2.416	2.379	2.340	2.298	2.272	2.255	2.218	2.189	2.159
	0.05	3.137	3.073	3.006	2.936	2.893	2.864	2.803	2.756	2.707
	0.01	5.257	5.111	4.962	4.808	4.713	4.649	4.517	4.415	4.311
10	0.10	2.323	2.284	2.244	2.201	2.174	2.155	2.117	2.087	2.055
	0.05	2.978	2.913	2.845	2.774	2.730	2.700	2.637	2.588	2.538
	0.01	4.849	4.706	4.558	4.405	4.311	4.247	4.115	4.014	3.909
11	0.10	2.248	2.209	2.167	2.123	2.095	2.076	2.036	2.005	1.972
	0.05	2.854	2.788	2.719	2.646	2.601	2.570	2.507	2.457	2.404
	0.01	4.539	4.397	4.251	4.099	4.005	3.941	3.810	3.708	3.602
12	0.10	2.188	2.147	2.105	2.060	2.031	2.011	1.970	1.938	1.904
	0.05	2.753	2.687	2.617	2.544	2.498	2.466	2.401	2.350	2.296
	0.01	4.296	4.155	4.010	3.858	3.765	3.701	3.569	3.467	3.361
13	0.10	2.138	2.097	2.053	2.007	1.978	1.958	1.915	1.882	1.846
	0.05	2.671	2.604	2.533	2.459	2.412	2.380	2.314	2.261	2.206
	0.01	4.100	3.960	3.815	3.665	3.571	3.507	3.375	3.272	3.165
14	0.10	2.095	2.054	2.010	1.962	1.933	1.912	1.869	1.834	1.797
	0.05	2.602	2.534	2.463	2.388	2.341	2.308	2.241	2.187	2.131
	0.01	3.939	3.800	3.656	3.505	3.412	3.348	3.215	3.112	3.004
15	0.10	2.059	2.017	1.972	1.924	1.894	1.873	1.828	1.793	1.755
	0.05	2.544	2.475	2.403	2.328	2.280	2.247	2.178	2.123	2.066
	0.01	3.805	3.666	3.522	3.372	3.278	3.214	3.081	2.977	2.868
16	0.10	2.028	1.985	1.940	1.891	1.860	1.839	1.793	1.757	1.718
	0.05	2.494	2.425	2.352	2.276	2.227	2.194	2.124	2.068	2.010
	0.01	3.691	3.553	3.409	3.259	3.165	3.101	2.967	2.863	2.753
17	0.10	2.001	1.958	1.912	1.862	1.831	1.809	1.763	1.726	1.686
	0.05	2.450	2.381	2.308	2.230	2.181	2.148	2.077	2.020	1.960
	0.01	3.593	3.455	3.312	3.162	3.068	3.003	2.869	2.764	2.653
18	0.10	1.997	1.933	1.887	1.837	1.805	1.783	1.736	1.698	1.657
	0.05	2.412	2.342	2.269	2.191	2.141	2.107	2.035	1.978	1.917
	0.01	3.508	3.371	3.227	3.077	2.983	2.919	2.784	2.678	2.566
19	0.10	1.956	1.912	1.865	1.814	1.782	1.759	1.711	1.673	1.631
	0.05	2.378	2.308	2.237	2.155	2.106	2.071	1.999	1.940	1.878
	0.01	3.434	3.297	3.153	3.003	2.909	2.844	2.709	2.602	2.489

					v_1					
v_2	α	10	12	15	20	25	30	50	100	∞
20	0.10	1.937	1.892	1.845	1.794	1.761	1.738	1.690	1.650	1.607
	0.05	2.348	2.278	2.203	2.124	2.074	2.039	1.966	1.907	1.843
	0.01	3.368	3.231	3.088	2.928	2.843	2.778	2.643	2.535	2.421
21	0.10	1.920	1.875	1.827	1.776	1.742	1.719	1.670	1.630	1.586
	0.05	2.321	2.250	2.176	2.096	2.045	2.010	1.936	1.876	1.812
	0.01	3.310	3.173	3.030	2.880	2.785	2.720	2.584	2.475	2.360
22	0.10	1.904	1.859	1.811	1.759	1.726	1.702	1.652	1.611	1.657
	0.05	2.297	2.226	2.151	2.071	2.020	1.984	1.909	1.849	1.783
	0.01	3.258	3.121	2.978	2.827	2.733	2.667	2.531	2.422	2.305
23	0.10	1.890	1.845	1.796	1.744	1.710	1.686	1.636	1.594	1.549
	0.05	2.275	2.204	2.128	2.048	1.996	1.961	1.885	1.823	1.757
	0.01	3.211	3.074	2.931	2.781	2.686	2.620	2.483	2.373	2.256
24	0.10	1.877	1.832	1.783	1.730	1.696	1.672	1.621	1.579	1.533
	0.05	2.255	2.183	2.108	2.027	1.975	1.939	1.863	1.800	1.733
	0.01	3.168	3.032	2.889	2.738	2.643	2.577	2.440	2.329	2.211
25	0.10	1.866	1.820	1.771	1.718	1.683	1.659	1.607	1.565	1.518
	0.05	2.236	2.165	2.089	2.007	1.955	1.912	1.842	1.779	1.711
	0.01	3.129	2.993	2.850	2.699	2.604	2.538	2.400	2.289	2.169
26	0.10	1.855	1.809	1.760	1.706	1.671	1.647	1.594	1.551	1.504
	0.05	2.220	2.148	2.072	1.990	1.938	1.901	1.823	1.760	1.691
	0.01	3.094	2.958	2.815	2.664	2.569	2.603	2.364	2.252	2.131
27	0.10	1.845	1.799	1.749	1.695	1.660	1.636	1.583	1.539	1.491
	0.05	2.204	2.132	2.056	1.974	1.921	1.884	1.806	1.742	1.672
	0.01	3.062	2.926	2.783	2.632	2.536	2.470	2.330	2.218	2.097
28	0.10	1.836	1.790	1.740	1.685	1.650	1.625	1.572	1.528	1.478
	0.05	2.190	2.118	2.041	1.959	1.906	1.869	1.790	1.725	1.654
	0.01	3.032	2.896	2.753	2.602	2.506	2.440	2.300	2.187	2.064
29	0.10	1.827	1.781	1.731	1.676	1.640	1.616	1.562	1.517	1.467
	0.05	2.177	2.104	2.027	1.945	1.891	1.854	1.775	1.710	1.638
	0.01	3.005	2.868	2.726	2.574	2.478	2.412	2.271	2.158	2.034
30	0.10	1.819	1.773	1.722	1.667	1.632	1.606	1.552	1.507	1.456
	0.05	2.165	2.092	2.015	1.932	1.878	1.841	1.761	1.695	1.622
	0.01	2.979	2.843	2.700	2.549	2.453	2.386	2.245	2.131	2.006
35	0.10	1.787	1.739	1.688	1.632	1.595	1.569	1.513	1.465	1.411
	0.05	2.114	2.041	1.963	1.878	1.824	1.786	1.703	1.635	1.558
	0.01	2.876	2.740	2.597	2.445	2.348	2.281	2.137	2.020	1.891
40	0.10	1.763	1.715	1.662	1.605	1.568	1.541	1.483	1.434	1.377
	0.05	2.077	2.003	1.924	1.839	1.783	1.744	1.660	1.589	1.509
	0.01	2.801	2.665	2.522	2.369	2.271	2.203	2.058	1.938	1.805
45	0.10	1.744	1.695	1.643	1.585	1.546	1.519	1.460	1.409	1.349
	0.05	2.049	1.974	1.895	1.808	1.752	1.713	1.626	1.554	1.470
	0.01	2.743	2.608	2.464	2.311	2.213	2.144	1.997	1.875	1.737

v_2	α	10	12	15	v_1 20	25	30	50	100	∞
50	0.10	1.729	1.680	1.627	1.568	1.529	1.502	1.441	1.388	1.327
	0.05	2.026	1.952	1.871	1.784	1.727	1.687	1.599	1.525	1.438
	0.01	2.698	2.562	2.419	2.265	2.167	2.098	1.949	1.825	1.683
60	0.10	1.707	1.657	1.603	1.543	1.504	1.476	1.413	1.358	1.291
	0.05	1.993	1.917	1.836	1.748	1.690	1.649	1.559	1.481	1.389
	0.01	2.632	2.496	2.352	2.198	2.098	2.028	1.877	1.749	1.601
70	0.10	1.691	1.641	1.587	1.526	1.486	1.457	1.392	1.335	1.265
	0.05	1.969	1.839	1.812	1.722	1.664	1.622	1.530	1.450	1.353
	0.01	2.585	2.450	2.306	2.150	2.050	1.980	1.826	1.695	1.540
80	0.10	1.680	1.629	1.574	1.513	1.472	1.443	1.377	1.318	1.245
	0.05	1.951	1.875	1.793	1.703	1.644	1.602	1.508	1.426	1.325
	0.01	2.551	2.415	2.271	2.115	2.015	1.944	1.788	1.655	1.494
90	0.10	1.670	1.620	1.564	1.503	1.461	1.432	1.365	1.304	1.228
	0.05	1.938	1.861	1.779	1.688	1.629	1.586	1.491	1.407	1.302
	0.01	2.524	2.389	2.244	2.088	1.987	1.916	1.759	1.623	1.457
100	0.10	1.663	1.612	1.557	1.494	1.453	1.423	1.355	1.293	1.214
	0.05	1.927	1.850	1.768	1.676	1.616	1.573	1.477	1.392	1.283
	0.01	2.503	2.368	2.223	2.067	1.965	1.893	1.735	1.598	1.427
120	0.10	1.652	1.601	1.545	1.482	1.440	1.409	1.340	1.277	1.193
	0.05	1.910	1.834	1.750	1.659	1.598	1.554	1.457	1.369	1.254
	0.01	2.472	2.336	2.192	2.035	1.932	1.860	1.700	1.559	1.381
150	0.10	1.642	1.590	1.533	1.470	1.427	1.396	1.325	1.259	1.169
	0.05	1.894	1.817	1.734	1.641	1.580	1.535	1.436	1.345	1.223
	0.01	2.441	2.305	2.160	2.003	1.900	1.827	1.665	1.520	1.331
200	0.10	1.631	1.579	1.522	1.458	1.414	1.383	1.310	1.242	1.144
	0.05	1.878	1.801	1.717	1.623	1.561	1.516	1.415	1.321	1.189
	0.01	2.411	2.275	2.129	1.971	1.868	1.794	1.629	1.481	1.279
500	0.10	1.612	1.559	1.501	1.435	1.391	1.358	1.282	1.209	1.087
	0.05	1.850	1.772	1.686	1.592	1.528	1.482	1.376	1.275	1.113
	0.01	2.356	2.220	2.075	1.915	1.810	1.735	1.566	1.408	1.164
1000	0.10	1.605	1.552	1.494	1.428	1.383	1.350	1.273	1.197	1.060
	0.05	1.840	1.762	1.676	1.581	1.517	1.471	1.363	1.260	1.078
	0.01	2.339	2.203	2.056	1.897	1.791	1.716	1.544	1.383	1.112

Appendix H

The Wilcoxon Rank-Sum Test

This appendix contains the exceedance probabilities $P(T_1 \geq t_1)$ for different values of n_1 and n_2, and for different values of t_1.

				$n_1 = 1$					
t_1	$n_2 = 3$	$n_2 = 4$	$n_2 = 5$	$n_2 = 6$	$n_2 = 7$	$n_2 = 8$	$n_2 = 9$	$n_2 = 10$	$n_2 = 11$
3	0.500	0.600							
4	0.250	0.400	0.500	0.571					
5		0.200	0.333	0.429	0.500	0.556			
6			0.167	0.286	0.375	0.444	0.500	0.545	
7				0.143	0.250	0.333	0.400	0.455	0.500
8					0.125	0.222	0.300	0.364	0.417
9						0.111	0.200	0.273	0.333
10							0.100	0.182	0.250
11								0.091	0.167
12									0.083

				$n_1 = 1$					
t_1	$n_2 = 12$	$n_2 = 13$	$n_2 = 14$	$n_2 = 15$	$n_2 = 16$	$n_2 = 17$	$n_2 = 18$	$n_2 = 19$	$n_2 = 20$
7	0.528								
8	0.462	0.500	0.533						
9	0.385	0.429	0.467	0.500	0.529				
10	0.308	0.357	0.400	0.438	0.471	0.500	0.526		
11	0.231	0.286	0.333	0.375	0.412	0.444	0.474	0.500	0.524
12	0.154	0.214	0.267	0.312	0.353	0.389	0.421	0.450	0.476
13	0.077	0.143	0.133	0.250	0.294	0.333	0.368	0.400	0.429
14		0.071	0.067	0.180	0.235	0.278	0.263	0.350	0.381
15				0.125	0.176	0.222	0.211	0.300	0.333
16				0.062	0.118	0.167	0.158	0.250	0.286
17					0.059	0.111	0.105	0.200	0.238
18						0.056	0.053	0.150	0.190
19								0.100	0.143
20								0.050	0.095
21									0.048

Statistics with JMP: Hypothesis Tests, ANOVA and Regression, First Edition. Peter Goos and David Meintrup.
© 2016 John Wiley & Sons, Ltd. Published 2016 by John Wiley & Sons, Ltd.
Companion Website: http://www.wiley.com/go/goosandmeintrup/JMP

| | | | $n_1 = 2$ | | | | | |
t_1	$n_2 = 3$	$n_2 = 4$	$n_2 = 5$	$n_2 = 6$	$n_2 = 7$	$n_2 = 8$	$n_2 = 9$	$n_2 = 10$	$n_2 = 11$
6	0.600								
7	0.400	0.600							
8	0.200	0.400	0.571						
9	0.100	0.267	0.429	0.571					
10		0.133	0.286	0.429	0.556				
11		0.067	0.190	0.321	0.444	0.556			
12			0.095	0.214	0.333	0.444	0.545		
13			0.048	0.143	0.250	0.356	0.455	0.545	
14				0.071	0.167	0.267	0.364	0.455	0.538
15				0.036	0.111	0.200	0.291	0.379	0.462
16					0.056	0.133	0.218	0.303	0.385
17					0.028	0.089	0.164	0.242	0.321
18						0.044	0.109	0.182	0.256
19						0.022	0.073	0.136	0.205
20							0.036	0.091	0.154
21							0.018	0.061	0.115
22								0.300	0.077
23								0.150	0.051
24									0.026
25									0.013

| | | | $n_1 = 2$ | | | | | |
t_1	$n_2 = 12$	$n_2 = 13$	$n_2 = 14$	$n_2 = 15$	$n_2 = 16$	$n_2 = 17$	$n_2 = 18$	$n_2 = 19$	$n_2 = 20$
15	0.538								
16	0.462	0.533							
17	0.396	0.467	0.533						
18	0.330	0.400	0.467	0.529					
19	0.275	0.343	0.408	0.471	0.529				
20	0.220	0.286	0.350	0.412	0.471	0.526			
21	0.176	0.238	0.300	0.360	0.418	0.474	0.526		
22	0.132	0.190	0.250	0.309	0.366	0.421	0.474	0.524	
23	0.099	0.152	0.208	0.265	0.320	0.327	0.426	0.476	0.524
24	0.066	0.114	0.167	0.221	0.275	0.287	0.376	0.129	0.476
25	0.044	0.086	0.133	0.184	0.235	0.246	0.337	0.386	0.433
26	0.022	0.057	0.100	0.147	0.196	0.211	0.295	0.343	0.390
27	0.011	0.038	0.075	0.118	0.163	0.175	0.258	0.305	0.351
28		0.019	0.050	0.088	0.131	0.146	0.221	0.267	0.312
29		0.100	0.033	0.066	0.105	0.117	0.189	0.233	0.277
30			0.017	0.044	0.078	0.094	0.158	0.200	0.242
31			0.008	0.029	0.059	0.070	0.132	0.171	0.212
32				0.015	0.039	0.053	0.105	0.143	0.182
33				0.007	0.026	0.035	0.084	0.119	0.156
34					0.013	0.023	0.063	0.095	0.130
35					0.007	0.012	0.047	0.057	0.108

t_1	$n_2 = 12$	$n_2 = 13$	$n_2 = 14$	$n_2 = 15$	$n_2 = 16$	$n_2 = 17$	$n_2 = 18$	$n_2 = 19$	$n_2 = 20$
						$n_1 = 2$			
36						0.006	0.032	0.043	0.087
37							0.021	0.029	0.069
38							0.011	0.019	0.052
39							0.005	0.010	0.039
40								0.005	0.026
41									0.017
42									0.009
43									0.004

t_1	$n_2 = 3$	$n_2 = 4$	$n_2 = 5$	$n_2 = 6$	$n_2 = 7$	$n_2 = 8$	$n_2 = 9$	$n_2 = 10$	$n_2 = 11$
						$n_1 = 3$			
11	0.500								
12	0.350	0.571							
13	0.200	0.429							
14	0.100	0.314	0.500						
15	0.050	0.200	0.393	0.548					
16		0.114	0.286	0.452					
17		0.057	0.196	0.357	0.500				
18		0.029	0.125	0.274	0.417	0.539			
19			0.071	0.190	0.333	0.461			
20			0.036	0.131	0.258	0.388	0.500		
21			0.018	0.083	0.192	0.315	0.432	0.531	
22				0.048	0.133	0.248	0.364	0.469	
23				0.024	0.092	0.188	0.300	0.406	0.500
24				0.012	0.058	0.097	0.241	0.346	0.442
25					0.033	0.067	0.186	0.287	0.385
26					0.017	0.042	0.141	0.234	0.330
27					0.008	0.024	0.105	0.185	0.277
28						0.012	0.073	0.143	0.228
29						0.006	0.050	0.108	0.184
30							0.032	0.080	0.146
31							0.018	0.056	0.113
32							0.009	0.038	0.085
33							0.005	0.024	0.063
34								0.014	0.044
35								0.007	0.030
36								0.003	0.019
37									0.011
38									0.005
39									0.003

$n_1 = 3$

t_1	$n_2 = 12$	$n_2 = 13$	$n_2 = 14$	$n_2 = 15$	$n_2 = 16$	$n_2 = 17$	$n_2 = 18$	$n_2 = 19$	$n_2 = 20$
24	0.527								
25	0.473								
26	0.420	0.500							
27	0.367	0.450	0.524						
28	0.316	0.400	0.476						
29	0.268	0.352	0.429	0.500					
30	0.224	0.305	0.384	0.456	0.521				
31	0.182	0.261	0.338	0.412	0.179				
32	0.147	0.220	0.296	0.369	0.138	0.500			
33	0.116	0.182	0.254	0.327	0.396	0.461	0.519		
34	0.090	0.148	0.216	0.287	0.346	0.421	0.481		
35	0.068	0.120	0.181	0.249	0.317	0.382	0.444	0.500	
36	0.051	0.095	0.150	0.213	0.280	0.345	0.407	0.464	0.517
37	0.035	0.073	0.122	0.180	0.244	0.308	0.370	0.429	0.483
38	0.024	0.055	0.099	0.151	0.211	0.273	0.335	0.394	0.449
39	0.015	0.041	0.078	0.125	0.180	0.239	0.300	0.359	0.415
40	0.009	0.029	0.060	0.102	0.152	0.208	0.267	0.325	0.382
41	0.004	0.020	0.046	0.082	0.127	0.179	0.235	0.293	0.349
42	0.002	0.012	0.034	0.065	0.105	0.153	0.206	0.262	0.317
43		0.007	0.024	0.050	0.086	0.129	0.178	0.320	0.286
44		0.004	0.016	0.038	0.055	0.108	0.153	0.204	0.257
45		0.002	0.010	0.028	0.042	0.089	0.131	0.178	0.229
46			0.006	0.200	0.032	0.073	0.111	0.154	0.202
47			0.003	0.013	0.024	0.059	0.092	0.132	0.177
48			0.001	0.009	0.017	0.046	0.077	0.113	0.155
49				0.005	0.011	0.036	0.062	0.095	0.134
50				0.002	0.007	0.027	0.050	0.080	0.115
51				0.001	0.004	0.020	0.040	0.066	0.098
52					0.002	0.014	0.031	0.054	0.083
53					0.001	0.010	0.023	0.044	0.069
54						0.006	0.017	0.034	0.058
55						0.004	0.012	0.027	0.047
56						0.002	0.008	0.020	0.038
57						0.001	0.005	0.015	0.030
58							0.003	0.010	0.023
59							0.002	0.007	0.018
60							0.001	0.005	0.013
61								0.003	0.009
62								0.001	0.006
63								0.001	0.004
64									0.002
65									0.001
66									0.001

				$n_1 = 4$				
t_1	$n_2 = 4$	$n_2 = 5$	$n_2 = 6$	$n_2 = 7$	$n_2 = 8$	$n_2 = 9$	$n_2 = 10$	$n_2 = 11$
18	0.557							
19	0.443							
20	0.343	0.548						
21	0.243	0.452						
22	0.171	0.365	0.543					
23	0.100	0.278	0.049					
24	0.057	0.206	0.381	0.536				
25	0.029	0.143	0.305	0.046				
26	0.014	0.095	0.238	0.394	0.533			
27		0.056	0.176	0.324	0.467			
28		0.032	0.129	0.264	0.404	0.530		
29		0.016	0.086	0.206	0.341	0.470		
30		0.008	0.057	0.158	0.285	0.413	0.527	
31			0.033	0.082	0.230	0.355	0.473	
32			0.019	0.055	0.184	0.302	0.420	0.525
33			0.010	0.036	0.141	0.252	0.367	0.475
34			0.005	0.021	0.107	0.207	0.318	0.426
35				0.012	0.077	0.165	0.270	0.377
36				0.006	0.055	0.130	0.227	0.330
37				0.003	0.036	0.099	0.187	0.286
38					0.024	0.074	0.152	0.245
39					0.014	0.053	0.120	0.206
40					0.008	0.038	0.094	0.171
41					0.004	0.025	0.071	0.140
42						0.002	0.017	0.053
43						0.010	0.038	0.089
44						0.006	0.027	0.069
45						0.003	0.018	0.052
46						0.001	0.012	0.039
47							0.007	0.028
48							0.004	0.020
49							0.002	0.013
50							0.001	0.009
51								0.005
52								0.003
53								0.001
54								0.001

				$n_1 = 4$				
t_1	$n_2 = 12$	$n_2 = 13$	$n_2 = 14$	$n_2 = 15$	$n_2 = 16$	$n_2 = 17$	$n_2 = 18$	$n_2 = 19$	$n_2 = 20$
34	0.524								
35	0.476								
36	0.431	0.522							
37	0.385	0.478							
38	0.342	0.435	0.521						

				$n_1 = 4$				
t_1	$n_2 = 12$	$n_2 = 13$	$n_2 = 14$	$n_2 = 15$	$n_2 = 16$	$n_2 = 17$	$n_2 = 18$	$n_2 = 19$	$n_2 = 20$
39	0.299	0.392	0.479						
40	0.260	0.352	0.439	0.519					
41	0.223	0.312	0.399	0.481					
42	0.190	0.274	0.360	0.443	0.518				
43	0.158	0.239	0.323	0.405	0.482				
44	0.131	0.206	0.287	0.368	0.446	0.517			
46	0.085	0.148	0.221	0.298	0.375	0.449	0.516		
47	0.066	0.123	0.191	0.265	0.341	0.415	0.484		
48	0.052	0.101	0.164	0.235	0.308	0.381	0.451	0.516	
49	0.039	0.082	0.139	0.205	0.277	0.349	0.419	0.484	
50	0.029	0.065	0.116	0.179	0.247	0.318	0.387	0.453	0.515
51	0.021	0.051	0.096	0.154	0.219	0.287	0.356	0.422	0.485
52	0.015	0.039	0.079	0.131	0.192	0.258	0.326	0.392	0.455
53	0.010	0.030	0.063	0.110	0.168	0.231	0.297	0.363	0.426
54	0.007	0.022	0.051	0.092	0.145	0.205	0.269	0.334	0.390
55	0.004	0.016	0.040	0.076	0.124	0.181	0.242	0.306	0.368
56	0.002	0.011	0.031	0.062	0.106	0.158	0.217	0.028	0.341
57	0.001	0.008	0.023	0.050	0.089	0.138	0.193	0.253	0.314
58	0.001	0.005	0.017	0.040	0.074	0.119	0.171	0.228	0.288
59		0.003	0.012	0.031	0.061	0.101	0.150	0.205	0.262
60		0.002	0.009	0.024	0.050	0.086	0.131	0.183	0.239
61		0.001	0.006	0.018	0.040	0.072	0.113	0.162	0.216
62		0.000	0.004	0.014	0.320	0.060	0.098	0.143	0.194
63			0.002	0.010	0.025	0.049	0.083	0.125	0.174
64			0.001	0.007	0.019	0.040	0.070	0.109	0.155
65			0.001	0.005	0.015	0.032	0.059	0.094	0.137
66			0.000	0.003	0.011	0.026	0.049	0.081	0.120
67				0.002	0.008	0.020	0.040	0.069	0.105
68				0.001	0.006	0.016	0.033	0.058	0.091
69				0.001	0.004	0.012	0.027	0.049	0.079
70				0.000	0.002	0.009	0.021	0.041	0.067
71					0.001	0.006	0.017	0.033	0.057
72					0.001	0.005	0.013	0.027	0.048
73					0.000	0.003	0.010	0.022	0.041
74					0.000	0.002	0.007	0.018	0.034
75						0.001	0.005	0.014	0.028
76						0.001	0.004	0.011	0.023
77						0.000	0.002	0.008	0.018
78						0.000	0.002	0.006	0.015
79							0.001	0.004	0.011
80							0.001	0.003	0.009
81							0.000	0.002	0.007
82							0.000	0.001	0.005
83								0.001	0.004

				$n_1 = 4$					
t_1	$n_2 = 12$	$n_2 = 13$	$n_2 = 14$	$n_2 = 15$	$n_2 = 16$	$n_2 = 17$	$n_2 = 18$	$n_2 = 19$	$n_2 = 20$
84								0.000	0.003
85								0.000	0.002
86								0.000	0.001
87									0.001
88									0.000
89									0.000
90									0.000

			$n_1 = 5$			
x	$n_2 = 5$	$n_2 = 6$	$n_2 = 7$	$n_2 = 8$	$n_2 = 9$	$n_2 = 10$
28	0.500					
29	0.421					
30	0.345	0.535				
31	0.274	0.465				
32	0.210	0.396				
33	0.155	0.331	0.500			
34	0.111	0.268	0.438			
35	0.075	0.214	0.378	0.528		
36	0.048	0.165	0.319	0.472		
37	0.028	0.123	0.265	0.416	0.500	
38	0.016	0.089	0.216	0.362	0.449	
39	0.008	0.063	0.172	0.311	0.399	
40	0.004	0.041	0.134	0.262	0.350	0.523
41		0.026	0.101	0.218	0.303	0.477
42		0.015	0.074	0.177	0.259	0.430
43		0.009	0.053	0.142	0.219	0.384
44		0.004	0.037	0.111	0.182	0.339
45		0.002	0.024	0.085	0.149	0.297
46			0.015	0.064	0.120	0.270
47			0.001	0.047	0.095	0.220
48			0.005	0.033	0.073	0.185
49			0.003	0.023	0.056	0.155
50			0.001	0.015	0.041	0.127
51				0.009	0.030	0.103
52				0.005	0.021	0.082
53				0.003	0.014	0.065
54				0.002	0.009	0.050
55				0.001	0.006	0.038
56					0.003	0.028
57					0.002	0.020
58					0.001	0.014
59					0.000	0.010

			$n_1 = 5$			
x	$n_2 = 5$	$n_2 = 6$	$n_2 = 7$	$n_2 = 8$	$n_2 = 9$	$n_2 = 10$
60						0.006
61						0.004
62						0.002
63						0.001
64						0.001
65						0.000

			$n_1 = 6$		
t_1	$n_2 = 6$	$n_2 = 7$	$n_2 = 8$	$n_2 = 9$	$n_2 = 10$
39	0.531				
40	0.469				
41	0.409				
42	0.350	0.527			
43	0.294	0.473			
44	0.242	0.418			
45	0.197	0.365	0.525		
46	0.155	0.314	0.475		
47	0.120	0.267	0.426		
48	0.090	0.223	0.377	0.523	
49	0.066	0.183	0.331	0.477	
50	0.047	0.147	0.286	0.432	
51	0.032	0.117	0.245	0.388	0.521
52	0.021	0.090	0.207	0.344	0.479
53	0.013	0.069	0.172	0.303	0.437
54	0.008	0.051	0.141	0.264	0.396
55	0.004	0.037	0.114	0.228	0.356
56	0.002	0.026	0.091	0.194	0.318
57	0.001	0.017	0.071	0.164	0.281
58		0.011	0.054	0.136	0.246
59		0.007	0.041	0.112	0.214
60		0.004	0.030	0.091	0.184
61		0.002	0.021	0.072	0.157
62		0.001	0.015	0.057	0.132
63		0.001	0.010	0.044	0.110
64			0.006	0.033	0.090
65			0.004	0.025	0.074
66			0.002	0.018	0.059
67			0.001	0.013	0.047
68			0.001	0.001	0.036
69			0.000	0.006	0.028
70				0.004	0.021
71				0.002	0.016
72				0.001	0.011
73				0.001	0.008

	$n_1 = 6$				
t_1	$n_2 = 6$	$n_2 = 7$	$n_2 = 8$	$n_2 = 9$	$n_2 = 10$
74				0.000	0.005
75				0.000	0.004
76					0.002
77					0.001
78					0.011
79					0.000
80					0.000
81					0.000

	$n_1 = 7$					$n_1 = 8$		
t_1	$n_2 = 7$	$n_2 = 8$	$n_2 = 9$	$n_2 = 10$	t_1	$n_2 = 8$	$n_2 = 9$	$n_2 = 10$
53	0.500				68	0.520		
54	0.451				69	0.480		
55	0.402				70	0.439		
56	0.355	0.522			71	0.399		
57	0.310	0.478			72	0.360	0.519	
58	0.267	0.433			73	0.323	0.481	
59	0.228	0.389			74	0.287	0.444	
60	0.191	0.347	0.500		75	0.253	0.407	
61	0.159	0.306	0.459		76	0.210	0.371	0.517
62	0.130	0.268	0.419		77	0.191	0.336	0.483
63	0.104	0.232	0.379	0.519	78	0.164	0.303	0.448
64	0.082	0.198	0.340	0.481	79	0.139	0.271	0.414
65	0.064	0.168	0.303	0.443	80	0.117	0.240	0.381
66	0.049	0.140	0.268	0.406	81	0.097	0.212	0.348
67	0.036	0.116	0.235	0.370	82	0.080	0.185	0.317
68	0.027	0.095	0.204	0.335	83	0.065	0.161	0.286
69	0.019	0.076	0.176	0.300	84	0.052	0.138	0.257
70	0.013	0.060	0.150	0.268	85	0.041	0.118	0.230
71	0.009	0.047	0.126	0.237	86	0.032	0.100	0.204
72	0.006	0.036	0.105	0.209	87	0.025	0.008	0.180
73	0.003	0.027	0.087	0.182	88	0.019	0.069	0.158
74	0.002	0.020	0.071	0.157	89	0.014	0.057	0.137
75	0.001	0.014	0.057	0.135	90	0.010	0.046	0.118
76	0.001	0.010	0.045	0.115	91	0.007	0.037	0.102
77	0.000	0.007	0.036	0.097	92	0.005	0.030	0.086
78		0.005	0.027	0.081	93	0.003	0.023	0.073
79		0.003	0.021	0.067	94	0.002	0.018	0.061
80		0.002	0.016	0.054	95	0.001	0.014	0.051
81		0.001	0.011	0.044	96	0.001	0.010	0.042
82		0.001	0.008	0.035	97	0.001	0.008	0.034
83		0.000	0.006	0.028	98	0.000	0.006	0.027
84		0.000	0.004	0.022	99	0.000	0.004	0.022
85			0.003	0.017	100	0.000	0.003	0.017

| | $n_1 = 7$ | | | | $n_1 = 8$ | | |
t_1	$n_2 = 7$	$n_2 = 8$	$n_2 = 9$	$n_2 = 10$	t_1	$n_2 = 8$	$n_2 = 9$	$n_2 = 10$
86			0.002	0.012	101		0.002	0.013
87			0.001	0.009	102		0.001	0.010
88			0.001	0.007	103		0.001	0.008
89			0.000	0.005	104		0.000	0.006
90			0.000	0.003	105		0.000	0.004
91			0.000	0.002	106		0.000	0.003
92				0.002	107		0.000	0.002
93				0.001	108		0.000	0.002
94				0.001	109			0.001
95				0.000	110			0.001
96				0.000	111			0.000
97				0.000	112			0.000
98				0.000	113			0.000
					114			0.000
					115			0.000
					116			0.000

| | $n_1 = 9$ | | | $n_1 = 10$ | |
t_1	$n_2 = 9$	$n_2 = 10$		t_1	$n_2 = 10$
86	0.500			105	0.515
87	0.466			106	0.485
88	0.432			107	0.456
89	0.398			108	0.427
90	0.365	0.516		109	0.398
91	0.333	0.484		110	0.370
92	0.302	0.452		111	0.342
93	0.273	0.421		112	0.315
94	0.245	0.390		113	0.289
95	0.218	0.360		114	0.264
96	0.193	0.330		115	0.241
97	0.170	0.302		116	0.218
98	0.149	0.274		117	0.197
99	0.129	0.248		118	0.176
100	0.111	0.223		119	0.157
101	0.095	0.200		120	0.140
102	0.081	0.178		121	0.124
103	0.068	0.158		122	0.109
104	0.057	0.139		123	0.095
105	0.047	0.121		124	0.083
106	0.039	0.106		125	0.072
107	0.031	0.091		126	0.062
108	0.025	0.078		127	0.053
109	0.020	0.067		128	0.045
110	0.016	0.056		129	0.038

| | $n_1 = 9$ | | | $n_1 = 10$ | |
t_1	$n_2 = 9$	$n_2 = 10$		t_1	$n_2 = 10$
111	0.012	0.047		130	0.032
112	0.009	0.039		131	0.026
113	0.007	0.033		132	0.022
114	0.005	0.027		133	0.018
115	0.004	0.022		134	0.014
116	0.003	0.017		135	0.012
117	0.002	0.014		136	0.009
118	0.001	0.011		137	0.007
119	0.001	0.009		138	0.006
120	0.001	0.007		139	0.004
121	0.000	0.005		140	0.003
122	0.000	0.004		141	0.003
123	0.000	0.003		142	0.002
124	0.000	0.002		143	0.001
125	0.000	0.001		144	0.001
126	0.000	0.001		145	0.001
127		0.001		146	0.001
128		0.000		147	0.000
129		0.000		148	0.000
130		0.000		149	0.000
131		0.000		150	0.000
132		0.000		151	0.000
133		0.000		152	0.000
134		0.000		153	0.000
135		0.000		154	0.000
				155	0.000

Appendix I

The Studentized Range or Q-Distribution

This appendix contains the critical values $Q_{\alpha;g;v}$ as function of the parameters g and v. A critical value corresponds to the exceedance probability

$$P(Q_{g;v} \geq Q_{\alpha;g;v}) = \alpha.$$

For example, $Q_{0.10;9;3} = 7.062$.

Statistics with JMP: Hypothesis Tests, ANOVA and Regression, First Edition. Peter Goos and David Meintrup.
© 2016 John Wiley & Sons, Ltd. Published 2016 by John Wiley & Sons, Ltd.
Companion Website: http://www.wiley.com/go/goosandmeintrup/JMP

$$Q_{0.10; g, v}$$

v	2	3	4	5	6	7	8	9	10	11	12	13	14	15	16	17	18	19	20
1	8.929	13.437	16.358	18.488	20.150	21.504	22.642	23.621	24.477	25.237	25.918	26.536	27.100	27.618	28.097	28.542	28.958	29.347	29.713
2	4.129	5.733	6.772	7.538	8.139	8.633	9.049	9.409	9.725	10.006	10.259	10.488	10.698	10.891	11.070	11.237	11.392	11.538	11.676
3	3.328	4.467	5.199	5.738	6.162	6.511	6.806	7.062	7.287	7.487	7.667	7.831	7.982	8.120	8.248	8.368	8.479	8.584	8.683
4	3.015	3.976	4.586	5.035	5.388	5.679	5.926	6.139	6.327	6.494	6.645	6.783	6.909	7.025	7.132	7.233	7.326	7.414	7.497
5	2.850	3.717	4.264	4.664	4.979	5.238	5.458	5.648	5.816	5.965	6.100	6.223	6.336	6.439	6.536	6.626	6.710	6.788	6.863
6	2.748	3.558	4.065	4.435	4.726	4.966	5.168	5.344	5.499	5.637	5.762	5.875	5.979	6.075	6.164	6.247	6.325	6.398	6.466
7	2.679	3.451	3.931	4.280	4.555	4.780	4.971	5.137	5.283	5.413	5.530	5.637	5.735	5.826	5.910	5.988	6.061	6.130	6.195
8	2.630	3.374	3.834	4.169	4.431	4.646	4.829	4.987	5.126	5.250	5.362	5.464	5.558	5.644	5.724	5.799	5.869	5.935	5.997
9	2.592	3.316	3.761	4.084	4.337	4.545	4.721	4.873	5.007	5.126	5.234	5.333	5.423	5.506	5.583	5.655	5.722	5.786	5.845
10	2.563	3.270	3.704	4.018	4.264	4.465	4.636	4.783	4.913	5.029	5.134	5.229	5.316	5.397	5.472	5.542	5.607	5.668	5.726
11	2.540	3.234	3.658	3.965	4.205	4.401	4.567	4.711	4.838	4.951	5.053	5.145	5.231	5.309	5.382	5.450	5.514	5.573	5.630
12	2.521	3.204	3.621	3.921	4.156	4.349	4.511	4.652	4.776	4.886	4.986	5.076	5.160	5.236	5.308	5.374	5.436	5.495	5.550
13	2.504	3.179	3.589	3.885	4.116	4.304	4.464	4.602	4.724	4.832	4.930	5.019	5.100	5.175	5.245	5.310	5.371	5.429	5.483
14	2.491	3.158	3.563	3.854	4.081	4.267	4.424	4.560	4.679	4.786	4.882	4.969	5.050	5.124	5.192	5.256	5.316	5.372	5.426
15	2.479	3.140	3.540	3.828	4.052	4.235	4.390	4.524	4.641	4.746	4.841	4.927	5.006	5.079	5.146	5.209	5.268	5.324	5.376
16	2.469	3.124	3.520	3.804	4.026	4.207	4.360	4.492	4.608	4.712	4.805	4.890	4.968	5.040	5.106	5.169	5.227	5.282	5.333
17	2.460	3.110	3.503	3.784	4.003	4.182	4.334	4.464	4.579	4.681	4.774	4.857	4.934	5.005	5.071	5.133	5.190	5.244	5.295
18	2.452	3.098	3.487	3.766	3.984	4.161	4.310	4.440	4.553	4.654	4.746	4.829	4.905	4.975	5.040	5.101	5.158	5.211	5.262
19	2.445	3.087	3.474	3.751	3.966	4.142	4.290	4.418	4.530	4.630	4.721	4.803	4.878	4.948	5.012	5.072	5.129	5.182	5.232
20	2.439	3.077	3.462	3.736	3.950	4.124	4.271	4.398	4.510	4.609	4.699	4.780	4.855	4.923	4.987	5.047	5.103	5.155	5.205
21	2.433	3.069	3.451	3.724	3.936	4.109	4.255	4.380	4.491	4.590	4.678	4.759	4.833	4.901	4.965	5.024	5.079	5.131	5.180
22	2.428	3.061	3.441	3.712	3.923	4.095	4.239	4.364	4.474	4.572	4.660	4.740	4.814	4.882	4.944	5.003	5.058	5.109	5.158
23	2.424	3.054	3.432	3.701	3.911	4.082	4.226	4.350	4.459	4.556	4.644	4.723	4.796	4.863	4.926	4.984	5.038	5.089	5.138
24	2.420	3.047	3.423	3.692	3.900	4.070	4.213	4.336	4.445	4.541	4.628	4.707	4.780	4.847	4.909	4.966	5.020	5.071	5.119

25	2.416	3.041	3.416	3.683	3.890	4.059	4.201	4.324	4.432	4.528	4.614	4.693	4.765	4.831	4.893	4.950	5.004	5.055	5.102
26	2.412	3.036	3.409	3.675	3.881	4.049	4.191	4.313	4.420	4.515	4.601	4.680	4.751	4.817	4.878	4.936	4.989	5.039	5.086
27	2.409	3.030	3.402	3.667	3.873	4.040	4.181	4.302	4.409	4.504	4.590	4.667	4.739	4.804	4.865	4.922	4.975	5.025	5.072
28	2.406	3.026	3.396	3.660	3.865	4.032	4.172	4.293	4.399	4.493	4.579	4.656	4.727	4.792	4.853	4.909	4.962	5.012	5.058
29	2.403	3.021	3.391	3.654	3.858	4.024	4.163	4.284	4.389	4.484	4.568	4.645	4.716	4.781	4.841	4.897	4.950	4.999	5.046
30	2.400	3.017	3.386	3.648	3.851	4.016	4.155	4.275	4.381	4.474	4.559	4.635	4.706	4.770	4.830	4.886	4.939	4.988	5.034
31	2.398	3.013	3.381	3.642	3.845	4.009	4.148	4.268	4.372	4.466	4.550	4.626	4.696	4.760	4.820	4.876	4.928	4.977	5.023
32	2.396	3.010	3.376	3.637	3.839	4.003	4.141	4.260	4.365	4.458	4.541	4.617	4.687	4.751	4.811	4.866	4.918	4.967	5.013
33	2.393	3.006	3.372	3.632	3.833	3.997	4.135	4.253	4.357	4.450	4.533	4.609	4.679	4.743	4.802	4.857	4.909	4.957	5.003
34	2.391	3.003	3.368	3.627	3.828	3.991	4.129	4.247	4.351	4.443	4.526	4.602	4.671	4.734	4.794	4.849	4.900	4.949	4.994
35	2.389	3.000	3.364	3.623	3.823	3.986	4.123	4.241	4.344	4.436	4.519	4.594	4.663	4.727	4.786	4.841	4.892	4.940	4.986
36	2.388	2.998	3.361	3.619	3.819	3.981	4.117	4.235	4.338	4.430	4.512	4.588	4.656	4.720	4.778	4.833	4.884	4.932	4.978
37	2.386	2.995	3.357	3.615	3.814	3.976	4.112	4.230	4.332	4.424	4.506	4.581	4.650	4.713	4.771	4.826	4.877	4.925	4.970
38	2.384	2.992	3.354	3.611	3.810	3.972	4.107	4.224	4.327	4.418	4.500	4.575	4.643	4.706	4.765	4.819	4.870	4.918	4.963
39	2.383	2.990	3.351	3.608	3.806	3.967	4.103	4.220	4.322	4.413	4.495	4.569	4.637	4.700	4.758	4.812	4.863	4.911	4.956
40	2.381	2.988	3.348	3.605	3.802	3.963	4.099	4.215	4.317	4.408	4.490	4.564	4.632	4.694	4.752	4.806	4.857	4.904	4.949
48	2.372	2.973	3.330	3.583	3.778	3.937	4.070	4.185	4.285	4.375	4.455	4.528	4.595	4.656	4.713	4.766	4.816	4.863	4.907
60	2.363	2.959	3.312	3.562	3.755	3.911	4.042	4.155	4.254	4.342	4.421	4.493	4.558	4.619	4.675	4.727	4.775	4.821	4.864
80	2.353	2.945	3.294	3.541	3.731	3.885	4.014	4.125	4.223	4.309	4.387	4.457	4.521	4.581	4.636	4.687	4.735	4.780	4.822
120	2.344	2.930	3.276	3.520	3.707	3.859	3.986	4.096	4.191	4.276	4.353	4.422	4.485	4.543	4.597	4.647	4.694	4.738	4.779
240	2.335	2.916	3.258	3.499	3.684	3.834	3.959	4.066	4.160	4.244	4.319	4.386	4.448	4.505	4.558	4.607	4.653	4.696	4.737
∞	2.326	2.902	3.240	3.478	3.661	3.808	3.931	4.037	4.129	4.211	4.285	4.351	4.412	4.468	4.519	4.568	4.612	4.654	4.694

$$Q_{0.05;g,v}$$

v	\multicolumn{19}{c}{g}
	2	3	4	5	6	7	8	9	10	11	12	13	14	15	16	17	18	19	20
1	17.969	26.976	32.819	37.082	40.408	43.119	45.397	47.357	49.071	50.592	51.957	53.194	54.323	55.361	56.32	57.212	58.044	58.824	59.558
2	6.085	8.331	9.798	10.881	11.734	12.435	13.027	13.539	13.988	14.389	14.749	15.076	15.375	15.65	15.905	16.143	16.365	16.573	16.769
3	4.501	5.910	6.825	7.502	8.037	8.478	8.852	9.177	9.462	9.717	9.946	10.155	10.346	10.522	10.686	10.838	10.98	11.114	11.24
4	3.926	5.040	5.757	6.287	6.706	7.053	7.347	7.602	7.826	8.027	8.208	8.373	8.524	8.664	8.793	8.914	9.027	9.133	9.233
5	3.635	4.602	5.218	5.673	6.033	6.330	6.582	6.801	6.995	7.167	7.323	7.466	7.596	7.716	7.828	7.932	8.030	8.122	8.208
6	3.460	4.339	4.896	5.305	5.628	5.895	6.122	6.319	6.493	6.649	6.789	6.917	7.034	7.143	7.244	7.338	7.426	7.508	7.586
7	3.344	4.165	4.681	5.060	5.359	5.606	5.815	5.997	6.158	6.302	6.431	6.550	6.658	6.759	6.852	6.939	7.020	7.097	7.169
8	3.261	4.041	4.529	4.886	5.167	5.399	5.596	5.767	5.918	6.053	6.175	6.287	6.389	6.483	6.571	6.653	6.729	6.801	6.869
9	3.199	3.948	4.415	4.755	5.024	5.244	5.432	5.595	5.738	5.867	5.983	6.089	6.186	6.276	6.359	6.437	6.510	6.579	6.643
10	3.151	3.877	4.327	4.654	4.912	5.124	5.304	5.460	5.598	5.722	5.833	5.935	6.028	6.114	6.194	6.269	6.339	6.405	6.467
11	3.113	3.820	4.256	4.574	4.823	5.028	5.202	5.353	5.486	5.605	5.713	5.811	5.901	5.984	6.062	6.134	6.202	6.265	6.325
12	3.081	3.773	4.199	4.508	4.750	4.950	5.119	5.265	5.395	5.510	5.615	5.710	5.797	5.878	5.953	6.023	6.089	6.151	6.209
13	3.055	3.734	4.151	4.453	4.690	4.884	5.049	5.192	5.318	5.431	5.533	5.625	5.711	5.789	5.862	5.931	5.995	6.055	6.112
14	3.033	3.701	4.111	4.407	4.639	4.829	4.990	5.130	5.253	5.364	5.463	5.554	5.637	5.714	5.785	5.852	5.915	5.973	6.029
15	3.014	3.673	4.076	4.367	4.595	4.782	4.940	5.077	5.198	5.306	5.403	5.492	5.574	5.649	5.719	5.785	5.846	5.904	5.958
16	2.998	3.649	4.046	4.333	4.557	4.741	4.896	5.031	5.150	5.256	5.352	5.439	5.519	5.593	5.662	5.726	5.786	5.843	5.896
17	2.984	3.628	4.020	4.303	4.524	4.705	4.858	4.991	5.108	5.212	5.306	5.392	5.471	5.544	5.612	5.675	5.734	5.790	5.842
18	2.971	3.609	3.997	4.276	4.494	4.673	4.824	4.955	5.071	5.173	5.266	5.351	5.429	5.501	5.567	5.629	5.688	5.743	5.794
19	2.960	3.593	3.977	4.253	4.468	4.645	4.794	4.924	5.037	5.139	5.231	5.314	5.391	5.462	5.528	5.589	5.647	5.701	5.752
20	2.950	3.578	3.958	4.232	4.445	4.620	4.768	4.895	5.008	5.108	5.199	5.282	5.357	5.427	5.492	5.553	5.610	5.663	5.714
21	2.941	3.565	3.942	4.213	4.424	4.597	4.743	4.870	4.981	5.081	5.170	5.252	5.327	5.396	5.460	5.520	5.576	5.629	5.679
22	2.933	3.553	3.927	4.196	4.405	4.577	4.722	4.847	4.957	5.056	5.144	5.225	5.299	5.368	5.431	5.491	5.546	5.599	5.648
23	2.926	3.542	3.914	4.180	4.388	4.558	4.702	4.826	4.935	5.033	5.121	5.201	5.274	5.342	5.405	5.464	5.519	5.571	5.620
24	2.919	3.532	3.901	4.166	4.373	4.541	4.684	4.807	4.915	5.012	5.099	5.179	5.251	5.319	5.381	5.439	5.494	5.545	5.594

25	2.913	3.523	3.890	4.153	4.358	4.526	4.667	4.789	4.897	4.993	5.079	5.158	5.230	5.297	5.359	5.417	5.471	5.522	5.570
26	2.907	3.514	3.880	4.141	4.345	4.511	4.652	4.773	4.880	4.975	5.061	5.139	5.211	5.277	5.339	5.396	5.450	5.500	5.548
27	2.902	3.506	3.870	4.130	4.333	4.498	4.638	4.758	4.864	4.959	5.044	5.122	5.193	5.259	5.320	5.377	5.430	5.480	5.528
28	2.897	3.499	3.861	4.120	4.322	4.486	4.625	4.745	4.850	4.944	5.029	5.106	5.177	5.242	5.302	5.359	5.412	5.462	5.509
29	2.892	3.493	3.853	4.111	4.311	4.475	4.613	4.732	4.837	4.930	5.014	5.091	5.161	5.226	5.286	5.342	5.395	5.445	5.491
30	2.888	3.486	3.845	4.102	4.301	4.464	4.601	4.720	4.824	4.917	5.001	5.077	5.147	5.211	5.271	5.327	5.379	5.429	5.475
31	2.884	3.481	3.838	4.094	4.292	4.454	4.591	4.709	4.812	4.905	4.988	5.064	5.134	5.198	5.257	5.313	5.365	5.414	5.460
32	2.881	3.475	3.832	4.086	4.284	4.445	4.581	4.698	4.802	4.894	4.976	5.052	5.121	5.185	5.244	5.299	5.351	5.400	5.445
33	2.877	3.470	3.825	4.079	4.276	4.436	4.572	4.689	4.791	4.883	4.965	5.040	5.109	5.173	5.232	5.287	5.338	5.386	5.432
34	2.874	3.465	3.820	4.072	4.268	4.428	4.563	4.680	4.782	4.873	4.955	5.030	5.098	5.161	5.220	5.275	5.326	5.374	5.420
35	2.871	3.461	3.814	4.066	4.261	4.421	4.555	4.671	4.773	4.863	4.945	5.020	5.088	5.151	5.209	5.264	5.315	5.362	5.408
36	2.868	3.457	3.809	4.060	4.255	4.414	4.547	4.663	4.764	4.855	4.936	5.010	5.078	5.141	5.199	5.253	5.304	5.352	5.397
37	2.865	3.453	3.804	4.054	4.249	4.407	4.540	4.655	4.756	4.846	4.927	5.001	5.069	5.131	5.189	5.243	5.294	5.341	5.386
38	2.863	3.449	3.799	4.049	4.243	4.400	4.533	4.648	4.749	4.838	4.919	4.993	5.060	5.122	5.180	5.234	5.284	5.331	5.376
39	2.861	3.445	3.795	4.044	4.237	4.394	4.527	4.641	4.741	4.831	4.911	4.985	5.052	5.114	5.171	5.225	5.275	5.322	5.367
40	2.858	3.442	3.791	4.039	4.232	4.388	4.521	4.634	4.735	4.824	4.904	4.977	5.044	5.106	5.163	5.216	5.266	5.313	5.358
48	2.843	3.420	3.764	4.008	4.197	4.351	4.481	4.592	4.690	4.777	4.856	4.927	4.993	5.053	5.109	5.161	5.210	5.256	5.299
60	2.829	3.399	3.737	3.977	4.163	4.314	4.441	4.550	4.646	4.732	4.808	4.878	4.942	5.001	5.056	5.107	5.154	5.199	5.241
80	2.814	3.377	3.711	3.947	4.129	4.277	4.402	4.509	4.603	4.686	4.761	4.829	4.892	4.949	5.003	5.052	5.099	5.142	5.183
120	2.800	3.356	3.685	3.917	4.096	4.241	4.363	4.468	4.560	4.641	4.714	4.781	4.842	4.898	4.950	4.998	5.043	5.086	5.126
240	2.786	3.335	3.659	3.887	4.063	4.205	4.324	4.427	4.517	4.596	4.668	4.733	4.792	4.847	4.897	4.944	4.988	5.030	5.069
∞	2.772	3.314	3.633	3.858	4.030	4.170	4.286	4.387	4.474	4.552	4.622	4.685	4.743	4.796	4.845	4.891	4.934	4.974	5.012

$$Q_{0.01;g;v}$$

v	g 2	3	4	5	6	7	8	9	10	11	12	13	14	15	16	17	18	19	20
1	90.024	135.041	164.258	185.575	202.21	215.769	227.166	236.966	245.542	253.151	259.979	266.165	271.812	277.003	281.803	286.263	290.426	294.328	297.997
2	14.036	19.019	22.294	24.717	26.629	28.201	29.530	30.679	31.689	32.589	33.398	34.134	34.806	35.426	36.000	36.534	37.034	37.502	37.943
3	8.260	10.619	12.170	13.324	14.241	14.998	15.641	16.199	16.691	17.130	17.526	17.887	18.217	18.522	18.805	19.068	19.315	19.546	19.765
4	6.511	8.120	9.173	9.958	10.583	11.101	11.542	11.925	12.264	12.567	12.840	13.090	13.318	13.530	13.726	13.909	14.081	14.242	14.394
5	5.702	6.976	7.804	8.421	8.913	9.321	9.669	9.971	10.239	10.479	10.696	10.894	11.076	11.244	11.400	11.545	11.682	11.811	11.932
6	5.243	6.331	7.033	7.556	7.972	8.318	8.612	8.869	9.097	9.300	9.485	9.653	9.808	9.951	10.084	10.208	10.325	10.434	10.538
7	4.949	5.919	6.542	7.005	7.373	7.678	7.939	8.166	8.367	8.548	8.711	8.860	8.997	9.124	9.242	9.353	9.456	9.553	9.645
8	4.745	5.635	6.204	6.625	6.959	7.237	7.474	7.680	7.863	8.027	8.176	8.311	8.436	8.552	8.659	8.760	8.854	8.943	9.027
9	4.596	5.428	5.957	6.347	6.657	6.915	7.134	7.325	7.494	7.646	7.784	7.910	8.025	8.132	8.232	8.325	8.412	8.495	8.573
10	4.482	5.270	5.769	6.136	6.428	6.669	6.875	7.054	7.213	7.356	7.485	7.603	7.712	7.812	7.906	7.993	8.075	8.153	8.226
11	4.392	5.146	5.621	5.970	6.247	6.476	6.671	6.841	6.992	7.127	7.250	7.362	7.464	7.560	7.648	7.731	7.809	7.883	7.952
12	4.320	5.046	5.502	5.836	6.101	6.320	6.507	6.670	6.814	6.943	7.060	7.166	7.265	7.356	7.441	7.520	7.594	7.664	7.730
13	4.260	4.964	5.404	5.726	5.981	6.192	6.372	6.528	6.666	6.791	6.903	7.006	7.100	7.188	7.269	7.345	7.417	7.484	7.548
14	4.210	4.895	5.322	5.634	5.881	6.085	6.258	6.409	6.543	6.663	6.772	6.871	6.962	7.047	7.125	7.199	7.268	7.333	7.394
15	4.167	4.836	5.252	5.556	5.796	5.994	6.162	6.309	6.438	6.555	6.660	6.756	6.845	6.927	7.003	7.074	7.141	7.204	7.264
16	4.131	4.786	5.192	5.489	5.722	5.915	6.079	6.222	6.348	6.461	6.564	6.658	6.744	6.823	6.897	6.967	7.032	7.093	7.151
17	4.099	4.742	5.140	5.430	5.659	5.847	6.007	6.147	6.270	6.380	6.480	6.572	6.656	6.733	6.806	6.873	6.937	6.997	7.053
18	4.071	4.703	5.094	5.379	5.603	5.787	5.944	6.081	6.201	6.309	6.407	6.496	6.579	6.655	6.725	6.791	6.854	6.912	6.967
19	4.046	4.669	5.054	5.334	5.553	5.735	5.889	6.022	6.141	6.246	6.342	6.430	6.510	6.585	6.654	6.719	6.780	6.837	6.891
20	4.024	4.639	5.018	5.293	5.510	5.688	5.839	5.970	6.086	6.190	6.285	6.370	6.449	6.523	6.591	6.654	6.714	6.770	6.823
21	4.004	4.612	4.986	5.257	5.470	5.646	5.794	5.924	6.038	6.140	6.233	6.317	6.395	6.467	6.534	6.596	6.655	6.710	6.762
22	3.986	4.588	4.957	5.225	5.435	5.608	5.754	5.882	5.994	6.095	6.186	6.269	6.346	6.417	6.482	6.544	6.602	6.656	6.707
23	3.970	4.566	4.931	5.195	5.403	5.573	5.718	5.844	5.955	6.054	6.144	6.226	6.301	6.371	6.436	6.497	6.553	6.607	6.658
24	3.955	4.546	4.907	5.168	5.373	5.542	5.685	5.809	5.919	6.017	6.105	6.186	6.261	6.330	6.394	6.453	6.510	6.562	6.612

25	3.942	4.527	4.885	5.144	5.347	5.513	5.655	5.778	5.886	5.983	6.070	6.150	6.224	6.292	6.355	6.414	6.469	6.522	6.571
26	3.930	4.510	4.865	5.121	5.322	5.487	5.627	5.749	5.856	5.951	6.038	6.117	6.190	6.257	6.319	6.378	6.432	6.484	6.533
27	3.918	4.495	4.847	5.101	5.300	5.463	5.602	5.722	5.828	5.923	6.008	6.087	6.158	6.225	6.287	6.344	6.399	6.450	6.498
28	3.908	4.481	4.830	5.082	5.279	5.441	5.578	5.697	5.802	5.896	5.981	6.058	6.129	6.195	6.256	6.314	6.367	6.418	6.465
29	3.898	4.467	4.814	5.064	5.260	5.420	5.556	5.674	5.778	5.871	5.955	6.032	6.103	6.168	6.228	6.285	6.338	6.388	6.435
30	3.889	4.455	4.799	5.048	5.242	5.401	5.536	5.653	5.756	5.848	5.932	6.008	6.078	6.142	6.202	6.258	6.311	6.361	6.407
31	3.881	4.443	4.786	5.032	5.225	5.383	5.517	5.633	5.736	5.827	5.910	5.985	6.055	6.119	6.178	6.234	6.286	6.335	6.381
32	3.873	4.433	4.773	5.018	5.210	5.367	5.500	5.615	5.716	5.807	5.889	5.964	6.033	6.096	6.155	6.211	6.262	6.311	6.357
33	3.865	4.423	4.761	5.005	5.195	5.351	5.483	5.598	5.698	5.789	5.870	5.944	6.013	6.076	6.134	6.189	6.240	6.289	6.334
34	3.859	4.413	4.750	4.992	5.181	5.336	5.468	5.581	5.682	5.771	5.852	5.926	5.994	6.056	6.114	6.169	6.220	6.268	6.313
35	3.852	4.404	4.739	4.980	5.169	5.323	5.453	5.566	5.666	5.755	5.835	5.908	5.976	6.038	6.096	6.150	6.200	6.248	6.293
36	3.846	4.396	4.729	4.969	5.156	5.310	5.439	5.552	5.651	5.739	5.819	5.892	5.959	6.021	6.078	6.132	6.182	6.229	6.274
37	3.840	4.388	4.720	4.959	5.145	5.298	5.427	5.538	5.637	5.725	5.804	5.876	5.943	6.004	6.061	6.115	6.165	6.212	6.256
38	3.835	4.381	4.711	4.949	5.134	5.286	5.414	5.526	5.623	5.711	5.790	5.862	5.928	5.989	6.046	6.099	6.148	6.195	6.239
39	3.830	4.374	4.703	4.940	5.124	5.275	5.403	5.513	5.611	5.698	5.776	5.848	5.914	5.974	6.031	6.084	6.133	6.179	6.223
40	3.825	4.367	4.695	4.931	5.114	5.265	5.392	5.502	5.599	5.685	5.764	5.835	5.900	5.961	6.017	6.069	6.118	6.165	6.208
48	3.793	4.324	4.644	4.874	5.052	5.198	5.322	5.428	5.522	5.606	5.681	5.750	5.814	5.872	5.926	5.977	6.024	6.069	6.111
60	3.762	4.282	4.594	4.818	4.991	5.133	5.253	5.356	5.447	5.528	5.601	5.667	5.728	5.784	5.837	5.886	5.931	5.974	6.015
80	3.732	4.241	4.545	4.763	4.931	5.069	5.185	5.284	5.372	5.451	5.521	5.585	5.644	5.698	5.749	5.796	5.840	5.881	5.920
120	3.702	4.200	4.497	4.709	4.872	5.005	5.118	5.214	5.299	5.375	5.443	5.505	5.561	5.614	5.662	5.708	5.750	5.790	5.827
240	3.672	4.160	4.450	4.655	4.814	4.943	5.052	5.145	5.227	5.300	5.366	5.426	5.480	5.530	5.577	5.621	5.661	5.699	5.735
∞	3.643	4.120	4.403	4.603	4.757	4.882	4.987	5.078	5.157	5.227	5.290	5.348	5.400	5.448	5.493	5.535	5.574	5.611	5.645

Appendix J

The Two-Tailed Dunnett Test

This appendix contains the requird quantiles or percentiles for a two-tailed Dunnett test (and corresponding confidence intervals) in the context of a one-way analysis of variance. The first table assumes a significance level α of 5%. The second table is based on a significance level α of 1%.

	$\alpha = 0.05$							
	$g - 1$							
$n - g$	1	2	3	4	5	6	7	8
1	12.706	17.369	20.034	21.850	23.208	24.284		
2	4.303	5.418	6.065	6.513	6.852	7.123	7.349	7.540
3	3.182	3.866	4.263	4.538	4.748	4.916	5.056	5.176
4	2.776	3.310	3.618	3.832	3.994	4.125	4.235	4.328
5	2.571	3.030	3.293	3.476	3.615	3.727	3.821	3.900
6	2.447	2.863	3.099	3.263	3.388	3.489	3.573	3.644
7	2.365	2.752	2.971	3.123	3.238	3.331	3.408	3.475
8	2.306	2.673	2.880	3.023	3.131	3.219	3.292	3.354
9	2.262	2.614	2.812	2.948	3.052	3.135	3.205	3.264
10	2.228	2.568	2.759	2.890	2.990	3.070	3.137	3.194
11	2.201	2.532	2.717	2.845	2.941	3.019	3.084	3.139
12	2.179	2.502	2.683	2.807	2.901	2.977	3.040	3.094
13	2.160	2.478	2.655	2.776	2.868	2.942	3.003	3.056
14	2.145	2.457	2.631	2.750	2.840	2.913	2.973	3.024
15	2.132	2.439	2.610	2.727	2.816	2.887	2.947	2.997
16	2.120	2.424	2.592	2.708	2.795	2.866	2.924	2.974
17	2.110	2.410	2.577	2.691	2.777	2.847	2.904	2.953
18	2.101	2.399	2.563	2.676	2.762	2.830	2.887	2.935
19	2.093	2.388	2.551	2.663	2.747	2.815	2.871	2.919
20	2.086	2.379	2.540	2.651	2.735	2.802	2.857	2.905

(continued)

Statistics with JMP: Hypothesis Tests, ANOVA and Regression, First Edition. Peter Goos and David Meintrup.
© 2016 John Wiley & Sons, Ltd. Published 2016 by John Wiley & Sons, Ltd.
Companion Website: http://www.wiley.com/go/goosandmeintrup/JMP

				$\alpha = 0.05$				
				$g - 1$				
$n - g$	1	2	3	4	5	6	7	8
21	2.080	2.370	2.531	2.640	2.723	2.790	2.845	2.892
22	2.074	2.363	2.522	2.631	2.713	2.779	2.834	2.880
23	2.069	2.356	2.514	2.622	2.704	2.769	2.824	2.870
24	2.064	2.349	2.507	2.614	2.695	2.760	2.814	2.860
25	2.060	2.344	2.500	2.607	2.688	2.752	2.806	2.852
26	2.056	2.338	2.494	2.600	2.680	2.745	2.798	2.843
27	2.052	2.333	2.488	2.594	2.674	2.738	2.791	2.836
28	2.048	2.329	2.483	2.588	2.668	2.731	2.784	2.829
29	2.045	2.325	2.478	2.583	2.662	2.725	2.778	2.823
30	2.042	2.321	2.474	2.578	2.657	2.720	2.772	2.817
35	2.030	2.305	2.455	2.558	2.635	2.697	2.749	2.792
40	2.021	2.293	2.441	2.543	2.619	2.680	2.731	2.774
50	2.009	2.276	2.422	2.522	2.597	2.657	2.707	2.749
60	2.000	2.265	2.410	2.508	2.582	2.642	2.691	2.733
80	1.990	2.252	2.394	2.491	2.564	2.623	2.671	2.712
100	1.984	2.244	2.385	2.481	2.554	2.611	2.659	2.700
∞	1.960	2.212	2.349	2.442	2.511	2.567	2.613	2.652

				$\alpha = 0.01$				
				$g - 1$				
$n - g$	1	2	3	4	5	6	7	8
1	63.655	86.954	100.270	109.345	116.134	121.513	125.943	
2	9.925	12.387	13.825	14.824	15.581	16.187	16.690	17.119
3	5.841	6.974	7.638	8.103	8.459	8.745	8.984	9.188
4	4.604	5.363	5.809	6.121	6.361	6.554	6.716	6.854
5	4.032	4.627	4.975	5.218	5.405	5.556	5.683	5.791
6	3.707	4.212	4.506	4.711	4.869	4.997	5.104	5.196
7	3.499	3.948	4.208	4.389	4.529	4.642	4.736	4.817
8	3.355	3.766	4.002	4.167	4.294	4.396	4.483	4.557
9	3.250	3.633	3.853	4.006	4.124	4.219	4.298	4.367
10	3.169	3.531	3.739	3.883	3.994	4.084	4.158	4.223
11	3.106	3.452	3.649	3.787	3.893	3.978	4.049	4.110
12	3.054	3.387	3.577	3.710	3.811	3.892	3.960	4.019
13	3.012	3.334	3.518	3.646	3.744	3.822	3.888	3.944
14	2.977	3.290	3.468	3.592	3.687	3.763	3.827	3.882
15	2.947	3.253	3.426	3.547	3.639	3.713	3.776	3.829
16	2.921	3.221	3.390	3.508	3.598	3.671	3.731	3.783
17	2.898	3.192	3.359	3.474	3.563	3.634	3.693	3.744

(continued)

| | $\alpha = 0.01$ | | | | | | | |
| | $g - 1$ | | | | | | | |
$n - g$	1	2	3	4	5	6	7	8
18	2.879	3.168	3.331	3.445	3.531	3.601	3.659	3.709
19	2.861	3.146	3.307	3.419	3.504	3.572	3.630	3.679
20	2.845	3.127	3.285	3.395	3.479	3.547	3.603	3.652
21	2.831	3.109	3.266	3.375	3.457	3.524	3.580	3.627
22	2.819	3.094	3.249	3.356	3.438	3.503	3.558	3.605
23	2.807	3.080	3.233	3.339	3.420	3.485	3.539	3.585
24	2.797	3.067	3.219	3.323	3.403	3.468	3.521	3.567
25	2.788	3.055	3.205	3.309	3.388	3.452	3.505	3.551
26	2.779	3.044	3.193	3.296	3.375	3.438	3.491	3.536
27	2.771	3.034	3.182	3.285	3.362	3.425	3.477	3.522
28	2.763	3.025	3.172	3.274	3.351	3.413	3.465	3.509
29	2.756	3.017	3.163	3.263	3.340	3.402	3.453	3.497
30	2.750	3.009	3.154	3.254	3.330	3.391	3.442	3.486
35	2.724	2.976	3.118	3.215	3.289	3.349	3.398	3.441
40	2.705	2.952	3.091	3.187	3.259	3.317	3.366	3.408
50	2.678	2.920	3.054	3.147	3.218	3.274	3.321	3.362
60	2.660	2.898	3.030	3.121	3.191	3.246	3.292	3.332
80	2.639	2.871	3.001	3.090	3.157	3.211	3.256	3.295
100	2.626	2.856	2.983	3.071	3.137	3.191	3.235	3.273
∞	2.576	2.794	2.915	2.998	3.060	3.111	3.152	3.188

Appendix K

The One-Tailed Dunnett Test

This appendix contains the required quantiles or percentiles for a one-tailed Dunnett test (and the corresponding confidence intervals) in the context of a one-way analysis of variance. The first table assumes a significance level α of 5%. The second table is based on a significance level α of 1%.

	$\alpha = 0.05$							
	$g - 1$							
$n - g$	1	2	3	4	5	6	7	8
1	6.314	8.649	9.982	10.890	11.569	12.108		
2	2.920	3.721	4.182	4.500	4.740	4.932	5.091	5.228
3	2.353	2.912	3.232	3.453	3.621	3.755	3.866	3.961
4	2.132	2.598	2.863	3.046	3.185	3.296	3.389	3.468
5	2.015	2.433	2.669	2.832	2.956	3.055	3.137	3.207
6	1.943	2.332	2.551	2.701	2.815	2.906	2.982	3.047
7	1.895	2.264	2.470	2.612	2.720	2.806	2.877	2.938
8	1.860	2.215	2.413	2.548	2.651	2.733	2.802	2.860
9	1.833	2.178	2.369	2.500	2.599	2.679	2.745	2.801
10	1.813	2.149	2.335	2.463	2.559	2.636	2.700	2.755
11	1.796	2.126	2.308	2.433	2.527	2.602	2.664	2.718
12	1.782	2.107	2.286	2.408	2.500	2.574	2.635	2.687
13	1.771	2.091	2.267	2.387	2.478	2.550	2.611	2.662
14	1.761	2.078	2.252	2.370	2.459	2.531	2.590	2.640
15	1.753	2.066	2.238	2.355	2.443	2.514	2.572	2.622
16	1.746	2.056	2.226	2.342	2.429	2.499	2.557	2.606
17	1.740	2.048	2.216	2.331	2.417	2.486	2.543	2.592
18	1.734	2.040	2.207	2.321	2.406	2.475	2.531	2.579
19	1.729	2.033	2.199	2.312	2.397	2.464	2.521	2.569
20	1.725	2.027	2.192	2.304	2.388	2.455	2.511	2.559

(*continued*)

Statistics with JMP: Hypothesis Tests, ANOVA and Regression, First Edition. Peter Goos and David Meintrup.
© 2016 John Wiley & Sons, Ltd. Published 2016 by John Wiley & Sons, Ltd.
Companion Website: http://www.wiley.com/go/goosandmeintrup/JMP

				$\alpha = 0.05$				
				$g - 1$				
$n - g$	1	2	3	4	5	6	7	8
21	1.721	2.021	2.185	2.297	2.380	2.447	2.503	2.550
22	1.717	2.016	2.179	2.290	2.373	2.440	2.495	2.542
23	1.714	2.012	2.174	2.284	2.367	2.433	2.488	2.534
24	1.711	2.008	2.169	2.279	2.361	2.427	2.481	2.528
25	1.708	2.004	2.165	2.274	2.356	2.421	2.476	2.522
26	1.706	2.000	2.161	2.269	2.351	2.416	2.470	2.516
27	1.703	1.997	2.157	2.265	2.347	2.411	2.465	2.511
28	1.701	1.994	2.153	2.261	2.342	2.407	2.461	2.506
29	1.699	1.991	2.150	2.258	2.338	2.403	2.456	2.502
30	1.697	1.989	2.147	2.254	2.335	2.399	2.452	2.498
35	1.690	1.978	2.135	2.240	2.320	2.383	2.436	2.480
40	1.684	1.970	2.125	2.230	2.309	2.372	2.424	2.468
50	1.676	1.959	2.112	2.216	2.294	2.355	2.407	2.450
60	1.671	1.952	2.104	2.206	2.283	2.345	2.395	2.438
80	1.664	1.943	2.093	2.195	2.271	2.331	2.381	2.424
100	1.660	1.938	2.087	2.188	2.263	2.323	2.373	2.415
∞	1.645	1.916	2.062	2.160	2.234	2.292	2.340	2.381

				$\alpha = 0.01$				
				$g - 1$				
$n - g$	1	2	3	4	5	6	7	8
1	31.815	43.453	50.133	54.673	58.069	60.760	62.976	
2	6.963	8.711	9.730	10.437	10.973	11.401	11.757	12.061
3	4.541	5.449	5.979	6.350	6.633	6.861	7.050	7.213
4	3.747	4.396	4.774	5.039	5.241	5.404	5.542	5.658
5	3.365	3.895	4.202	4.416	4.581	4.713	4.824	4.919
6	3.143	3.604	3.871	4.057	4.199	4.314	4.410	4.493
7	2.998	3.417	3.657	3.824	3.953	4.056	4.143	4.217
8	2.896	3.285	3.508	3.663	3.781	3.876	3.956	4.025
9	2.821	3.189	3.398	3.543	3.654	3.744	3.819	3.883
10	2.764	3.114	3.314	3.452	3.558	3.643	3.714	3.775
11	2.718	3.056	3.247	3.380	3.481	3.563	3.631	3.689
12	2.681	3.008	3.193	3.321	3.419	3.498	3.563	3.620
13	2.650	2.969	3.149	3.273	3.368	3.444	3.508	3.563
14	2.625	2.936	3.111	3.233	3.325	3.400	3.462	3.515
15	2.603	2.908	3.080	3.198	3.289	3.361	3.422	3.474
16	2.584	2.884	3.052	3.169	3.257	3.329	3.388	3.439
17	2.567	2.862	3.028	3.143	3.230	3.300	3.358	3.408

(*continued*)

	\(\alpha = 0.01\)							
	\(g - 1\)							
\(n - g\)	1	2	3	4	5	6	7	8
18	2.552	2.844	3.007	3.120	3.206	3.275	3.332	3.382
19	2.540	2.828	2.989	3.100	3.185	3.253	3.309	3.358
20	2.528	2.813	2.972	3.082	3.166	3.233	3.289	3.337
21	2.518	2.800	2.957	3.066	3.149	3.215	3.270	3.318
22	2.508	2.788	2.944	3.052	3.133	3.199	3.254	3.301
23	2.500	2.777	2.932	3.039	3.119	3.184	3.239	3.285
24	2.492	2.767	2.921	3.027	3.107	3.171	3.225	3.271
25	2.485	2.758	2.911	3.016	3.095	3.159	3.212	3.258
26	2.479	2.750	2.902	3.006	3.085	3.148	3.201	3.246
27	2.473	2.743	2.893	2.996	3.075	3.138	3.190	3.235
28	2.467	2.736	2.885	2.988	3.066	3.128	3.181	3.225
29	2.462	2.729	2.878	2.980	3.057	3.120	3.171	3.216
30	2.457	2.723	2.871	2.973	3.050	3.111	3.163	3.207
35	2.438	2.698	2.843	2.942	3.018	3.078	3.128	3.171
40	2.423	2.680	2.822	2.920	2.994	3.053	3.103	3.145
50	2.403	2.655	2.794	2.889	2.962	3.019	3.068	3.109
60	2.390	2.638	2.775	2.869	2.940	2.997	3.044	3.085
80	2.374	2.618	2.752	2.844	2.914	2.970	3.016	3.055
100	2.364	2.605	2.738	2.829	2.898	2.953	2.999	3.038
\(\infty\)	2.326	2.558	2.685	2.772	2.837	2.889	2.933	2.970

Appendix L

The Kruskal–Wallis Test

This appendix contains the quantiles or percentiles h_α of the probability distribution of the test statistic H in the Kruskal–Wallis test for significance levels α of about 0.10, 0.05, and 0.01. These quantiles can be used as critical values. Source: Meyer and Seaman, 2008. Reproduced with permission from http://faculty.virginia.edu/kruskal-wallis/.

n_1	n_2	n_3	$\alpha \approx 0.10$		$\alpha \approx 0.05$		$\alpha \approx 0.01$	
			$h_{0.10}$	p-value	$h_{0.05}$	p-value	$h_{0.01}$	p-value
1	1	1						
2	1	1						
2	2	1						
2	2	2	4.57143	0.06667				
3	1	1						
3	2	1	4.28571	0.10000				
3	2	2	4.46429	0.08571	4.71429	0.04762		
3	3	1	4.57143	0.10000	5.14286	0.04286		
3	3	2	4.55556	0.10000	5.36111	0.03214		
3	3	3	4.62222	0.10000	5.60000	0.05000	6.48889	0.00833
4	1	1						
4	2	1	4.50000	0.07619				
4	2	2	4.45833	0.10000	5.33333	0.03333		
4	3	1	4.05556	0.09286	5.20833	0.05000		
4	3	2	4.51111	0.09841	5.44444	0.04603	6.44444	0.00794
4	3	3	4.70000	0.09881	5.79091	0.04571	6.74546	0.01000
4	4	1	4.16667	0.08254	4.96667	0.04762	6.66667	0.00952
4	4	2	4.55455	0.09778	5.45455	0.04571	7.03636	0.00571
4	4	3	4.54546	0.09905	5.59849	0.04866	7.14394	0.00970

(*continued*)

Statistics with JMP: Hypothesis Tests, ANOVA and Regression, First Edition. Peter Goos and David Meintrup.
© 2016 John Wiley & Sons, Ltd. Published 2016 by John Wiley & Sons, Ltd.
Companion Website: http://www.wiley.com/go/goosandmeintrup/JMP

n_1	n_2	n_3	$\alpha \approx 0.10$		$\alpha \approx 0.05$		$\alpha \approx 0.01$	
			$h_{0.10}$	p-value	$h_{0.05}$	p-value	$h_{0.01}$	p-value
4	4	4	4.65385	0.09662	5.69231	0.04866	7.65385	0.00762
5	1	1						
5	2	1	4.20000	0.09524	5.00000	0.04762		
5	2	2	4.37333	0.08995	5.16000	0.03439	6.53333	0.00794
5	3	1	4.01778	0.09524	4.87111	0.04960		
5	3	2	4.65091	0.09127	5.25091	0.04921	6.82182	0.00913
5	3	3	4.53333	0.09697	5.64849	0.04892	7.07879	0.00866
5	4	1	3.98727	0.09841	4.98546	0.04444	6.95455	0.00794
5	4	2	4.54091	0.09841	5.27273	0.04877	7.20455	0.00895
5	4	3	4.54872	0.09892	5.65641	0.04863	7.44487	0.00974
5	4	4	4.66813	0.09817	5.65714	0.04906	7.76044	0.00946
5	5	1	4.10909	0.08586	5.12727	0.04618	7.30909	0.00938
5	5	2	4.62308	0.09704	5.33846	0.04726	7.33846	0.00962
5	5	3	4.54506	0.09965	5.62637	0.04906	7.57802	0.00969
5	5	4	4.52286	0.09935	5.66571	0.04931	7.79143	0.01000
5	5	5	4.56000	0.09952	5.78000	0.04878	8.00000	0.00946
6	1	1						
6	2	1	4.20000	0.09524	4.82222	0.04762		
6	2	2	4.43636	0.09841	5.34546	0.03810	6.65455	0.00794
6	3	1	3.90909	0.09524	4.85455	0.05000	6.58182	0.00952
6	3	2	4.68182	0.08528	5.34849	0.04632	6.96970	0.00909
6	3	3	4.53846	0.09913	5.61539	0.04968	7.19231	0.00796
6	4	1	4.03788	0.09437	4.94697	0.04675	7.08333	0.00996
6	4	2	4.49359	0.09986	5.26282	0.04964	7.33974	0.00967
6	4	3	4.60440	0.09997	5.60440	0.04888	7.46703	0.00976
6	4	4	4.52381	0.10000	5.66667	0.04967	7.79524	0.00990
6	5	1	4.12821	0.09271	4.98974	0.04726	7.18205	0.00974
6	5	2	4.59560	0.09807	5.31868	0.04920	7.37582	0.00982
6	5	3	4.53524	0.09932	5.60191	0.04956	7.59048	0.00999
6	5	4	4.52250	0.09974	5.66083	0.04991	7.93583	0.00998
6	5	5	4.54706	0.09835	5.69853	0.04988	8.02794	0.00988
6	6	1	4.00000	0.09774	4.85714	0.04945	7.06593	0.00982
6	6	2	4.43810	0.09824	5.40952	0.04993	7.46667	0.00982
6	6	3	4.55833	0.09948	5.62500	0.04999	7.72500	0.00985
6	6	4	4.54779	0.09982	5.72427	0.04950	8.00000	0.00998
6	6	5	4.54248	0.09987	5.76471	0.04993	8.11895	0.00995
6	6	6	4.53801	0.09985	5.71930	0.04944	8.22222	0.00994
7	1	1	4.26667	0.08333				
7	2	1	4.20000	0.10000	4.70649	0.05000		
7	2	2	4.52597	0.09899	5.14286	0.04444	7.00000	0.00707
7	3	1	4.17316	0.09546	4.95238	0.04697	6.64935	0.00909
7	3	2	4.58242	0.09874	5.35714	0.04950	6.83883	0.00985
7	3	3	4.60283	0.09814	5.62009	0.04924	7.22763	0.00991
7	4	1	4.12088	0.09141	4.98626	0.04798	6.98626	0.00960

(*continued*)

n_1	n_2	n_3	$\alpha \approx 0.10$		$\alpha \approx 0.05$		$\alpha \approx 0.01$	
			$h_{0.10}$	p-value	$h_{0.05}$	p-value	$h_{0.01}$	p-value
7	4	2	4.54945	0.09992	5.37598	0.04670	7.30455	0.00991
7	4	3	4.52721	0.09973	5.62313	0.04930	7.49864	0.00996
7	4	4	4.56250	0.09943	5.65000	0.04995	7.81429	0.00980
7	5	1	4.03517	0.09790	5.06374	0.04895	7.06060	0.00971
7	5	2	4.48490	0.09979	5.39265	0.04953	7.44980	0.00974
7	5	3	4.53524	0.09983	5.58857	0.04981	7.69714	0.00992
7	5	4	4.54160	0.09951	5.73277	0.04976	7.93109	0.00997
7	5	5	4.54006	0.09963	5.70756	0.04975	8.10084	0.00997
7	6	1	4.03265	0.09998	5.06667	0.04937	7.25442	0.00932
7	6	2	4.50000	0.09869	5.35714	0.04995	7.49048	0.00986
7	6	3	4.55042	0.09989	5.67227	0.04875	7.75630	0.00993
7	6	4	4.56163	0.09978	5.70588	0.04988	8.01634	0.00999
7	6	5	4.55973	0.09933	5.76993	0.04994	8.15673	0.01000
7	6	6	4.53008	0.09998	5.73008	0.04969	8.25714	0.00992
7	7	1	3.98571	0.09728	4.98571	0.04876	7.15714	0.00975
7	7	2	4.49055	0.09832	5.39811	0.04903	7.49055	0.00999
7	7	3	4.59010	0.09985	5.67694	0.04941	7.80952	0.00999
7	7	4	4.55890	0.09997	5.76566	0.04962	8.14160	0.00993
7	7	5	4.54556	0.09973	5.74556	0.04948	8.24481	0.00999
7	7	6	4.56803	0.09970	5.79252	0.04985	8.34150	0.00998
7	7	7	4.59369	0.09933	5.81818	0.04911	8.37848	0.00992
8	1	1	4.41818	0.06667				
8	2	1	4.01136	0.09293	4.90909	0.04444		
8	2	2	4.58654	0.09562	5.35577	0.04040	6.66346	0.00875
8	3	1	4.00962	0.09899	4.88141	0.04849	6.80449	0.00909
8	3	2	4.45055	0.09915	5.31593	0.04957	6.98626	0.00995
8	3	3	4.50476	0.09943	5.61667	0.04905	7.25476	0.00999
8	4	1	4.03846	0.09946	5.04396	0.04817	6.97253	0.00964
8	4	2	4.50000	0.09861	5.39286	0.04817	7.35000	0.00959
8	4	3	4.52917	0.09875	5.62292	0.04858	7.58542	0.00983
8	4	4	4.56066	0.09927	5.77941	0.04910	7.85294	0.00974
8	5	1	3.96714	0.09935	4.86857	0.04973	7.11000	0.00977
8	5	2	4.46625	0.09978	5.41500	0.04877	7.44000	0.00990
8	5	3	4.51434	0.09994	5.61434	0.04996	7.70552	0.00991
8	5	4	4.54902	0.09930	5.71765	0.04960	7.99216	0.00998
8	5	5	4.55526	0.09931	5.76930	0.04972	8.11579	0.00996
8	6	1	4.01458	0.09994	5.01458	0.04937	7.25625	0.00981
8	6	2	4.44118	0.09983	5.40441	0.04991	7.52206	0.00968
8	6	3	4.57353	0.09999	5.67811	0.04981	7.79575	0.00998
8	6	4	4.56287	0.09972	5.74269	0.04991	8.04532	0.00997
8	6	5	4.55026	0.09955	5.75026	0.04997	8.21026	0.01000
8	6	6	4.59881	0.09930	5.77024	0.04977	8.29405	0.00999
8	7	1	4.04543	0.09920	5.04123	0.04718	7.30777	0.00993
8	7	2	4.45098	0.09984	5.40336	0.04915	7.57143	0.00996

(*continued*)

n_1	n_2	n_3	$\alpha \approx 0.10$		$\alpha \approx 0.05$		$\alpha \approx 0.01$	
			$h_{0.10}$	p-value	$h_{0.05}$	p-value	$h_{0.01}$	p-value
8	7	3	4.55556	0.09939	5.69841	0.04994	7.82707	0.00996
8	7	4	4.54850	0.09984	5.75921	0.04997	8.11805	0.00998
8	7	5	4.55061	0.09975	5.77745	0.04991	8.24194	0.00999
8	7	6	4.55288	0.09987	5.78123	0.04990	8.33272	0.00999
8	7	7	4.57369	0.09931	5.79503	0.04993	8.35630	0.00996
8	8	1	4.04412	0.09978	5.03922	0.04929	7.31373	0.00999
8	8	2	4.50877	0.09736	5.40790	0.04953	7.65351	0.00987
8	8	3	4.55526	0.09975	5.73421	0.04975	7.88947	0.00998
8	8	4	4.57857	0.09957	5.74286	0.04984	8.16786	0.00989
8	8	5	4.57273	0.09915	5.76104	0.04998	8.29740	0.00999
8	8	6	4.57213	0.09914	5.77866	0.04988	8.36660	0.00996
8	8	7	4.57065	0.09981	5.79115	0.05000	8.41887	0.00999
8	8	8	4.59500	0.09933	5.80500	0.04973	8.46500	0.00991

Appendix M

The Rank Correlation Test

This appendix contains the quantiles or percentiles of the probability distribution of the rank correlation coefficient $R^{(s)}$ for significance levels α of about 0.10, 0.05, 0.02, and 0.01 for a two-tailed test, and approximately 0.05, 0.025, 0.01, and 0.005 for a right-tailed test. These quantiles can be used as critical values. The required quantiles for a left-tailed test are the opposites of the quantiles for a right-tailed test. The exact p-value corresponding to the specified quantile or percentile is less than or equal to the indicated value of α. Source: Zar, 1972. Reproduced with permission of Taylor and Francis.

	α for one-tailed tests			
	≈ 0.05	≈ 0.025	≈ 0.01	≈ 0.005
	α for two-tailed tests			
n	≈ 0.1	≈ 0.05	≈ 0.02	≈ 0.01
4	1.000			
5	0.900	1.000	1.000	
6	0.829	0.886	0.943	1.000
7	0.714	0.786	0.893	0.929
8	0.643	0.738	0.833	0.881
9	0.600	0.700	0.783	0.833
10	0.564	0.648	0.745	0.794
11	0.536	0.618	0.709	0.755
12	0.503	0.587	0.671	0.727
13	0.484	0.560	0.648	0.703
14	0.464	0.538	0.622	0.675
15	0.443	0.521	0.604	0.654
16	0.429	0.503	0.582	0.635
17	0.414	0.485	0.566	0.615

(continued)

Statistics with JMP: Hypothesis Tests, ANOVA and Regression, First Edition. Peter Goos and David Meintrup.
© 2016 John Wiley & Sons, Ltd. Published 2016 by John Wiley & Sons, Ltd.
Companion Website: http://www.wiley.com/go/goosandmeintrup/JMP

	α for one-tailed tests			
	≈ 0.05	≈ 0.025	≈ 0.01	≈ 0.005
	α for two-tailed tests			
n	≈ 0.1	≈ 0.05	≈ 0.02	≈ 0.01
18	0.401	0.472	0.550	0.600
19	0.391	0.460	0.535	0.584
20	0.380	0.447	0.520	0.570
21	0.370	0.435	0.508	0.556
22	0.361	0.425	0.496	0.544
23	0.353	0.415	0.486	0.532
24	0.344	0.406	0.476	0.521
25	0.337	0.398	0.466	0.511
26	0.331	0.390	0.457	0.501
27	0.324	0.382	0.448	0.491
28	0.317	0.375	0.440	0.483
29	0.312	0.368	0.433	0.475
30	0.306	0.362	0.425	0.467
31	0.301	0.356	0.418	0.459
32	0.296	0.350	0.412	0.452
33	0.291	0.345	0.405	0.446
34	0.278	0.340	0.399	0.439
35	0.283	0.335	0.394	0.433
36	0.279	0.330	0.388	0.427
37	0.275	0.325	0.383	0.421
38	0.271	0.321	0.378	0.415
39	0.267	0.317	0.373	0.410
40	0.264	0.313	0.368	0.405
41	0.261	0.309	0.364	0.400
42	0.257	0.305	0.359	0.395
43	0.254	0.301	0.355	0.391
44	0.251	0.298	0.351	0.386
45	0.248	0.294	0.347	0.382
46	0.246	0.291	0.343	0.378
47	0.243	0.288	0.340	0.374
48	0.240	0.285	0.336	0.370
49	0.238	0.282	0.333	0.366
50	0.235	0.279	0.329	0.363

Index

Statistics with JMP: Hypothesis Tests, ANOVA and Regression, First Edition. Peter Goos and David Meintrup.
© 2016 John Wiley & Sons, Ltd. Published 2016 by John Wiley & Sons, Ltd.
Companion Website: http://www.wiley.com/go/goosandmeintrup/JMP